DSP Software Development Techniques for Embedded and Real-Time Systems

DSP Software Development Techniques for Embedded and Real-Time Systems

by Robert Oshana

ELSEVIER

AMSTERDAM • BOSTON • HEIDELBERG • LONDON
NEW YORK • OXFORD • PARIS • SAN DIEGO
SAN FRANCISCO • SINGAPORE • SYDNEY • TOKYO

Newnes is an imprint of Elsevier

Newnes

Newnes is an imprint of Elsevier
30 Corporate Drive, Suite 400, Burlington, MA 01803, USA
Linacre House, Jordan Hill, Oxford OX2 8DP, UK

Library of Congress Cataloging-in-Publication Data

Application submitted.

British Library Cataloguing-in-Publication Data
A catalogue record for this book is available from the British Library.

ISBN-13: 978-0-7506-7759-2
ISBN-10: 0-7506-7759-7

For information on all Newnes publications,
visit our website at www.books.elsevier.com.

06 07 08 09 10 10 9 8 7 6 5 4 3 2

Printed in the United States of America.

Dedicated to
Susan, Sam, and Noah

Table of Contents

Acknowledgments

This book has been written with significant technical and emotional support from my family, friends, and co-workers. It is not possible to list everyone who helped to sustain me throughout this project. I apologize for any omissions.

My editorial staff was great. Tiffany Gasbarrini, you have been a pleasure to work with; it was an honor to be associated with you on this endeavor. Carol Lewis, I won't forget you. Thanks for getting me started with Elsevier. To Kelly Johnson of Borrego Publishing, thanks for all of the hard work and support.

Thanks to Frank Coyle, my academic and personal mentor at Southern Methodist University, who was the initial inspiration for this project. Thanks for everything Frank!

I would like to recognize those who provided me with significant input and support for this project: Gene Frantz, Gary Swoboda, Oliver Sohm, Scott Gary, Dennis Kertis, Bob Frankel, Leon Adams, Eric Stotzer, George Mock, Jonathan Humphreys, Gerald Watson, the many outstanding technical writers from the TI technical training group, and the many unnamed authors whose excellent application notes I have used and referenced in this project. Also, special thanks to Cathy Wicks, Suzette Harris, Lisa Ferrara, Christy Brunton, and Sarah Gonzales for your support, dedication, and humor.

Thanks to my management for giving me the opportunity to work on this project: Greg Delagi, David Peterman, Hasan Khan, Ed Morgan—thanks!

Thanks to the reviewers. I have attempted to incorporate all feedback received into this project, and I will continue to appreciate any additional feedback. Many thanks to those who granted me permission to use several of the figures in this book. These figures have added to the quality of this material.

I also want to thank my family and friends who offered their support and patience, as this book project consumed time ordinarily spend with them. To Susan, Sam, and Noah—thanks and it's great to have you with me!

Go DSP!

Introduction: Why Use a DSP?

In order to understand the usefulness of programmable Digital Signal Processing, I will first draw an analogy and then explain the special environments where DSPs are used.

A DSP is really just a special form of microprocessor. It has all of the same basic characteristics and components; a CPU, memory, instruction set, buses, etc. The primary difference is that each of these components is customized slightly to perform certain operations more efficiently. We'll talk about specifics in a moment, but in general, a DSP has hardware and instruction sets that are optimized for high-speed numeric processing applications and rapid, real-time processing of analog signals from the environment. The CPU is slightly customized, as is the memory, instruction sets, buses, and so forth.

I like to draw an analogy to society. We, as humans, are all processors (cognitive processors) but each of us is specialized to do certain things well; engineering, nursing, finance, and so forth. We are trained and educated in certain fields (specialized) so that we can perform certain jobs efficiently. When we are specialized to do a certain set of tasks, we expend less energy doing those tasks. It is not much different for microprocessors. There are hundreds to choose from and each class of microprocessor is specialized to perform well in certain areas. A DSP is a specialized processor that does signal processing very efficiently. And, like our specialty in life, because a DSP specializes in signal processing, it expends less energy getting the job done. DSPs, therefore, consume less time, energy and power than a general-purpose microprocessor when carrying out signal processing tasks.

When you specialize a processor, it is important to specialize those areas that are commonly and frequently put to use. It doesn't make sense to make something efficient at doing things that are hardly ever needed! Specialize those areas that result in the biggest bang for the buck!

But before I go much further, I need to give a quick summary of what a processor must do to be considered a digital signal processor. It must do two things very well. First, it must be good at math and be able to do millions (actually billions) of multiplies and adds per second. This is necessary to implement the algorithms used in digital signal processing.

The second thing it must do well is to guarantee real time. Let's go back to our real-life example. I took my kids to a movie recently and when we arrived, we had to wait in line to purchase our tickets. In effect, we were put into a queue for processing, standing in line behind other moviegoers. If the line stays the same length and doesn't continue to get longer and longer, then the queue is real-time in the sense that the same number of customers are being processed as there are joining the queue. This queue of people may get shorter or grow a bit longer but does not grow in an unbounded way. If you recall the evacuation from Houston as Hurricane Rita approached, that was a queue that was growing in an unbounded way! This queue was definitely not real time and it grew in an unbounded way, and the system (the evacuation system) was considered a failure. Real-time systems that cannot perform in real time are failures.

If the queue is really big (meaning, if the line I am standing in at the movies is really long) but not growing, the system may still not work. If it takes me 50 minutes to move to the front of the line to buy my ticket, I will probably be really frustrated, or leave altogether before buying my ticket to the movies (my kids will definitely consider this a failure). Real-time systems also need to be careful of large queues that can cause the system to fail. Real-time systems can process information (queues) in one of two ways: either one data element at a time, or by buffering information and then processing the "queue." The queue length cannot be too long or the system will have significant latency and not be considered real time.

If real time is violated, the system breaks and must be restarted. To further the discussion, there are two aspects to real time. The first is the concept that for every sample period, one input piece of data must be captured, and one output piece of data must be sent out. The second concept is latency. Latency means that the delay from the signal being input into the system and then being output from the system must be preserved as immediate.

Keep in mind the following when thinking of real-time systems: producing the correct answer too late is wrong! If I am given the right movie ticket and charged the correct amount of money after waiting in line, but the movie has already started, then the system is still broke (unless I arrived late to the movie to begin with). Now go back to our discussion.

So what are the "special" things that a DSP can perform? Well, like the name says, DSPs do signal processing very well. What does "signal processing" mean? Really, it's a set of algorithms for processing signals in the digital domain. There are analog equivalents to these algorithms, but processing them digitally has been proven to be more efficient. This has been a trend for many many years. Signal processing algorithms are the basic building blocks for many applications in the world; from cell phones to MP3 players, digital still cameras, and so on. A summary of these algorithms is shown in the following table.

Algorithm	Equation
Finite Impulse Response Filter	$y(n)=\sum_{k=0}^{M} a_k x(n-k)$
Infinite Impulse Response Filter	$y(n)=\sum_{k=0}^{M} a_k x(n-k)+\sum_{k=1}^{N} b_k y(n-k)$
Convolution	$y(n)=\sum_{k=0}^{N} x(k)h(n-k)$
Discrete Fourier Transform	$X(k)=\sum_{n=0}^{N-1} x(n)\exp[-j(2\pi/N)nk]$
Discrete Cosine Transform	$F(u)=\sum_{x=0}^{N-1} c(u).f(x).\cos\left[\frac{\pi}{2N}u(2x+1)\right]$

One or more of these algorithms are used in almost every signal processing application. Finite Impulse Response Filters and Infinite Impulse Response Filters are used to remove unwanted noise from signals being processed, convolution algorithms are used for looking for similarities in signals, discrete Fourier transforms are used for representing signals in formats that are easier to process, and discrete cosine transforms are used in image processing applications. We'll discuss the details of some of these algorithms later, but there are some things to notice about this entire list of algorithms. First, they all have a summing operation, the function. In the computer world, this is equivalent to an accumulation of a large number of elements which is implemented using a "for" loop. DSPs are designed to have large accumulators because of this characteristic. They are specialized in this way. DSPs also have special hardware to perform the "for" loop operation so that the programmer does not have to implement this in software, which would be much slower.

The algorithms above also have multiplications of two different operands. Logically, if we were to speed up this operation, we would design a processor to accommodate the multiplication and accumulation of two operands like this very quickly. In fact, this is what has been done with DSPs. They are designed to support the multiplication and accumulation of data sets like this very quickly; for most processors, in just one cycle. Since these algorithms are very common in most DSP applications, tremendous execution savings can be obtained by exploiting these processor optimizations.

There are also inherent structures in DSP algorithms that allow them to be separated and operated on in parallel. Just as in real life, if I can do more things in parallel, I can get more done in the same amount of time. As it turns out, signal processing algorithms have this characteristic as well. So, we can take advantage of this by putting multiple orthogonal (nondependent) execution units in our DSPs and exploit this parallelism when implementing these algorithms.

DSPs must also add some reality to the mix of these algorithms shown above. Take the IIR filter described above. You may be able to tell just by looking at this algorithm that there is a feedback component that essentially feeds back previous outputs into the calculation of the current output. Whenever you deal with feedback, there is always an inherent stability issue. IIR filters can become unstable just like other feedback systems. Careless implementation of feedback systems like the IIR filter can cause the output to oscillate instead of asymptotically decaying to zero (the preferred approach). This problem is compounded in the digital world where we must deal with finite word lengths, a key limitation in all digital systems. We can alleviate this using saturation checks in software or use a specialized instruction to do this for us. DSPs, because of the nature of signal processing algorithms, use specialized saturation underflow/overflow instructions to deal with these conditions efficiently.

There is more I can say about this, but you get the point. Specialization is really all it's about with DSPs; these devices are specifically designed to do signal processing really well. DSPs may not be as good as other processors when dealing with nonsignal processing centric algorithms (that's fine; I'm not any good at medicine either). So it's important to understand your application and pick the right processor.

With all of the special instructions, parallel execution units and so on designed to optimize signal processing algorithms, there is not much room left to perform other types of general-purpose optimizations. General-purpose processors contain optimization logic such as branch prediction and speculative execution, which provide performance improvements in other types of applications. But some of these optimizations don't work as well for signal processing applications. For example, branch prediction works really well when there are a lot of branches in the application. But DSP algorithms do not have a lot of branches. Much signal processing code consists of well defined functions that execute off a single stimulus, not complicated state machines requiring a lot of branch logic.

Digital signal processing also requires optimization of the software. Even with the fancy hardware optimizations in a DSP, there is still some heavy-duty tools support required—specifically, the compiler—that makes it all happen. The compiler is a nice tool for taking a language like C and mapping the resultant object code onto this specialized microprocessor. Optimizing compilers perform a very complex and difficult task of producing code that fully "entitles" the DSP hardware platform. We'll talk a lot about optimizing compilers later on in the book.

There is no black magic in DSPs. As a matter of fact, over the last couple of years, the tools used to produce code for these processors have advanced to the point where you can write much of the code for a DSP in a high level language like C or C++ and let the compiler map and optimize the code for you. Certainly, there will always be special things you can do, and certain hints you need to give the compiler to produce the optimal code, but it's really no different from other processors. As a matter of fact, we'll spend a couple of chapters talking about how to optimize DSP code to achieve optimal performance, memory, and power.

The environment in which a DSP operates is important as well; not just the types of algorithms running on the DSP. Many (but not all) DSP applications are required to interact with the real world. This is a world that has a lot of stuff going on; voices, light, temperature, motion, and more. DSPs, like other embedded processors, have to *react* in certain ways within this real world. Systems like this are actually referred to as *reactive* systems. When a system is reactive, it needs to respond and control the *real* world, not too surprisingly, in *real*-time. Data and signals coming in from the real world must be processed in a timely way. The definition of timely varies from application to application, but it requires us to keep up with what is going on in the environment.

Because of this timeliness requirement, DSPs, as well as other processors, must be designed to respond to real-world events quickly, get data in and out quickly, and process the data quickly. We have already addressed the processing part of this. But believe it or not, the bottleneck in many real-time applications is not getting the data processed, but getting the data in and out of the processor quickly enough. DSPs are designed to support this real-world requirement. High speed I/O ports, buffered serial ports, and other peripherals are designed into DSPs to accommodate this. DSPs are, in fact, often referred to as data pumps, because of the speed in which they can process streams of data. This is another characteristic that makes DSPs unique.

DSPs are also found in many embedded applications. I'll discuss the details of embedded systems in Chapter 2. However, one of the constraints of an embedded application is scarce resources. Embedded systems, by their very nature, have scarce resources. The main resources I am referring to here are processor cycles, memory, power and I/O. It has always been this way, and always will. Regardless of how fast embedded processors run, how much memory can be fit on chip, and so on, there will always be applications that consume all available resources and then look for more! Also, embedded applications are very application-specific, not like a desktop application that is much more general-purpose.

At this point, we should now understand that a DSP is like any other programmable processor, except that it is specialized to perform signal processing really efficiently. So now the only question should be; why program anything at all? Can't I do all this signal processing stuff in hardware? Well, actually you can. There is a fairly broad spectrum of DSP implementation techniques, with corresponding trade-offs in flexibility, as well as cost, power, and a few other parameters. The graph below summarizes two of the main trade-offs in the programmable vs. fixed-function decision; flexibility and power.

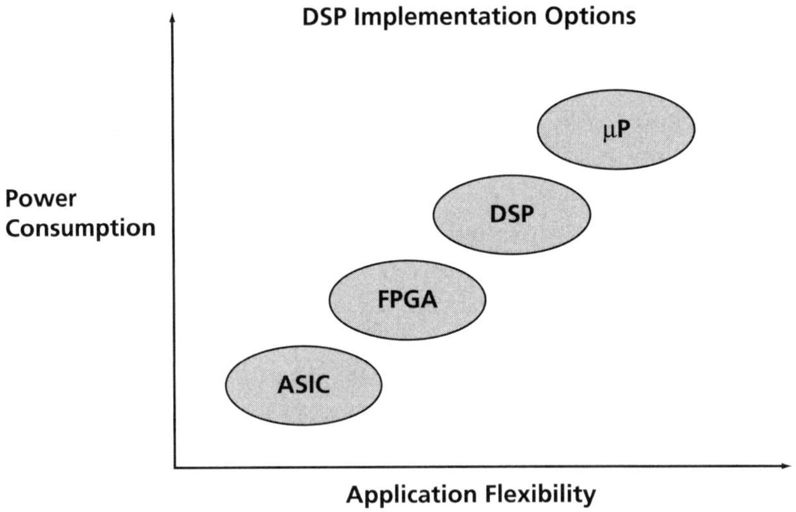

An application-specific integrated circuit (ASIC) is a hardware only implementation option. These devices are programmed to perform a fixed-function or set of functions. Being a hardware only solution, an ASIC does not suffer from some of the programmable von Neumann-like limitations such as loading and storing of instructions and data. These devices run exceedingly fast in comparison to a programmable solution, but they are not as flexible. Building an ASIC is like building any other microprocessor, to some extent. It's a rather complicated design process, so you have to make sure the algorithms you are designing into the ASIC work and won't need to be changed for a while! You cannot simply recompile your application to fix a bug or change to a new wireless standard. (Actually, you could, but it will cost a lot of money and take a lot of time). If you have a stable, well-defined function that needs to run really fast, an ASIC may be the way to go.

Field-programmable gate arrays (FPGAs) are one of those in-between choices. You can program them and re-program them in the field, to a certain extent. These devices are not as flexible as true programmable solutions, but they are more flexible than an ASIC. Since FPGAs are hardware they offer similar performance advantages to other hardware-based solutions. An FPGA can be "tuned" to the precise algorithm, which is great for performance. FPGAs are not truly application specific, unlike an ASIC. Think of an FPGA as a large sea of gates where you can turn on and off different gates to implement your function. In the end, you get your application implemented, but there are a lot of spare gates laying around, kind of going along for the ride. These take up extra space as well as cost, so you need to do the trade-offs; are the cost, physical area, development cost and performance all in line with what you are looking for?

DSP and μP (microprocessor): We have already discussed the difference here, so there is no need to rehash it. Personally, I like to take the flexible route: programmability. I make a lot of mistakes when I develop signal processing systems; it's very complicated technology! So, I like to know that I have the flexibility to make changes when I need to in order to fix a bug, perform an additional optimization to increase performance or reduce power (we talk a lot about this as well in this book), or change to the next standard. The entire signal processing field is growing and changing so quickly—witness the standards that are evolving and changing all the time—that I prefer to make the rapid and inexpensive upgrades and changes that only a programmable solution can afford.

The general answer, as always, lies somewhere in between. In fact, many signal processing solutions are partitioned across a number of different processing elements. Certain parts of the algorithm stream—those that have a pretty good probability of changing in the near future—are mapped to a programmable DSP. Signal processing functions that will remain fairly stable for the foreseeable future are mapped into hardware gates (either an ASIC, an FPGA, or other hardware acceleration). Those parts of the signal processing system that control the input, output, user interface and overall management of the system heartbeat may be mapped to a more general-purpose processor. Complicated signal processing systems need the right combination of processing elements to achieve true system performance/cost/power trade-offs. We'll spend more time on this later in the book as well.

Signal processing is here to stay. It's everywhere. Any time you have a signal that you want to know more about, communicate in some way, make better or worse, you need to process it. The digital part is just the process of making it all work on a computer of some sort. If it's an embedded application you must do this with the minimal amount of resources possible. Everything costs money; cycles, memory, power—so everything must be conserved. This is the nature of embedded computing; be application specific, tailor to the job at hand, reduce cost as much as possible, and make things as efficient as possible. This was the way things were done in 1982 when I started in this industry, and the same techniques and processes apply today. The scale has certainly changed; computing problems that required supercomputers in those days are on embedded devices today!

This book will touch on these areas and more as it relates to digital signal processing. There is a lot to discuss and I'll take a practical rather than theoretical approach to describe the challenges and processes required to do DSP well.

What's on the CD-ROM?

Test drive Code Composer Studio™ (CCStudio) Development Tools for 120 days absolutely free with the "Essential Guide to Getting Started with DSP" CD-ROM. Benchmark, write sample algorithms or just explore the rich feature set of the CCStudio IDE. For more information on TI DSP, visit www.ti.com/dsp.

1

Introduction to Digital Signal Processing

What is Digital Signal Processing

Digital signal processing (DSP) is a method of processing signals and data in order to enhance or modify those signals, or to analyze those signals to determine specific information content. It involves the processing of real-world signals that are converted into and represented by sequences of numbers. These signals are then processed using mathematical techniques in order to extract certain information from the signal or to transform the signal in some (preferably beneficial) way.

The "digital" term in DSP requires processing using discrete signals to represent the data in the form of numbers that can be easily manipulated. In other words, the signal is represented numerically. This type of representation implies some form of quantization of one or more properties of the signal, including time.

This is just one type of digital data; other types include ASCII numbers and letters.

The "signal" term in DSP refers to a variable parameter. This parameter is treated as information as it flows through an electronic circuit. The signal usually[1] starts out in the analog world as a constantly changing piece of information. Examples of real world signals include:

- air temperature
- sound
- humidity
- speed
- position
- flow
- light
- pressure
- volume

The signal is essentially a voltage that varies among a theoretically infinite number of values. This represents patterns of variation of physical quantities. Other examples of signals are sine waves, the waveforms representing human speech, and the signals from a conventional television camera. A signal is a detectable physical quantity. Messages or information can be transmitted based on these signals.

---

[1] Usually because some signals may already be in a discrete form. An example of this would be a switch, which is represented discretely as being either open or closed.

A signal is called *one-dimensional* (1-D) when it describes variations of a physical quantity as a function of a single independent variable. An audio/speech signal is one-dimensional because it represents the continuing variation of air pressure as a function of time.

Finally, the "processing" term in DSP relates to the processing of data using software programs as opposed to hardware circuitry. A digital signal processor is a device or a system that performs signal processing functions on signals from the real (analog) world using primarily software programs to manipulate the signals. This is an advantage in the sense that the software program can be changed relatively easily to modify the performance or behavior of the signal processing. This is much harder to do with analog circuitry.

Since DSPs interact with signals in the environment, the DSP system must be "reactive" to the environment. In other words, the DSP must keep up with changes in the environment. This is the concept of "real-time" processing and we will talk about it shortly.

A Brief History of Digital Signal Processing

Some of the first digital signal processing solutions were TTL[2] medium scale integration (MSI) silicon chips. Up to 100 of these chips were used to form cascadable ALU sections and standalone multipliers. These early systems were large, expensive and hot.

The first single-chip DSP solution appeared in 1982. This was the TMS32010 DSP from Texas Instruments. NEC came out with the uPD7720 not long after. These processors had performance close to 5 MIPS[3]. These early single-chip solutions had very small RAM memory and sold for about $600[4]. These solutions were able to reduce overall system chip count, as well as provide lower power requirements and more reliable systems due to reduced manufacturing complexity and cost. Most of these DSPs used NMOS technology[5].

As the market for DSP devices continued to grow, vendors began adding more integration, as well as internal RAM, ROM, and EPROM. Advanced addressing functionality including FFT bit-reversed addressing and circular buffer addressing were developed (these are two common DSP-centric addressing modes and will be

---

[2] Transistor-transistor logic, a common type of digital circuit in which the output is derived from two transistors. The first semiconductors using TTL were developed by Texas Instruments in 1965.

[3] The number of MIPS (millions of instructions per second) is a general measure of computing performance and, by implication, the amount of work a larger computer can do. Historically, the cost of computing measured in the number of MIPS per dollar has been reduced by half on an annual basis for a number of years (Moore's Law).

[4] A similar device sells for under $2 today.

[5] Acronym for negative-channel metal-oxide semiconductor. This is a type of semiconductor that is negatively charged so that transistors are turned on or off by the movement of electrons. In contrast, PMOS (positive-channel MOS) works by moving electron vacancies. NMOS is faster than PMOS, but also more expensive to produce.

discussed in more detail later). Serial ports for fast data transfer were added. Other architectural enhancements to these second generation devices included timers, direct memory access (DMA) controllers, interrupt systems including shadow registers, and integrated analog-to-digital (ADC), and digital-to-analog (DAC) converters.

Floating-point DSPs were introduced in 1988. The DSP32 was introduced by AT&T. The Texas Instruments TMS320C30 was introduced during the same time period. These devices were easier to program and provided features such as automatic scaling. Because of the larger silicon area to support the floating-point architecture, these devices cost more than the traditional fixed-point processors. They also used more power and tended to be lower in processing speed.

In the early 1990s, parallel processing DSP support began to emerge. Single processor DSPs with advanced communication support, such as the Texas Instruments TMS320C40, appeared. Multiple processing elements were designed into a single integrated circuit (such as the TMS320C80).

Today, there are many advanced DSP architecture styles. We will be studying several of them in this book. Architectural advances have included multiple functional units, very long instruction word (VLIW) architectures, and specialized functional units to perform specific tasks very quickly (such as echo cancellation in a cell phone).

Advantages of DSP

There are many advantages of using a digital signal processing solution over an analog solution. These include:

- *Changeability* – It is easy to reprogram digital systems for other applications or to fine tune existing applications. A DSP allows for easy changes and updates to the application.
- *Repeatability* – Analog components have characteristics that may change slightly over time or temperature variances. A programmable digital solution is much more repeatable due to the programmable nature of the system. Multiple DSPs in a system, for example, can also run the exact same program and be very repeatable. With analog signal processing, each DSP in the system would have to be individually tuned.
- *Size, weight, and power* – A DSP solution that requires mostly programming means the DSP device itself consumes less overall power than a solution using all hardware components.
- *Reliability* – Analog systems are reliable only to the extent that the hardware devices function properly. If any of these devices fail due to physical conditions, the entire system degrades or fails. A DSP solution implemented in software will function properly as long as the software is implemented correctly.

- *Expandability* – To add more functionality to the system, the engineer must add more hardware. This may not be possible. Adding the same functionality to a DSP involves adding software, which is much easier.

Figure 1.1 shows an example of an analog signal plotted as amplitude over time. A signal like this may represent a noise source such as white noise plus a speech signal or maybe an acoustic echo. The change for a signal processing system would be to eliminate or filter out the noise signal and keep the speech signal. A hands-free cell phone car kit would be a system where this type of noise and acoustic echo removal would be implemented. The time domain is where a large part of digital signal processing occurs. As you can see, this domain is primarily concerned with the value of a signal over time. This is natural, since that is the way many of these signals are produced from the source anyway; a continuous stream of signal over time. We will see later that it makes sense, at times, to represent this same signal in other domains to enable more efficient processing of the signal.

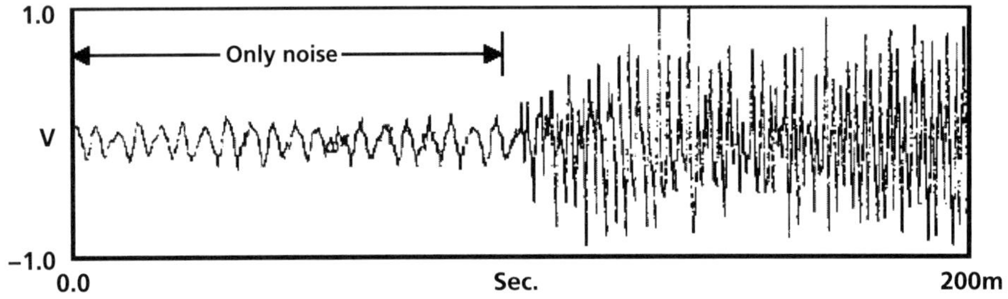

Figure 1.1 An example of an analog signal plotted over time (This figure comes from the application note SPRA095, Integrated Automotive Signal Processing and Audio System Using the TMS320C3x from Texas Instruments)

DSP Systems

The signals that a DSP processor uses come from the real world. Because a DSP must respond to signals in the real world, it must be capable of changing based on what it sees in the real world. We live in an analog world in which the information around us changes, sometimes very quickly. A DSP system must be able to process these analog signals and respond in a timely manner. A typical DSP system (Figure 1.2) consists of the following:

- *Signal source* – Something that is producing the signal, such as a microphone, a radar sensor, or a flow gauge.
- *Analog signal processing (ASP)* – Circuitry to perform some initial signal amplification or filtering.
- *Analog-to-digital conversion (ADC)* – An electronic process in which a continuously variable signal is changed, without altering its essential content, into a multilevel (digital) signal. The output of the ADC has defined levels or states. The number of states is almost always a power of two—that is, 2, 4, 8, 16, and so on. The simplest digital signals have only two states, and are called binary.

- *Digital signal processing (DSP)* – The various techniques used to improve the accuracy and reliability of modern digital communications. DSP works by clarifying, or standardizing, the levels or states of a digital signal. A DSP system is able to differentiate, for example, between human-made signals, which are orderly, and noise, which is inherently chaotic.
- *Computer* – If additional processing is required in the system, additional computing resources can be applied if necessary. For example, if the signals being processed by the DSP are to be formatted for display to a user, an additional computer can be used to perform these tasks.
- *Digital-to-analog conversion (DAC)* – The process in which signals having a few (usually two) defined levels or states (digital) are converted into signals having a theoretically infinite number of states (analog). A common example is the processing, by a modem, of computer data into audio-frequency (AF) tones that can be transmitted over a twisted pair telephone line.
- *Output* – A system for realizing the processed data. This may be a terminal display, a speaker, or another computer.

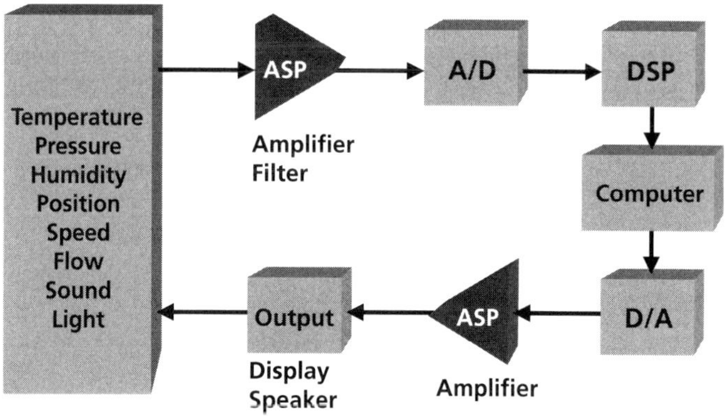

Figure 1.2 A DSP system

Systems operate on signals to produce new signals. For example, microphones convert air pressure to electrical current and speakers convert electrical current to air pressure.

Analog-to-Digital Conversion

The first step in a signal processing system is getting the information from the real world into the system. This requires transforming an analog signal to a digital representation suitable for processing by the digital system. This signal passes through a device called an analog-to-digital converter (A/D or ADC). The ADC converts the analog signal to a digital representation by sampling or measuring the signal at a periodic rate. Each sample is assigned a digital code (Figure 1.3). These digital codes can then be processed by the DSP. The number of different codes or states is almost always a power of two

(2, 4, 8, 16, etc.) The simplest digital signals have only two states. These are referred to as binary signals.

Examples of analog signals are waveforms representing human speech and signals from a television camera. Each of these analog signals can be converted to digital form using ADC and then processed using a programmable DSP.

Digital signals can be processed more efficiently than analog signals. Digital signals are generally well-defined and orderly, which makes them easier for electronic circuits to distinguish from *noise*, which is chaotic. Noise is basically unwanted information. Noise can be background noise from an automobile, or a scratch on a picture that has been converted to digital. In the analog world, noise can be represented as electrical or electromagnetic energy that degrades the quality of signals and data. Noise, however, occurs in both digital and analog systems. Sampling errors (we'll talk more about this later) can degrade digital signals as well. Too much noise can degrade all forms of information including text, programs, images, audio and video, and telemetry. Digital signal processing provides an effective way to minimize the effects of noise by making it easy to filter this "bad" information out of the signal.

Figure 1.3 Analog-to-digital conversion for signal processing

As an example, assume that the analog signal in Figure 1.3 needs to be converted into a digital signal for further processing. The first question to consider is how often to *sample* or measure the analog signal in order to represent that signal accurately in the digital domain. The sample rate is the number of samples of an analog event (like sound) that are taken per second to represent the event in the digital domain. Let's assume that we are going to sample the signal at a rate of T seconds. This can be represented as:

Sampling period (T) = 1 / Sampling Frequency (fs)

where the sampling frequency is measured in *hertz*[6]. If the sampling frequency is 8 kilohertz (KHz), this would be equivalent to 8000 cycles per second. The sampling period would then be:

T = 1 / 8000 = 125 microseconds = 0.000125 seconds

This tells us that, for a signal being sampled at this rate, we would have 0.000125 seconds to perform all the processing necessary before the next sample arrived (remember, these samples are arriving on a continuous basis and we cannot fall behind

[6] Hertz is a unit of frequency (change in state or cycle in a sound wave, alternating current, or other cyclical waveform) of one cycle per second. The unit of measure is named after Heinrich Hertz, a German physicist.

in processing them). This is a common restriction for *real-time* systems, which we will discuss shortly.

Since we now know the time restriction, we can determine the processor speed required to keep up with this sampling rate. Processor "speed" is measured not by how fast the clock rate is for the processor, but how fast the processor executes instructions. Once we know the processor instruction cycle time, we can determine how many instructions we have available to process the sample:

Sampling period (T) / Instruction cycle time = number of instructions per sample

For a 100 MHz processor that executes one instruction per cycle, the instruction cycle time would be 1/100 MHz = 10 nanoseconds

125 μs / 10 ns = 12,500 instructions per sample

125 μs / 5 ns = 25,000 instructions per sample (for a 200 MHz processor)

125 ms / 2 ns = 62,500 instruction per sample (for a 500 MHz processor)

As this example demonstrated, the higher the processor instruction cycle execution, the more processing we can do on each sample. If it were this easy, we could just choose the highest processor speed available and have plenty of processing margin. Unfortunately, it is not as easy as this. Many other factors including cost, accuracy and power limitations must be considered. Embedded systems have many constraints such as these as well as size and weight (important for portable devices). For example, how do we know how fast we should sample the input analog signal to represent it accurately in the digital domain? If we do not sample often enough, the information we obtain will not be representative of the true signal. If we sample too much we may be "over designing" the system and overly constrain ourselves.

Digital-to-Analog Conversion

In many applications, a signal must be sent back out to the real world after being processed, enhanced and/or transformed while inside the DSP. Digital-to-analog conversion (DAC) is a process in which signals having a few (usually two) defined levels or states (digital) are converted into signals having a very large number of states (analog).

Both the DAC and the ADC are of significance in many applications of digital signal processing. The fidelity of an analog signal can often be improved by converting the analog input to digital form using a DAC, clarifying or enhancing the digital signal and then converting the enhanced digital impulses back to analog form using an ADC (A single digital output level provides a DC output voltage).

Figure 1.4 shows a digital signal passing through another device called a digital-to-analog (D/A or DAC) converter which transforms the digital signal into an analog signal and outputs that signal to the environment.

Figure 1.4 Digital-to-analog conversion

Applications for DSPs

In this section, we will explore some common applications for DSPs. Although there are many different DSP applications, I will focus on three categories:

- Low cost, good performance DSP applications.
- Low power DSP applications.
- High performance DSP applications.

Low-Cost DSP Applications

DSPs are becoming an increasingly popular choice as low-cost solutions in a number of different areas. One popular area is electronic motor control. Electric motors exist in many consumer products, from washing machines to refrigerators. The energy consumed by the electric motor in these appliances is a significant portion of the total energy consumed by the appliance. Controlling the speed of the motor has a direct effect on the total energy consumption of the appliance[7]. In order to achieve the performance improvements necessary to meet energy consumption targets for these appliances, manufacturers use advanced three phase variable speed drive systems. DSP based motor control systems have the bandwidth required to enable the development of more advanced motor drive systems for many domestic appliance applications.

As performance requirements have continued to increase in the past few years, the need for DSPs has increased as well (Figure 1.5).

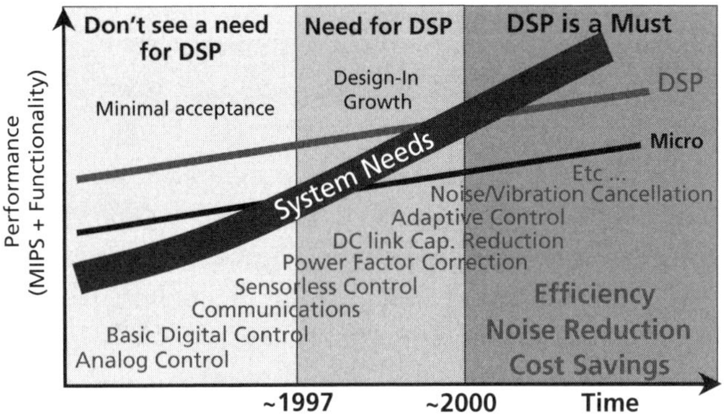

Figure 1.5 Low-cost, high-performance DSP motor control applications (courtesy of Texas Instruments)

[7] Many of today's energy efficient compressors require the motor speed to be controlled in the range from 1200 rpm to 4000 rpm.

Application complexity has continued to grow as well, from basic digital control to advanced noise and vibration cancellation applications. As shown in Figure 1.5, as the complexity of these applications has grown, there has also been a migration from analog to digital control. This has resulted in an increase in reliability, efficiency, flexibility and integration, leading to overall lower system cost.

Many of the early control functions used what is called a *microcontroller* as the basic control unit. A microcontroller is an integrated microprocessor which includes a CPU, a small amount of RAM and/or ROM, and a set of specialized peripherals, all on the same chip. As the complexity of the algorithms in motor control systems increased, the need also grew for higher performance and more programmable solutions (Figure 1.6). Digital signal processors provide much of the bandwidth and programmability required for such applications[8]. DSPs are now finding their way into some of the more advanced motor control technologies:

- Variable speed motor control.
- Sensorless control.
- Field-oriented control.
- Motor modeling in software.
- Improvements in control algorithms.
- Replacement of costly hardware components with software routines.

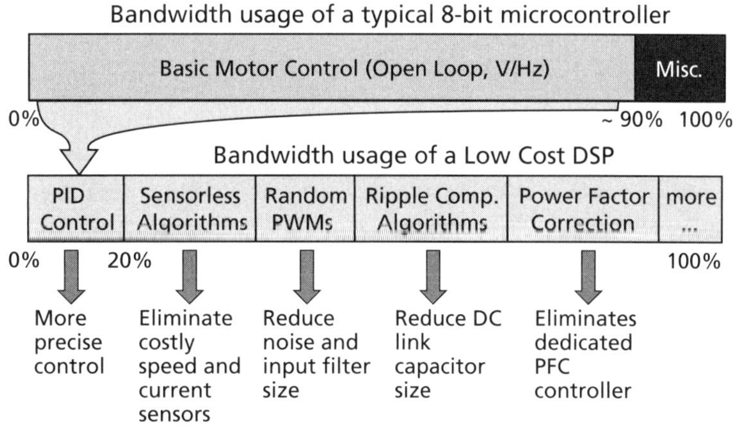

Figure 1.6 Microcontrollers vs. DSPs in motor control (courtesy of Texas Instruments)

[8] For example, one of the trends on motor control has been the conversion from brush motors to brushless motors. DSP-based control has facilitated this conversion. Eliminating the brushes provides improvements. First, since there is no brush drag, the overall efficiency of the motor is higher. Second, there is far less electrical noise generated to interfere with the remote control. Third, there is no required maintenance on the brushless motor and there is no deterioration of performance over the life of the motor.

The typical motor control model is shown in Figure 1.7. In this example, the DSP is used to provide fast and precise PWM switching of the converter. The DSP also provides the system with fast, accurate feedback of the various analog motor control parameters such as current, voltage, speed, temperature, etc. There are two different motor control approaches; open-loop control and closed-loop control. The open-loop control system is the simplest form of control. Open-loop systems have good steady state performance and the lack of current feedback limits much of the transient performance (Figure 1.8). A low-cost DSP is used to provide variable speed control of the three phase induction motor, providing improved system efficiency.

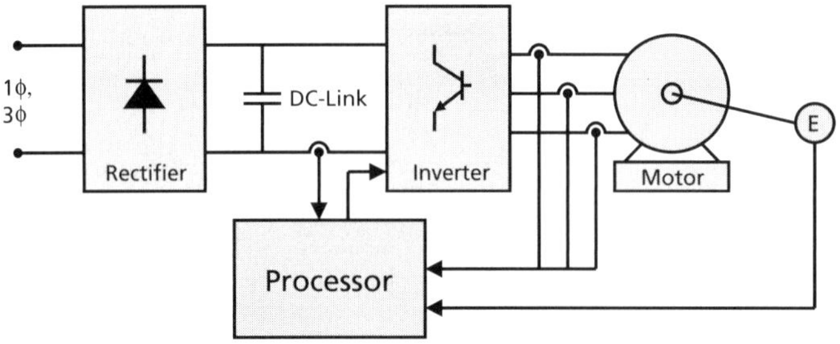

Figure 1.7 Simplified DSP controlled motor control system (courtesy of Texas Instruments)

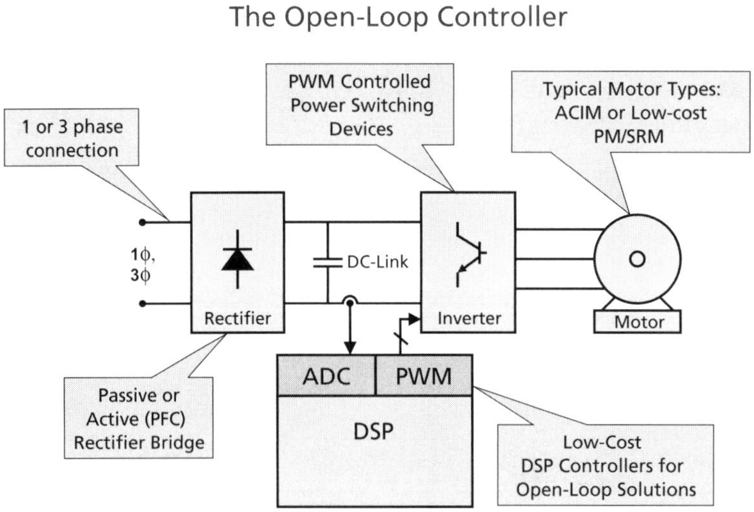

Figure 1.8 Open-loop controller (courtesy of Texas Instruments)

A closed-loop solution (Figure 1.9) is more complicated. A higher performance DSP is used to control current, speed, and position feedback, which improves the transient response of the system and enables tighter velocity/position control. Other, more sophisticated, motor control algorithms can also implemented in the higher performance DSP.

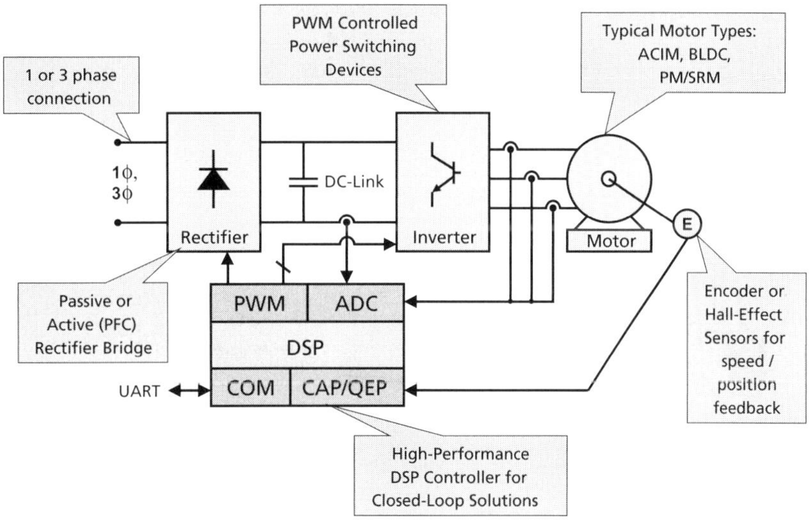

Figure 1.9 Closed-loop controller (courtesy of Texas Instruments)

There are many other applications using low-cost DSPs (Figure 1.10). Refrigeration compressors, for example, use low-cost DSPs to control variable speed compressors that dramatically improve energy efficiency. Low-cost DSPs are used in many washing machines to enable variable speed controls which eliminate the need for mechanical gearing. DSPs also provide sensorless control for these devices, which eliminates the need for speed and current sensors. Improved off balance detection and control enable higher spin speeds, which gets clothes dryer with less noise and vibration. Heating, ventilating and air conditioning (HVAC) systems use DSPs in variable speed control of the blower and inducer, which increases furnace efficiency and improves comfort level.

Figure 1.10 There are many applications of low-cost DSPs in the motor control industry, including refrigeration, washing machines, and heating, ventilation, and air conditioning systems. (courtesy of Texas Instruments)

Power Efficient DSP Applications

We live in a portable society. From cell phones to personal digital assistants, we work and play on the road! These systems are dependent on the batteries that power them. The longer the battery life can be extended the better. So it makes sense for the designers of these systems to be sensitive to processor power. Having a processor that consumes less power enables longer battery life, and makes these systems and applications possible.

As a result of reduced power consumption, systems dissipate lower heat. This results in the elimination of costly hardware components like heat sinks to dissipate the heat effectively. This leads to overall lower system cost as well as smaller overall system size because of the reduced number of components. Continuing along this same line of reasoning, if the system can be made less complex with fewer parts, designers can bring these systems to market more quickly.

Low power devices also give the system designer a number of new options, such as potential battery back-up to enable uninterruptible operation as well as the ability to do more with the same power (as well as cost) budget to enable greater functionality and/or higher performance.

There are several classes of systems that make them suitable for low power DSPs. Portable consumer electronics (Figure 1.11) use batteries for power. Since the average consumer of these devices wants to minimize the replacement of batteries, the longer they can go on the same batteries, the better off they are. This class of customer also cares about size. Consumers want products they can carry with them, clip onto their belts or carry in their pockets.

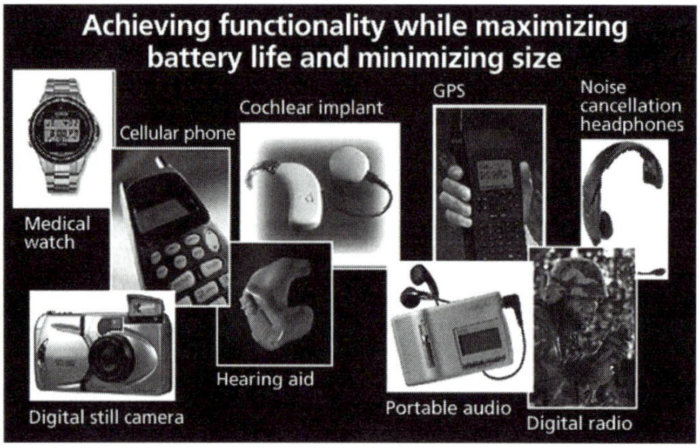

Figure 1.11 Battery operated products require low power DSPs (courtesy of Texas Instruments)

Certain classes of systems require designers to adhere to a strict power budget. These are systems that have a fixed power budget, such as systems that operate on limited line power, battery back-up, or with fixed power source (Figure 1.12). For this class of systems, designers aim to deliver functionality within the constraints imposed by the power supply. Examples of these systems include many defense and aerospace systems.

These systems have very tight size, weight, and power restrictions. Low power processors give designers more flexibility in all three of these important constraints.

Figure 1.12 Low power DSPs allow designers to meet strict size, weight, and power constraints. (courtesy of Texas Instruments)

Another important class of power-sensitive systems are high density systems (Figure 1.13). These systems are often high performance system or multiprocessor systems. Power efficiency is important for these systems, not only because of the power supply constraints, but also because of heat dissipation concerns. These systems contain very dense boards with a large number of components per board. There may also be several boards per system in a very confined area. Designers of these systems are concerned about reduced power consumption as well as heat dissipation. Low power DSPs can lead to higher performance and higher density. Fewer heat sinks and cooling systems enable lower cost systems that are easier to design. The main concerns for these systems are:

- creating more functions per channel;
- achieving more functions per square inch;
- avoiding cooling issues (heat sinks, fans, noise);
- reducing overall power consumption.

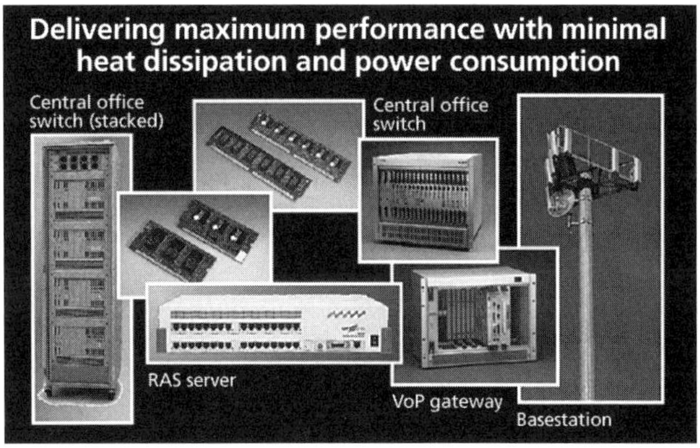

Figure 1.13 Low power DSPs allow designers to deliver maximum performance and higher density systems (courtesy of Texas Instruments)

Power is the limiting factor in many systems today. Designers must optimize the system design for power efficiency at every step. One of the first steps in any system design is the selection of the processor. A processor should be selected based on an architecture and instruction set optimized for power efficient performance[9]. For signal processing intensive systems, a common choice is a DSP (Figure 1.14).

ALGORITHM	BASIC FUNCTION
Voice compression Phase detection	FIR filter
DTMF, Graphic EQ	IIR filter
Echo cancellation; high bit-rate modems; motion detectors	Adaptive filter
Audio decoder (MP3, AP3)	Inverse modified DCT (FFT)
Forward error correction	Viterbi

Figure 1.14 Many of today's complex algorithms are composed from basic function signal processing blocks that DSPs are very efficient at computing

As an example of a low power DSP solution, consider a solid-state audio player like the one shown in Figure 1.15. This system requires a number of DSP-centric algorithms to perform the signal processing necessary to produce high fidelity music quality sound. Figure 1.16 shows some of the important algorithms required in this system. A low power DSP can handle the decompression, decryption and processing of audio data. This data may be stored on external memory devices which can be interchanged like individual CD's. These memory devices can be reprogrammed as well. The user interface functions can be handled by a microcontroller. The memory device which holds the audio data may be connected to the micro which reads it and transfers to the DSP. Alternately, data might be downloaded from a PC or Internet site and played directly or written onto blank memory devices. A digital-to-analog (DAC) converter translates the digital audio output of the DSP into an analog form to be played on user headphones. The entire system must be powered from batteries (for example, two AA batteries).

[9] Process technology also has a significant effect on power consumption. By integrating memory and functionality (for example, DTMF and V.22) onto the processor, you can lower your system power level. This is discussed more in the chapter on DSP architectures.

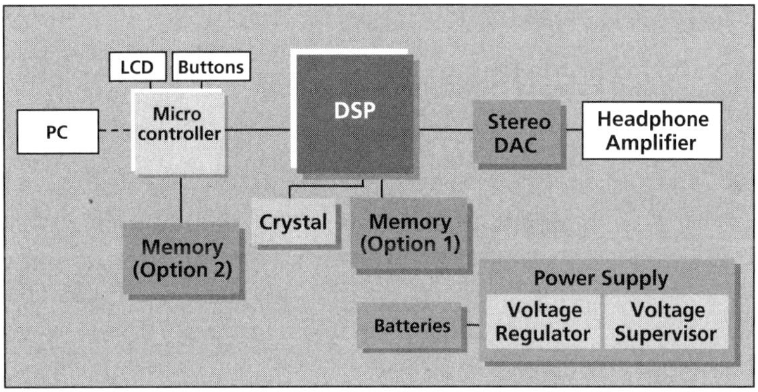

Figure 1.15 Block diagram of a solid-state low-power audio music player (courtesy of Texas Instruments)

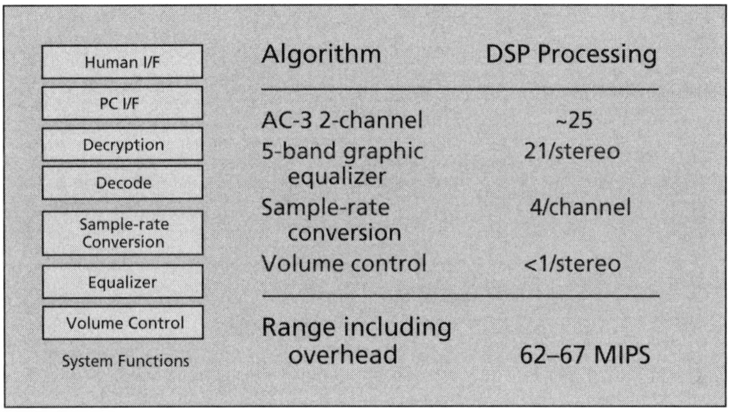

Figure 1.16 Common signal processing algorithms for a solid-state audio music player (courtesy of Texas Instruments)

For this type of product, a key design constraint would be power. Customers do not like replacing the batteries in their portable devices. Thus, battery life, which is directly related to system power consumption, is a key issue. By not having any moving parts, a solid-state audio player uses less power than previous generation players (tape and CD). Since this is a portable product, size and weight are also obviously concerns. Solid-state devices such as the one described here are also size efficient because of fewer parts in the overall system.

To the system designer, programmability is a key concern. With a programmable DSP solution, this portable audio player can be updated with the newest decompression, encryption and audio processing algorithms instantly from the World Wide Web or from memory devices. A low power DSP-based system solution like the one described here could have system power consumption as low as 200 mW. This will allow the portable audio player will have three times the battery life of a CD player on the same two AA battery supply.

High Performance DSP Applications

At the high end of the performance spectrum, DSPs utilize advanced architectures to perform signal processing at high rates. Advanced architectures such as very long instruction word (VLIW) use extensive parallelism and pipelining to achieve high performance. These advanced architectures take advantage of other technologies such as optimizing compilers to achieve this performance. There is a growing need for high performance computing (Figure 1.21). Applications include:

- DSL modems
- base station transceivers
- wireless LAN
- multimedia gateway
- professional audio
- networked camera
- security identification
- industrial scanner
- high speed printer
- advanced encryption

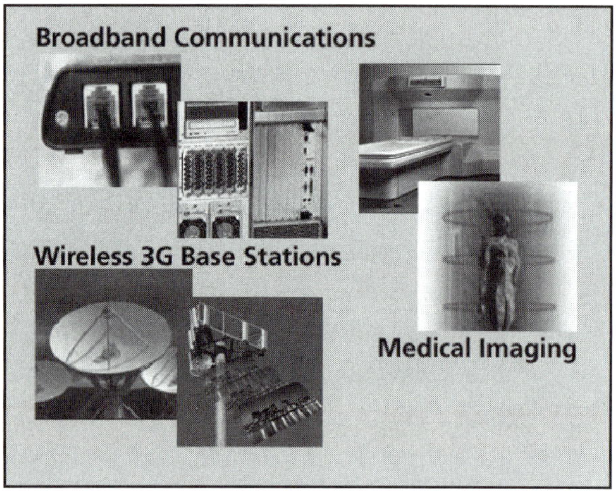

Figure 1.17 There is a growing need for high performance signal processing (courtesy of Texas Instruments)

Conclusion

Though analog signals can also be processed using analog hardware (that is, electrical circuits containing active and passive elements), there are several advantages to digital signal processing:

- Analog hardware is usually limited to linear operations; digital hardware can implement nonlinear operations.
- Digital hardware is programmable, which allows for easy modification of the signal processing procedure in both real-time and non real-time modes of operation.
- Digital hardware is less sensitive than analog hardware to variations such as temperature, and so forth.

These advantages lead to lower cost, which is the main reason for the ongoing shift from analog to digital processing in wireless telephones, consumer electronics, industrial controllers and numerous other applications.

The discipline of signal processing, whether analog or digital, consists of a large number of specific techniques. These can be roughly categorized into two families:

- Signal-analysis/feature-extraction techniques, which are used to extract useful information from a signal. Examples include speech recognition, location and identification of targets from radar signals, detection and characterization of changes in meteorological or seismographic data.
- Signal filtering/shaping techniques, which are used to improve the quality of a signal. Sometimes this is done as an initial step before analysis or feature extraction. Examples of these techniques include the removal of noise and interference using filtering algorithms, separating a signal into simpler components, and other time-domain and frequency-domain averaging.

A complete signal processing system usually consists of many components and incorporates multiple signal processing techniques.

References

Digital Signal Processing Demystified, James D. Broesch, Newnes, 1997

Digital Audio Processing, Doug Coulter, R&D Books, 2000

Software Engineering, Ian Sommerville, Addison Wesley, 1997

2

Overview of Embedded Systems and Real-Time Systems

Introduction

Nearly all real-world DSP applications are part of an embedded real-time system. While this book will focus primarily on the DSP-specific portion of such a system, it would be naive to pretend that the DSP portions can be implemented without concern for the real-time nature of DSP or the embedded nature of the entire system.

This chapter will highlight some of special design considerations that apply to embedded real-time systems. I will look first at real-time issues, then some specific embedded issues, and finally, at trends and issues that commonly apply to both real-time and embedded systems.

Real-Time Systems

A real-time system is a system that is required to react to stimuli from the environment (including the passage of physical time) within time intervals dictated by the environment. The Oxford Dictionary defines a real-time system as "any system in which the time at which output is produced is significant." This is usually because the input corresponds to some movement in the physical world, and the output has to relate to that same movement. The lag from input time to output time must be sufficiently small for acceptable timeliness. Another way of thinking of real-time systems is any information processing activity or system which has to respond to externally generated input stimuli within a finite and specified period. Generally, real-time systems are systems that maintain a *continuous timely* interaction with their environment (Figure 2.1).

Types of real-time systems—soft and hard

Correctness of a computation depends not only upon its results but also upon the time at which its outputs are generated A real-time system must satisfy response time constraints or suffer significant system consequences. If the consequences consist of a degradation of performance, but not failure, the system is referred to as a soft real-time system. If the consequences are system failure, the system is referred to as a hard real-time system. (for instance, anti-lock braking systems in an automobile).

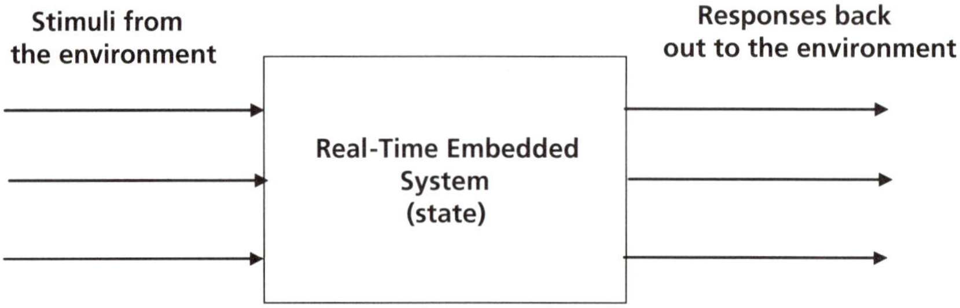

Figure 2.1 A real-time system reacts to inputs from the environment and produces outputs that affect the environment

Hard Real-Time and Soft Real-Time Systems

Hard real-time and soft real-time systems introduction

A system function (hardware, software, or a combination of both) is considered hard real-time if, and only if, it has a hard deadline for the completion of an action or task. This deadline must always be met, otherwise the task has failed. The system may have one or more hard real-time tasks as well as other nonreal-time tasks. This is acceptable, as long as the system can properly schedule these tasks in such a way that the hard real-time tasks always meet their deadlines. Hard real-time systems are commonly also embedded systems.

Differences between real-time and time-shared systems

Real-time systems are different from time shared systems in the three fundamantal areas (Table 1). These include predictably fast response to urgent events:

High degree of schedulability – Timing requirements of the system must be satisfied at high degrees of resource usage,

Worst-case latency – Ensuring the system still opertates under worst-case response time to events,

Stability under transient overload – When the system is overloaded by events and it is impossible to meet all deadlines, the deadlines of selected critical tasks must still be guaranteed.

Characteristic	Time-shared systems	Real-time systems
System capacity	High throughput	Schedulability and the ability of system tasks to meet all deadlines
Responsiveness	Fast average response time	Ensured worst-case latency, which is the worst-case response time to events
Overload	Fairness to all	Stability – When the system is overloaded, important tasks must meet deadlines while others may be starved

Table 2.1 Real-time systems are fundamentally different from time-shared systems

DSP Systems are Hard Real-Time

Usually, DSP systems qualify as hard real-time systems. As an example, assume that an analog signal is to be processed digitally. The first question to consider is how often to *sample* or measure an analog signal in order to represent that signal accurately in the digital domain. The sample rate is the number of samples of an analog event (like sound) that are taken per second to represent the event in the digital domain. Based on a signal processing rule called the Nyquist rule, the signal must be sampled at a rate at least equal to twice the highest frequency that we wish to preserve. For example, if the signal contains important components at 4 kilohertz (kHZ), then the sampling frequency would need to be at least 8 KHz. The sampling period would then be:

T = 1 / 8000 = 125 microseconds = 0.000125 seconds

Based on signal sample, time to perform actions before next sample arrives

This tells us that, for this signal being sampled at this rate, we would have 0.000125 seconds to perform *all* the processing necessary before the next sample arrives. Samples are arriving on a continuous basis, and the system cannot fall behind in processing these samples and still produce correct results—it is *hard real-time.*

Hard real-time systems

The collective timeliness of the hard real-time tasks is binary—that is, either they will all always meet their deadlines (in a correctly functioning system), or they will not (the system is infeasible). In all hard real-time systems, collective timeliness is deterministic. This determinism does not imply that the actual individual task completion times, or the task execution ordering, are necessarily known in advance.

A computing system being hard real-time says nothing about the magnitudes of the deadlines. They may be microseconds or weeks. There is a bit of confusion with regards to the usage of the term "hard real-time." Some relate hard real-time to response time magnitudes below some arbitrary threshold, such as 1 msec. This is not the case. Many of these systems actually happen to be soft real-time. These systems would be more accurately termed "real fast" or perhaps "real predictable." But certainly not hard real-time.

The feasibility and costs (for example, in terms of system resources) of hard real-time computing depend on how well known a priori are the relevant future behavioral characteristics of the tasks and execution environment. These task characteristics include:

- timeliness parameters, such as arrival periods or upper bounds
- deadlines
- worst-case execution times
- ready and suspension times
- resource utilization profiles
- precedence and exclusion constraints
- relative importances, and so on

There are also pertinent characteristics relating to the execution environment:

- system loading
- resource interactions
- queuing disciplines
- arbitration mechanisms
- service latencies
- interrupt priorities and timing
- caching, and so on

Deterministic collective task timeliness in hard (and soft) real-time computing requires that the future characteristics of the relevant tasks and execution environment be deterministic—that is, known absolutely in advance. The knowledge of these characteristics must then be used to pre-allocate resources so all deadlines will always be met.

Usually, the task's and execution environment's future characteristics must be adjusted to enable a schedule and resource allocation that meets all deadlines. Different algorithms or schedules that meet all deadlines are evaluated with respect to other factors. In many real-time computing applications, it is common that the primary factor is maximizing processor utilization.

Allocation for hard real-time computing has been performed using various techniques. Some of these techniques involve conducting an offline enumerative search for a static schedule that will deterministically always meet all deadlines. Scheduling algorithms include the use of priorities that are assigned to the various system tasks. These priorities can be assigned either offline by application programmers, or online by the application or operating system software. The task priority assignments may either be static (fixed), as with rate monotonic algorithms[1] or dynamic (changeable), as with the earliest deadline first algorithm[2].

Real-Time Event Characteristics

Real-time event categories

Real-time events fall into one of three categories: asynchronous, synchronous, or isochronous.

Asynchronous events are entirely unpredictable. An example of this is a cell phone call arriving at a cellular base station. As far as the base station is concerned, the action of making a phone call cannot be predicted.

Synchronous events are predictable and occur with precise regularity. For example, the audio and video in a camcorder take place in synchronous fashion.

Isochronous events occur with regularity within a given window of time. For example, audio data in a networked multimedia application must appear within a window of time when the corresponding video stream arrives. Isochronous is a sub-class of asynchronous.

[1] Rate monotonic analysis (RMA) is a collection of quantitative methods and algorithms that allow engineers to specify, understand, analyze, and predict the timing behavior of real-time software systems, thus improving their dependability and evolvability.

[2] A strategy for CPU or disk access scheduling. With EDF, the task with the earliest deadline is always executed first.

In many real-time systems, task and future execution environment characteristics are hard to predict. This makes true hard real-time scheduling infeasible. In hard real-time computing, deterministic satisfaction of the collective timeliness criterion is the driving requirement. The necessary approach to meeting that requirement is static (that is, a priori[3]) scheduling of deterministic task and execution environment characteristic cases. The requirement for advance knowledge about each of the system tasks and their future execution environment to enable offline scheduling and resource allocation significantly restricts the applicability of hard real-time computing.

Efficient Execution and the Execution Environment

Efficiency overview

Real-time systems are time critical, and the efficiency of their implementation is more important than in other systems. Efficiency can be categorized in terms of processor cycles, memory or power. This constraint may drive everything from the choice of processor to the choice of the programming language. One of the main benefits of using a higher level language is to allow the programmer to abstract away implementation details and concentrate on solving the problem. This is not always true in the embedded system world. Some higher level languages have instructions that be an order of magnitude slower than assembly language. However, higher level languages can be used in real-time systems effectively, using the right techniques. We will be discussing much more about this topic in the chapter on optimizing source code for DSPs.

Resource management

A system operates in real time as long as it completes its time-critical processes with acceptable timeliness. *Acceptable timeliness* is defined as part of the behavioral or "nonfunctional" requirements for the system. These requirements must be objectively quantifiable and measurable (stating that the system must be "fast," for example, is not quantifiable). A system is said to be real-time if it contains some model of real-time resource management (these resources must be explicitly managed for the purpose of operating in real time). As mentioned earlier, resource management may be performed statically, offline, or dynamically, online.

Real-time resource management comes at a cost. The degree to which a system is required to operate in real time cannot necessarily be attained solely by hardware over-capacity (such as, high processor performance using a faster CPU). To be cost effective, there must exist some form of real-time resource management. Systems that must operate in real time consist of both real-time resource management and hardware resource capacity. Systems that have interactions with physical devices require higher degrees of real-time resource management. These computers are referred to as

[3] Relating to or derived by reasoning from self-evident propositions (formed or conceived beforehand), as compared to a posteriori that is presupposed by experience (www.wikipedia.org).

embedded systems, which we spoke about earlier. Many of these embedded computers use very little real-time resource management. The resource management that is used is usually static and requires analysis of the system prior to it executing in its environment. In a real-time system, physical time (as opposed to logical time) is necessary for real-time resource management in order to relate events to the precise moments of occurrence. Physical time is also important for action time constraints as well as measuring costs incurred as processes progress to completion. Physical time can also be used for logging history data.

All real-time systems make trade-offs of scheduling costs vs. performance in order to reach an appropriate balance for attaining acceptable timeliness between the real-time portion of the scheduling optimization rules and the offline scheduling performance evaluation and analysis.

Types of real-time systems—reactive and embedded

There are two types of real-time systems: reactive and embedded. A reactive real-time system has constant interaction with its environment (such as a pilot controlling an aircraft). An embedded real-time system is used to control specialized hardware that is installed within a larger system (such as a microprocessor that controls anti-lock brakes in an automobile).

Challenges in Real-Time System Design

Designing real-time systems poses significant challenges to the designer. One of these challenges comes from the fact that real-time systems must interact with the environment. The environment is complex and changing and these interactions can become very complex. Many real-time systems don't just interact with one, but many different entities in the environment, with different characteristics and rates of interaction. A cell phone base station, for example, must be able to handle calls from literally thousands of cell phone subscribers at the same time. Each call may have different requirements for processing and be in different sequences of processing. All of this complexity must be managed and coordinated.

Response Time

Real-time systems must respond to external interactions in the environment within a predetermined amount of time. Real-time systems must produce the correct result and produce it in a timely way. This implies that response time is as important as producing correct results. Real-time systems must be engineered to meet these response times. Hardware and software must be designed to support response time requirements for these systems. Optimal partitioning of the system requirements into hardware and software is also important.

Real-time systems must be architected to meet system response time requirements. Using combinations of hardware and software components, engineering makes architecture decisions such as interconnectivity of the system processors, system link speeds, processor speeds, memory size, I/O bandwidth, etc. Key questions to be answered include:

Is the architecture suitable? – To meet the system response time requirements, the system can be architected using one powerful processor or several smaller processors. Can the application be partitioned among the several smaller processors without imposing large communication bottlenecks throughout the system. If the designer decides to use one powerful processor, will the system meet its power requirements? Sometimes a simpler architecture may be the better approach—more complexity can lead to unnecessary bottlenecks which cause response time issues.

Are the processing elements powerful enough? – A processing element with high utilization (greater than 90%) will lead to unpredictable run time behavior. At this utilization level, lower priority tasks in the system may get starved. As a general rule, real-time systems that are loaded at 90% take approximately twice as long to develop, due to the cycles of optimization and integration issues with the system at these utilization rates. At 95% utilization, systems can take three times longer to develop, due to these same issues. Using multiple processors will help, but the inter-processor communication must be managed.

Are the communication speeds adequate? – Communication and I/O are a common bottleneck in real-time embedded systems. Many response time problems come not from the processor being overloaded but in latencies in getting data into and out of the system. On other cases, overloading a communication port (greater than 75%) can cause unnecessary queuing in different system nodes and this causes delays in message passing throughout the rest of the system.

Is the right scheduling system available? – In real-time systems, tasks that are processing real-time events must take higher priority. But, how do you schedule multiple tasks that are all processing real-time events? There are several scheduling approaches available, and the engineer must design the scheduling algorithm to accommodate the system priorities in order to meet all real-time deadlines. Because external events may occur at any time, the scheduling system must be able to preempt currently running tasks to allow higher priority tasks to run. The scheduling system (or real-time operating system) must not introduce a significant amount of overhead into the real-time system.

Recovering from Failures

Real-time systems interact with the environment, which is inherently unreliable. Therefore, real-time systems must be able to detect and overcome failures in the environment. Also, since real-time systems are often embedded into other systems and may be hard to get at (such as a spacecraft or satellite) these systems must also be able to detect and overcome internal failures (there is no "reset" button in easy reach of the user!). Also since events in the environment are unpredictable, its almost impossible to test for every possible combination and sequence of events in the environment. This is a characteristic of real-time software that makes it somewhat nondeterministic in the sense that it is almost impossible in some real-time systems to predict the multiple paths of execution based on the nondeterministic behavior of the environment. Examples of internal and external failures that must be detected and managed by real-time systems include:

- Processor failures
- Board failures
- Link failures
- Invalid behavior of external environment
- Interconnectivity failure

Distributed and Multiprocessor Architectures

Real-time systems are becoming so complex that applications are often executed on multiprocessor systems distributed across some communication system. This poses challenges to the designer that relate to the partitioning of the application in a multiprocessor system. These systems will involve processing on several different nodes. One node may be a DSP, another node a more general-purpose processor, some specialized hardware processing elements, etc. This leads to several design challenges for the engineering team:

Initialization of the system – Initializing a multiprocessor system can be very complicated. In most multiprocessor systems, the software load file resides on the general-purpose processing node. Nodes that are directly connected to the general-purpose processor, for example a DSP, will initialize first. After these nodes complete loading and initialization, other nodes connected to them may then go through this same process until the system completes initialization.

Processor interfaces – When multiple processors must communicate with each other, care must be taken to ensure that messages sent along interfaces between the processors are well defined and consistent with the processing elements. Differences in message protocol, including endianness, byte ordering and other padding rules, can complicate system integration, especially if there is a system requirement for backwards compatibility.

Load distribution – As mentioned earlier, multiple processors lead to the challenge of distributing the application, and possibly developing the application to support efficient partitioning of the application among the processing elements. Mistakes in partitioning the application can lead to bottlenecks in the system and this degrades the full capability of the system by overloading certain processing elements and leaving others under utilized. Application developers must design the application to be partitioned efficiently across the processing elements.

Centralized Resource Allocation and Management – In systems of multiple processing elements, there is still a common set of resources including peripherals, cross bar switches, memory, etc that must be managed. In some cases the operating system can provide mechanisms like semaphores to manage these shared resources. In other cases there may be dedicated hardware to manage the resources. Either way, important shared resources in the system must be managed in order to prevent more system bottlenecks.

Embedded Systems

An embedded system is a specialized computer system that is usually integrated as part of a larger system. An embedded system consists of a combination of hardware and software components to form a computational engine that will perform a specific function. Unlike desktop systems which are designed to perform a general function, embedded systems are constrained in their application. Embedded systems often perform in reactive and time-constrained environments as desribed earlier. A rough partitioning of an embedded system consists of the hardware which provides the performance necessary for the application (and other system properties like security) and the software, which provides a majority of the features and flexibility in the system. A typical embedded system is shown in Figure 2.3.

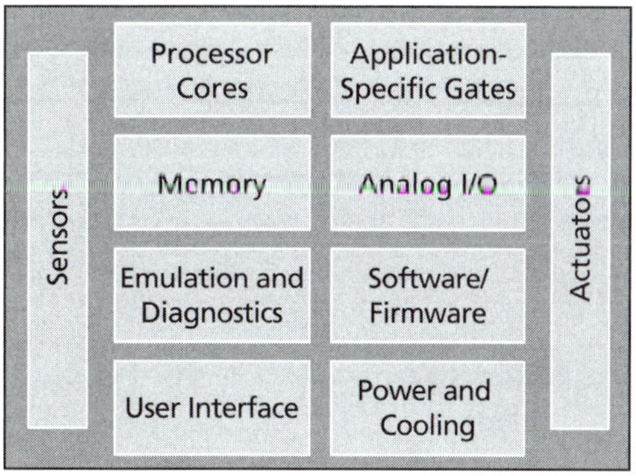

Figure 2.3 Typical embedded system components

- *Processor core* – At the heart of the embedded system is the processor core(s). This can be a simple inexpensive 8 bit microcontroller to a more complex 32 or 64 bit microprocessor. The embedded designer must select the most cost sensitive device for the application that can meet all of the functional and nonfunctional (timing) requirements.

- *Analog I/O* – D/A and A/D converters are used to get data from the environment and back out to the environment. The embedded designer must understand the type of data required from the environment, the accuracy requirements for that data, and the input/output data rates in order to select the right converters for the application. The external environment drives the reactive nature of the embedded system. Embedded systems have to be at least fast enough to keep up with the environment. This is where the analog information such as light or sound pressure or acceleration are sensed and input into the embedded system (see Figure 2.4 below).

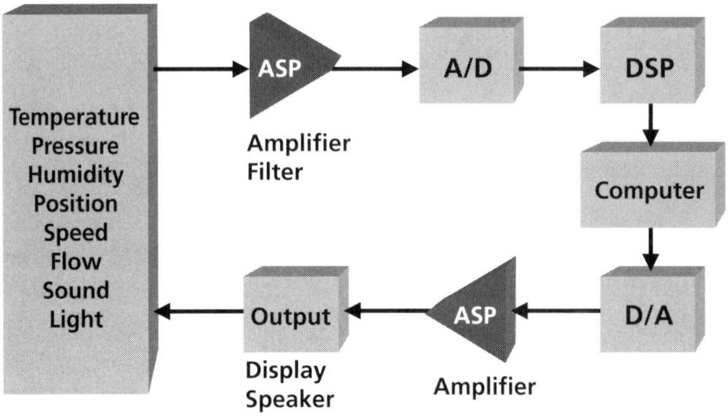

Figure 2.4 Analog information of various types is processed by embedded system

- *Sensors and Actuators* – Sensors are used to sense analog information from the environment. Actuators are used to control the environment in some way.
- *Embedded systems* also have user interfaces. These interfaces may be as simple as a flashing LED to a sophisticated cell phone or digital still camera interface.
- *Application-specific gates* – Hardware acceleration like ASICs or FPGA are used for accelerating specific functions in the application that have high performance requirements. The embedded designer must be able to map or partition the application appropriately using available accelerators to gain maximum application performance.
- *Software* is a significant part of embedded system development. Over the last several years, the amount of embedded software has grown faster than Moore's law, with the amount doubling approximately every 10 months. Embedded software is usually optimized in some way (performance, memory, or power). More and more embedded software is written in a high level language like C/C++ with some of the more performance critical pieces of code still written in assembly language.
- *Memory* is an important part of an embedded system and embedded applications can either run out of RAM or ROM depending on the application. There are many

types of volatile and nonvolatile memory used for embedded systems and we will talk more about this later.

- *Emulation and diagnostics* – Many embedded systems are hard to see or get to. There needs to be a way to interface to embedded systems to debug them. Diagnostic ports such as a JTAG (joint test action group) port are used to debug embedded systems. On-chip emulation is used to provide visibility into the behavior of the application. These emulation modules provide sophisticated visibility into the runtime behavior and performance, in effect replacing external logic analyzer functions with on board diagnostic capabilities.

Embedded systems are reactive systems

A typical embedded system responds to the environment via sensors and controls the environment using actuators (Figure 2.5). This imposes a requirement on embedded systems to achieve performance consistent with that of the environment. This is why embedded systems are referred to as reactive systems. A reactive system must use a combination of hardware and software to respond to events in the environment, within defined constraints. Complicating the matter is the fact that these external events can be periodic and predictable or aperiodic and hard to predict. When scheduling events for processing in an embedded system, both periodic and aperiodic events must be considered and performance must be guranteed for worst-case rates of execution. This can be a significant challenge. Consider the example in Figure 2.6. This is a model of an automobile airbag deployment system showing sensors including crash severity and occupant detection. These sensors monitor the environment and could signal the embedded system at any time. The embedded control unit (ECU) contains accelerometers to detect crash impacts. Also, rollover sensors, buckle sensors and weight sensors (Figure 2.8) are used to determine how and when to deploy airbags. Figure 2.7 shows the actuators in this same system. These include Thorax bags actuators, pyrotechnic buckle pretensioner with load limiters and the central airbag control unit. When an impact occurs, the sensors must detect and send a signal to the ECU, which must deploy the appropriate airbags within a hard real-time deadline for this system to work properly.

The previous example demonstrates several key characteristics of embedded systems:

- *Monitoring and reacting to the environment* – Embedded systems typically get input by reading data from input sensors. There are many different types of sensors that monitor various analog signals in the evironment, including temperature, sound pressure, and vibration. This data is processed using embedded system algorithms. The results may be displayed in some format to a user or simply used to control actuators (like deploying the airbags and calling the police).
- *Control the environment* – Embedded systems may generate and transmit commands that control actuators such as airbags, motors, and so on.

- *Processing of information* – Embedded systems process the data collected from the sensors in some meaningful way, such as data compression/decompression, side impact detection, and so on.
- *Application-specific* – Embedded systems are often designed for applications such as airbag deploment, digital still cameras or cell phones. Embedded systems may also be designed for processing control laws, finite state machines, and signal processing algorithms. Embedded systems must also be able to detect and react appropriately to faults in both the internal computing environment as well as the surrounding systems.

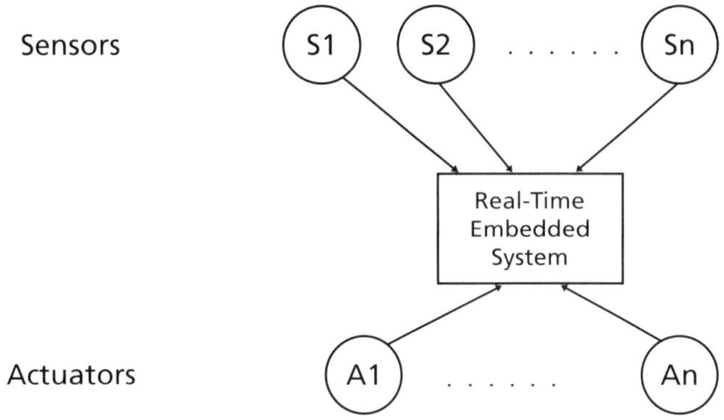

Figure 2.5 A model of sensors and actuators in embedded systems

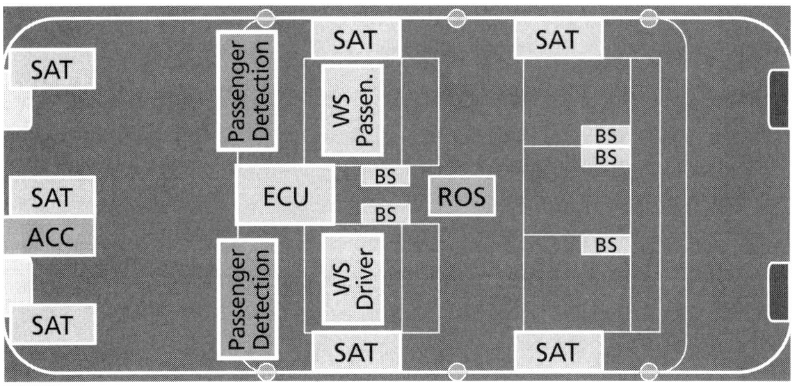

Figure 2.6 Airbag system: possible sensors (including crash severity and occupant detection) (courtesy of Texas Instruments)

SAT = satellite with serial communication interface

ECU = central airbag control unit (including accelerometers)

ROS = roll over sensing unit

WS = weight sensor

BS = buckle switch

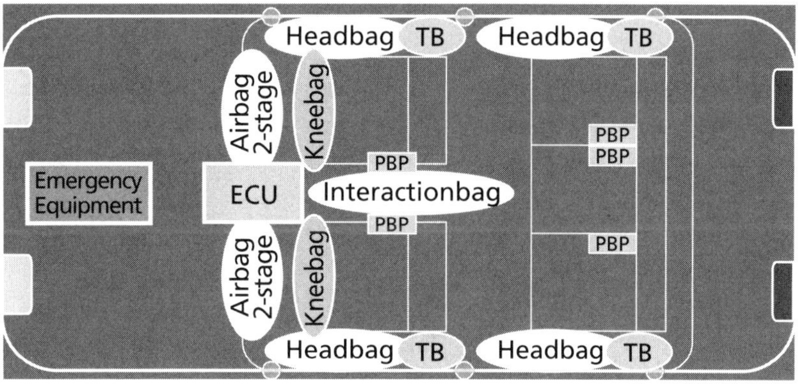

Figure 2.7 Airbag system: possible sensors (including crash severity and occupant detection) (courtesy of Texas Instruments)

TB = thorax bag

PBP = pyrotechnic buckle pretensioner with load limiter

ECU = central airbag control unit

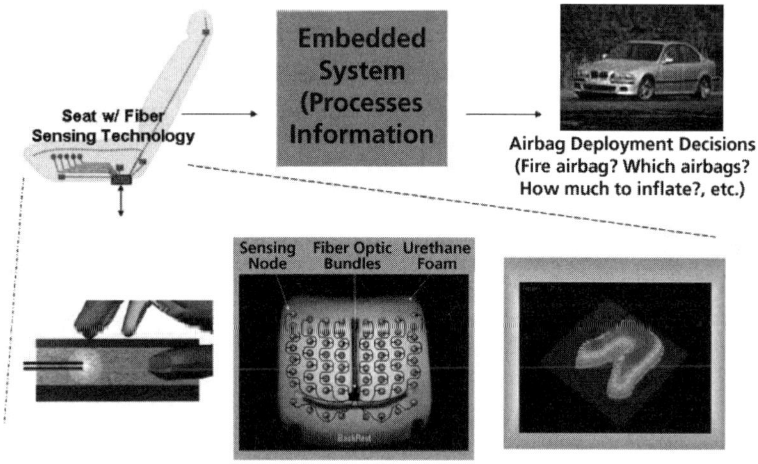

Figure 2.8 Automotive seat occupancy detection (courtesy of Texas Instruments)

Figure 2.9 shows a block diagram of a digital still camera (DSC). A DSC is an example of an embedded system. Referring back to the major components of an embedded system shown in Figure 2.3 we can see the following components in the DSC:

- The charge-coupled device analog front-end (CCD AFE) acts as the primary sensor in this system.
- The digital signal processor is the primary processor in this system.
- The battery management module controls the power for this system.
- The preview LCD screen is the user interface for this system.

- The Infrared port and serial ports are actuators in this system that interface to a computer.
- The graphics controller and picture compression modules are dedicated application-specific gates for processing accleration.
- The signal processing software runs on the DSP.

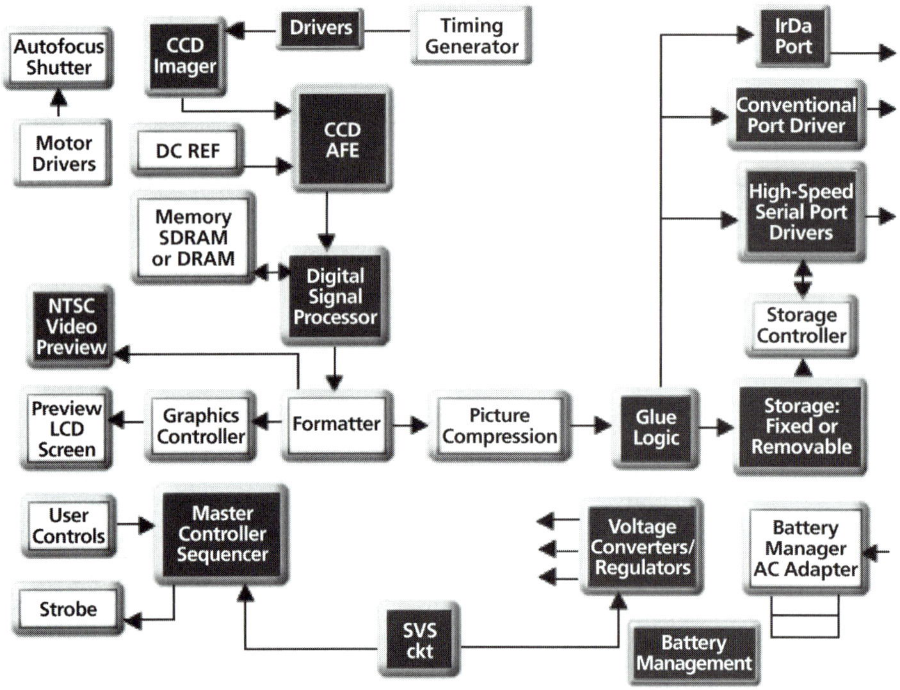

Figure 2.9 Block diagram of a digital still camera (courtesy of Texas Instruments)

Figure 2.10 shows another example of an embedded system. This is a block diagram of a cell phone. In this diagram, the major components of an embedded system are again obvious:

- The antenna is one of the sensors in this system. The microphone is another sensor. The keyboard also provides aperiodic events into the system.
- The voice codec is an application-specific acceleration in hardware gates.
- The DSP is one of the primary processor cores which runs most of the signal processing algorithms.
- The ARM processor is the other primary system processor running the state machines, controlling the user interface, and other components in this system.
- The battery/temp monitor controls the power in the system along with the supply voltage supervisor.
- The display is the primary user interface in the system.

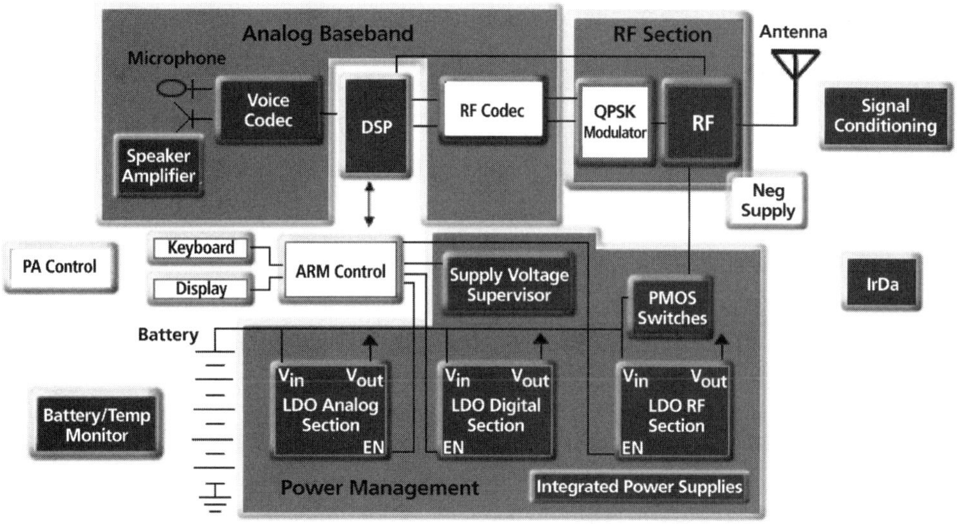

Figure 2.10 Block diagram of a cell phone (courtesy of Texas Instruments)

Summary

Many of the items that we interface with or use on a daily basis contain an embedded system. An embedded system is a system that is "hidden" inside the item we interface with. Systems such as cell phones, answering machines, microwave ovens, VCRs, DVD players, video game consoles, digital cameras, music synthesizers, and cars all contain embedded processors. A late model car contains more than 60 embedded microprocessors. These embedded processors keep us safe and comfortable by controlling such tasks as antilock braking, climate control, engine control, audio system control, and airbag deployment.

Embedded systems have the added burden of reacting quickly and efficiently to the external "analog" environment. That may include responding to the push of a button, a sensor to trigger an air bag during a collision, or the arrival of a phone call on a cell phone. Simply put, embedded systems have deadlines that can be hard or soft. Given the "hidden" nature of embedded systems, they must also react to and handle unusual conditions without the intervention of a human.

DSPs are useful in embedded systems principally for one reason; signal processing. The ability to perform complex signal processing functions in real time gives DSP the advantage over other forms of embedded processing. DSPs must respond in real time to analog signals from the environment, convert them to digital form, perform value added processing to those digital signals, and, if required, convert the processed signals back to analog form to send back out to the environment.

We will discuss the special architectures and techniques that allow DSPs to perform these real-time embedded tasks so quickly and efficiently. These topics are discussed in the coming chapters.

Programming embedded systems requires an entirely different approach from that used in desktop or mainframe programming. Embedded systems must be able to respond to external events in a very predictable and reliable way. Real-time programs must not only execute correctly, they must execute on time. A late answer is a wrong answer. Because of this requirement, we will be looking at issues such as concurrency, mutual exclusion, interrupts, hardware control and processing, and more later in the book because these topics become the dominant considerations. Multitasking, for example, has proven to be a powerful paradigm for building reliable and understandable real-time programs.

3

Overview of Embedded Systems Development Life Cycle Using DSP

Embedded Systems

As mentioned earlier, an embedded system is a specialized computer system that is integrated as part of a larger system. Many embedded systems are implemented using digital signal processors. The DSP will interface with the other embedded components to perform a specific function. The specific embedded application will determine the specific DSP to be used. For example, if the embedded application is one that performs video processing, the system designer may choose a DSP that is customized to perform media processing, including video and audio processing. An example of an application specific DSP for this function is shown in Figure 3.1. This device contains dual channel video ports that are software configurable for input or output, as well as video filtering and automatic horizontal scaling and support of various digital TV formats such as HDTV, multichannel audio serial ports, multiple stereo lines, and an Ethernet peripheral to connect to IP packet networks. It is obvious that the choice of a DSP "system" depends on the embedded application.

In this chapter we will discuss the basic steps to develop an embedded application using DSP.

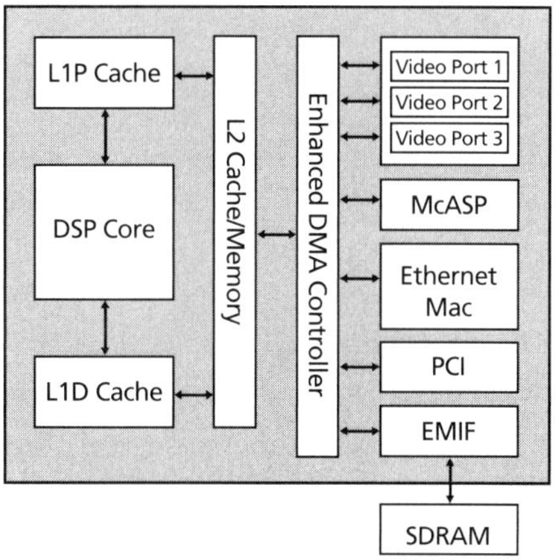

Figure 3.1 Example of a DSP-based "system" for embedded video applications

The Embedded System Life Cycle Using DSP

In this section we will overview the general embedded system life cycle using DSP. There are many steps involved in developing an embedded system—some are similar to other system development activities and some are unique. We will step through the basic process of embedded system development, focusing on DSP applications.

Step 1—Examine the Overall Needs of the System

Choosing a design solution is a difficult process. Often the choice comes down to emotion or attachment to a particular vendor or processor, inertia based on prior projects and comfort level. The embedded designer must take a positive logical approach to comparing solutions based on well defined selection criteria. For DSP, specific selection criteria must be discussed. Many signal processing applications will require a mix of several system components as shown in Figure 3.2.

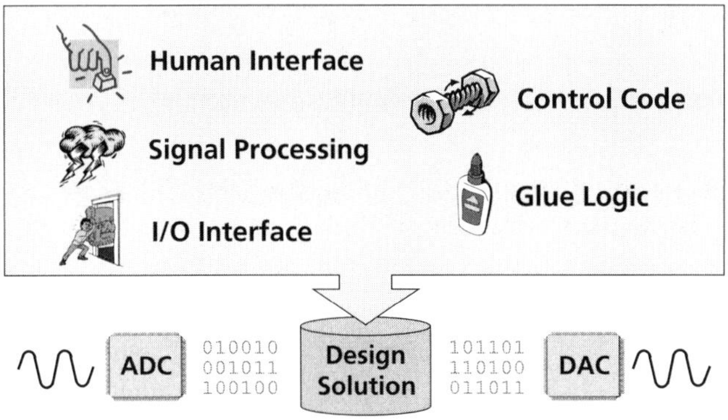

Figure 3.2 Most signal processing applications will require a mix of various system components (courtesy of Texas Instruments)

What is a DSP solution?

A typical DSP product design uses the digital signal processor itself, analog/mixed signal functions, memory, and software, all designed with a deep understanding of overall system function. In the product, the analog signals of the real world, signals representing anything from temperature to sound and images, are translated into digital bits—zeros and ones—by an analog/mixed signal device. Then the digital bits or signals are processed by the DSP. Digital signal processing is much faster and more precise than traditional analog processing. This type of processing speed is needed for today's advanced communications devices where information requires instantaneous processing, and in many portable applications that are connected to the Internet.

There are many selection criteria for embedded DSP systems. Some of these are shown in Figure 3.3. These are the major selection criteria defined by Berkeley Design Technology Incorporated (bdti.com). Other selection criteria may be "ease of use," which is closely linked to "time-to-market" and also "features." Some of the basic rules to consider in this phase are:

- For a fixed cost, maximize performance.
- For a fixed performance, minimize cost.

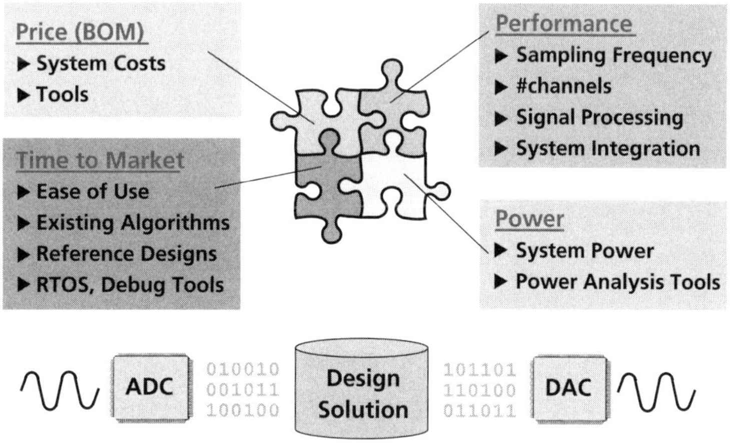

Figure 3.3 The design solution will be influenced by these major criteria and others (courtesy of Texas Instruments)

Step 2—Select the Hardware Components Required for the System

In many systems, a general-purpose processor (GPP), field-programmable gate array (FPGA), microcontroller (mC) or DSP is not used as a single-point solution. This is because designers often combine solutions, maximizing the strengths of each device (Figure 3.4).

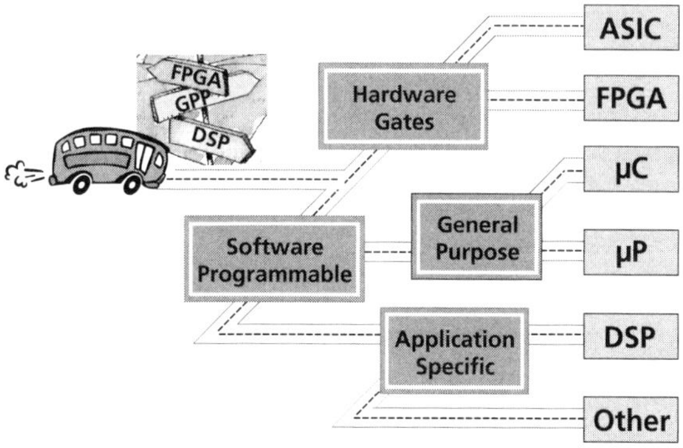

Figure 3.4 Many applications, multiple solutions (courtesy of Texas Instruments)

One of the first decisions that designers often make when choosing a processor is whether they would like a software-programmable processor in which functional blocks are developed in software using C or assembly, or a hardware processor in which functional blocks are laid out logically in gates. Both FPGAs and application specific integrated circuits (ASICs) may integrate a processor core (very common in ASICs).

Hardware Gates

Hardware gates are logical blocks laid out in a flow, therefore any degree of parallelization of instructions is theoretically possible. Logical blocks have very low latency, therefore FPGAs are more efficient for building peripherals than "bit-banging" using a software device.

If a designer chooses to design in hardware, he or she may design using either an FPGA or ASIC. FPGAs are termed "field programmable" because their logical architecture is stored in a nonvolatile memory and booted into the device. Thus, FPGAs may be reprogrammed in the field simply by modifying the nonvolatile memory (usually FLASH or EEPROM). ASICs are not field-programmable. They are programmed at the factory using a mask which cannot be changed. ASICs are often less expensive and/or lower power. They often have sizable nonrecurring engineering (NRE) costs.

Software-Programmable

In this model, instructions are executed from memory in a serial fashion (that is, one per cycle). Software-programmable solutions have limited parallelization of instructions; however, some devices can execute multiple instructions in parallel in a single cycle. Because instructions are executed from memory in the CPU, device functions can be changed without having to reset the device. Also, because instructions are executed from memory, many different functions or routines may be integrated into a program without the need to lay out each individual routine in gates. This may make a software-programmable device more cost efficient for implementing very complex programs with a large number of subroutines.

If a designer chooses to design in software, there are many types of processors available to choose from. There are a number of general-purpose processors, but in addition, there are processors that have been optimized for specific applications. Examples of such application specific processors are graphics processors, network processors and digital signal processors (DSPs). Application specific processors usually offer higher performance for a target application, but are less flexible than general-purpose processors.

General-Purpose Processors

Within the category of general-purpose processors are microcontrollers (μC) and microprocessors (μP) (Figure 3.5).

Microcontrollers usually have control-oriented peripherals. They are usually lower cost and lower performance than microprocessors. Microprocessors usually have communications-oriented peripherals. They are usually higher cost and higher performance than microcontrollers.

Strengths	▸ Familiar Design Environment (tools, s/w, emulation) ▸ Robust communication peripherals ▸ Ability to use higher-end O/Ss (control code) ▸ Great for compiling generic (non-tuned) C code
Signal Processing	▸ Fair to good
Home Run Apps	▸ PC, PDA

Figure 3.5 General-purpose processor solutions (courtesy of Texas Instruments)

Note that some GPPs have integrated MAC units. It is not a "strength" of GPPs to have this capability because all DSPs have MACs—but, it is worth noting because a student might mention it. Regarding performance of the GPP's MAC, it is different for each one.

Microcontrollers

A microcontroller is a highly integrated chip that contains many or all of the components comprising a controller. This includes a CPU, RAM and ROM, I/O ports, and timers. Many general-purpose computer are designed the same way. But a microcontroller is usually designed for very specific tasks in embedded systems. As the name implies, the specific task is to control a particular system, hence the name microcontroller. Because of this customized task, the device's parts can be simplified, which makes these devices very cost effective solutions for these types of applications.

PIC12	68HC11/16	MCS51	MSP430

Strengths	▸ Good control peripherals ▸ May have ability to use mid-range O/Ss ▸ Very low cost ▸ Integrated FLASH ▸ Can be very low power
Signal Processing	▸ Poor to fair
Home Run Apps	▸ Embedded control, small home appliances

Figure 3.6 Microcontroller solutions (courtesy of Texas Instruments)

Some microcontrollers can actually do a multiply and accumulate (MAC) in a single cycle. But that does not necessarily make it a DSP. True DSPs can allow two 16x16 MACS in a single cycle including bringing the data in over the buses, and so on. It is this that truly makes the part a DSP. So, devices with hardware MACs might get a "fair" rating. Others get a "poor" rating. In general, microcontrollers can do DSP but they will generally do it slower.

FPGA Solutions

An FPGA is an array of logic gates that are hardware-programmed to perform a user-specified task. FPGAs are arrays of programmable logic cells interconnected by a matrix of wires and programmable switches. Each cell in an FPGA performs a simple logic function. These logic funcctions are defined by an engineer's program. FPGA contain large numbers of these cells (1000–100,000) available to use as building blocks in DSP applications.The advantage of using FPGAs is that the engineer can create special purpose functional units that can perform limited tasks very efficiently. FPGAs can be reconfigured dynamically as well (usually 100–1,000 times per second depending on the device). This makes it possible to optimize FPGAs for complex tasks at speeds higher than what can be achieved using a general-purpose processor. The ability to manipulate logic at the gate level means it is possible to construct custom DSP-centric processors that efficiently implement the desired DSP function. This is possible by simultaneously performing all of the algorithm's subfunctions. This is where the FPGA can achieve performance gains over a programmable DSP processor.

The DSP designer must understand the trade-offs when using an FPGA (Figure 3.7). If the application can be done in a single programmable DSP, that is usually the best way to go since talent for programming DSPs is usually easier to find than FPGA designers. Also, software design tools are common, cheap and sophisticated, which improves development time and cost. Most of the common DSP algorithms are also available in well packaged software components. Its harder to find these same algorithms implemented and available for FPGA designs.

An FPGA is worth considering, however, if the desired performance cannot be achieved using one or two DSPs, or when there may be significant power concerns (although a DSP is also a power efficient device—benchmarking needs to be performed) or when there may be significant programmatic issues when developing and integrating a complex software system.

Typical applications for FPGAs include radar/sensor arrays, physical system and noise modeling, and any really high I/O and high-bandwidth application.

FLEX 10K/ Stratix   Spartan - 3/ Virtex - II

Strengths	▶ Fastest possible computation ▶ Excellent design support tools ▶ Some kind of PLD usually required in design ▶ Ability to synthesize almost any peripheral ▶ Easy to develop with ▶ Flexible features – field reprogrammable
Signal Processing	▶ Excellent for hi-speed & parallel signal processing
Home Run Apps	▶ Glue logic, radar/sensor arrays

Figure 3.7 FPGA solutions for DSP (courtesy of Texas Instruments)

Digital Signal Processors

A DSP is a specialized microprocessor used to perform calculations efficiently on digitized signals that are converted from the analog domain. One of the big advantages of DSP is the programmability of the processor, which allows important system parameters to be changed easily to accommodate the application. DSPs are optimized for digital signal manipulations.

DSPs provide ultra-fast instruction sequences such as shift and add, and multiply and add. These instruction sequences are common in many math-intensive signal processing applications. DSPs are used in devices where this type of signal processing is important, such as sound cards, modems, cell phones, high-capacity hard disks and digital TVs (Figure 3.8).

ANALOG DEVICES   TEXAS INSTRUMENTS

BlackFin/Sharc   C2000/C5000/C6000   DSP56xxx/StarCore

Strengths	▶ Architecture optimized for computing DSP algorithms ▶ Excellent MIP/mW/$$ tradeoff ▶ Efficient compilers – can program entire app in C ▶ Some have a real-time O/S (for task scheduling) ▶ Can be very low power
Signal Processing	▶ Good to excellent
Home Run Apps	▶ Cell phones, telecom infrastructure, digital cameras ▶ DSL/cable/modems, audio/video, multimedia

Figure 3.8 DSP processor solutions (courtesy of Texas Instruments)

A General Signal Processing Solution

The solution shown in Figure 3.9 allows each device to perform the tasks it's best at, achieving a more efficient system in terms of cost/power/performance. For example, in Figure 3.9, the system designer may put the system control software (state machines and other communication software) on the general-purpose processor or microcontroller, the high performance, single dedicated fixed functions on the FPGA and the high I/O signal processing functions on the DSP.

When planning the embedded product development cycle, there are multiple opportunities to reduce cost and/or increase functionality using combinations of GPP/uC, FPGA, and DSP. This becomes more of an issue in higher-end DSP applications. These are applications which are computationally intensive and performance critical. These applications require more processing power and channel density than can be provided by GPPs alone. For these high-end applications, there are software/hardware alternatives that the system designer must consider. Each alternative provides different degrees of performance benefits and must be also be weighed against other important system parameters including cost, power consumption and time-to-market.

The system designer may decide to use an FPGA in a DSP system for the following reasons:

- A decision to extend the life of a generic, lower-cost microprocessor or DSP by offloading computationally intensive work to a FPGA.
- A decision to reduce or eliminate the need for a higher-cost, higher performance DSP processor.
- To increase computational throughput. If the throughput of an existing system must increase to handle higher resolutions or larger signal bandwidths, an FPGA may be an option. If the required performance increases are computational in nature, an FPGA may be an option.
- For prototyping new signal processing algorithms; since the computational core of many DSP algorithms can be defined using a small amount of C code, the system designer can quickly prototype new algorithmic approaches on FPGAs before committing to hardware or other production solution like an ASIC.
- For implementing "glue" logic; various processor peripherals and other random or "glue" logic are often consolidated into a single FPGA. This can lead to reduced system size, complexity and cost.

By combining the capabilities of FPGAs and DSP processors, the system designer can increase the scope of the system design solution. Combinations of fixed hardware and programmable processors are a good model for enabling flexibility, programmability, and computational acceleration of hardware for the system.

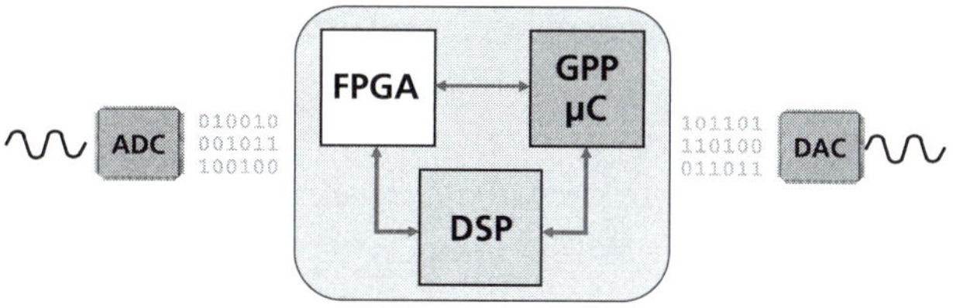

Figure 3.9 General signal processing solution (courtesy of Texas Instruments)

DSP Acceleration Decisions

In DSP system design, there are several things to consider when determining whether a functional component should be implemented in hardware or software:

Signal processing algorithm parallelism – Modern processor architectures have various forms of instruction level parallelism (ILP). One example is the 64x DSP which has a very long instruction word (VLIW) architecture (more about this in Chapter 5). The 64x DSP exploits ILP by grouping multiple instructions (adds, multiplies, loads and stores) for execution in a single processor cycle. For DSP algorithms that map well to this type of instruction parallelism, significant performance gains can be realized. But not all signal processing algorithms exploit such forms of parallelism. Filtering algorithms such as finite impulse response (FIR) algorithms are recursive and are sub-optimal when mapped to programmable DSPs. Data recursion prevents effective parallelism and ILP. As an alternative, the system designer can build dedicated hardware engines in an FPGA.

Computational complexity – Depending on the computational complexity of the algorithms, these may run more efficiently on a FPGA instead of a DSP. It may make sense, for certain algorithmic functions, to implement in a FPGA and free up programmable DSP cycles for other algorithms. Some FPGAs have multiple clock domains built into the fabric, which can be used to separate different signal processing hardware blocks into separate clock speeds based on their computational requirements. FPGAs can also provide flexibility by exploiting data and algorithm parallelism using multiple instantiations of hardware engines in the device.

Data locality – The ability to access memory in a particular order and granularity is important. Data access takes time (clock cycles) due to architectural latency, bus contention, data alignment, direct memory access (DMA) transfer rates, and even the type of memory being used in the system. For example, static RAM (SRAM) which is very fast but much more expensive than dynamic RAM (DRAM), is often used as cache memory due to its speed. Synchronous DRAM (SDRAM), on the other hand, is directly dependent on the clock speed of the entire system (that's why they call it synchronous). It basically works at the same speed as the system bus. The overall performance of the system is driven in part by which type of memory is being used. The physical interfaces between the data unit and the arithmetic unit are the primary drivers of the data locality issue.

Data parallelism – Many signal processing algorithms operate on data that is highly capable of parallelism, such as many common filtering algorithms. Some of the more advanced high-performance DSPs have single instruction multiple data (SIMD) capability in the architectures and/or compilers that implement various forms of vector processing operations. FPGA devices are also good at this type of parallelism. Large amounts of RAM are used to support high bandwidth requirements. Depending on the DSP processor being used, an FPGA can be used to provide this SIMD processing capability for certain algorithms that have these characteristics.

A DSP-based embedded system could incorporate one, two or all three of these devices depending on various factors:

- # signal processing tasks/channels
- Sampling rate
- Memory/peripherals needed
- Power requirements
- Availability of desired algorithms
- Amount of control code
- Development environment
- Operating system (O/S or RTOS)
- Debug capabilities
- Form factor, system cost

The trend in embedded DSP development is moving more towards programmable solutions as shown in Figure 3.10. There will always be a trade-off depending on the application but the trend is moving towards software and programmable solutions.

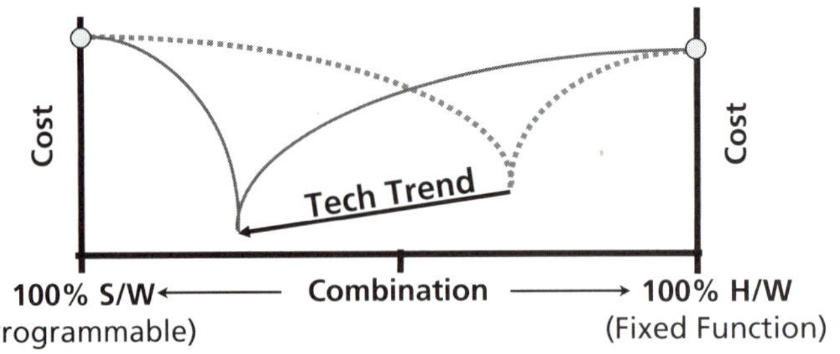

Figure 3.10 Hardware /software mix in an embedded system; the trend is towards more software (courtesy of Texas Instruments)

"Cost" can mean different things to different people. Sometimes, the solution is to go with the lowest "device cost." However, if the development team then spends large amounts of time re-doing work, the project may be delayed; the "time-to-market" window may extend, which, in the long run, costs more than the savings of the low-cost device.

The first point to make is that a 100% software or hardware solution is usually the most expensive option. A combination of the two is the best. In the past, more functions were done in hardware and less in software. Hardware was faster, cheaper (ASICs) and good C compilers for embedded processors just weren't available. However, today, with better compilers, faster and lower-cost processors available, the trend is toward more of a software-programmable solution. A software-only solution is not (and most likely never will be) the best overall cost. Some hardware will still be required. For example, let's say you have ten functions to perform and two of them require extreme speed. Do you purchase a very fast processor (which costs 3–4x the speed you need for the other eight functions) or do you spend 1x on a lower-speed processor and purchase an ASIC or FPGA to do only those two critical functions? It's probably best to choose the combination.

- **Cost can be defined by as a combination of the following:**

▸ Device Cost	▸ Power Dissipation
▸ NRE	▸ Time to Market
▸ Manufacturing Cost	▸ Weight
▸ Opportunity Cost	▸ Size

A combination of software and hardware always gives the lowest cost system design.

Step 3—Understand DSP Basics and Architecture

One compelling reason to choose a DSP processor for an embedded system application is performance. Three important questions to understand when deciding on a DSP are:

- What makes a DSP a DSP?
- How fast can it go?
- How can I achieve maximum performance without writing in assembly?

In this section we will begin to answer these questions. We know that a DSP is really just an application specific microprocessor. They are designed to do a certain thing, signal processing, very efficiently. We mentioned the types of signal processing algorithms that are used in DSP. They are shown again in Figure 3.11 for reference.

Algorithm	Equation
Finite Impluse Response Filter	$y(n)=\sum_{k=0}^{M} a_k x(n-k)$
Infinite Impulse Response Filter	$y(n)=\sum_{k=0}^{M} a_k x(n-k)+\sum_{k=1}^{N} b_k y(n-k)$
Convolution	$y(n)=\sum_{k=0}^{N} x(k)h(n-k)$
Discrete Fourier Transform	$X(k)=\sum_{n=0}^{N-1} x(n)\exp[-j(2\pi/N)nk]$
Discrete Cosine Transform	$F(u)=\sum_{x=0}^{N-1} c(u).f(x).\cos\left[\frac{\pi}{2N}u(2x+1)\right]$

Figure 3.11 Typical DSP algorithms (courtesy of Texas Instruments)

Notice the common structure of each of the algorithms:

- They all accumulate a number of computations.
- They all sum over a number of elements.
- They all perform a series of multiplies and adds.

These algorithms all share some common characteristics; they perform multiplies and adds over and over again. This is generally referred to as the sum of products (SOP).

DSP designers have developed hardware architectures that allow the efficient execution of algorithms to take advantage of this algorithmic specialty in signal processing. For example, some of the specific architectural features of DSPs accommodate the algorithmic structure described in Figure 3.11.

As an example, consider the FIR diagram in Figure 3.12 as an example DSP algorithm which clearly shows the multiply/accumulate and shows the need for doing MACs very fast, along with reading at least two data values. As shown in Figure 3.12, the filter algorithm can be implemented using a few lines of C source code. The signal flow diagram shows this algorithm in a more visual context. Signal flow diagrams are used to show overall logic flow, signal dependencies, and code structure. They make a nice addition to code documentation.

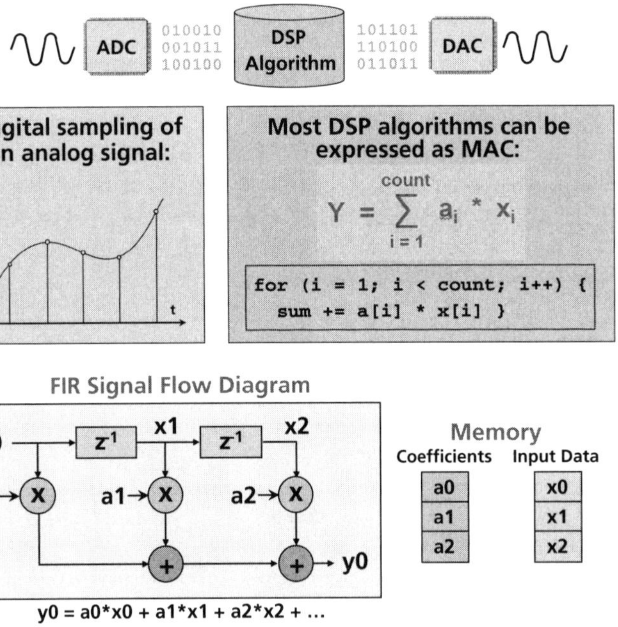

Figure 3.12 DSP filtering using a FIR filter (courtesy of Texas Instruments)

To execute at top speed, a DSP needs to:

- read *at least* two values from memory (minimum),
- multiply coeff * data,
- accumulate (+) answer (an * xn) to running total…,
- …and do all of the above in a single cycle (or less).

DSP architectures support the requirements above (Figure 3.13):

- High-speed memory architectures support multiple accesses/cycle.
- Multiple read buses allow two (or more) data reads/cycle from memory.
- The processor pipeline overlays CPU operations allowing one-cycle execution.

All of these things work together to result in the highest possible performance when executing DSP algorithms. A deeper discussion of DSP architectures is given in Chapter 5).

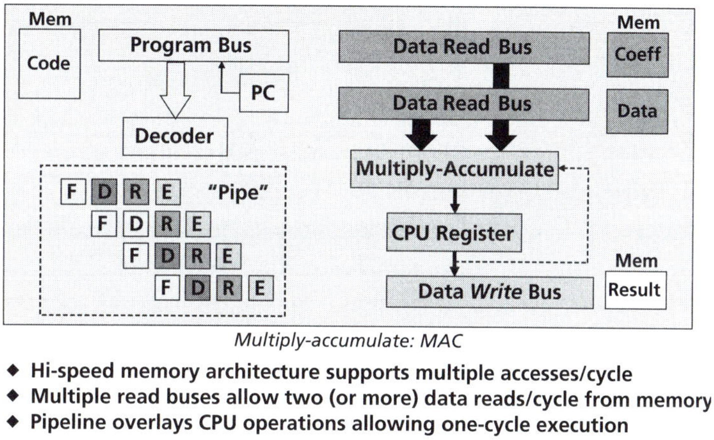

Figure 3.13 Architectural block diagram of a DSP (courtesy of Texas Instruments)

Other DSP architectural features are summarized in Figure 3.14.

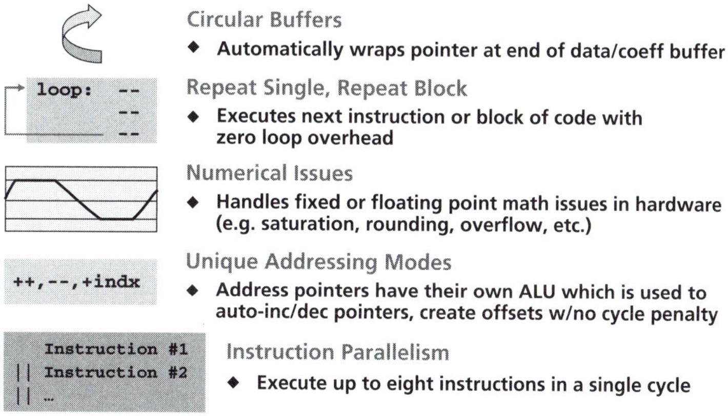

Figure 3.14 DSP CPU architectural highlights (courtesy of Texas Instruments)

Models of DSP Processing

There are two types of DSP processing models—*single sample* model and *block processing* model. In a single sample model of signal processing (Figure 3.15a), the output must result before next input sample. The goal is minimum latency (in-to-out time). These systems tend to be interrupt intensive; interrupts drive the processing for the next sample. Example DSP applications include motor control and noise cancellation.

In the block processing model (Figure 3.15b), the system will output a buffer of results before the next input buffer fills. DSP systems like this use the DMA to transfer samples to the buffer. There is increased latency in this approach as the buffers are filled before processing. However, these systems tend to be computationally efficient. The main types of DSP applications that use block processing include cellular telephony, video, and telecom infrastructure.

An example of stream processing is averaging data sample. A DSP system that must average the last three digital samples of a signal together and output a signal at the same rate as what is being sampled must do the following:

- Input a new sample and store it.
- Average the new sample with the last two samples.
- Output the result.

These three steps must complete before the next sample is taken. This is an example of stream processing. The signal must be processed in real time. A system that is sampling at 1000 samples per second has one thousandth of a second to complete the operation in order to maintain real-time performance.

Block processing, on the other hand, accumulates a large number of samples at a time and processes those samples while the next buffer of samples is being collected. Algorithms such as the fast Fourier transform (FFT) operate in this mode.

Block processing (processing a block of data in a tight inner loop) can have a number of advantages in DSP systems:

- If the DSP has an instruction cache, this cache will optimize instructions to run faster the second (or subsequent) time through the loop.
- If the data accesses adhere to a locality of reference (which is quite common in DSP systems) the performance will improve. Processing the data in stages means the data in any given stage will be accessed from fewer areas, and therefore less likely to thrash the data caches in the device.
- Block processing can often be done in simple loops. These loops have stages where only one kind of processing is taking place. In this manner there will be less thrashing from registers to memory and back. In many cases, most if not all of the intermediate results can be kept in registers or in level one cache.

- By arranging data access to be sequential, even data from the slowest level of memory (DRAM) will be much faster because the various types of DRAM assume sequential access.

DSP designers will use one of these two methods in their system. Typically, control algorithms will use single-sample processing because they cannot delay the output very long such as in the case of block processing. In audio/video systems, block processing is typically used—because there can be some delay tolerated from input to output.

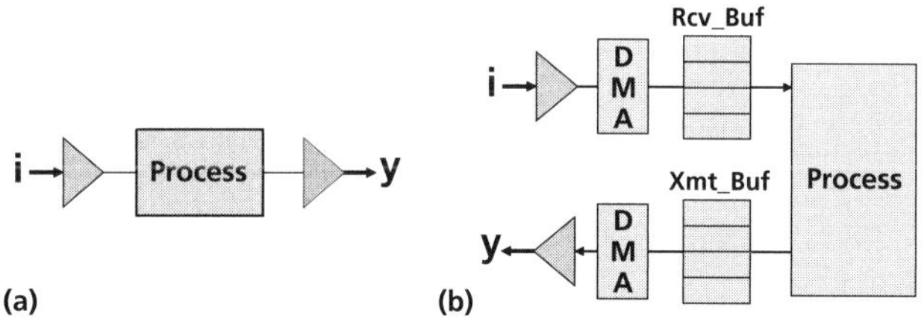

Figure 3.15 Single sample (a) and block processing (b) models of DSP

Input/Output Options

DSPs are used in many different systems including motor control applications, performance-oriented applications and power sensitive applications. The choice of a DSP processor is dependent on not just the CPU speed or architecture but also the mix of peripherals or I/O devices used to get data in and out of the system. After all, much of the bottleneck in DSP applications is not in the compute engine but in getting data in and out of the system. Therefore, the correct choice of peripherals is important in selecting the device for the application. Example I/O devices for DSP include:

GPIO – A flexible parallel interface that allows a variety of custom connections.

UART – Universal asynchronous receiver-transmitter. This is a component that converts parallel data to serial data for transmission and also converts received serial data to parallel data for digital processing.

CAN – Controller area network. The CAN protocol is an international standard used in many automotive applications.

SPI – Serial peripheral interface. A three-wire serial interface developed by Motorola.

USB – Universal serial bus. This is a standard port that enables the designer to connect external devices (digital cameras, scanners, music players, etc) to computers. The USB standard supports data transfer rates of 12 Mbps (million bits per second).

McBSP – Multichannel buffered serial port. These provide direct full-duplex serial interfaces between the DSP and other devices in a system.

HPI – Host port interface. This is used to download data from a host processor into the DSP.

A summary of I/O mechanisms for DSP application class is shown in Figure 3.16.

Motor	•12-bit ADC •PWM DAC •McBSP •UART	•CAN 2.0B •GPIO •EMIF	•SPI •SCI •I^2C
Power	•USB •McBSP •HPI	•EMIF •GPIO •10-bit ADC	•MMC/SD serial ports •UART •I^2C
Perf	•PCI •McBSP •HPI •Utopia SP	•EMIF •GPIO •I2C	•Video ports •Audio ports •McASP •Ethernet 10/100 MAC

Figure 3.16 Input/output options (courtesy of Texas Instruments)

Calculating DSP Performance

Before choosing a DSP processor for a specific application, the system designer must evaluate three key system parameters as shown below:

- **Maximum CPU Performance**
  "What is the maximum number of times the CPU can execute your algorithm? (max # channels)
- **Maximum I/O Performance**
  "Can the I/O keep up with this maximum #channels?"
- **Available Hi-Speed Memory**
  "Is there enough hi-speed internal memory?"

With this knowledge, the system designer can scale the numbers to meet the application's needs and then determine:

- CPU load (% of maximum CPU).
- At this CPU load, what other functions can be performed?

The DSP system designer can use this process for any CPU they are evaluating. The goal is the find the "weakest link" in terms of performance so that you know what the system constraints are. The CPU might be able to process numbers at sufficient rates, but if the CPU cannot be fed with data fast enough, then having a fast CPU doesn't really matter. The goal is to determine the maximum number of channels that can be processed given a specific algorithm and then work that number down based on other constraints (maximum input/output speed and available memory).

As an example, consider the process shown in Figure 3.17. The goal is to determine the maximum number of channels that this specific DSP processor can handle given a specific algorithm. To do this, we must first determine the benchmark of the chosen algorithm (in this case, a 200-tap FIR filter). The relevant documentation for an algorithm like this (from a library of DSP functions) gives us the benchmark with two variables: nx (size of buffer) and nh (# coeffs)—these are used for the first part of the computation. This FIR routine takes about 106K cycles per frame. Now, consider the sampling frequency. A key question to answer at this point is "How many times is a frame FULL per second?" To answer this, divide the sampling frequency (which specifies how often a new data item is sampled) by the size of the buffer. Performing this calculation determines that we fill about 47 frames per second. Next, is the most important calculation—how many MIPS does this algorithm require of a processor? We need to find out how many cycles this algorithm will require per second. Now we multiply frames/second * cycles/frame and perform the calculation using these data to get a throughput rate of about 5 MIPs. Assuming this is the only computation being performed on the processor, the channel density (how many channels of simultaneous processing can be performed by a processor) is a maximum of 300/5 = 60 channels. This completes the CPU calculation. This result can not be used in the I/O calculation.

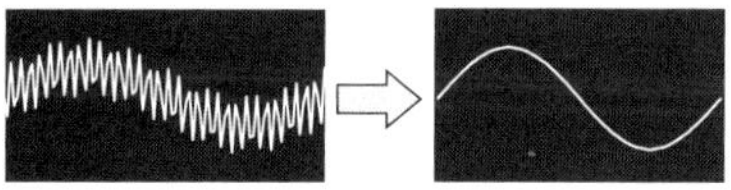

Algorithm: 200-tap (nh) low-pass FIR filter
Frame size: 256 (nx) 16-bit elements
Sampling frequency: 48KHz

How many channels can the DSP handle given this algorithm?

C	FIR benchmark:	(nx/2) (nh+7) = 128 * 207 =	26496 cyc/frm
	#times frm full/s:	(samp freq / frm size) = 48000/256 =	187.5 frm/s
P	MIP calc:	(frm/s) (cyc/frm) = 187.5 * 26496 =	4.97M cyc/s
	Conclusion:	FIR takes ~5MIPs on a C5502	
U	Max #channels:	60 @300MHz	

Max # channels: does not include overhead for interrupts, control code, RTOS, etc.

Are the I/O and memory capable of handling this many channels?

I	Required I/O rate:	48Ksamp/s * #Ch = 48000 * 16 * 60 =	46.08 Mbps
	DSP SP rate:	serial port is full duplex	50.00 Mbps ✓
/	DMA Rate:	(2x16-bit xfrs/cycle) * 300MHz =	9600 Mbps ✓
	Req'd Data Mem	(60 * 200) + (60 * 4 * 256) + (60 * 2 * 199) = 97K x 16-bit	
O	Avail int'l mem:		32K x 16-bit X

Required memory assumes: 60 different filters, 199 element delay buffer, double buffering rcv/xmt

Figure 3.17 Example – performance calculation (courtesy of Texas Instruments)

The next question to answer is "Can the I/O interface feed the CPU fast enough to handle 60 channels?" Step one is to calculate the "bit rate" required of the serial port. To do this, the required sampling rate (48 KHz) is multiplied by the maximum channel density (60). This is then multiplied by 16 (assuming the word size is 16—which it is given the chosen algorithm). This calculation yields a requirement of 46 Mbps for 60 channels operating at 48 KHz. In this example what can the 5502 DSP serial port support? The specification says that the maximum bit rate is 50 Mbps (half the CPU clock rate up to 50 Mbps). This tells us that the processor can handle the rates we need for this chosen application. Can the DMA move these samples from the McBSP to memory fast enough? Again, the specification tells us that this should not be a problem.

The next step considers the issue of required data memory. This calculation is somewhat confusing and needs some additional explanation.

Assume that all 60 channels of this application are using different filters—that is, 60 different sets of coefficients and 60 double-buffers (this can be implemented using a ping-pong buffer on both the receive and transmit sides. This is a total of 4 buffers per channel hence the *4 + the delay buffers for each channel (only the receive side has delay buffers…) so the algorithm becomes:

Number of channels * 2 * delay buffer size
= 60 * 2 * 199

This is extremely conservative and the system designer could save some memory if this is not the case. But this is a worst-case scenario. Hence, we'll have 60 sets of 200 coefficients, 60 double-buffers (ping and pong on receive and transmit, hence the *4) and we'll also need a delay buffer of number of coefficients—1 which is 199 for each channel. So, the calculation is:

(#Channels * #coefficients) + (#Channels * 4 * frame size) + (#Channels * #delay_buffers * delay_buffer_size)

= (60 * 200) + (60 * 4 * 256) + (60 * 2 * 199) = 97320 bytes of memory

This results in a requirement of 97K of memory. The 5502 DSP only has 32K of on-chip memory, so this is a limitation. Again, you can redo the calculation assuming only one type of filter is used, or look for another processor.

	DSP	FIR Benchmark	cyc/frm	frm/s	cyc/s	%CPU	Max Ch
C	C2812	(nx/2)(nh+12)+ â	27712	187.5	5.20M	3.5	28
P	C5502	(nx/2)(nh+7)	26496	187.5	4.97M	1.7	60
U	C6416	(nx/4+15)(nh+11)	16669	187.5	3.13M	0.4	230

â = 36nx/16 = additional time to transfer 16 samples to memory

	DSP	#Ch	Req'd IO rate	Avail SP rate	Avail DMA Rate	Req'd Memory	Avail Int Mem
I	C2812	28	21.5 Mbps	50Mbps ✓	None	46K	18K X
/	C5502	60	46.1 Mbps	50Mbps ✓	9.6 Gbps	97K	32K X
O	C6416	230	176.6 Mbps	100Mbps X	46.1 Gbps	373K	512K ✓

- Bandwidth calculations help determine processor's capability
- Limiting factors: I/O rate, available memory, CPU performance
- Use your system needs (such as 8 Ch) to calculate CPU loading (for example, 3%).

CPU load can help guide your system design...

Figure 3.18 Performance calculation analysis (courtesy of Texas Instruments)

Now we extend the calculations to the 2812 and the 6416 processors (Figure 3.18). A couple of things to note:

The 2812 is best used in a single-sample processing mode, so using a block FIR application on a 2812 is not the best fit. But for example purposes it is done this way to benchmark one processor vs. another. Where block processing hurts the 2812 is in relation to getting the samples into on-chip memory. There is no DMA on the 2812 because in single-sample processing, it is not required. The term "beta" in the calculation is the time it takes to move (using CPU cycles) the incoming sampled signals from the A/D to memory. This would be performed by an interrupt service routine and it must be accounted for. Notice that the benchmarks for the 2812 and 5502 are very close.

The 6416 is a high performance machine when doing 16-bit operations—it can do 269 channels given the specific FIR used in this example. Of course, the I/O (on one serial port) can't keep up with this, but it could with 2 serial ports in operation.

Once you've done these calculations, you can "back off" the calculation to the exact number of channels your system requires, determine an initial theoretical CPU load that is expected and then make some decisions about what to do with any additional bandwidth that is left over (Figure 3.19).

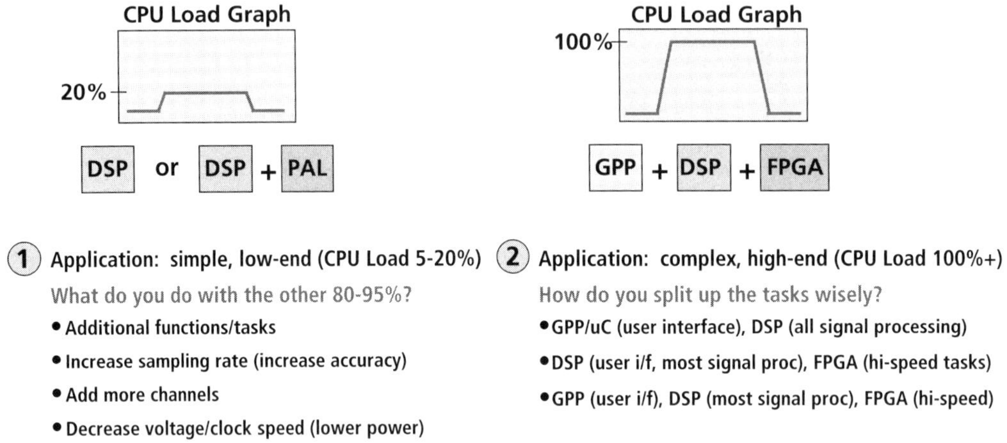

Figure 3.19 Determining what to do based on available CPU bandwith (courtesy of Texas Instruments)

Two sample cases that help drive discussion on issues related to CPU load are shown in Figure 3.19. In the first case, the entire application only takes 20% of the CPU's load. What do you do with the extra bandwidth? The designer can add more algorithmic processing, increase the channel density, increase the sampling rate to achieve higher resolution or accuracy, or decrease the clock/voltage so that the CPU load goes up and you save lots of power. It is up to the system designer to determine the best strategy here based on the system requirements.

The second example application is the other side of the fence—where the application takes more processing power than the CPU can handle. This leads the designer to consider a combined solution. The architecture of this again depends on the application's needs.

DSP Software

DSP software development is primarily focused on achieving the performance goals of the system. Its more efficient to develop DSP software using a high-level language like C or C++ but it is not uncommon to see some of the high performance, MIPS intensive algorithms written at least partially in assembly language. When generating DSP algorithm code, the designer should use one or more of the following approaches:

- Find existing algorithms (free code).
- Buy or license algorithms from vendors. These algorithms may come bundled with tools or may be classes of libraries for specific applications (Figure 3.20).
- Write the algorithms in house. If using this approach, implement as much of the algorithm as possible in C/C++. This usually results in faster time-to-market and requires a common skill found in the industry. It is much easier to find a C

programmer than a 5502 DSP assembly language programmer. DSP compiler efficiency is fairly good and significant performance can be achieved using a compiler with the right techniques. There are several tuning techniques used to generate optimal code and these will be discussed in later chapters.

- **May be bundled with tools and contains:**
  - **C-callable highly-optimized assembly routines**
  - **Documentation on each algorithm**
  - **Examples: FIR, IIR, FFT, convolution, min/max, log, etc.**
- **Other libraries available for specific DSPs:**
  - **Image libraries**
  - **Other control-specific free libraries**
- **Use third parties**
  - **Lists application software by platform, algorithm, and third party**
  - **Includes specs such as data/code size, performance, licensing fees**

Figure 3.20 Reuse opportunities–using DSP libraries and third parties

To fine-tune code and get the highest efficiency possible, the system designer needs to know three things:

- The architecture.
- The algorithms.
- The compiler.

Figure 3.21 shows some ways to help the compiler generate efficient code. These techniques will be discussed in more detail in Chapter 6. Compilers are pessimistic by nature, so the more information that can be provided about the system algorithms, where data is in memory, and so on, the better. The C6000 compiler can achieve 100% efficiency vs. hand-coded assembly if the right techniques are used. There are pros and cons to writing DSP algorithms in assembly language as well, so if this must be done, these must be understood from the beginning (Figure 3.22).

- Pragmas : **target-specific instructions/hints to the compiler**

```
#pragma DATA_SECTION (buffer, "buffer_sect");
int buffer[32];
// places buffer in specific location in memory map
```

- Intrinsics : **C function call to access specific ASM instructions**

  C: `y = a*b;` Intrinsic: `y = smpy (a,b); //saturated multiply`

- Compiler options: **affect efficiency of compiler**
  - Optimization levels
  - Amount of debug info
  - Target-specific options
  - Many, many more...

Figure 3.21 Compiler optimization techniques for producing high performance code (courtesy of Texas Instruments)

Pros	• Can result in highest possible performance • Access to native instruction set (including application-specific instructions)
Cons	• Usually difficult learning curve (often increases development time) • Usually not portable
Conclusions	• Write in C when possible (most of the time, assembly is not required) • Don't reinvent the wheel– make full use of libraries, third parties, etc.

Figure 3.22 Pros and cons of writing DSP code in assembly language (courtesy of Texas Instruments)

DSP Frameworks

All DSP systems have some basic needs—basic requirements for processing high performance algorithms. These include:

Input/Output

- Input consists of analog information being converted to digital data.
- Output consists of digital data converted back out to analog format.
- Device drivers to talk to the actual hardware.

Processing

- Algorithms that are applied to the digitized data, for example an algorithm to encrypt secure data streams or to decode an MP3 file for playback.

Control

- Control structures with the ability to make system level decisions, for example to stop or play an MP3 file.

A DSP framework must be developed to connect device drivers and algorithms for correct data flow and processing (Figure 3.23).

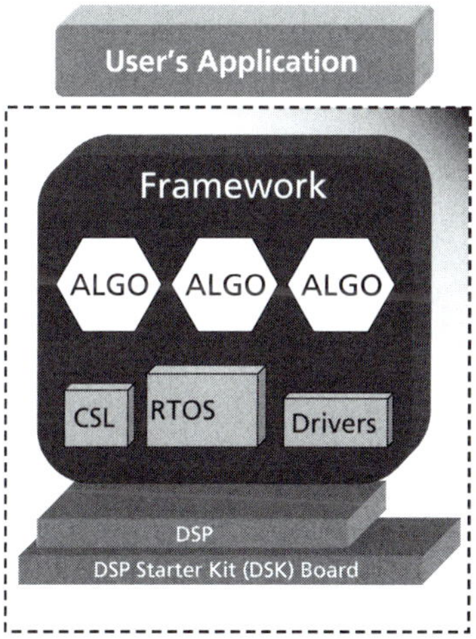

Figure 3.23 A model of a DSP framework for signal processing

A DSP framework can be custom developed for the application, reused from another application, or even purchased or acquired from a vendor. Since many DSP systems have similar processing frameworks as described above, reuse is a viable option. A framework is system software that uses standardized interfaces to algorithms and software. This includes algorithms as well as hardware drivers. The benefits of using a DSP framework include:

- The development does not have to start from scratch.
- The framework can be used as a starting point for many applications.
- The software components within a framework have well defined interfaces and work well together.
- The DSP designer can focus on the application layer which is usually the main differentiator in the product being developed. The framework can be reused.

An example DSP reference framework is shown in Figure 3.24. This DSP framework consists of:

- I/O drivers for input/output.
- Two processing threads with generic algorithms.
- Split/join threads used to simulate/utilize a stereo codec.

This reference framework has two channels by default. The designer can add and remove channels to suit the applications needs.

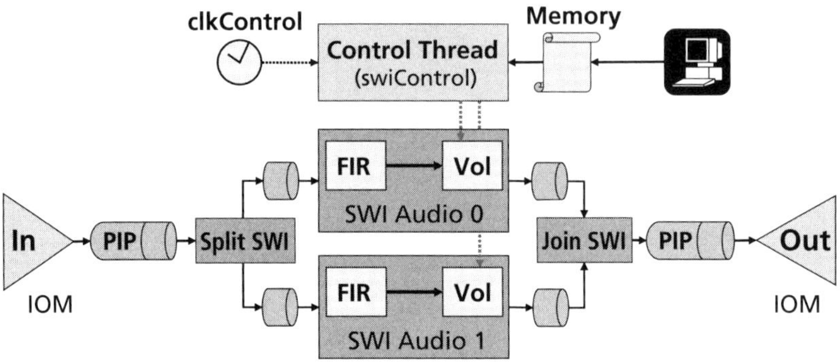

Figure 3.24 An example DSP reference framework (courtesy of Texas Instruments)

An example complete DSP Solution is shown in Figure 3.25. There is the DSP as the central processing element. There are mechanisms to get data into and out of the system (the ADC and DAC components). There is a power control module for system power management, a data transmission block with several possible peripherals including USB, FireWire®, and so on, some clock generation components and a sensor for the RF component. Of course, this is only one example, but many DSP applications follow a similar structure.

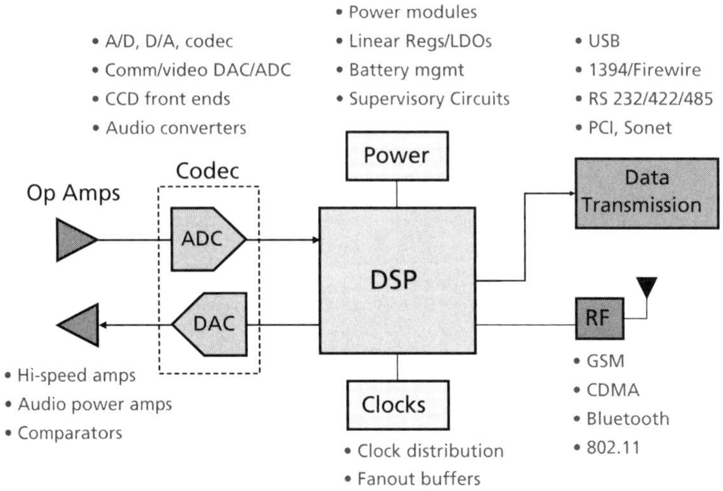

Figure 3.25 An example DSP application with major building blocks (courtesy of Texas Instruments)

4

Overview of Digital Signal Processing Algorithms

Definition of an Algorithm

An algorithm is a formula or set of steps for solving a particular problem[1]. Think of an algorithm as consisting of a set of rules that are unambiguous and have a clear stopping point. Algorithms can be expressed in many forms. For example, natural language, or a programming language like C or C++.

Algorithms are used in many situations. The recipe for baking a cake is an example of an algorithm. Most computer programs consist of various combinations of algorithms. One of the primary challenges in programming, especially digital signal processing programming, is the invention of algorithms that are simple and require the fewest steps possible.

From a DSP perspective, an algorithm is a set of rules that specify the order and form of arithmetic operations that are used on a specified set of data. Examples of arithmetic operations include rounding rules, logical decisions, a sine function or other specific formula.

To be effective in computer programming, algorithms must be precise, effective and finite. Each step in an algorithm must have a clearly defined meaning. Algorithms must also be effective, in the sense that the task should always be performed as required. Finally, algorithms must have a finite number of steps and must ultimately terminate. Although many embedded systems application are wrapped in a "while (1)" loop, effectively running "forever," the specific algorithms within these application loops must terminate.

Computer algorithms consist of a combination of different logic constructs. The basic constructs that make up an algorithm running on a computer are:

Sequence – Steps must follow each other in a logical sequence.

Decision – There may be alternative steps that may be taken subject to a particular condition.

Repetition – Certain steps may be repeated while, or until, a certain condition is true.

Given the real-time nature of digital signal processing, it is important to analyze the algorithms running on a DSP or in a DSP system based on various performance measures such as execution time (how long does it take the algorithm to run).

---

[1] The word algorithm is named after a mathematician from Baghdad named Mohammed-Musa Al-Khowarizmi from the ninth century. His work is also the likely source for the word "algebra."

Analyzing the run time performance of algorithms is important. In real-time embedded applications, it is not always reasonable to solve the problem of a slow running algorithm by using a faster processor. There are many reasons for this. Faster processors cost more money, and this is a unit cost expense. In other words, every system built with a faster processor costs more money. If the engineer can spend the time to optimize an algorithm to run on a smaller processor, this is a nonrecurring cost which is only borne once. For a high volume product, this approach makes a lot of sense. Faster processors also use more power, which will result in larger battery requirements for portable applications, or a larger fan to cool the processor. Some DSP algorithms cost less to develop (usually) and can have significant qualitative improvements (for example, the FFT reduces the amount of time to perform a DFT transform from N^2 operations to N log N operations). Keep in mind that a faster algorithm may help the specific problem but may not have a dramatic effect on other parts of the system; more about this later.

For real-time DSP applications, it is important to estimate the run time performance of an algorithm. This requires analysis of several aspects of the problem:

- The problem size.
- The input size.
- The number of primitive operations used (add, multiply, compare and so on).

Development of efficient algorithms will always be important. Despite the fact that processors are getting faster all the time, more and more algorithms are required to develop the advanced applications customers demand today. An inefficient algorithm, regardless of the speed of the processor, will take a relatively long time to execute and use more than its share of processor cycles, memory, or power. DSP engineers should always perform an analysis of system algorithms to estimate the time the system will take to complete its task; they should also find and eliminate wasteful code. Identifying inefficient algorithms allows the designer to improve overall system design.

DSP algorithm analysis starts with high level algorithm execution analysis and proceeds through a number of increasingly more detailed estimation steps to determine the run time performance of an application.

One method used to estimate the efficiency of an algorithm is called *big-OH notation* or O(n), where "n" is the function being estimated. This notation will not produce an exact run time for an algorithm. For example, a constant multiply is ignored when analyzing an algorithm using O(n) since this approach is only concerned with the order. Big OH estimates how fast the overall execution time grows as the size of the input grows. This is an important analysis step, because many algorithms can be implemented in multiple ways. In many cases, the simplest, most obvious algorithm will also be the slowest. A Fourier transform algorithm is a perfect example. The discrete Fourier transform (DFT), one of the more important algorithms for signal processing, has a growth rate of n^2, or O(n) = n^2. The fast Fourier transform (FFT) is a faster implementation of the DFT with a growth rate of n log n, or O(n) = n log n. As the number of input samples, n, grows, this difference in computation can become significant. Figure 4.1 shows a plot of the DFT vs. the FFT. This figure shows the difference in computation cycles in a logarithmic scale. The difference is very significant as "n" grows. Some common running times for algorithms are:

Constant growth rate: O(n) = c

Linear growth rate: O(n) = c*n

Logarithmic growth rate: O(n) = c*log n

Quadratic growth rate: O(n) = c*n2

Exponential growth rate: O(n) = c*2n

Linear growth rate is the most efficient and preferred running time.

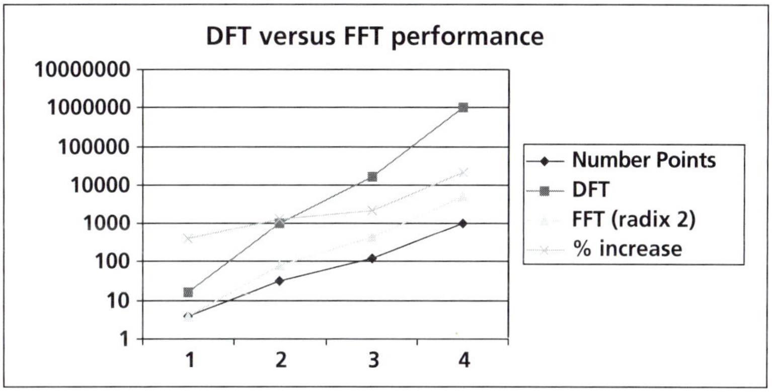

Figure 4.1 Difference in algorithmic efficiency in compute cycles for a linear growth rate algorithm (DFT) vs. a logarithmic growth rate algorithm (FFT)

Signal processing performs a variety of functions in many real-time embedded systems. Some of the most popular are:

- *Filtering* – Removing unwanted frequencies from a signal.
- *Spectrum Analysis* – Determining the frequencies in a signal.
- *Synthesis* – Generating complex signals such as speech.
- *System Identification* – Identifying the properties of a system by numerical analysis.
- *Compression* – Reducing the memory or bandwidth it takes to store a signal, such as audio or video.

Digital signal processing engineers use a toolbox of algorithms to implement complex signal processing functions. Complex system applications can be built using combinations of supporting algorithms that themselves are composed of combinations of underlying signal processing functions (Figure 4.2). This chapter will introduce some of these underlying functions that are the building blocks of complex signal processing functions.

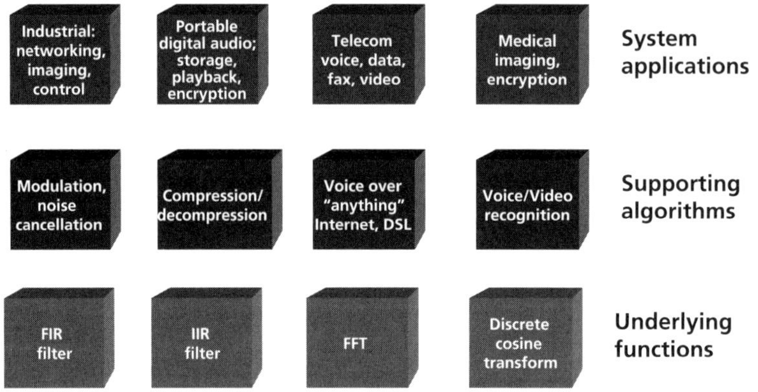

Figure 4.2 System building blocks; system applications, supporting algorithms, and underlying functions

Systems and signals

Before we start talking about DSP systems, it is necessary to introduce a few concepts that are common to all systems. However, emphasis will be placed on how these concepts are used in DSP systems. The systems generally used in DSP applications are linear time-invariant systems. The property of linearity is important for a couple of reasons. The most important reason is that the system is not dependant on the order in which processes are applied. For example, it does not matter if we scale our input before or after we run it through a filter; the output is the same. As a result, a complex system can be divided into multiple systems. For example, an eighth order system can be divided into four-second order systems and still produce the same output. Time Invariance is important because we need to know that our system will react in the same way to an input, in other words the output is not dependant on when the input is applied. These two qualities make our system very predictable.

Signals and systems are usually graphed to show how the input and output relate to each other. There are two typical ways of viewing the data: in the time domain and in the frequency domain. The time domain is especially handy for applications such as control systems, where response times are important. The frequency domain is useful for viewing filter results to see which frequencies will pass and which will be attenuated.

We usually describe our systems in the time domain in terms of impulse response. An impulse is a stimulus of infinite magnitude and zero duration; its integral is equal to one. Since it is only a single instant of time, and spans all frequencies, the impulse makes a good test to show how a system will react to an input. The response to an impulse input, called the *impulse response*, can be used to describe the system. In the digital domain, an impulse is an input signal whose magnitude is 1 at time 0 and 0 everywhere else, as shown in Figure 4.3. The way a system reacts to an impulse input can also be considered the system's transfer function. The transfer function (or impulse response) can provide us with everything we need to determine how a system will react to an input in the time or the frequency domain.

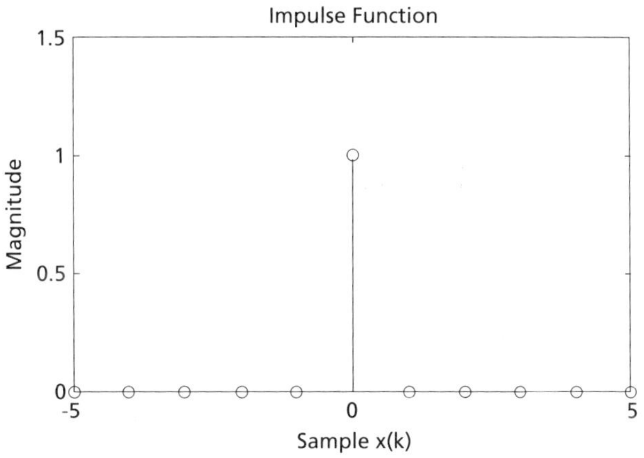

Figure 4.3 An impulse response function

A system response is usually plotted in the frequency domain to show how it will affect signals of different frequencies. When plotted in the frequency domain, there are two characteristics to be concerned about: magnitude and phase. Magnitude is the ratio of the output's strength compared to the strength of its input; for example, how much of a radio signal will pass through a filter designed to pass a specific frequency. The phase is how much a frequency of the signal will be changed by the filter, usually lagging or leading. While not important in all applications, sometimes the phase can be particularly important, such as in music or speech applications.

DSP Systems

DSPs are often used in applications that manipulate some sort of continuous phenomena, like voice, radio frequency signals, or motor currents. DSPs, like other microprocessors, are *digital* devices, running algorithms that manipulate *discrete* data. Discrete data implies the DSP must process a sampled representation (or approximation) of the actual phenomenon, and not the real thing. Whether the sampled approximation will adequately represent the original depends on how the sample was obtained.

How are analog signals converted to digital signals? Analog signals are converted to digital signals through a process called sampling. Sampling is the process of taking an analog signal and converting it to discrete numbers. The sampling frequency (or sampling rate) is how many times per second the signal will be sampled. This is important because it restricts the highest frequency that can be present in a signal. Any signal greater than half the sampling rate will be folded into a lower frequency that is less than half the sampling rate. This is known as aliasing and it will be discussed in more detail in the next section.

The inverse of the sampling frequency is known as the sampling period. The sampling period is the amount of time that elapses between the samples of the analog signal. This conversion is done by hardware that takes the best approximation of the voltage level and converts it to the nearest digital level that can be represented by the computer. The loss of information during the conversion is referred to as quantization.

Analog-to-Digital Conversion

The first step in a signal processing system is getting the information that must be processed from the real world into the system. This implies transforming an analog signal to a digital representation for processing by the system. The analog signal passes through a device called an analog-to-digital converter (A/D or ADC). The ADC converts the analog signal to a digital form by sampling or measuring the signal periodically and assigning a digital code to the measurement. Once the signal is in digital form, it can be processed by the DSP.

The input to an analog-to-digital converter (ADC) consists of a voltage that varies continuously over time. Examples are waveforms representing human speech, or the signals from a conventional television camera. The output of the ADC has defined levels or states. The number of states is usually a power of two; that is, the output is normally chosen to be represented conveniently in binary. The maximum number of bits in the binary representation defines the "resolution" of the ADC.

An Audio Example

When used in an audio context, analog refers to a signal that directly represents the sound waves traveling through the air. A simple audio tone, such as a sine wave, causes the air to produce evenly spaced ripples of alternating high and low pressure. When these waves reach an eardrum, or a microphone, they cause it to move evenly back and forth at the same rate. You can measure the voltage coming from the microphone vibrating in this fashion. If plotted over time this signal will look similar to Figure 4.4.

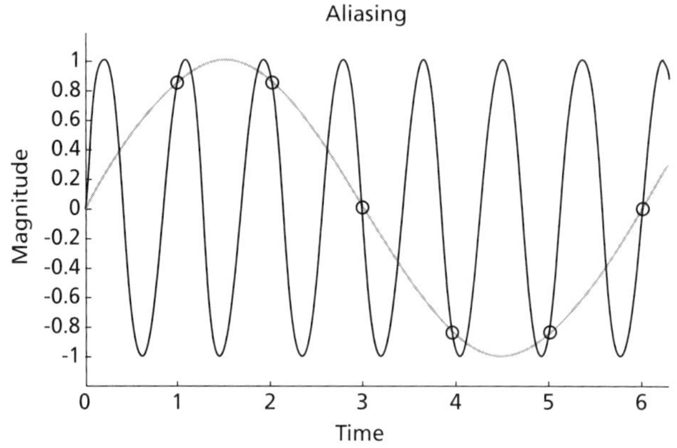

Figure 4.4 Analog data (a simple sine wave) plotted over time and an aliased signal which is a misrepresentation of the original signal

This analog data must be stored, edited, manipulated or transmitted using digital equipment. Therefore it must be converted to digital form first. This requires sampling. The incoming voltage levels are converted into binary numbers using analog-to-digital conversion. The two primary constraints when performing this operation are sampling frequency (how often the voltage is measured) and resolution (the size of numbers used to measure the voltage). When an analog waveform is digitized we are effectively taking continuous "snapshots" of the waveform at intervals. We call this interval the sampling frequency. The samples are stored one after another as binary numbers. When the waveform is reconstructed from the list of sampled numbers, the result will be a "stairstep" approximation of what we started with (Figure 4.5).

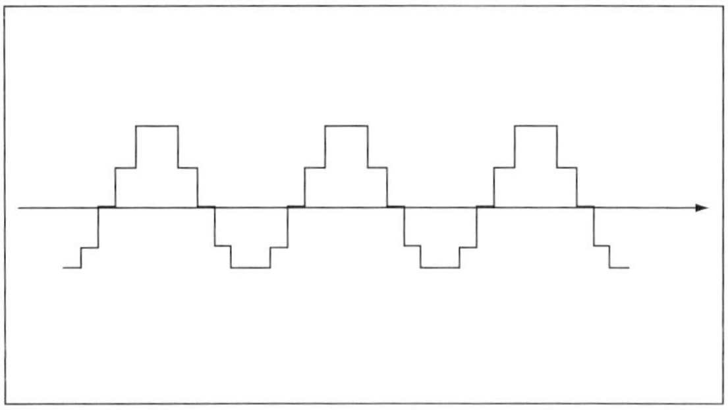

Figure 4.5 Resultant analog signal reconstructed from the digitized samples

When the digitized samples are converted back into voltages, the stair-step approximation is "smoothed" using a filter. This results in the output looking similar to the input. The constraints on sampling are very important because if the sampling frequency, resolution, or both are too low, the reconstructed waveform will be of lower quality.

The sample rate sets a hard real-time performance requirement for the system. Failing to acquire a sample on time or to produce a result on time has the same effect on the result as a missed calculation in a hard real-time system such as a CD player or cell phone. Therefore the sample rate together with the algorithm complexity sets a lower bound for the necessary processor throughput. This can be generalized to:

(Number of Instructions to process * Sample Rate) < Fclk * Instructions/cycle (MIPS),

where Fclk is the clock frequency of the DSP device. The DSP engineer must understand these constraints early in the development process in order to set the right nonfunctional requirements for timeliness in the system. Timing requirements are a form of nonfunctional requirement.

The Nyquist Criteria

One of the most important rules of sampling is called the Nyquist Theorem[2]. This theorem states that the highest frequency which can be represented accurately is one half of the sampling rate. The Nyquist rate specifies the minimum sampling rate that fully describes a given signal; in other words a sampling rate that enables the signal's accurate reconstruction from the samples. In reality, the sampling rate required to reconstruct the original signal must be somewhat higher than the Nyquist rate, because of quantization errors[3] introduced by the sampling process.

As an example, humans can detect or hear frequenies in the range of 20 Hz to 20,000 Hz. If we were to store sound, like music, to a CD, the audio signal must be sampled at a rate of at least 40,000 Hz to reproduce the 20,000 Hz signal. A standard CD is sampled at 44,100 times per second, or 44.1 kHz.

The required sampling rate depends on the signal frequencies processed by the application. Radar signal processing sampling rates are on the order of one to several Gigahertz while video applications are sampling at or near 10 megahertz. Audio applications are sampled in the 40 to 60 kilohertz range and modeling applications such as weather or financial modeling systems are sampled at much lower rates, sometimes less than once per second.

The Nyquist Criteria sets a lower bound for the sampling rate. In practice, algorithm complexity may set an upper bound. The more complex the algorithm, the more instruction cycles are required to compute the result, and the lower the sampling rate must be to accommodate the time for processing these complex algorithms. This is another reason why efficient algorithms must be designed and implemented in order to meet the required sampling rates to achieve the right application performance or resolution.

Aliasing

If an analog signal is not sampled at the minimum Nyquist rate, the data sampled will not accurately represent the true signal. Consider the sine wave in Figure 4.4 again.

The circles in this diagram indicate the sampled points on the waveform during the ADC process. The sampling rate in Figure 4.4 is below the Nyquist frequency, which means that we will not be able to reconstruct the signal accurately as shown in Figure 4.4.

The resultant signal looks nothing like the input signal. This misrepresentation is called *aliasing*. In applications that perform sound and image generation, aliasing is the generation of a false (alias) frequency along with the correct one when doing frequency sampling.

---

[2] Named in 1933 after scientist Harry Nyquist.

[3] Quantization errors result from the rounding or truncation of a continuous signal to the nearest defined reference value.

Aliasing is a sampling-related phenomenon that occurs when measurable frequency content exists above 1/2 the sampling rate. When this happens, energy or power in a signal above the Nyquist frequency is actually "mirrored" back into the represented region from 0 to 1/2 the sample rate. For example, a 4500 Hz signal sampled at 6000 samples/sec appears as a 1500 Hz signal. This is an "aliased" signal.

Aliasing is due to the fact that mathematically, the following two equations are equal:

$$X(n) = \sin(2 * pi * fo * n * ts) = \sin(2 * pi * fo * n * ts) + 2 * pi * k$$

These equations can be re-written as

$$X(n) = \sin (2 * pi * fo * n * ts) = \sin (2 * pi * fo + k * fs) * n * ts)$$

To avoid this ambiguity, we must restrict all frequencies in our signal to be in the range of 0 to $f_s/2$. The frequency $f_s/2$ is called the *Nyquist frequency* (or *Nyquist rate*). While this provides a limit to the frequencies available in our system, there is no other choice but to do this. The way this is achieved is by building an analog (anti-aliasing) filter and placing it before our digital-to-analog (D/A) converter.

Basic DSP system

The basic DSP system pictured in Figure 4.6 consists of an analog-to-digital converter (A/D), a digital signal processor (DSP) and a digital-to-analog converter (D/A). Typically, systems have analog filters before and after the converters to make the signals more pure. Let's discuss each component in detail.

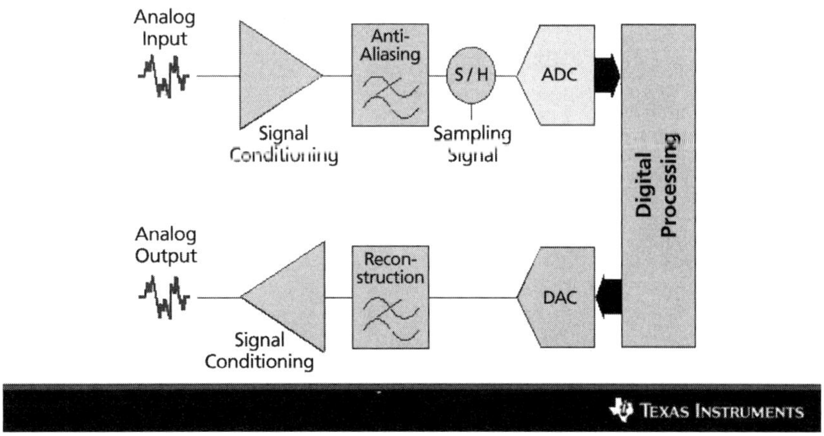

Figure 4.6 A basic DSP system (courtesy of Texas Instruments)

Since the signal can contain no frequencies above the Nyquist frequency, steps must be taken to ensure that the high frequencies are eliminated from the signal. This is done with an analog low pass filter with a cutoff rate set around the Nyquist frequency. This analog filter is known as the anti-aliasing filter. The signal is then used as input into an A/D converter where the signal is converted to a digital signal so that the DSP can handle and process it. The DSP will perform the actions required of it, such as filtering, and then pass the new signal to the D/A. The D/A then converts the digital output back to an analog signal. This analog output usually contains high frequencies introduced by the D/A, so a low pass filter is need to smooth the waveform back to its intended shape. This filter is known as a reconstruction filter.

Anti-Aliasing Filter

In digitized images, aliasing will produce a jagged edge, or stair-step effect. In an audio application with a sampled sound signal, aliasing will produce a buzz . In order to eliminate this buzz, A/D converters use lowpass filtering to remove all signals above the Nyquist frequency. This is referred to as "antialiasing." It is the smoothing of an image, sound, or other signal roughness caused by aliasing. The approach may vary from signal to signal. An image, for example, may apply a technique that adjusts pixel positions or sets pixel intensities so that there is a more gradual transition between the color of a line and the background color. With sound signals, aliases are removed by eliminating frequencies above half the sampling frequency.

Low pass filtering is also used to eliminate unwanted high-frequency noise and interference introduced prior to sampling. Anti-aliasing filters actually help reduce system cost, as well as acquisition storage requirements and analysis time by enabling a lower sampling rate. Low-pass filters serve as an important element of any data acquisition system in which the accuracy of the acquired data is essential. We will talk more about low pass filters later in the chapter.

Sample Rate and Processor Speed

We have discussed the ADC process shown in Figure 4.7. To gain a better understanding of the processor speeds required by a typical application, consider the example of a simple modem. A modem converts outgoing digital signals from a computer or other digital device to analog signals. Modems also demodulate incoming analog signals and convert them to digital signals for the computer.

Figure 4.7 Analog-to-digital conversion for signal processing

For this example, assume that the analog signal has no frequency content above about 3500 Hz. If we therefore sample the signal at 8 KHz, we should be safe as far as the Nyquist rule indicates. The sampling period T would then be:

T = 1 / 8000 = 125 microseconds = 0.000125 seconds

From a processing requirement perspective, for a signal being sampled at this rate, we would have 0.000125 seconds to perform all the processing necessary before the next sample arrives (remember, these samples are arriving on a continuous basis and we cannot fall behind in processing them.). This is an example of a hard real-time requirement.

Whether we can meet this real-time requirement is dependent on the algorithm performance and the processor speed that is executing these algorithms. Knowing the sampling rate, we can determine how many instructions the processor can perform during each sample period:

Sampling period (T) / Instruction cycle time = number of instructions per sample

For a 100 MHz processor that executes one instruction per cycle, the instruction cycle time would be 1/100 MHz = 10 nanoseconds

125 µs / 10 ns = 12,500 instructions per sample

125 µs / 5 ns = 25,000 instructions per sample (for a 200 MHz processor)

125 µs / 2 ns = 62,500 instruction per sample (for a 500 MHz processor)

As this example demonstrated, the higher the processor instruction cycle execution, the more processing we can do on each sample. If speed were free, we could just choose the highest processor speed available and have plenty of processing margin. Unfortunately, it's not that easy. Speed costs money and consumes more power and requires more cooling, and so on. You get the picture!

A to D Converters

Analog-to-digital converters convert analog signals from the environment to digital representations that can be processed by a digital computer like a DSP. DSP engineers have a choice of several different analog-to-digital converter types. The choice depends on the accuracy and resolution required in the application as well as the speed required to acquire the samples. Digital numbers produced by the ADC represent the input voltage in discrete steps with a finite resolution. The resolution of an ADC is determined by the number of bits used to represent the digital number. In general, an n-bit converter will have a resolution of 1 part in 2^n. The general description of this is:

1 LSB = Full scale range / 2n –1 for an n-bit converter

For example, a 10-bit converter has a resolution of 1 part in 1024 (2^{10} = 1024). 10-bit resolution corresponds to 4.88mV per step for a 5V range. If that resolution is not good enough, using a 12-bit ADC will produce a resolution of 5V / (2^{12}) = 1.22 mV per step. Given these characteristics, the only way to increase the resolution of an ADC without reducing the dynamic range is to use an ADC with more bits. Much of this analysis must be performed early in the DSP development life cycle. Improper analysis may lead to the choice of a ADC that does not meet application requirements for video resolution, voice quality, codec resolution and so forth.

Some of the more popular types of these forms of data converters are:

- *Flash converters* – These converters are probably the simplest ADC implementations. They use a circuit called a voltage divider consisting of resistors and comparators to "divide" down the reference voltage into a series of steps that form the resolution of the converter. Flash converters are generally very fast but have limited resolution due to increased circuitry requirements. An 8-bit ADC, for example, will require 256 comparators (28 = 256). All these comparators take up space and power.
- *Successive approximation* – Successive approximation ADCs determine the digital value of an analog signal by performing a series of approximations that get more and more accurate. These successive approximations are done in series, so successive approximation ADCs are slower than flash converters, but they are relatively inexpensive.
- *Voltage-to-frequency* – These ADCs operate by producing an output pulse train with a frequency proportional to the input voltage. The output frequency is determined by counting pulses over a fixed time interval. This form of data converter is used when a high degree of noise rejection is required, such as signals that are relatively slow and noisy, and in applications such as automotive where we find sensors in noisy environments. By converting the analog signal to a frequency pulse representation, signals can be transferred over long distances without worrying about transmission noise.
- *Sigma delta* – A sigma-delta converter uses a very fast signal sampling rate to produce a very accurate conversion of an analog signal. The accuracy of these data converters is directly related to the clock rate used to sample the signal. The higher the clock rate, the more accurate the conversion. This approach bypasses the use of analog components such as resistor networks to produce the approximation, but it may be slower than other converters due to the fast clock required to oversample the input.

Digital-to-Analog Conversion

A digital-to-analog converter (DAC), as the name implies, is a data converter which generates an analog output from a digital input. A DAC converts a limited number of discrete digital codes to a corresponding number of discrete analog output values.

Because of the finite precision of any digitized value, the finite word length is a source of error in the analog output. This is referred to as quantization error. Any digital value is really only an approximation of the real world analog signal. The more digital bits represented by the DAC, the more accurate the analog output signal. Basically, one LSB of the converter will represent the height of one step in the successive analog output. You can think of a DAC as a digital potentiometer that produces an analog output that is a fraction of the full scale analog voltage determined by the value of the digital code applied to the converter. Similar to ADCs, the performance of a DAC is determined by the number of samples it can process and the number of bits used in the conversion process. For example, a three bit converter as shown in Figure 4.8 will have less performance than the four bit converter shown in Figure 4.9.

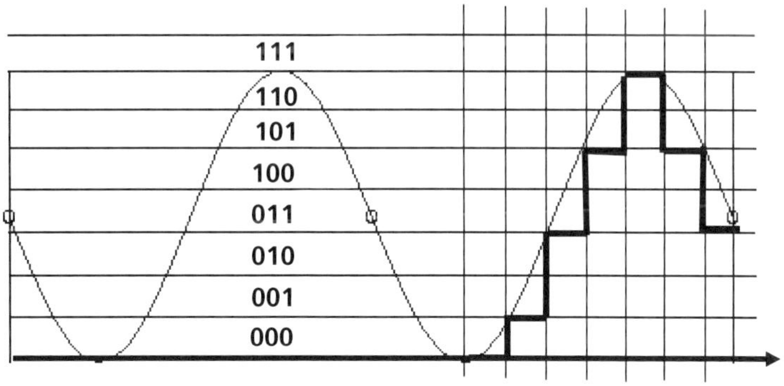

Figure 4.8 A three bit DAC (create a similar image)

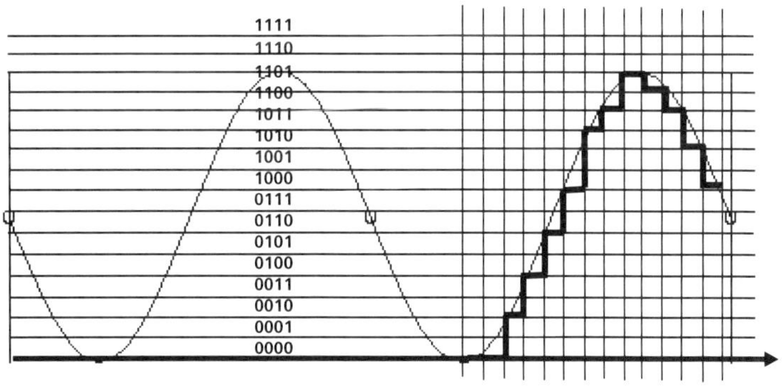

Figure 4.9 A four bit DAC (create a similar image)

We now know that a DAC generates an analog output based on a train of digital values. This operation produces an analog waveform that is discrete, not continuous. The discrete output can be integrated on the receiving device, or in the DAC itself. If the integration is done on the DAC, circuitry called a *sample and hold* is used to perform the integration. Sample and hold logic will "hold" the signal from the DAC until the next pulse comes along. Regardless of where the integration is performed, the resultant analog waveform is filtered using a low pass filter. This operation "smooths" the waveform by removing the high frequency energy introduced during the DAC operation.

Multirate Applications

In many real-time signal processing applications, time domain processing requires changing the sampling rate of the system signals at various processing stages. This is referred to as multirate sampling. Multirate sampling techniques use more than one sampling rate in the same system. The sample spacing can vary from point to point. This approach is based on the Nyquist criteria, in which a signal only needs to be sampled at twice its maximum frequency component. Oversampling may provide increased system benefits, but if oversampling a signal provides no significant value and the system processing burden can be reduced by sampling at a lower rate, it makes sense to do so. There are other benefits to doing this. The more you sample a signal, the more memory is required to hold the sampled data, as well as the additional processing required to process all these samples. So oversampling a signal comes with a processing and storage burden that cannot be ignored. Multirate processing, therefore, often results in increased efficiency in signal processing since the sample rates can be "tuned" to the specific task and be kept as small as possible. Multirate sampling and signal processing uses techniques called decimation and interpolation, or downsampling and upsampling, respectively.

Decimation is defined by the relation y(n) = x (Mn), where M is some integer. This relation implies that the ouput at a time "n" is equal to the input at time Mn. In other words, only the input samples at integer multiples of M are retained. This is effectively an algorithmic form of sample rate reduction where the input samples are "decimated" by a selector factor. This reduction in sampling rate is comparable with a reduction in bandwidth using filtering.

An interpolator function performs the opposite. Instead of "throwing away" all but a selected set of samples in an input train, an interpolator inserts samples in between the input samples. This operation effectively creates a "streched" sample train. This relation can be described as y(n) = x (n / M) where n is a multiple of M and y(n) = 0 otherwise. This algorithm results in an increase in the sampling rate. Another term for this process is called "sample rate expansion." Interpolation and decimation are used in applications such as software defined radios as well as radar systems.

Summary of Sampling

To exploit the power of digital processing, we must convert analog signals from the environment into a digital form for efficient processing. This involves an analog-to-digital conversion process. In order to characterize analog signals from the environment accurately, we must sample them often enough during this process to represent them in the processing chain. Not sampling these signals often enough leads to improper characterization and representation. Sampling too often can lead to undue processing costs and memory use. Not only must we sample appropriately, we must also use the right accuracy to represent the sampled signal. The bit width of the conversion process is just as important as how often we sample the signal. Once we are through with the signal processing steps, the resultant waveform(s) may need to be sent back out to the environment to control something such as a motor, TV volume, and so on. Digital-to-analog converters are used for this conversion process.

Introduction to Filters

Introduction

We use filters for many things in our daily lives; we use coffee filters to make our coffee in the morning; we wear sunglasses to filter harmful rays from the environment; we use our digital cell phone to filter the appropriate phone call and we filter out other speech and noise to listen for our name to be called at a busy restaurant. We also use filters at work to search for certain useful information stored on our company mainframes and computers.

Signals are composed of different sinusoidal components. Signals are composed of many combinations of sinusoidal periodic functions and these can be used to approximate many different signals that are very useful is the signal processing field. For example, a square wave can be approximated from a combination of the following sinusoidals:

$$y = \sin(wt) + 1/3\sin(3wt) + 1/5\ \sin(5wt) + 1/7\ \sin(7wt)$$

The successive approximations are shown in Figure 4.10. The first harmonic, or fundamental harmonic, is simply a sine wave with a frequency equal to the frequency of the desired square wave. Each additional term produces a closer approximation to the desired square wave. Filtering can also be used on signals like this. A filtering operation could be used to eliminate one of more of these sinusoids from the signal. This is often done to remove noise or other unwanted sinusoids from a given signal.

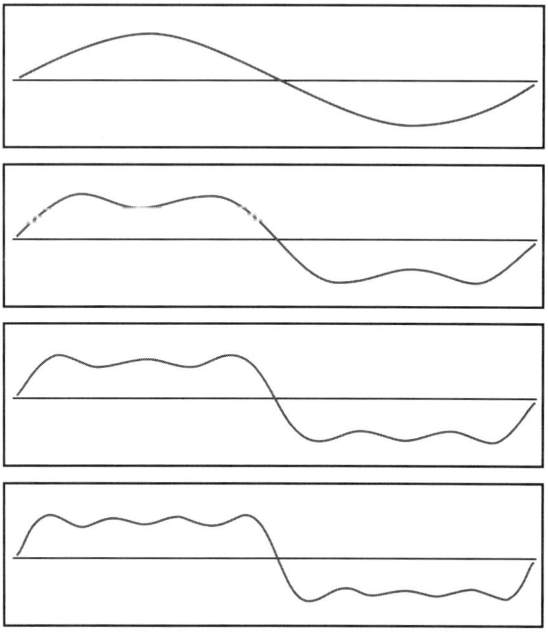

Figure 4.10 Creating a square wave from a series of sine waves y = sin(wt) + 1/3sin(3wt) + 1/5 sin(5wt) + 1/7 sin(7wt)

This section will introduce the main concepts of filters, discuss the advantages of digital filters over analog filters, and provide a brief overview of the characteristics of digital filters.

What is a filter?

A filter can be thought of as an operation that accepts a certain type of data as input, transforms it in some manner, and then outputs the transformed data.

A filter can be a device, or even a material, that is used either to suppress or minimize signals of certain frequencies. The analog world is filled with unwanted noise such as sound, light, and so on. In digital signal processing, the main function of a filter is to remove unwanted parts of the signal, which is usually random noise from the environment, and keep the useful parts of the signal, which usually corresponds to a specific range of frequencies.

Broadly, the notion of a filter is independent of its implementation. While filters are a broad concept (as described above), most of the time DSP filters process some analog signal in the environment.

Filters are used for two main functions; signal separation and signal restoration. Signal separation is used when a signal has been corrupted with noise or some other type of unwanted interference. Signal restoration is used when a signal has been distorted in some way and needs to be processed to better represent the signal as it actually occurred.

Usually DSP filters are applied to signals to select certain frequencies or certain information while rejecting other frequencies or information. Consider the successive diagrams shown in Figure 4.11. Figure 4.11a shows a simple sine wave. This sine wave could represent a simple tone frequency that we are interested in processing. This signal may come from the environment where it must compete with many other "unwanted" signals, referred to as noise. Noise may be generated by natural sources or be man made. Regardless, noise is an unwanted disturbance within the band of information that we are interested in. Noise corrupts our tone signal in an additive way. The undesired signal is added to the desired signal producing a resultant signal that may not look anything like the signal we are "looking" for. Part b shows additional noise added to the sine wave, producing a resulting waveform that is corrupted. Many electronic systems must handle this form of noise disturbance in one way or another, and the filtering of noise from the desired signal is very important to the successful operation of the system. In fact, one of the most important metrics used in electronic systems is the *signal to noise ratio* (SNR), which is the ratio of the amplitude of the desired signal to the amplitude of noise signals at a given point in time.

Whereas we tend to think naturally in the spatial sense (the passage of time, and so on), digital computers require a different way of thinking—that of the frequency domain. The frequency domain is primarily concerned with periodicity (frequency) and phase rather than with duration and timing. Figure 4.11c shows the same signal in the frequency domain. The sinusoidal time-based signal is represented as a set of frequency components. We have simply transformed an analog based view of the signal to a frequency based view of the signal. Figure 4.11c shows the signal consisting of frequency components at two discrete locations in the frequency spectrum (ignore the negative representation for the moment). This concept of transformation between variables of time and frequency (periodicity) has proven to be extremely important in signal processing. As we will see later, there are efficient mechanisms to convert signals from the spatial domain into the frequency domain and vice-versa.

If we were now to think about how to keep the useful information and attenuate or eliminate the noise, the analysis becomes a bit simpler. In this example, the useful signal is the large "spike" and the noise is the smaller "spike." If I were to draw a box around that part of this signal that I wanted to keep, the box would look something like the dotted line in Figure 4.11d.

Figure 4.11e represents an actual digital filter designed to approximate the ideal filter shown in Figure 4.11d. You can see that the actual filter does not have the ideal characteristics of the desired filter. An ideal filter has a infinitely sharp cutoff (called the transition ratio). This is extremely hard to achieve in real world applications. The steeper the cutoff, the higher the quality factor of the filter. To achieve a high quality factor, a more complex filter design is required (Figure 4.11f). As we will see later, high quality factor filter designs can also make the filter unstable, which leads to other problems. The key to selecting a filter is to understand and characterize the frequencies and the corresponding amplitudes of all interfering signals and design appropriately. If you are designing a filter for a cell phone, this means you must understand the worst case amplitude as well as the location of adjacent signals, and design the filter to prevent these adjacent signals from interfering with the desired signal. Finally, Figure 4.11g represents the signal in 4.11b after filtering out the high frequency noise. This looks like the signal we wanted to extract from the noise (4.11a). Note that Figures 4.11a–4.11g were generated using "DSP Calculator" software from James Broesch, 1997.

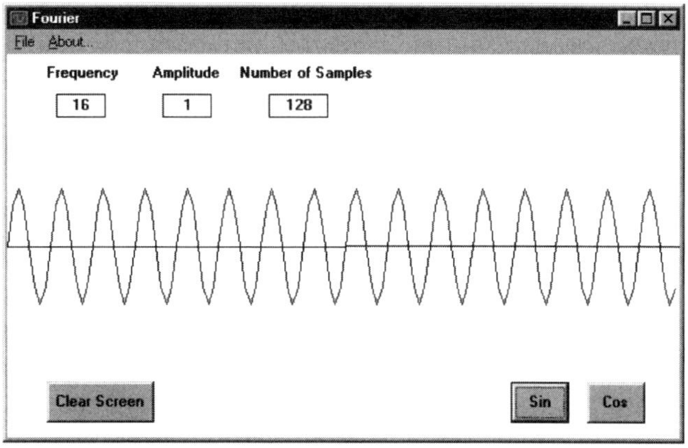

Figure 4.11a The desired sine wave

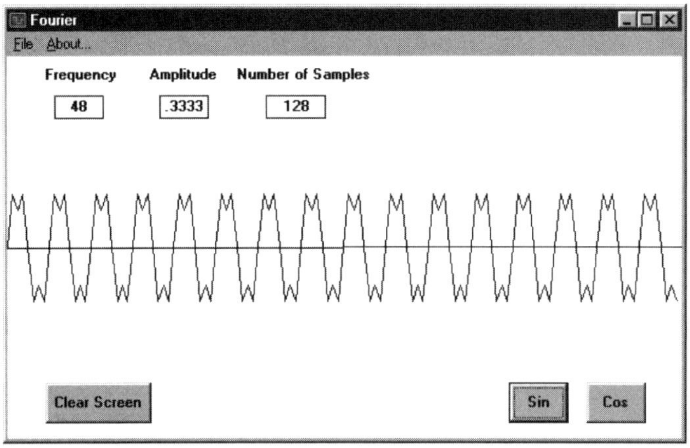

Figure 4.11b The desired sine wave corrupted with noise

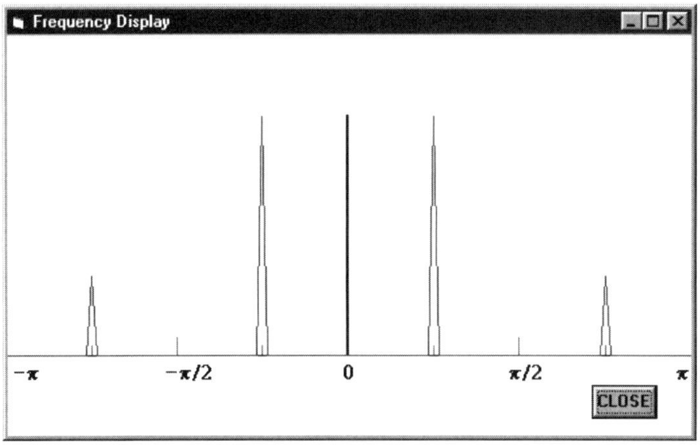

Figure 4.11c The Frequency plot for the resultant signal showing signal energy at the two main frequencies of the signal

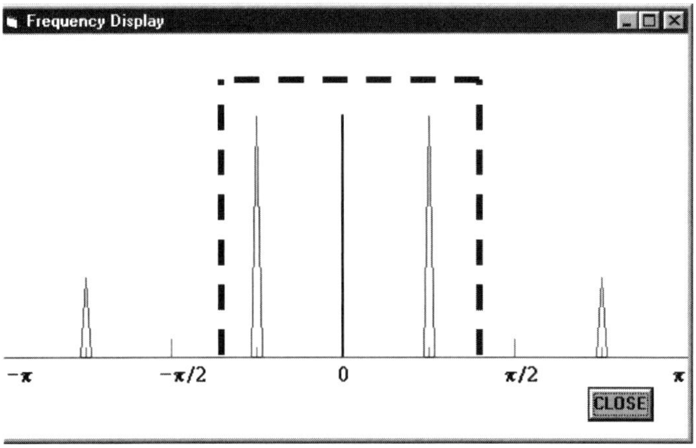

Figure 4.11d The desired filtering operation to keep the desired signal and attenuate the undesirable signal

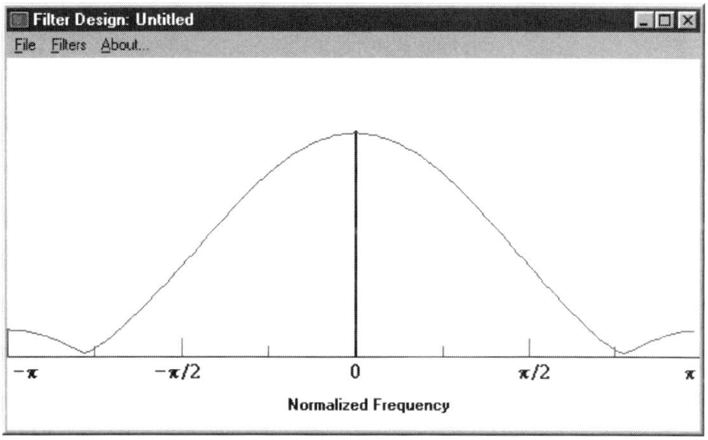

Figure 4.11e Actual digital filter designed to approximate the ideal filter shown in Part d

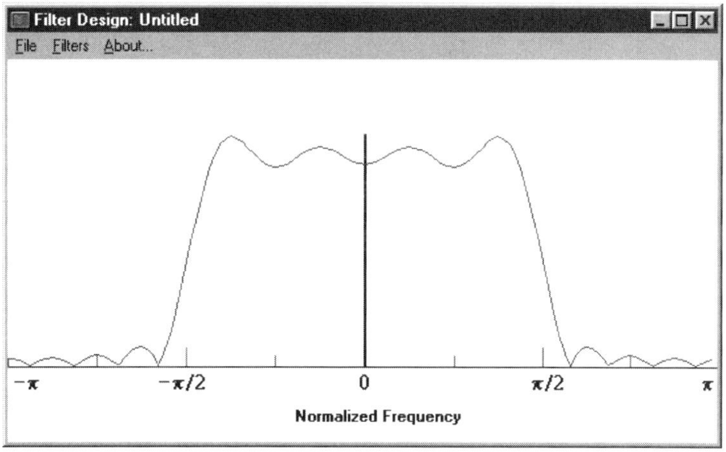

Figure 4.11f A more complex filter with sharper rolloff

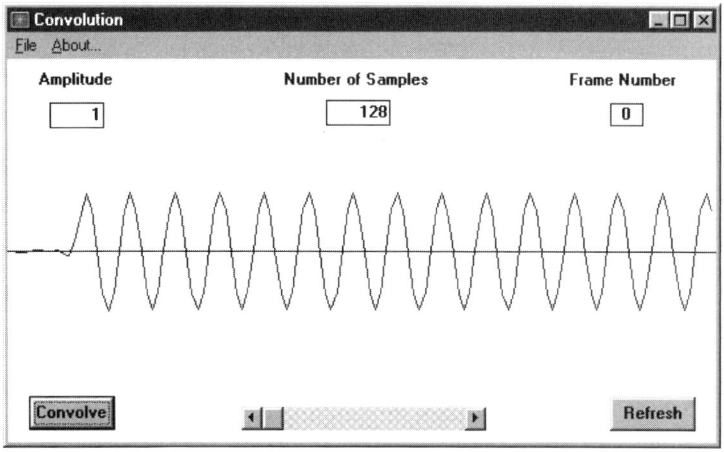

Figure 4.11g The resulting waveform after the filtering operation

Simple R/C filter

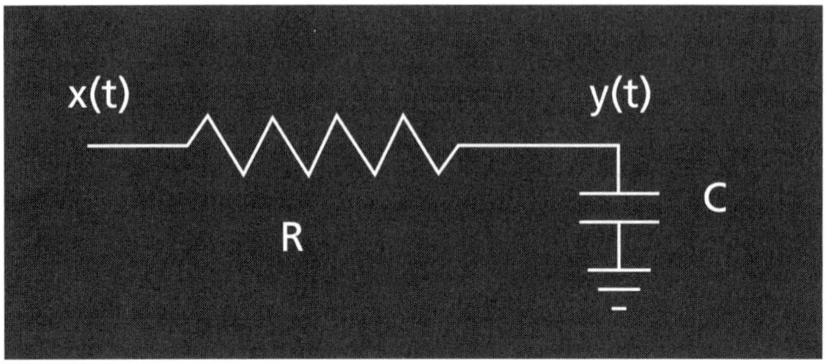

Figure 4.12 A simple analog low pass filter using a resistor and a capacitor

Filters can be implemented in hardware or software. One of the advantages of a DSP is being able to implement a filter in software efficiently. But before DSP, what did we do? Consider the simple analog low pass filter shown in Figure 4.12. This simple circuit consists of a resistor and a capacitor. In a filter like this, the capacitor C is a short to ground for high frequencies (high frequencies are attenuated). Low frequencies are allowed to pass on. The frequency response of this simple filter looks similar to that of Figure 4.11e. The frequency response of a filter can be defined as the spectrum of the output signal (range of frequencies) divided by the spectrum of the input signal. The frequency response shows how information represented in the frequency domain is being changed.

More selective filters

The filter shown in Figure 4.12 can be improved by adding more analog components. If we were doing this using hardware, one approach is to cascade the elements of the filter as shown in Figure 4.13. As you can see, creating a more complex filter in hardware requires adding more discrete components, which adds to overall circuit complexity. The alternative approach, using software algorithms, is easier—but it also requires additional computational complexity and, therefore, more processor cycles and memory. Nothing is free !

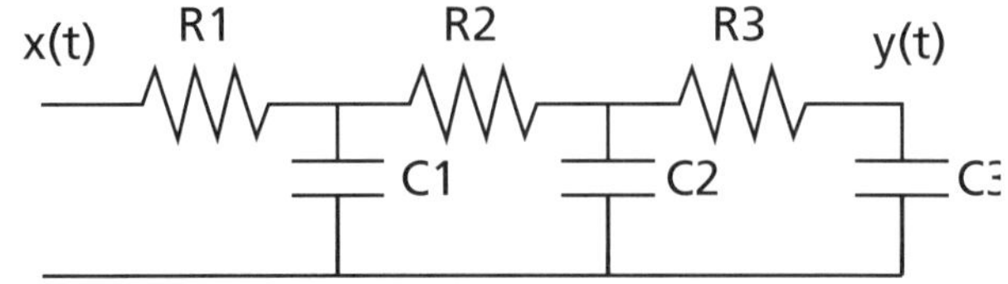

Figure 4.13 Another analog low pass filter with more filter components

There are other types of selective filters that are useful in the DSP world. A *bandpass* filter, for example, is a filter that allows signals between two specific frequencies to pass, and attenuates (eliminates) signals at other frequencies (see Figure 14a below). Bandpass filters are used in many applications including wireless transmitters to limit the bandwidth of the output signal to the minimum necessary. These filters are also used in receivers to allow signals within a certain selected range of frequencies to be processed while preventing signals at other unwanted frequencies from getting into the system.

Another form of filter is called *a high-pass filter.* These filters transmit energy above a certain frequency only (Figure 4.14b below). All other frequencies are attenuated. These filters are common in image processing applications, where high pass filtering is used to eliminate the low frequency, slowly changing areas of the image and amplify the high frequency details in the image. For example, if we were to high pass filter a picture of a house we would only see an outline of the house. The edge of the house is the only place in the picture where the neighboring pixels are different from one another.

We have already talked about low pass filters. Low pass filters transmit energy below a certain frequency (Figure 4.14c below). High frequency energy is attenuated. Low pass filters are common in speech processing. When processing speech signals, the steps are first to filter all frequencies above about 20 KHz (the upper limit of audible speech) and retain those frequencies in the frequency band below 20 kHz.

Figure 4.14d is an example of a band-stop filter. These filters only transmit energy outside a specific frequency band. These are also called *notch* filters. The goal of a filter of this type is effectively to pass all frequencies except for a predefined band of frequencies. This band is attenuated. These filters are primarily used for the removal of undesirable frequency components from an otherwise desired signal.

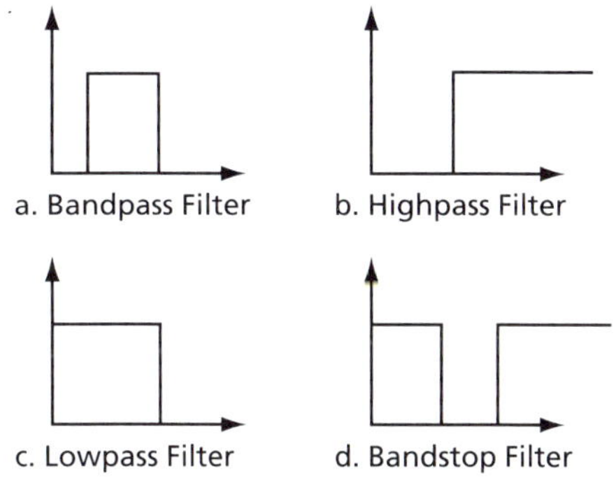

Figure 4.14 An analog filter design with cascading sections

Phase response

All filters, regardless of which frequencies they attenuate or amplify, introduce a delay between the input signal and the output signal. This is referred to as a simple time delay or transmission delay if we are in the time domain. In the frequency domain this same delay is called a *phase shift*. If this transmission delay was the same for all frequencies in the filter, this would not be a big problem. However, a potential problem in filtering is that this transmission delay may not be the same for all frequencies operated on by the filter. Some filters with sharp rolloff can be designed this way. This condition is referred to as *phase distortion*. If you view a filtered signal with phase distortion in the time domain, the signal will have preshoot, overshoot and ringing, especially if the input signal is changing rapidly. Phase distortion is not always a bad thing, because many applications do not require a linear phase response.

Some filters are designed to have constant transmission delay for all frequencies. These filters have the characteristic of a phase shift that is a linear function of frequency. These filters are sometimes called *linear-phase* filters. These filters work well for certain input conditions, such as a fast input transient signal. The downside is that they do not do as good a job at attenuating all of the undesirable frequencies unless a more complex filter is designed. Linear phase is an important requirement for applications such as digital data modems.

Summary of Filter Types

There are many types of filter types available to solve a variety of problems. We will talk about some of the fundamental filter types next. Here is a brief sampling of some of the common filter types available:

- *Boxcar* – Boxcar filters are filters in which each multiplicative coefficient is 1.0. For an N-tap boxcar filter, the output is just the sum of the past N samples.
- *Hilbert Transform* – Hilbert Transform filters shift the phase of the signal by 90 degrees. These filters are used to create the imaginary part of a complex signal from its real part.
- *Nyquist Filters* – These filters are used in multirate applications. In a Nyquist filter, one of every L coefficients is zero, which acts like a downsampling operation. This effectively reduces the number of multiply-accumulate operations required to implement the filter.
- *Raised-Cosine* – This filter is used for digital data applications. The filter has a frequency response in the passband that is a cosine shape which has been "raised" by a constant.
- *Butterworth* – This type of filter eliminates unwanted low frequency noise generated due to conditions such as time derivative calculations or noisy sensor inputs. The Butterworth filter is a good general-purpose filter which is also simple to understand. Butterworth filters aere used in applications such as audio processing.
- *Chebyshev* – This type of filter minimizes the difference between the ideal and the actual frequency response. A Chebyshev filter provides a much steeper rolloff, but also has a pass-band ripple that makes it unusable for audio systems. This filter is good for applications where the pass band includes only one frequency of interest. An example is when you need to produce a sine wave from a square wave by filtering out a number of other harmonics.

Finite Impulse Response (FIR) Filters

We now know that there are many types of filters. In computer programming applications, a filter, in its most general form, is a program that examines input or output requests (examples are data samples, files, and so forth) for certain qualifying criteria. The filter will then perform some type of processing on these requests and forward the processed request accordingly (This "filter" term is a common operation in UNIX systems and is now used in other operating systems as well). Generally, a filter is a "pass-through" function that proceeds through the following steps:

- Examine a set of input data;
- Make a specific decision(s) about the input data;
- Transform the data if necessary;
- Send the resultant data set to another processing function.

In DSP applications like telecommunications, filters are functions that selectively sort various signals, pass through a desired range of signals, and suppress the others. These types of filtering operations are used often to suppress noise and to separate signals into different bandwidth channels.

Remember that the environment in which a DSP interacts is analog in nature. Signals such as light, pressure, and so on are changing constantly. This continuously varying signal can be represented in the form:

$$V = x(t)$$

where t represents time in the analog world.

This "raw" analog data must now be converted to digital form using an analog-to-digital converter. In order to do this, the signal must be sampled at a specific time interval. This is the sampling interval. At each time interval in which the signal is sampled, a specific value is assigned to the signal. We can represent this as:

$$x_i = x(ih)$$

where x_i is the sampled value at time t = ih and i is the counting variable.

After ADC processing, the analog signal is now just a sequence of numbers inside the DSP:

$$x_0, x_1, x_2, x_3, \ldots x_n$$

This sequence represents the values of the signal at the various sampling intervals of time t = 0, h, 2h, 3h, 4h, ... nh, and so on.

The DSP will presumably perform some operations on these signal samples. This processing may be to improve the signal in some way, such as by removing noise or enhancing the signal. The DSP may then outputs an improved signal back to the environment. In this case the output samples must be converted back to analog form using a digital-to-analog converter:

$$(x_0, x_1, x_2, x_3, ..., x_n) \text{ -> some transformation by the DSP -> } y_0, y_1, y_2, y_3, ..., y_n$$

$y_0, y_1, y_2, y_3, ..., y_n$ are the enhanced or improved output sequence after processing by the DSP. The specific algorithm that defines how the output samples are calculated from the input samples is the filtering characteristic of the digital filter. There are many ways to perform this transformation, and we will now discuss several basic approaches.

A finite impulse response (FIR) filter is really just a weighted moving average of a given input signal. This algorithm can be implemented in custom-designed, dedicated hardware, or as a software algorithm running on a programmable platform like a DSP or even a field programmable gate array (FPGA). This section will focus on the structure of these filtering algorithms.

FIR Filters as Moving Averages

A Simple FIR

As a simple example of a filtering operation, consider the monthly recorded temperature in Houston, Texas. This is represented over a two-year period in Figure 4.15 by the "temperature" data plot. This same data can be filtered, or averaged, to show the average temperature over a six month period. The basic algorithm for this is to accumulate enough samples (six in this case), compute the average of the six samples, add a new sample and compute the "moving" average by repetitively summing the last six samples received. This produces a filtered average of the temperature over the period of interest. This is represented by the "semiannual average" in Figure 4.15. Notice the "phase delay" effect. The semiannual average has the same shape as the "temperature" plot but is "delayed" due to the averaging, and it is smoothed as well. Notice that the peaks and valleys are not as pronounced in the averaged or smoothed data.

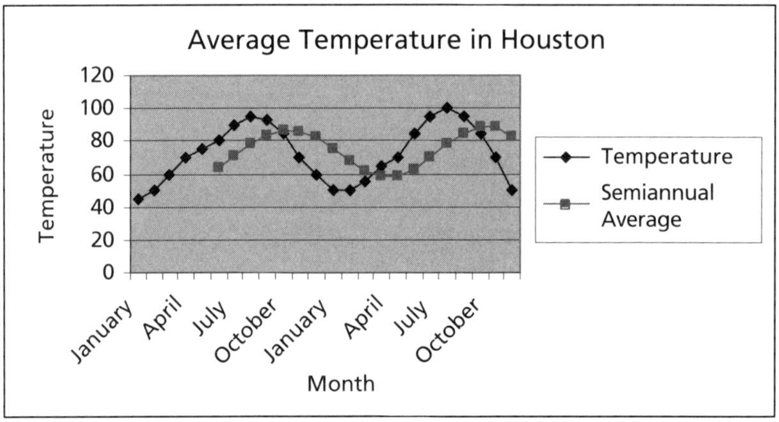

Figure 4.15 A simple filter to compute the running average of temperature

There are some observations that can be made about the example above. The first is the number of temperatures or "samples" used to perform the filter operation. In this example, the order is six, since we are computing a six-month moving average. This is referred to as the "order" of the filter. The *order* of a digital filter is defined as the number of previous input used to calculate the current output. These previous inputs are stored in a DSP memory when performing filtering operations on real signals.

For example, a zero order filter can be shown as:

$$y_n = x_n$$

In this case, the current output y_n depends only on the current input x_n and not on any other previous inputs. A zero order filter can also be scaled:

$$y_n = Kx_n$$

This uses a multiplicative factor K to scale the output sample by some constant value.

An example of a first order digital filter is shown as:

$$y_n = x_n - x_{n-1}$$

This filter uses one previous input value (x_{n-1}) as well as the current sample, x_n, to produce the current output y_n.

Likewise, the transformation shown below is a second order filter, since two previous input samples are required to produce the output sample.

$$y_n = (x_n + x_{n-1} + x_{n-2}) / 3$$

In the filters discussed so far, the current output sample, y_n, is calculated solely from the current and previous input values (such as x_n, x_{n-1}, x_{n-2}, ...). This type of filter is said to be nonrecursive. FIR filters are classified as nonrecursive filters.

Generalizing the Idea

We have been discussing simple filters this far, basically boxcar filters. While this simple unweighted or uniformly-weighted moving average produces acceptable results in many low-pass applications, we can get greater control over filter response characteristics by varying the weight assigned to past and present inputs. The resultant procedure is still the same; for each input sample, the present and previous inputs are all multiplied by the corresponding weights (also referred to as coefficients), and these products are added together to produce one output sample. Algebraically we can show this as:

$$y(n) = a0 \times x(n) + a1 \times x(n-1) + a2 \times x(n-2) + a3 \times x(n-3) + a4 \times x(n-4)$$

Or;

$$Y0 = a0 \times X0 + a1 \times x1 + a2 \times x2 + a3 \times x3 + a4 \times x4$$

The equation above can be generalized as follows:

$$Y = \sum_{n=1}^{40} a_n * x_n$$

Hardware Implementation (or Flow Diagram)

This computation can be implemented directly in dedicated hardware or in software. Structurally, hardware FIR filters consist of just two things: a sample delay line and a set of coefficients as shown in Figure 4.16. The delay lines are referred to in the figure as "z^–1." The coefficient or weights are shown in the figure as a0, a1, a2, a3 a4, and the input samples are the x values.

The sample delay line implies memory elements, required to store the previous signal samples. These are also referred to as delay elements and are mathematically represented as z**–1, where z**–1 is referred to as the unit delay. The unit delay operator produces the previous value in a sequence. The unit delay introduces a delay of one sampling interval. So if we apply the operator z–1 to an input value xn we get the previous input xn–1:

$$z{-}1\ x_n = x_{n-1}$$

A nonrecursive filter like the FIR filter shown above has a simple representation (called a transfer function) which does not contain any denominator terms. The transfer function of a second-order FIR filter is therefore:

$$y_n = a0x_n + a1x_{n-1} + a2x_{n-2}$$

where the a terms are the filter weights.

As we will see later, some filters do have a denominator term that can make them unstable (the output will never go to zero).

So now we know that the values a0, a1, and so on are referred to as the coefficients or weights of the operations and the values x(1), x(2), and so on are the data for the operation. Each coefficient/delay pair is called a *tap* (from the connection or "tap" between the delay elements).

The number of FIR taps, (often designated as "N") is an indication of:

- the amount of memory required to implement the filter,
- the number of calculations required, and
- the amount of "filtering" the filter can do; in effect, more taps means more stopband attenuation, less ripple, narrower filters, and so forth.

To operate the filter in Figure 4.16, the algorithmic steps are:

1. Put the input sample into the delay line.
2. Multiply each sample in the delay line by the corresponding coefficient and accumulate the result.
3. Shift the delay line by one sample to make room for the next input sample.

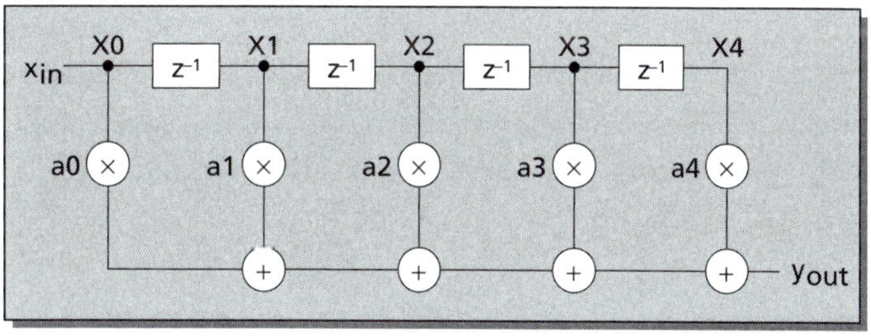

Figure 4.16 Structure of flow diagram of a finite impulse response filter (courtesy of Texas Instruments)

Basic Software Implementation

The implementation of a FIR is straightforward; it's just a weighted moving average. Any processor with decent arithmetic instructions or a math library can perform the necessary computations. The real constraint is the speed. Many general-purpose processors can't perform the calculations fast enough to generate real-time output from real-time input. This is why a DSP is used.

A dedicated hardware solution like a DSP has two major speed advantages over a general-purpose processor. A DSP has multiple arithmetic units, which can all be working in parallel on individual terms of the weighted average. A DSP architecture also has data paths that closely mirror the data movements used by the FIR filter. The delay line in a DSP automatically aligns the current window of samples with the appropriate coefficients, which increases throughput considerably. The results of the multiplications automatically flow to the accumulating adders, further increasing efficiency.

DSP architectures provide these optimizations and concurrent opportunities in a programmable processor. DSP processors have multiple arithmetic units that can be used in parallel, which closely mimics the parallelism in the filtering algorithm. These DSPs also tend to have special data movement operations. These operations can "shift" data among special purpose registers in the DSP. DSP processors almost always have special compound instructions (like a multiply and accumulate or MAC operation) that allow data to flow directly from a multiplier into an accumulator without explicit control intervention (Figure 4.17). This is why a DSP can perform one of these MAC operations in one clock cycle. A significant part of learning to use a particular DSP processor efficiently is learning how to exploit these special features.

In a DSP context, a "MAC" is the operation of multiplying a coefficient by the corresponding delayed data sample and accumulating the result. FIR filters usually require one MAC per tap.

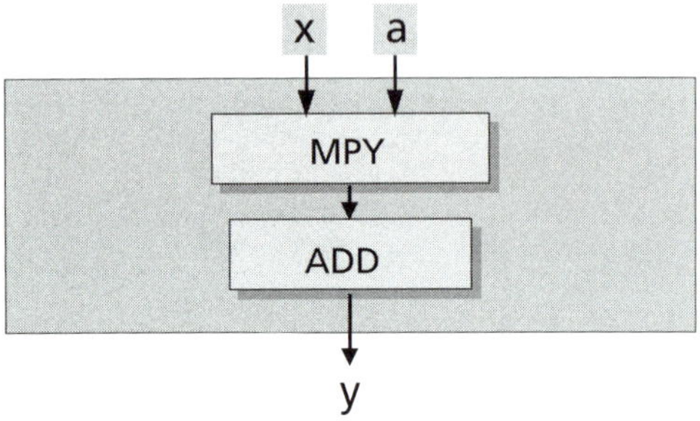

Figure 4.17 DSPs have optimized MAC instructions to perform multiply and accumulate operations very quickly

FIR Filter Characteristics

The "impulse response" of a FIR filter is just the set of FIR coefficients. In other words if you put an "impulse" into a FIR filter which consists of a "1" sample followed by a large number of "0" samples, the output of the filter will be simply the set of coefficients, as the 1 valued sample moves past each coefficient in turn to form the output.

We call the impulse response "finite" because there is no feedback loop in this form of filter. If you put in an impulse as described earlier, zeroes will eventually be output after the "1" valued sample has made its way in the delay line past all the filter coefficients. A more general way of stating this phenomenon is that, regardless of the type of signal input to the filter or how the long we apply the signal to the filter, the output will eventually go to zero. How long it takes for the output to go to zero is dependent on the filter length, which is defined by the number of taps (a multiplication of a delayed sample) as well as the sample rate (how quickly the taps are being computed) The time it takes for the FIR filter to compute all of the filter taps defines the delay from when a sample is input to the system and when a resultant sample is output. This is referred to as the phase delay of the filter.

If the coefficients are symmetrical in nature, the filter is called a *linear-phase* filter. Linear-phase filters delay the input signal, but don't distort its phase.

FIR filters are usually designed to be linear-phase, although they don't have to be. Like we discussed earlier, a FIR filter is linear-phase if (and only if) its coefficients are symmetrical around the center coefficient. This implies that the first coefficient is the same as the last, the second is the same as the next-to-last, and so on.

Calculating the delay in a FIR filter is straightforward. Given a FIR filter with N taps, the delay is computed as:

$$(N - 1) / Fs$$

where Fs is the sampling frequency. If we use a 21 tap linear-phase FIR filter operating at a 1 kHz rate, the delay is computed as:

(21 – 1) / 1 kHz = 20 milliseconds.

Adaptive FIR Filter

A common form of a FIR filter is called an *adaptive* filter. Adaptive filtering is used in cases where a speech signal must be extracted from a noisy environment. Assume a speech signal is buried in a very noisy environment with many periodic frequency components lying in the same bandwidth as the speech signal. An example might be an automotive application. An adaptive filtering system uses a noise cancellation model to eliminate as much of this noise as possible.

The adaptive filtering system uses two inputs. One input contains the speech signal corrupted by the noise. The second input is a noise reference input. The noise reference input contains noise related to that of the main input (like background noise). The adaptive system first filters the noise reference signal, which makes it more similar to that of the main input. The filtered version is then subtracted from the main input signal. The goal of this algorithm is to remove the noise and leave the speech signal intact. Although the noise may never be completely removed, it is reduced significantly.

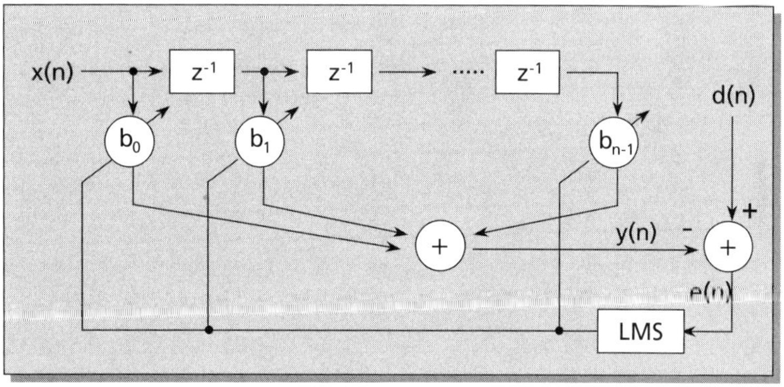

Figure 4.18 General structure of an adaptive FIR filter. Usually used in an adaptive algorithm since they are more tolerant of nonoptimal coefficients. (courtesy of Texas Instruments)

The filter to perform this adaptive algorithm could be any type, but a FIR filter is the most common because of its simplicity and stability (Figure 4.18). In this approach, there is a standard FIR filter algorithm which can use the MAC instruction to perform tap operations in one cycle. The "adaptation" process requires that calculations to tune the FIR filter match the characteristics of the desired response. To compute this, the output of the FIR is matched with the external system response. If the FIR output matches the system response, the filter is tuned and no further adaptation is required. If there are differences in the two values, this would indicate a need to tune the FIR filter coefficients further. This difference is referred to as the error term. This error term is used to adjust each of the coefficient values each time the filter is run.

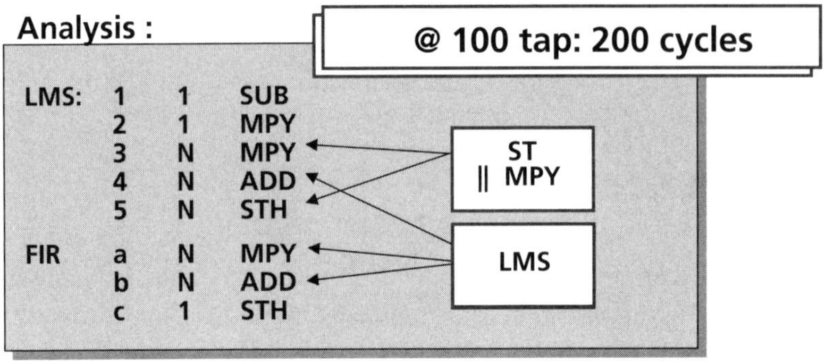

Each Iteration (only once)

1 - determine error: e(i) = d(i) - y(i)

2 - scale by "rate" term B: e´(i) = 2*B*e(i)

Each Term (N sets)

3 - Qualify error with signal strength: e´´(i) = x(i-k) * e´(i)

4 - Sum error with coefficient: b(i+1) = b(i) + e´´(i)

5 - Update coefficient: b(i) = b(i+1)

Figure 4.19 Analysis of the adaptive FIR algorithm (courtesy of Texas Instruments)

The basic process of computing an adaptive filter is shown in Figure 4.19. Each iteration of the filter operation requires the system first to determine the filter error and then scale it by a factor of how 'hot' or responsive the rate of adaptation must be. Since there is only one set of these to be performed each iteration, the overall system cost in cycles is very low.

The next step is to determine how much to tune each of the coefficients. All of the terms are not necessarily adjusted equally. The stronger the data was at a particular tap, the larger its contribution was to the error, and the more that term is scaled. (For example, a tap whose data happened to be a zero couldn't have had *any* influence on the results, and therefore would not be adapted in that iteration). This individual error term is added to the coefficient value, and written back to the coefficient memory location to "update," or "adapt" the filter. Analysis of the DSP load (Figure 4.18) shows an adaptation process which consists of 3 * N steps. This is larger than the FIR itself, which requires 2 * N processes to run. The adaptive filter requires 5N operations for each iteration of an N tap filter. This is many more cycles than a simple FIR. The overall load to compute this operation can be reduced using to 4N using the DSP MAC instruction. The DSP can also use parallel instruction execution that allows for a load or store to be performed while a separate math instruction is also being run.

In this example, a store in parallel with a multiply can absorb two steps of the LMS process, further reducing the overall load to 3N cycles per iteration. The LMS ADD needs the coefficient value, also being accessed during the MAC instruction. A specialized LMS instruction (another specialized instruction for DSP), merges the LMS ADD with the FIR's MAC. This can reduce the load to 2N cycles for an N tap adaptive filter. A 100th order system would run in about 200 cycles, vs. the expected 500. When selecting a DSP for use in your system, subtle performance issues like these can be seen to have a very significant effect on how many MIPS a given function will require.

Designing and Implementing FIRs Filters

A FIR filter is relatively straighforward to implement. They can be implemented easily using a high level language such as C:

```
y[n] = 0.0;
for (k = 0; k < n; k++)
{
y[n] = y[n] + c[k] * x[n–k];
}
```

y[n] represent the output samples,
c[k] represent the filter coefficients
x[n–k] represent the previous filter input samples

Although this code looks pretty simple, it may not run very efficiently on a DSP. The proper implementation or tuning is required or the algorithm will run very inefficiently, and not obtain the maximum "technology entitlement" from the DSP. For example, in the algorithm implemented above, there are some inefficiencies:

y[n] is accessed repeatably inside the loop, which is inefficient (memory accesses are significant overhead especially if the memory accesses are off chip). Even though DSP architectures (we will talk in detail about DSP architectures later) are designed to maximize access to several pieces of data from memory simultaneously, the programmer should take steps to minimize the number of memory accesses.

Accessing an array using indexes like c[k] is also inefficient; in many cases, an optimizing DSP compiler will generate more efficient code in terms of both memory and performance if a pointer is used instead of an array index (the array index computation could take several cycles and is very inefficient). This is especially true if a function like a FIR manipulates the same variable several times, or must step through the members of a large array. Using an index like in the example means that now the C compiler will only know the start address of the array. In order to read any array element, the compiler must first find the address of that element. When an array element is accessed by its array index [i], the compiler has to make a calculation and this takes time. In fact, the C language provides a 'pointer' type specifically to avoid the inefficiencies of accessing array elements by an index. Pointers can easily be modified to determine a new address without having to fetch it from memory, using simple and familiar operations such as *ptr++. The pointer has to be initialized, but only once.

Pointer use is generally more efficient than accessing an array index on any processor. For a DSP, this benefit is compounded because DSP processors are optimized for fast address arithmetic and simple address increments are often performed at no cost by using hardware support within the DSP, performed at the same time as the data access. When we discuss DSP architectures, it will become evident that a DSP can perform multiple data accesses at the same time and can, therefore, increment several addresses at the same time.

DSP processing is often hampered by the delay in getting data into and out of the processor. Memory accesses become the bottleneck for DSP processing. Even with the advantage of multiple access DSP architectures, the multiple accesses of data from memory to perform filtering operations can easily exceed the overall capacity of the DSP. This pattern of memory accesses is facilitated by the use of the fast DSP registers which allow the compiler to store often used memory values in registers as opposed to making expensive memory accesses (for example, the keyword "register" can be used to inform the compiler to, if at all possible, put the referenced variable in a fast access register).

The main drawback of a digital FIR filter is the time that it takes to execute. Since the filter has no feedback, many more coefficients are needed in the system equation, compared to an IIR filter, to meet the same requirements. For every extra coefficient, there is an extra multiply and extra memory requirements for the DSP. For a demanding system, the speed and memory requirements to implement an FIR system can make the system unfeasible.

We will discuss code optimization techniques in detail in a later chapter.

Basic FIR Optimizations for DSP Devices

FIR implementations can be made more efficient by not calculating things that don't need to be calculated (This applies to all algorithms!).

For example, if the filter has zero-valued coefficients, you don't actually have to calculate those taps; you can leave them out. A common case of this is a "half-band" filter, which has the property that every other coefficient is zero.

Also, if your filter is "symmetric" (linear phase), you can "pre-add" the samples which will be multiplied by the same coefficient value, prior to doing the multiply. Since this technique essentially trades an "add" for a "multiply," it isn't really useful in DSP microprocessors which can do a multiply in a single instruction cycle. However, it is useful in ASIC implementations (in which addition is usually much less expensive than multiplication); also, many DSP processors now offer special hardware and instructions to make use of this trick.

The process of symmetrical FIR implementation is first to add together the two data samples that share a common coefficient. In Figure 4.20, the first instruction is a dual operand ADD and it performs that function using registers AR2 and AR3. Registers AR2 and AR3 point to the first and last data values, allowing the A accumulator to hold their sum in a single cycle. The pointers to these data values are automatically (no extra cycles) incremented to point to the next pair of data samples for the subsequent ADD. The repeat instruction (RPTZ) instructs the DSP to implement the next instruction "N/2" times. The FIRS instruction implements the rest of the filter at one cycle per each two taps. FIRS takes the data sum from the A accumulator, multiplies it with the common coefficient drawn from the program bus, and adds the running filter sum to the B accumulator. In parallel to the MAC unit performing a multiply and accumulation operation (MAC), the ALU is being fed the next pair of data values via the C and D buses, and summing them into the A accumulator. The combination of the multiple buses, multiple math hardware, and instructions that task them in efficient ways allows the N tap FIR filter to be implemented in N/2 cycles. This process uses *three* input buses *every cycle*. This results in a lot of overall throughput (at 10nSec this amounts to 30 M words per sec—sustained, not 'burst').

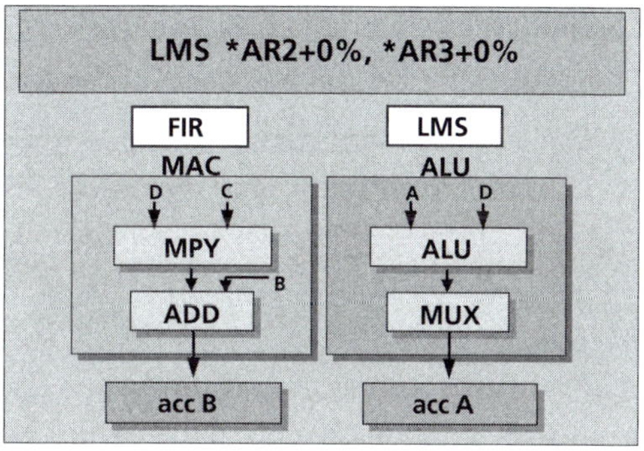

Figure 4.20 Symmetrical FIR implementation on a DSP (courtesy of Texas Instruments)

Designing an FIR filter

The simplest design of an FIR Filter is an averaging filter, where all the coefficients have the same value. However, this filter does not give a very desirable magnitude response. The trick to designing an FIR filter is getting the right coefficients. Today there are several good algorithms used to find these coefficients, and several software design programs to assist in calculating them. Once the coefficients are obtained, it is a fairly simple matter to place them in an algorithm to implement the filter. Let's talk about some of the techniques used in selecting these coefficients.

Parks-McClellan algorithm

One of the best "catch-all" algorithms used to determine the filter coefficients is the Parks-McClellan algorithm. Once the specifications are obtained (cutoff frequency, attenuation, band of filter), they can be supplied as parameters to the function, and the output of the function will be the coefficients for the filter. The program works by spreading out the error over the entire frequency response. So, an equal amount of "minimized" error will be present in the passband and stopband ripple. Also, the

Parks-McClellan algorithm isn't limited to the types of filters discussed earlier (low-pass, high-pass). It can have as many bands as are desired, and the error in each band can be weighted. This facilitates building filters of arbitrary frequency response. To design the filter, first calculate the order of the filter with the following equations:

$$\hat{M} = \frac{-20\log_{10}\sqrt{A\delta_1\delta_2} - 13}{14.6\Delta f} \; ; \qquad \Delta f = \frac{w_s - w_p}{2\pi}$$

where M is the order, w_s and w_p are the passband and stopband frequencies, and δ_1 and δ_2 are the ripple on the passband and stopband.

δ_1 and δ_2 are calculated from the desired passband ripple and stopband attenuation with the following formulas.

$$\delta_1 = 10^{Ap/20} - 1 \text{ and } \delta_2 = 10^{-As/20}$$

Once these values are obtained, the results can be plugged into the MATLAB function remez to get the coefficients. For example to obtain a filter that cuts off between .25 and .3 with a passband and stopband ripple of .2 and 50dB respectively, the following specifications can be plugged into the MATLAB script to get the filter coefficients:

```
% design specifications
wp = .23; ws = .27; ap = .025; as = 40;
%calculate deltas
d1 = 10^(ap/20) - 1; d2 = 10^(-as/20); df = ws - wp;
% calculate M
M = ((((-10 * log10(d1*d2)) - 13) / (14.6 * df)) + 1);
M = ceil(M);
% plug numbers into remez function for low pass filter
ht = remez(M-1, [0 wp ws 1], [1 1 0 0]);
```

ht will be a vector array containing the 35 (the value of M) coefficients. To get a graph of the frequency response, use the following MATLAB commands:

```
[h,w] = freqz(ht);      % Get frequency response
w = w/pi;               % normalize frequency
m = abs(h);             % calculate magnitude
plot(w,m);              % plot the graph
```

The graph is shown in Figure 4.21:

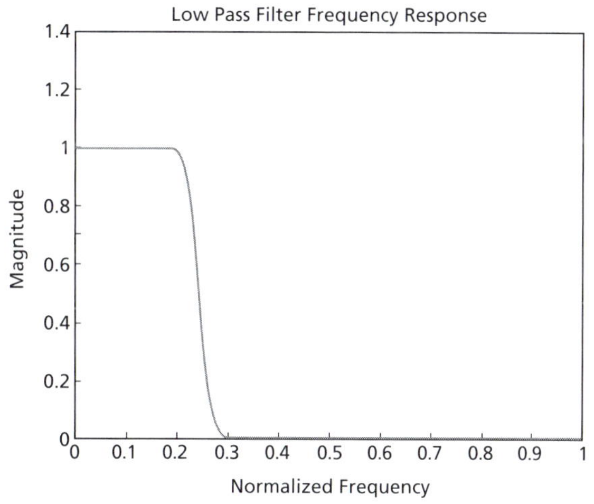

Figure 4.21 Low pass filter frequency response

Windowing

Another popular technique of the FIR filter is the ability to generate the frequency coefficients from an ideal impulse response. The time domain response of this ideal impulse response can then be used as coefficients for the filter. The problem with this approach is that the sharp transition of frequencies in the frequency domain will create a time domain response that is infinitely long. When the filter is truncated, ringing will be created around the cutoff frequency of the frequency domain due to the discontinuities in the time domain. To reduce this problem, a technique called windowing is used.

Windowing consists of multiplying the time domain coefficients by an algorithm to smooth the edges of the coefficients. The trade-off here is reducing the ringing but increasing the transition width. There are several windows discussed, each with a trade-off in transition width vs. stop band attenuation.

The following are several types of popular windows:

Rectangular – sharpest transition, least attenuation in the stopband (21 dB)

Hanning – Over 3x transition width of rectangular, but 30dB attenuation

Hamming – Winder transition, but 40 dB

Blackman – 6x transition of Rectangular, but 74 dB

Kaiser – Any (custom) window can be generated based on a stopband attenuation

When designing a filter using the window technique, the first step is to use response curves or trial and error and decide which window would be appropriate to use. Then, the desired number of filter coefficients is chosen. Once the length and type of window are determined, the window coefficients can be calculated. Then, the window coefficients are multiplied by the ideal filter response. Here is the code and frequency response for the same filter as before with a Blackman window:

```
%lowpass filter design using 67 coefficient hamming window

%design specificattions
ws = .25; wp = .3;
N = 67;

wc = (wp - ws) / 2 + ws %calculate cutoff frequency
%build filter coefficients ranges
n = -33:1:33;
hd = sin(2 * n * pi * wc) ./ (pi * n);    % ideal freq
hd(34) = 2 * pi * wc / pi;        %zero ideal freq
hm = hamming(N);                  % calculate window coefficients
hf = hd .* hm';                   %multiply window by ideal response
```

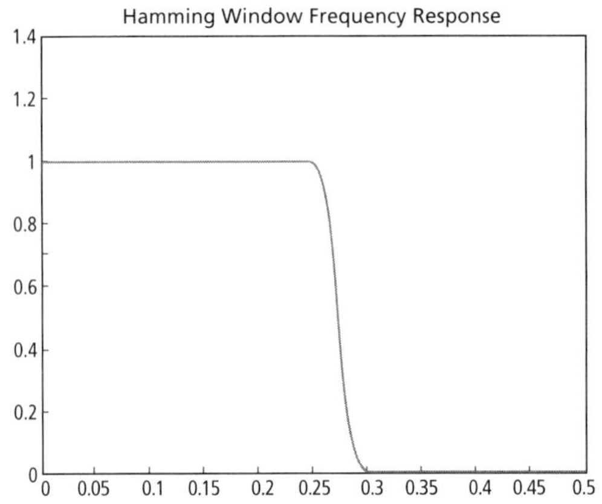

Figure 4.22 Hamming window frequency response

Summary of FIR Filters

In the digital filters discussed so far in this chapter, the current output sample, y_n, is calculated solely from the current and previous input values (such as x_n, x_{n-1}, x_{n-2}, ...). This type of filter is said to be nonrecursive. In fact, a FIR filter is also referred to as a *nonrecursive* filter.

FIR filters are easy to understand and implement. These filters are inherently stable, which also makes them easy to use. But FIR filters may require a significant number of filter taps to produce the desired filter characteristics. This may make the filter unusable for real-time applications where sample processing prohibits the use of more than a few filter taps. The use of many filter taps will also make the filter response characteristics somewhat imprecise because of the many stages of accumulating error buildup. This is especially true on integer fixed-point DSPs.

FIR filtering techniques are suitable for many audio applications. FIR filters in these applications can have significant consequences on the audio quality. The FIR linear-phase distortion is virtually inaudible, since all frequencies are effectively delayed by the same amount.

Despite the fact that recursive filters require the use of previous output values, there are fewer, not more, calculations to be performed in a recursive filter operation. Achieving a specific frequency response characteristic using a recursive filter generally requires a lower order filter, and therefore fewer terms to be evaluated by the DSP, than an equivalent nonrecursive filter.

Infinite Impulse Response Filters

When describing a FIR filter, we discussed the output as being a simple weighted average of a certain number of past input samples only. In other words, the computation of a FIR algorithm involves no feedback. In circuit theory, it is a well known fact that feedback can sometimes improve results. The same is true in digital filters. Feedback in DSP filters can also sometimes improve results. The IIR (infinite impulse response) filter includes a feedback component to the FIR structure we talked about earlier. IIR filters are more complicated to design, but thcy can sometimes produce better results with fewer taps.

IIR filters are also referred to as feedback filters or recursive filters. Recall that finite impulse response filters are considered nonrecursive because the current output (y_n) is calculated entirely from the current and previous input values (x_n, x_{n-1}, x_{n-2}, ...). A recursive filter is one which, in addition to current and previous input values, also uses previous output values to produce the result. Recursive, by definition, means "running back."

Recursive filter feeds back previously calculated output values when producing the latest output. These filters are classified as "infinite" because of this unique feedback mechanism. So, when describing a IIR filter, the current output from the filter depends on the previous outputs. In theory, IIR filters can use many (or an infinite) number of previous outputs, so this is where the term "infinite" comes from. This dependence on previous outputs also means that IIR filters do not have linear phase.

Like other feedback systems, the feedback mechanism of IIR filters can cause instability in the operation of the filter. Usually this filter instability can be managed with good filter design tools. In most cases, when an IIR filter becomes "unstable" it implies that the feedback in the filter is too large, similar to the causes of instability in other feedback systems.. If a IIR filter does become unstable, the output from the filter will cause oscillations that increase exponentially, as shown in Figure 4.23. This potential instability has software implications. Output oscillations can cause overflow in software calculations which, in turn, cause system exceptions to be generated or, worse, a system crash. Even if the system does not crash, incorrect answers will be generated, which may be difficult to detect and correct.

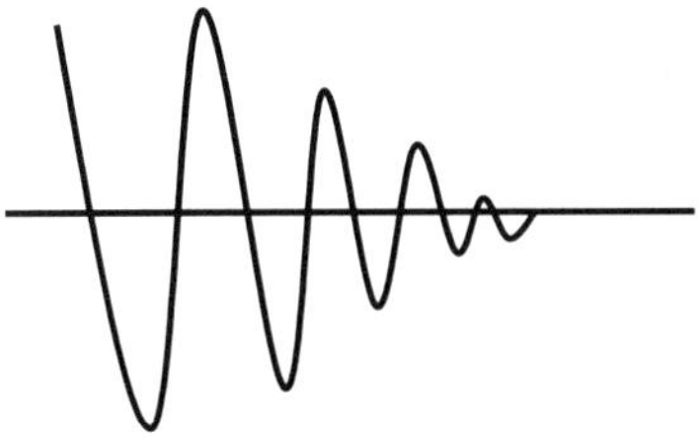

Figure 4.23a Feedback in an IIR filter that is controlled

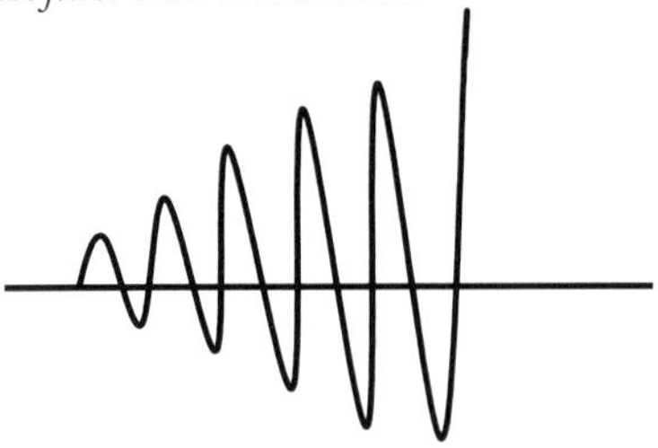

Figure 4.23b Feedback in an IIR filter that is out of control

If you think back to the moving average example used for FIR filters (the average temperature in Houston we discussed earlier), you'll remember that the output was computed using a number of previous input samples. If you add to that computation an average of a number of previous *output* samples, you will have a rough idea of how an IIR filter computes an output sample; using both previous input and output samples to compute the current output sample.

The description of a recursive filter contains input values (x_n, x_{n-1}, x_{n-2}, ...) as well as previous output values (y_{n-1}, y_{n-2},...). The simplest form of IIR filter is shown below:

$$y(n) = b0 * x(n) + a1 * y(n-1)$$

A simple example of a recursive filter is:

$$y_n = x_n + y_n - 1$$

In this expression the output, y_n, is equal to the current input, x_n, plus the previous output, y_{n-1}. If we expand this expression for n = 0, 1, 2, ... we now get additional terms shown below:

$$y_0 = x_0 + y_{-1}$$
$$y_1 = x_1 + y_0$$
$$y_2 = x_2 + y_1$$
... and so on.

Let's now do a simple comparison between recursive and non recursive filters. If we need to calculate the output of a filter at time t = 10h, a recursive filter will perform the calculation shown below:

Y10 = x10 + y9

To perform this same calculation using a nonrecursive filter the following calculation is performed;

Y10 = x10 + x9 + x8 + x7 + x6 + x5 + x4 + x3 + x2 + x1 + x0

Obviously, the nonrecursive filter requires significant addition operations. Many more terms must be stored in memory, and the computation will take longer to perform.

The C code for a simple IIR filter is shown below:

```
void iir(short *outPtr, short *inPtr, short *b, short *a, int M)
{
  int i, j, sum;
  for (i = 0; i < M; i++) {
    sum = b[0] *  inPtr[4+i]
          for (j = 1; j <= 4; j++)
            sum += b[j] * inPtr[4+i-j ]- a[j] * outPtr[4+i-j];
            outPtr[4+i] = (sum >> 15);
    }
}
```

IIR code, similar to FIR code, can be optimized using programming optimization techniques to take advantage of the specific DSP being used[4].

IIR As a Difference Equation

In many IIR filter representations, the feedback term is negative, which allows the IIR algorithm to be expressed as a difference equation. The general form is:

Difference Equation

$$y(n) = -\sum_{k=1}^{N} a(k) \bullet y(n-k) + \sum_{k=0}^{M} b(k) \bullet x(m-k)$$

[4] For an example, see Texas Instruments Application note SPRA517 which describes how to optimize an IIR filter to take advantage of a VLIW DSP architecture. Also see the reference "Source-Level Loop Optimization for DSP Code Generation" at http://www-acaps.cs.mcgill.ca/~jwang/ICASSP99.doc and the book *Analog and Digital Filter Design Using C* by Les Thede, Prentice Hall, 1996, ISBN 0-13-352627-5

The a(k) elements are the feedback coefficients. The y(n-k) element is the output. The b(k) elements are the feed forward elements. The x(m-k) element represents the input data stream. If all a(k) coefficients are zero, then the equation reduces to a FIR. In many IIR filters the number of feedback coefficients (N) and the number of feed forward elements (M) are the same (N=M) which simplifies programming somewhat by allowing us to effectively fold the delay line into the feed forward elements.

Figure 4.24 shows the general structure of an infinite impulse response (IIR) filter. This structure can map directly to hardware. You can notice the "b" coefficients supporting the feedforward line and the "a" coefficients on the feedback data line.

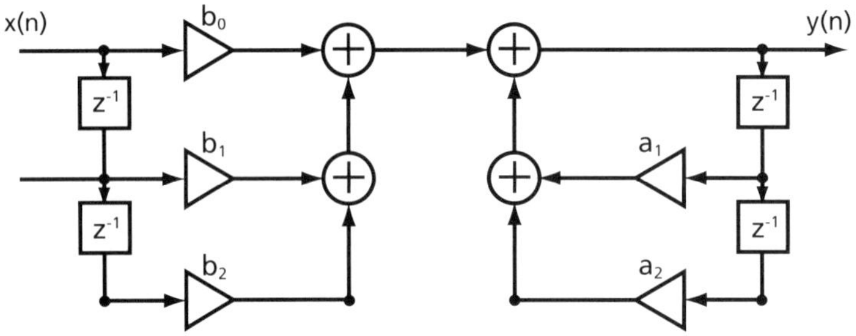

Figure 4.24 Difference equation-based circuit for a second order IIR filter

IIR As a Transfer Function

IIR filters can be described using a transfer function. A transfer function describes a filter using a convenient an d compact expression. This transfer function can be used to determine some important characteristics of these filters, such as the filter frequency response. The transfer function of an IIR filter is the ratio of two polynomials, as shown below.

Transfer Function

$$H(z) = \frac{b_0 + b_1 \bullet z^{-1} + b_2 \bullet z^{-2}}{1 + a_1 \bullet z^{-1} + a_2 \bullet z^{-2}}$$

The transfer function above is a characteristic of IIR filters that makes them more powerful than FIR filters, but can also subject them to instability. Like other expressions of this form, if the numerator of the transfer function goes to zero, the value of the entire transfer function becomes zero. In IIR filter design, the values that drive the numerator to zero like this are defined as "zeros" of the function. If the denominator goes to zero, we end up with a division by zero condition, and the value of the transfer function goes to (or approaches) infinity. The values that drive the transfer function to infinity are referred to as "poles" of the function. The goal in designing IIR filters is to select coefficients in order to prevent the filter from becoming unstable. This sounds difficult, but there are actually many good filter design packages available that can help the DSP engineer design a filter that meets the system requirements while remaining stable under the required operating conditions.

The form of a second order transfer function for an IIR filter is shown below.

Transfer Function

$$H(z) = \frac{b_0 + b_1 \bullet z^{-1} + b_2 \bullet z^{-2}}{1 + a_1 \bullet z^{-1} + a_2 \bullet z^{-2}}$$

Mapping this form directly onto hardware gives a much more compact circuit than the difference equation. See the second order filter shown in Figure 4.25.

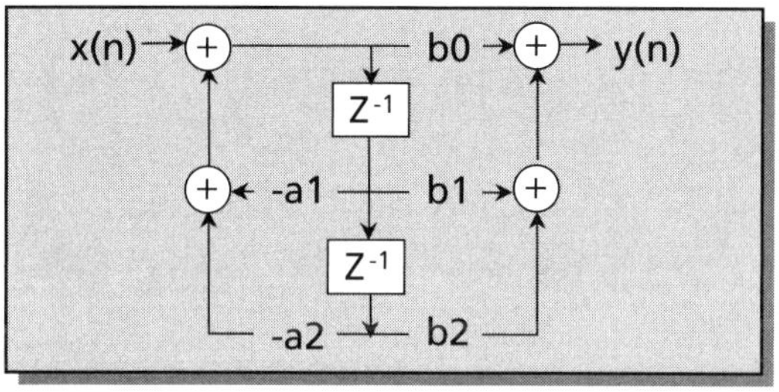

Figure 4.25 The structure of a second order recursive IIR filter (courtesy of Texas Instruments)

IIR Filter Design

Adding feedback to an IIR filter allows the equation to contain 5–10 times fewer coefficients than the FIR counterpart. However, it does mangle the phase and makes designing and implementing the filter more complicated.

While filters usually will be designed by software, it is a good idea to know the techniques involved in designing the filter so the designer has some idea of what the software is trying to accomplish, and what methods it is using to meet these goals. There are two primary techniques involved in designing IIR filters. They are direct and indirect design. Direct design does all its work in the z-domain (digital domain), while indirect design designs the filter in the s-domain (analog domain) and converts the results to the z-domain. Most of the time IIR filters are designed using analog techniques. While it may seem a less efficient way of doing things, analog methods for designing filters have been around a lot longer than digital design methods, and these proven techniques can be applied to digital filters in the same way.

In indirect design, the designer relies on optimized analog design techniques to develop the filter. Once the developer has an optimized solution for their analog filter, the problem lies in converting the analog solution to a digital solution. Since the analog domain can contain an infinite number of frequencies and the digital domain is limited to half the sampling rate, the two domains will not match up perfectly, and the frequencies must be mapped. There are two popular techniques used to accomplish this mapping. One is by wrapping the s-domain around the unit circle in the z-domain, and the other is done by compressing the s-domain into the unit circle.

There are several techniques which have been optimized for analog design over the years, most of which excel at one particular area or specification, such as passband ripple, transition, or phase. The most popular analog techniques and their useful characteristics are mentioned below.

- *Butterworth* – Use for a flat passband ripple. Also, the magnitude response will not increase as frequency increases.
- *Chebychev* – Sharper transition than Butterworth, with the cost of more ripple in the passband.
- *Chebychev II* – Monotonic passband but adds ripples to the stopband.
- *Bessel* – When phase is important in an IIR filter.
- *Eliptical* – Sharpest transition, but allows ripples in the stopband and passband.

Once the filter's poles and zeros are determined, they must be converted to the z-domain for use by the digital filter. The most popular technique for doing this is the Bilinear Transform. The Bilinear Transform method does this by mapping (or compressing) all the frequencies into the unit circle. It does this in a nonlinear manner, and to compensate for this "warping of frequencies", the frequencies must be "pre-warped" before the filter is designed.

So, to develop a filter using the Bilinear Transform technique, the designer should follow the instructions:

1. Determine the filter critical frequencies and sampling rate.
2. Pre-warp the filter's critical frequencies.
3. Design an analog filter using "classic" techniques with these pre-warped frequencies.
4. Convert the filter to the z-domain using the Bilinear Transform.

Another technique used to transform poles and zeros from the s-domain to the z-domain is called the *impulse invariance* method. The impulse invariance method takes the s-domain only up to half the sampling rate and converts it to the z-domain. For this reason, it is limited to low pass and band pass filters only. This has the benefit of creating an impulse response that is a sampled version of the s-domain impulse response.

There are MATLAB functions to assist in designing filters. The functions to design the popular analog filters are BUTTER, CHEB1AP, CHEB2AP and ELLIPAP. These functions will return the coefficients for the IIR filter; there are two additional functions for converting the analog coefficients to the digital domain: BILINEAR and IMPINVAR. The function BILINEAR does the necessary "pre-warping."

Typically, when designing IIR filters by hand, only low-pass filters would be used. They would then be converted to the appropriate specifications using complex formulas. However, when designing with a software package such as MATLAB, the user does not have to worry about this transformation.

IIR Trade-Offs

Advantages of recursive filter

There are trade-offs when using IIR filters. One of the advantages to using a recursive IIR filter structure is that these filters generally require a much lower order filter. This means that there will be fewer overall terms to be evaluated by the processor as compared to the equivalent nonrecursive filter. Recursive filters, on the other hand, calculate output terms more efficiently than the nonrecursive model. So if you needed to increase the filter structure for your signal processing application, you would need to add additional terms if you were using a FIR (nonrecursive) filter.

To achieve the same sharp rolloff using an IIR filter, you can implement it recursively with a small number of coefficients and still realize a sharp frequency cutoff in the output. The practical aspects can be substantial from a system design perspective. A recursive implementation leads to reduced hardware requirements in the system.

There are also some disadvantages to using IIR filters over FIR filters.

When using an IIR filter, the feedback component will also feed back the noise from the original signal. Because of this, the filter can possibly increase the amount of noise in the output. The amount of noise fed back into the system will actually increase as more stages are added to the filter design.

It is also important to keep in mind that IIR filters will exhibit nonlinear phase characteristics. This may make them a poor choice for some applications.

Just like with FIR filters, the DSP engineer must be aware of some implementation issues with IIR filters. For example, the precision of IIR filters is dependent on the accuracy (or quantization) of the filter coefficients. The precision of these coefficients is constrained by the word size of the DSP being used. When designing a filter like this, the engineer should use coefficients that are as large as possible in magnitude to take advantage of as many significant bits as possible in the coefficient word size.

Another implementation issue has to do with rounding. As you know, the accumulators of DSPs are large because of the iterative accumulations that must be done for the many looping constructs in signal processing algorithms. After these computations are complete, the result must then be stored back into a single word in processor memory. A common conversion may involve going from a 60-bit accumulator to a 32-bit result word somewhere in DSP memory. Because IIR filters are recursive, this same conversion problem can also exist.

Another potential issue is that of overflow. The same IIR filter feedback mechanism that can cause instability can also lead to overflow if not designed properly. There are a couple ways for the DSP engineer to alleviate this condition. The first approach is to scale the input and output of the filter operation. To do this, you must add additional software instructions (and therefore more cycles) to perform the task. This may or may not be a problem. It depends on the filter performance requirements as well as the available system processing resources (cycles, memory).

Another approach is to use the available DSP saturating arithmetic logic. More and more DSPs have this logic in the processor implementation. This logic is included in DSPs primarily for the purpose of preventing overflows (or underflows) without having to use software to do this. Overflow occurs when the result of a computationexceeds the available word length of the adder in the DSP. This results in an answer that can becomes negative (for overflow, positive for underflow). This will yield a result almost exactly opposite from the expected result. Saturating logic will prevent the adder circuitry from "wrapping" from highest positive to negative and will instead keep the result at the highest possible value (and likewise for underflow). This result is still wrong but a lot closer than the alternative.

Summary

IIR filters are implemented using the recursion described in this chapter to produce filters with sharp frequency cutoff characteristics. This can be achieved using a relatively small number of coefficients. The advantage of this approach is that the implementation lends itself to reduced memory and processing when implemented in software. However, these filters have nonlinear phase behavior and because of this characteristic, phase response must be of little concern to the DSP programmer. Amplitude response must be the primary concern when using these filters. This nonlinear phase characteristic makes IIR filters a poor choice for applications such as speech processing and stereo sound systems. Once these limitations are understood, the DSP engineer can choose the best filter for the application. The main drawback when using IIR filters implemented recursively is instability. Careful design techniques can avoid this.

DSP Architecture Optimization for Filter Implementation

Today's DSP architectures are made specifically to maximize throughput of DSP algorithms, such as a DSP filter. Some of the features of a DSP include:

- *On-chip memory* – Internal memory allows the DSP fast access to algorithm data such as input values, coefficients and intermediate values.
- *Special MAC instruction* – For performing a multiply and accumulate, the crux of a digital filter, in one cycle.
- *Separate program and data buses* – Allows the DSP to fetch code without affecting the performance of the calculations.
- *Multiple read buses* – For fetching all the data to feed the MAC instruction in one cycle.
- *Separate Write Buses* – For writing the results of the MAC instruction.
- *Parallel architecture* – DSPs have multiple instruction units so that more than one instruction can be executed per cycle.
- *Pipelined architecture* – DSPs execute instructions in stages so more than one instruction can be executed at a time. For example, while one instruction is doing

a multiply another instruction can be fetching data with other DSP chip.

- *Circular buffers* – To make pointer addressing easier when cycli ficients and maintaining past inputs.
- *Zero overhead looping* – Special hardware to take care of counter loops.
- *Bit-reversed addressing* – For calculating FFTs.

Number format

When converting an analog signal to digital format, the signal has to be truncated due to the limited precision of a DSP. DSPs come in fixed- and floating-point format. When working with a floating-point format, this truncation usually is not much of a factor due to its good mix of precision and dynamic range. However, implementing hardware to deal with floating-point formats is harder and more expensive, so most DSPs on the market today are fixed-point format. When working with fixed-point format a number of considerations have to be taken into account. For example, when two 16-bit numbers are multiplied, the result is a 32-bit number. Since we ultimately want to store the final result in 16-bit format, we need to handle this loss of data. Clearly, by just truncating the number we would lose a significant portion of the number. To deal with this issue we work with a fractional format called Q format. For example, in Q15 (or 1.15) format, the most significant digit is used to represent the sign and the rest of the digits represent the fractional part of the data. This allows for a dynamic range of between –1 and just less than 1. However, the results of a multiply will never be greater than one. So, if the lower 16 bits of the result are dropped, a very insignificant portion of the results is lost. One nuance of the multiply is that there are two sign bits, so the result will have to be shifted to the left one bit to eliminate the redundant information. Most processors will take care of this, so the designer doesn't have to waste cycles when doing many multiplications in a row.

Overflow and saturation

Two other problems that can occur when using fixed-point arithmetic are overflow and saturation. However, DSPs help the programmer deal with these problems. One way a DSP does this is by providing guard bits in the accumulator. In a normal 16-bit processor, the accumulator may be 40 bits; 32 bits for the results (keep in mind that a 16x16 bit multiplication can be up to 32 bits) and an extra 8 bits to guard against overflow (of multiple multiplies in a repeat block.)

Even with the extra guard bits, multiplications can provide overflow situations where the result contains more bits than the processor can hold. This situation is handled with a flag called an overflow bit. The processor will set this automatically when the results of a multiplication overflow the accumulator.

When an overflow occurs, the results in the accumulator usually become invalid. So what can be done? Another feature of DSPs can be used: saturation. When the saturate instruction on a DSP is executed, the processor sets the value in the accumulator to the largest positive or negative value the accumulator can handle. That way, instead of possibly flipping the result from a high positive number to a negative number, the result will be the highest positive number the processor can handle.

There is also a mode DSP processors have that will automatically saturate a result if the overflow flag gets set. This saves the code from having to check the flag and manually saturate the results.

Implementing an FIR filter

We will begin our discussion of implementing an algorithm on a DSP by examining the C code required to implement an FIR filter. The code is pretty straightforward and looks like this.

```
long temp;
int block_count;
int loop_count;


// loop through inputs
for (block_count=0;block_count<output_size;block_count++)
{
      temp=0;
      for (loop_count = 0;loop_count<coeff_size;loop_count++)
      {
            temp += (((long)x[block_count+loop_count]*(long)a[loop_count])
            << 1);
      }
      y[block_count] = (temp >> 16);
}
```

This code is a very simple sum of projects written in C with a few caveats. The caveats stem from the number format issues we discussed earlier. First of all, temp must be declared a long so that it is represented by 32 bits for temporary calculations. Also, the value of the MAC must be right-shifted to the left one bit because of the fact that multiplying two 1.15 numbers results in a 2.30 number, and we want our result in 1.15. Finally, the temp value is shifted 16 bits to the right before writing to the output so that we take the most significant bits of our result.

As you can see, this is a very simple algorithm to write, and central to most DSP applications. The problem with implementing the code in C is that it is too slow. A general rule in designing embedded DSP applications is known as the 90/10 rule. It is used to determine what to write in C and what to write in assembly. It states that a DSP application generally spends 90% of its time in 10% of the code. This 10% of the code should be written in assembly, and in this case the code is the FIR filter code.

Here is an example of a filter written in TMS320c5500 assembler language for a DSP:

```
fir:
          AMOV    #184, T1              ; block_count = 184
          AMOV    #x0, XAR3             ; init input pointer
          AMOV    #y, XAR4              ; init output pointer

                                        ;do
oloop:    SUB     #1, T1                ; block_count--
          AMOV    #16, T0               ; loop_count = 16
          AMOV    #a0, XAR2             ; init coefficient pointer
          MOV     #0, AC0               ; y[block_count] = 0
                                        ; do
loop:     SUB     #1,T0                 ; loop_count--
          MPYM    *AR2+, *AR3+, AC1     ; temp1 = x[] * a[]
          nop
          ADD     AC1, AC0              ; temp = temp1
          BCC     loop, T0 != #0        ; while (loop_count > 0)
          nop
          nop

          MOV     HI(AC0), *AR4+        ; y[block_count] = temp >> 16
          SUB     #15, AR3              ; adjust input pointer
          BCC     oloop, T1 != 0        ;while (block_count > 0)
          RET
```

This code does the same thing as the C code, except it is written in assembly. However, it does not take advantage of any of the DSP architecture. We will now start rewriting this code to take advantage of the DSP architecture.

Utilizing on-chip RAM

Typically, data such as filter coefficients are stored in ROM. However, when running an algorithm, the designer would not want to have to read the next coefficient value from ROM. Therefore, it is a good practice to copy the coefficients from ROM into internal RAM for faster execution. The following code is an example of how to do so.

```
copy: AMOV      #table, XAR2
      AMOV      #a0, XAR3
      RPT       #7
                MOV        dbl(*ar2+), dbl(*ar3+)
      RET
```

Special MAC instruction

All DSPs are built to do a multiply-accumulate (MAC) in one instruction cycle. There are a lot of things going on in the MAC instruction. If you notice, there is a multiply, an add, an increment of the pointers, and a load of the values for the next MAC, all in one cycle. Therefore, it is efficient to take advantage of this useful instruction in the core loop. The new code will look like this:

```
MAC *AR2+, *AR3+, AC0     ; temp += x[] * a[]
```

Block filtering

Typically, an algorithm is not performed one cycle at a time. Usually a block of data is processed. This is known as block filtering. In the example, looping was used to apply the filter algorithm on a hundred inputs rather than just one, thus, generating 100 outputs at a time. This technique allows us to use many of the optimizations we will now talk about.

Separate program and data buses

The 55x architecture has three read buses and two write buses, as shown in Figure 4.26. We will take advantage of all three read buses and both write buses in the filter by using what's called a coefficient data pointer and calculating two outputs at a time. Since the algorithm uses the same coefficients in every loop, one bus can be shared for the coefficient pointer and the other two buses can be used for the input pointer. This will also allow the use of the two output buses and two MAC units in each inner loop, allowing the values to be calculated over twice as fast. Here is the new code to optimize the MAC hardware unit and the buses:

```
AMOV #x0, XAR2              ; x[n]
AMOV #x0+1, XAR3            ; x[n+1]
AMOV #y, XAR4               ; y[n]
AMOV #a0, XCDP              ; a[n] coefficient pointer

MAC    AR2+, CDP+, AC0      ; y[n] = x[n] * a[n]
 :: MAC *AR3+, CDP+, AC1    ; y[n+1] = x[n+1] * a[n]

MOV   pair(hi(AC0)), dbl(*AR4+); move AC0 and AC1 into mem pointed to
      by AR4
```

Notice that a colon separates the two MAC instructions. This tells the processor to execute the instructions in parallel. By executing in parallel we take advantage of the fact that the processor has two MAC units in hardware, and the DSP is instructed to execute 2 MAC instructions in one cycle by using both hardware units.

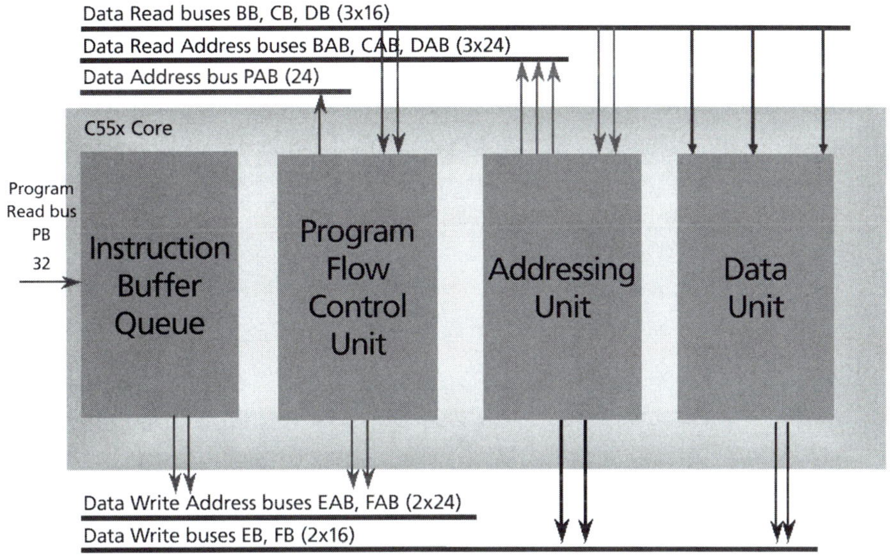

Figure 4.26 C55x architectural overview

Zero overhead looping

DSP processors have special hardware to take care of the overhead in looping. The designer need only set up a few registers and execute the RPT or RPTB (for a block of instructions) instruction and the processor will execute the loop the specified number of times. Here is the code taking advantage of zero overhead looping:

```
MOV        #92, BRC0      ;calculating 2 coefficients at a time, block loop
                           is 184/2
```

And here is the actual loop code:

```
RPTBlocal  endfir                   ; repeat to this label;, loop start
           MOV   #0, AC1            ; set outputs to zero
           MOV   #0, AC0
           MOV   #a0, XCDP          ; reset coefficient pointer

           RPT            #15       ; inner loop
                 MAC    *AR2+, *CDP+, AC0
                 :: MAC *AR3+, *CDP+, AC1
```

```
                    SUB     15, AR2          ; adjust input pointers
                    SUB     15, AR3
               MOV  pair(hi(AC0)), dbl(*AR4+) ; write y and y+1 output values
endfir:             nop
```

Circular buffers

Circular buffers are useful in DSP programming because most implementations include a loop of some sort. In the filter example, all the coefficients are processed, and then the coefficient pointer is reset when the loop is finished. Using circular buffering, the coefficient pointer will automatically wrap around to the beginning when the end of the loop is encountered. Therefore, the time that it takes to update the pointers is saved. Setting up circular buffers usually involves writing to some registers to tell the DSP the buffer start address, buffer size, and a bit to tell the DSP to use circular buffers. Here is the code to set up a circular buffer:

```
; setup coefficient circular buffer
        AMOV        #a0, XCDP       ; coefficient data pointer
        MOV         #a0, BSAC       ; starting address of circular buffer
        MOV         #16, BKC        ; size of circular buffer
        MOV         #0, CDP         ; starting offset for circular buffer
        BSET        CDPLC           ; set circular instead of linear
```

Another example where circular buffers are useful is when working with individual inputs and only saving the last N inputs. A circular buffer can be written so that when the end of the allocated input buffer is reached, the pointer automatically wraps around to the beginning of the buffer. Writing to the correct memory is then ensured. This saves the time of having to check for the end of the buffer and resetting the pointer if the end is reached.

System issues

After the filter code is set up, there are a few other things to take into consideration when writing the code. First, how does the DSP get the block of data? Typically, the A/D and D/A would be connected to serial ports built into the DSP. The serial ports will provide a common interface to the DSP, and will also handle many timing considerations. This will save the DSP a lot of cycles. Also, when the data comes in to the serial port, rather than having the DSP handle the serial port with an interrupt, a DMA can be configured to handle the data. A DMA is a peripheral designed for moving memory from one location to the other without hindering the DSP. This way, the DSP can concentrate on executing the algorithm and the DMA and serial port will worry about moving the data. The system block diagram for this type of implementation is shown in Figure 4.27.

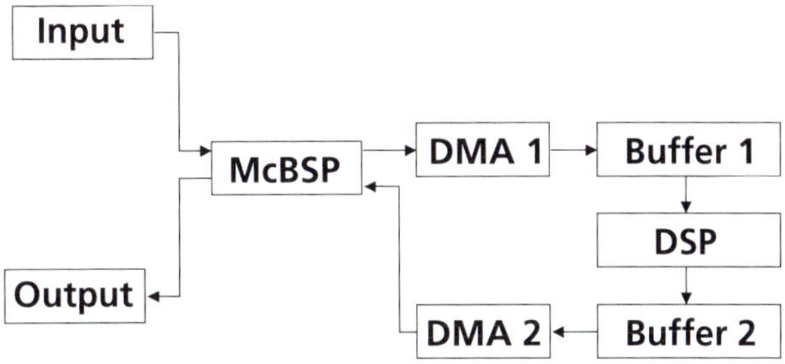

Figure 4.27 Using the DMA to bring data into and out of the DSP

Fast Fourier Transforms

One of the most common operations in digital signal processing involves a process called spectral analysis. Spectral analysis is a technique used to determine what frequencies are present in a signal. A good analogy is a filter that can be tuned to let just a narrow band of frequencies through (like tuning a radio). This approach would determine at what frequencies parts of the signal of interest would be exposed. The analysis could also reveal what sine wave frequencies should be added to create a duplicate of a signal of interest. This is similar to the function of a spectrum analyzer. This is an instrument that measures the frequency spectrum of a signal. These early tools were implemented using banks of analog filters and other components to separate the signal into the different frequency components. Modern systems use digital techniques to perform many of these operations. One of the most common techniques uses an algorithm called a fast Fourier transform. The fast Fourier transform (FFT) and other techniques give a picture of a signal in terms of frequency and the energy at the different frequencies of a particular signal.

In the signal processing field, a signal can be classified as a pure tone or a complex tone. A pure tone signal is composed of one single frequency and the wave form is a pure sine wave. A complex tone is not a pure sine wave but a complex tone can be periodic. Complex tones have underlying patterns that repeat. A sound may have a pattern that looks pretty much the same each time it occurs (for example, one dog bark sounds pretty much like any other). However, within the wave form itself, there is no long-term recognizable pattern.

The FFT is an algorithm used to decompose a signal in the time domain into all of its individual frequency components. The process of examining a time signal broken down into its individual frequency components like this is referred to as spectral analysis or harmonic analysis.

Time vs. Frequency

Jean Baptiste Fourier[5] discovered in the 1800s that any real world waveform can be generated by the addition of a number of different sinusoidal waveforms. Even a complex waveform like the one in Figure 4.28a can be recreated by summing a number of different sinusoids (Figure 4.28b and c). The work that Fourier did in the 1800's is still used today to decompose a signal that varies with time into components that vary only in frequency.

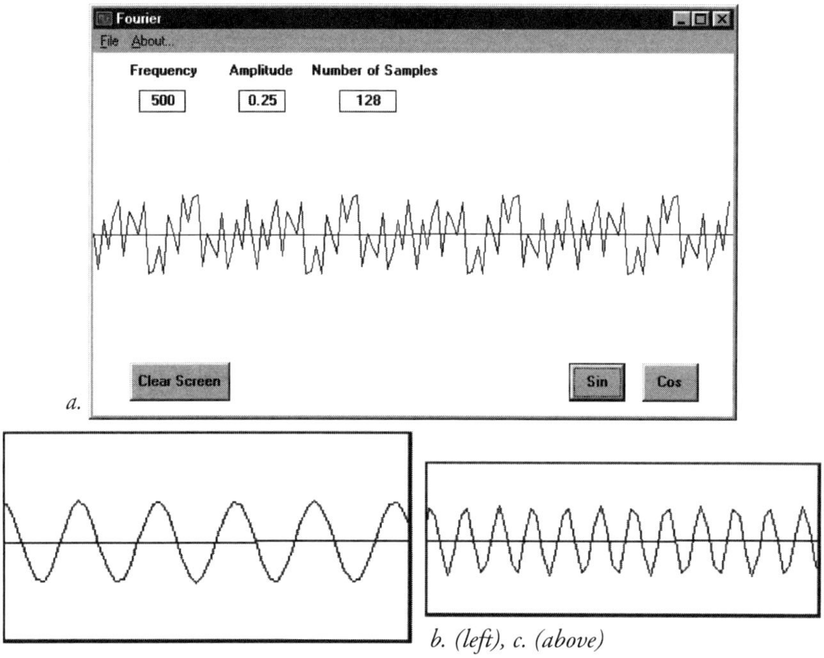

a.

b. (left), c. (above)

Figure 4.28 A complex signal is composed of the sum of a number of different sinusoids (generated using "DSP Calculator" software)

[5] Fourier published a prize winning essay, Théorie analytique de la chaleur, in1822. In this work he shows that any functions of a variable, whether continuous or discontinuous, can be expanded in a series of sines of multiples of the variable. This result is still used constantly today in modern signal analysis.

Relation between time and frequency

Before we begin to explain Fourier transforms, it is useful first to develop an understanding of the relationship between time and frequency domains. Many signals are easier to visualize in the frequency domain rather than the time domain. Other signals are actually easier to visualize in the time domain. The reason is simple. Some signals require less information to define them in one or the other domains. Consider a sine wave. This waveform requires a lot of information to define accurately in the time domain. In the frequency domain, however, there are only three pieces of information needed to accurately define this signal; the frequency, amplitude and phase.

The Fourier transform assumes a given signal is analyzed over long periods of time. Because of this assumption, there is no concept of time when analyzing signals in the frequency domain. This means that frequency cannot change with time in this analysis. In the time domain this is possible. When we analyze signals, we do not mix one representation with the other (we keep them orthogonal). We do, however, switch back and forth between the two domains, depending on the analysis we are performing on the signal.

Many real world signals have frequency components that change with time. A speech signal is a good example. When analyzing signals such as speech that could have, effectively, infinite duration, we can still perform an analysis on this signal by chopping the signal into shorter pieces and then using the Fourier transform to analyze each of these pieces. The resultant frequency spectrum of each piece of this speech signal describes the frequency content during that specific period. In many cases like this, when sampling a long sequence of related signals like speech, the average spectrum of the signal is often used for the analysis.

The Fourier transform operates under the assumption that any signal can be constructed by simply adding a series of sine waves of infinite duration. We know that a sine wave is a continuous and periodic signal. The Fourier transform will operate as if the data in the signal is also continuous and periodic.

The basic operation of a Fourier transform is as follows. For every frequency, the Fourier transform determines the contribution of a complex sinusoid at that frequency in the composition of the signal under analysis. Lets go back to the example of the spectrum analyzer. You can think of a Fourier transform as a spectrum analyzer that is composed of a filter sequence x(n) with a number of frequencies. Assume we run our input sequence through a very large number of these filters as shown in Figure 4.29. Assume each filter has a center frequency. The result of this operation is the sum of the magnitudes out of each of these filters. Of course, we don't want to use a spectrum analyzer to do this. A faster, cheaper way is to use an algorithm with a "big OH" execution time that is reasonable. This algorithm is the Fourier transform.

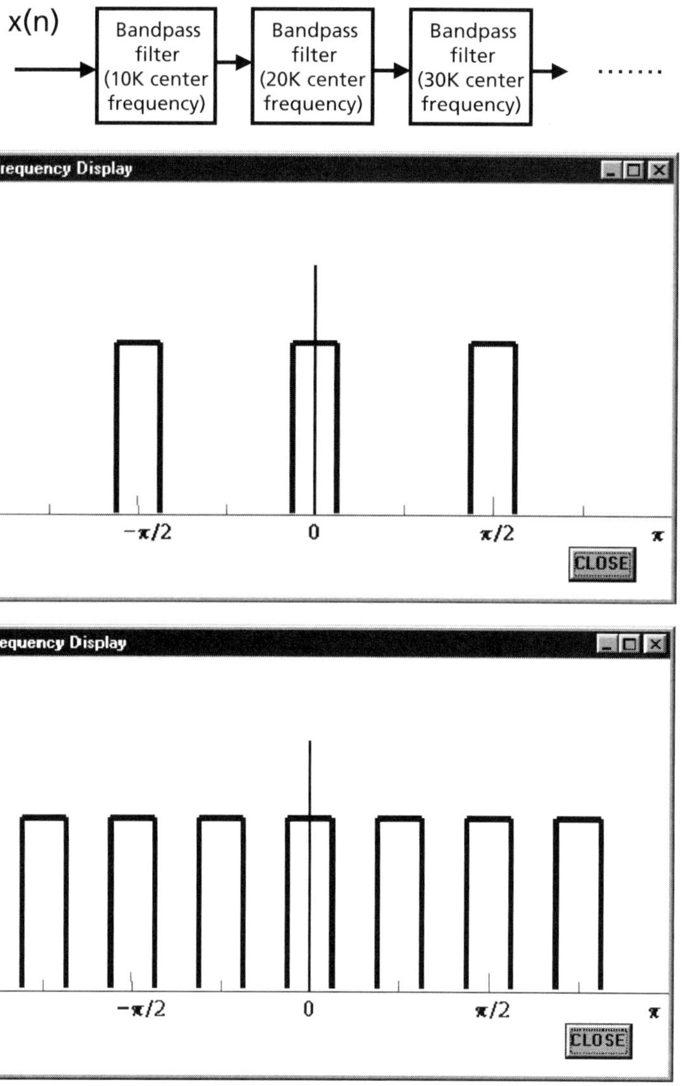

Figure 4.29 A Fourier transform is similar to passing an input sequence through a number of band-pass filters and summing the responses (generated using "DSP Calculator" software)

There are several different types of Fourier transforms. These can be summarized:

- The Fourier transform (FT) is a mathematical formula using integrals.
- The discrete Fourier transform (DFT) is a discrete numerical equivalent using sums instead of integrals.
- The fast Fourier transform (FFT) is a computationally faster way to calculate the DFT.

Since DSPs always work with discrete, sampled data, we will discuss only the discrete forms here (DFT and FFT).

The Discrete Fourier Transform (DFT)

We know from our previous discussion that in order for a computer to process a signal, the signal must be discrete. The signal must consist of a number of samples, usually from an ADC operating on a continuous signal. The "computer" form of a continuous Fourier transform is the discrete Fourier transform. DFTs are used on discrete input sample trains. The "continuous" or analog signal must be sampled at some rate (this is the Nyquist rate discussion we had earlier) to produce a representative number of samples for the computer. This sequence of N samples we call f(n). We can index this sequence from n = 0..N-1

The discrete Fourier transform (DFT) can now be defined as F(k), where k=0..N-1:

$$F(k) = \frac{1}{\sqrt{N}} \sum_{n-0}^{N-1} f(n) e^{-j2\pi kn/N}$$

F(k) are called the 'Fourier coefficients' (sometimes these are also called harmonics). This sequence operates on very long (maybe infinite) sequences so we accommodate this by dividing the result by N as shown above.

We can also go the other way. The sequence f(n) above can be calculated from F(k). This inverse transform is performed using the inverse discrete Fourier transform (IDFT):

$$f(n) = \frac{1}{\sqrt{N}} \sum_{k-0}^{N-1} F(k) e^{+j2\pi nk/N}$$

f(n) and F(k) are both complex numbers. You can see this by the fact that they are multiplied by the complex exponential (e^j2(pi)nk/N). This may or may not be an issue. If the signal being analyzed is a real signal (no imaginary part), these discrete transforms will make the result a complex number due to the multiplication by the complex exponential. When programming a DFT or IDFT, the DSP programmer must decide, based on the signal composition, how and whether to use the complex part of the result.

In DSP jargon, the complex exponential (e** -j (2 Pi n k) / N) is sometimes referred to as a "twiddle factor." In a DFT above, there are N twiddle factors. Each of these twiddle factors can be thought of as one of the bandpass filter "bins" referred to in Figure 4.14. When we talked about the Nyquist sampling rate earlier, we mentioned that sampling at twice the highest signal frequency was the only way to guarantee an accurate representation of the signal. In a Fourier transform, things work in a similar way. That is, the more samples you take of the signal, the more twiddle factors are required (think of this in terms of the bins analogy; you create more bins as you sample the signal more). With more bins to use in the analysis, the higher the resolution or accuracy of the signal in the frequency domain.

In the algorithms above, the sequence f(n) id referred to the time domain data and F(k) is referred to as the frequency domain data. The samples in f(n) do not necessarily need to be samples of a time dependant signal. They could also be spatial image samples (think of an MRI).

From an implementation perspective, a Fourier transform involves multiplying a vector of data (the input sequence) by a matrix. Since a Fourier transform can scale in terms of size of computation, for an N data point transform, the entry in the hth row and kth column of an N × N "Fourier matrix" is e^{2{\pi}ihk/N}.

Every entry in this Fourier matrix is nonzero. To perform a NXN multiplication (basically multiplying a matrix by a vector) is a very time consuming task. Using the big OH analysis, this involves a total of N^2 multiplications. From a big OH perspective, this is not that big an issue if N is small. But when N gets large (N > 512 or 1000), the N^2 multiplications become a significant computational bottleneck.

The Fast Fourier Transform (FFT)

The fast Fourier transform (FFT), as the name implies, is a fast version of the DFT. The FFT exploits the fact that the straightforward approach to computing the Fourier transform performs the many of the exact same multiplications repeatedly. The FFT algorithm organizes these redundant computations in a very efficient manner by taking advantage of the algebraic properties in the Fourier matrix. Specifically, the FFT makes use of periodicities in the sines that are multiplied to perform the calculation. Basically, the FFT takes the Fourier matrix and factorizes it into several sparse matrices. These sparse matrices have many entries that are equal to zero. Using sparse matrices reduces the total amount of calculations required. The FFT eliminates almost all of these redundant calculations, and this saves a significant amount of calculation, which makes the Fourier transform much more practical to use in many applications today.

The FFT algorithm takes a "divide and conquer" approach to solving problems. The FFT approach attempts to solve a series of smaller problems very quickly as opposed to trying to solve one big problem which is generally more difficult. A large data set is decomposed into smaller data sets and each of these smaller data sets may, in turn, be decomposed into still smaller data sets (depending on the size of the initial data set). An FFT of size 64 will first be decomposed into two data sets of 32. These data sets are then themselves decomposed into four data sets of 16. These 4 data sets of 16 are then decomposed into 8 data sets of 8, then 16 data sets of 4, and finally 32 data sets of 2. The calculation on a data set of size two is simple and inexpensive. The FFT then performs a DFT on these small data sets. The results of the transforms of these multiple stages of data are then combined to get the final result.

The DFT takes N^2 operations to calculate a N point DFT (using big OH nomenclature). A FFT on the same N point data set has log2(N) stages in the FFT operation. The total effort to perform the computation (the big OH) is proportional to N * log2(N). By this comparison, the FFT is N/log2(n) faster than the DFT. This says that a computation that originally required N^2 computations can now be done with only Nlog2 N computations. The speedup factor gets better as the number of data points gets larger (see Figure 4.1). Another benefit is that fewer computations means less chance for error in programming. Keep it simple is the motto here as well!

We'll talk more about this in the chapter on software optimization, but it needs to be mentioned here as well. Before beginning to optimize software to improve efficiency, the DSP engineer must understand the algorithms being run on the machine. Heavy duty code optimization is difficult and error prone, and much of the same performance improvement can be had by simply making algorithmic improvements, such as using a FFT instead of a DFT. These types of algorithmic improvements in many cases outpace other approaches to efficiency. You can speed up your existing algorithm by running it on a newer computer or processor that runs ten times faster than the one it replaced. This will give you a 10x improvement, but then all you've got is a 10x speedup. Faster algorithms like the FFT usually provide even bigger gains as the problem gets bigger. This should be the first approach in gaining efficiency in any complicated system like this. Focus first on algorithmic efficiency before diving into code efficiency.

The divide part of the FFT algorithm divides the input samples into a number of one-sample long signals. When this operation completes, the samples are re-ordered in what is referred to as bit-reversed order. Other sorting algorithms have this side affect as well. The actual FFT computations take place in the combine phase. The samples are combined by performing a complex multiplication on the data values of two groups that are merged. This computation is then followed by what is called a "butterfly" calculation. The butterfly operation calculates a complex conjugate operation. This is what gives the Fourier transform its symmetry.

The Butterfly Structure

The butterfly structure described in the previous section is a graph with a regular pattern. These butterfly structures exist in different sizes. A height three butterfly structure is shown in Figure 4.30. The FFT butterfly is a graphical method of representing the multiplications and additions required to process the data samples being transformed. The butterfly notation is shown as follows; each dot with entering arrows is an addition of the two values at the end of the arrows. This result is then multiplied by a constant term. Figure 4.31 shows an example of a simple FFT butterfly structure. The term W_N is the notation used to represent the twiddle factor discussed earlier.

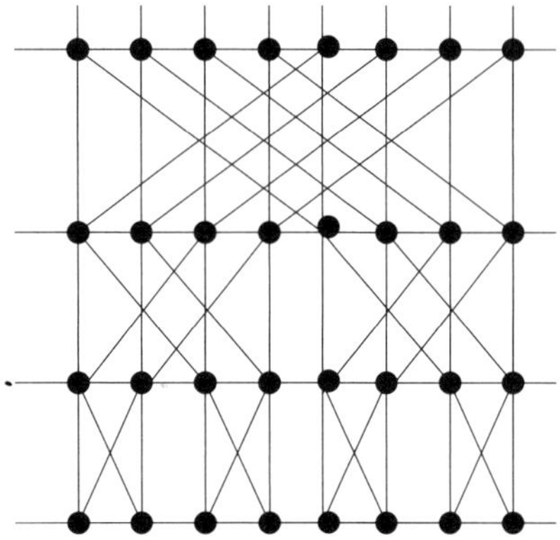

Figure 4.30 A butterfly structure of height three. Butterfly structures are used to compute FFT algorithms

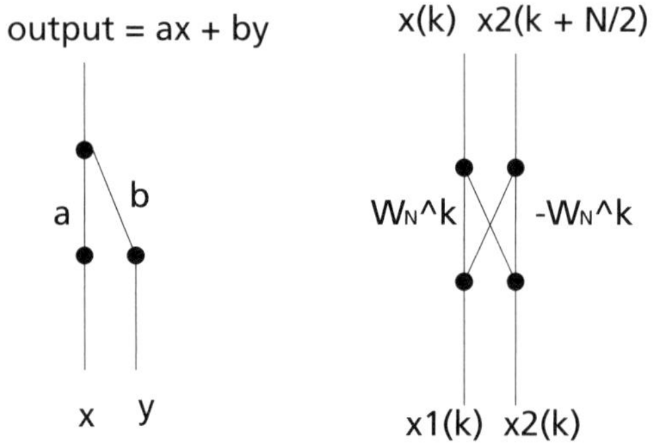

Figure 4.31 A simple butterfly computation and a FFT butterfly structure

Forms of the FFT Algorithm

There are two forms of the FFT algorithm. These are referred to as *decimation in time (DIT)* and *decimation in frequency (DIF)*. The differences in these two approaches involve how the terms of the DFT are grouped. From an algorithmic perspective, decimation refers to the process of decomposing something into its constituent parts. The DIT algorithm, therefore, involves decomposing a signal in the time domain into smaller signals. Each of these smaller signals then becomes easier to process. The DIF algorithm performs similar operation in the frequency domain. The DIF algorithm begins with a normal ordering of the input samples and generates bit-reversed order output. The DIT, on the other hand, begins with bit-reversed order input and generates normal order output (See Table 4.1). Many DSPs support what is called *bit-reversed addressing*, which makes accessing data samples in this order easy. The engineer does not have to write software to perform the bit-reversed addressing, which is very expensive computationally. If bit-reversed addressing modes exist on the DSP, the DIF and DIT can be used very interchangeably to perform forward and reverse transforms. An algorithm template for performing a FFT using this approach is as follows:

- Pad the input sequence of samples (where the number of samples = N) with zeros until the total number of samples is the nearest power of two (for example for N = 250, this means adding 6 zeros to get 256 total samples which is a power of two)
- Bit reverse the input sequence (if performing a decimation in time transform)
- Compute N / 2 two sample DFT's from the inputs.
- Compute N / 4 four sample DFT's from the two sample DFT's.
- Compute N / 8 eight sample DFT's from the four sample DFT's.
- Continue with this algorithm until the all the samples combine into one N-sample DFT

An eight point decimation in frequency structure is shown in Figure 4.32. A full listing of a FFT computation written in Java is shown in Figure 4.33.

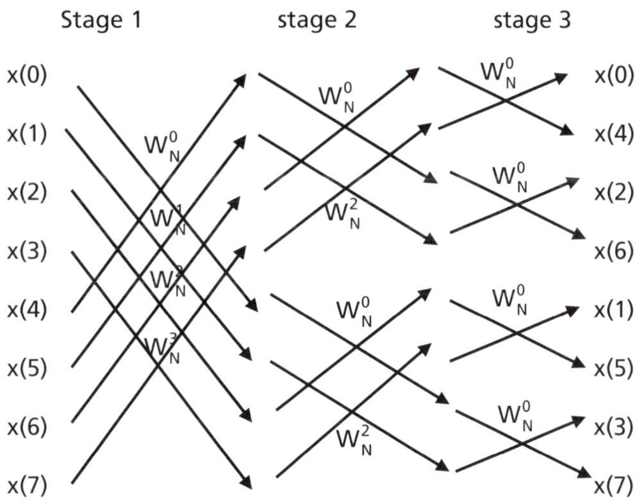

Figure 4.32 An 8 point decimation in frequency FFT structure

Original order	Decimal value	Final order	Decimal order
000	0	000	0
001	1	100	4
010	2	010	2
011	3	110	6
100	4	001	1
101	5	101	5
110	6	011	3
111	7	111	7

Table 4.1 Bit-reversed addressing in the FFT

```
import java.awt.*;

double[ ][ ] fft(  double[ ][ ] array )
{
        double  u_r,u_i, w_r,w_i, t_r,t_i;
        int     ln, nv2, k, l, le, le1, j, ip, i, n;

        n = array.length;
        ln = (int)( Math.log( (double)n )/Math.log(2) + 0.5 );
        nv2 = n / 2;
        j = 1;
        for (i = 1; i < n; i++ )
        {
                if (i < j)
                {
                        t_r = array[i - 1][0];
                        t_i = array[i - 1][1];
                        array[i - 1][0] = array[j - 1][0];
                        array[i - 1][1] = array[j - 1][1];
                        array[j - 1][0] = t_r;
                        array[j - 1][1] = t_i;
                }
                k = nv2;
                while (k < j)
                {
                        j = j - k;
                        k = k / 2;
                }
                j = j + k;
        }

        for (l = 1; l <= ln; l++) /* loops thru stages */
        {
                le = (int)(Math.exp( (double)l * Math.log(2) ) + 0.5 );
                le1 = le / 2;
                u_r = 1.0;
                u_i = 0.0;
                w_r =  Math.cos( Math.PI / (double)le1 );
                w_i = -Math.sin( Math.PI / (double)le1 );
                for (j = 1; j <= le1; j++) /* loops thru 1/2 twiddle values per stage */
                {
```

```
                for (i = j; i <= n; i += le) /* loops thru points per 1/2 twiddle */
                {
                    ip = i + le1;
                    t_r = array[ip - 1][0] * u_r - u_i * array[ip - 1][1];
                    t_i = array[ip - 1][1] * u_r + u_i * array[ip - 1][0];

                    array[ip - 1][0] = array[i - 1][0] - t_r;
                    array[ip - 1][1] = array[i - 1][1] - t_i;

                    array[i - 1][0] =  array[i - 1][0] + t_r;
                    array[i - 1][1] =  array[i - 1][1] + t_i;
                }
                t_r = u_r * w_r - w_i * u_i;
                u_i = w_r * u_i + w_i * u_r;
                u_r = t_r;
            }
        }
        return array;
} /* end of FFT_1d */
```

Figure 4.33 An FFT implementation in Java

Note that the DFT and the FFT involve a lot of multiplying and accumulating operations. This is typical of DSP operations and it is called a *multiply/accumulate* operation. We have seen this before. In fact, many DSP algorithms exhibit this characteristic. This is why so many DSP processors can do multiplications and additions in parallel. This is a huge efficiency boost to this specific signal processing algorithm class.

FFT Implementation Issues

Cache and memory issues

Cache-based DSP systems are becoming more common. The presence of cache in a processor can lead to nonintuitive effects regarding the performance of software, including signal processing algorithms such as FFT code. FFT algorithms can be implemented on programmable devices or in hardware gates. The inherent parallelism in FFT computations actually makes them good candidates for implementation in "seas of gates" like a FPGA or an ASIC, where multiple parallel channels of execution are easy to implement.

These same issues force the DSP designer to consider how to implement FFTs on programmable processors, including DSPs. One example of a possible inefficiency is the execution of the nested loop FFT. With this approach, each pass through the loop accesses and modifies the entire array of data being transformed. Depending on the size of this data set, it may or may not fit into the available processor cache. If the entire array of data does not fit inside the cache, then the entire array will be cycled through the cache during each pass, causing the cache to "thrash". When the cache thrashes like this, processor cycles are consumed in flushing the cache and loading it with new data (cache optimization is discussed in detail in Appendix C).

One way around this constraint is to perform the DIF and DIT algorithms "in place." This is a technique that treats the sub-transforms as a distinct computational segments which are completed before performing the next sub-transform. If these sub-transforms are small enough, they will fit in the cache and possible cache thrashing is minimized, resulting in a significant performance improvement. Of course, the actual performance improvement depends on factors such as the cache size, the cache line size and line replacement algorithm being used. It is difficult to determine the improvement precisely when using a cache based system, so the recommendation is to profile each experiment to determine which one performs best.

Many DSPs have small amounts of fast on-chip memory. This memory can be accessed in a single cycle. This can be referred to as a programmer managed cache. The advantage to this is determinism. The programmer/designer knows what is in cache at any point in time, and the application can be profiled much easier. This approach provides the advantages of performance and determinism. The general rule of thumb is that a DSP is good at loops but bad at stacks. The goal becomes to design FFT routines that perform as many sub-transforms as possible in internal memory.

One approach is to perform the first few passes of a large transform in external memory. Once the size of the sub-transform block is small enough, it is moved (via DMA) into on-chip memory. Each block is cycled into on-chip memory, transformed, and the result written back to external memory. The DIT transform can initially perform the first sub-transforms in on-chip memory and then complete the remaining passes in external memory. In both cases, the transfer to and from internal memory is performed using DMA, which allows for other useful computation to be performed in parallel.

Figure 4.34 describes how a 2-D FFT can be cycled into on-chip memory, one row at a time, processed in on-chip memory, and then cycled back out to external memory.

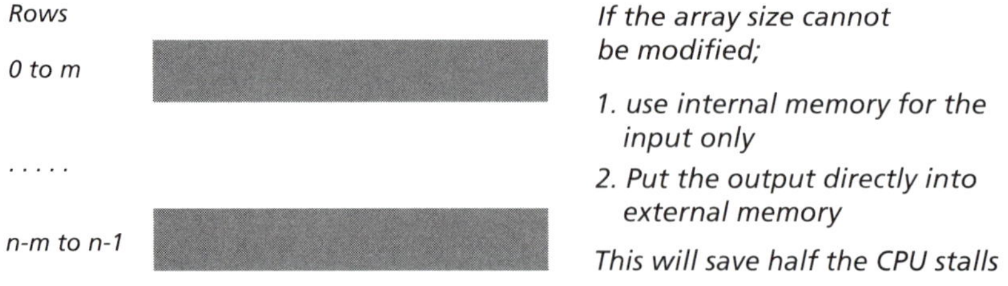

Figure 4.34 A 2-D FFT can be cycled into DSP on-chip memory one row at a time to improve overall performance

Summary

The discrete Fourier transform (DFT) was described as a useful tool employed to produce frequency analysis of discrete nonperiodic signals. The FFT is a more efficient method of performing the same operations, especially if the size of the transform is large. When signals become too complex to analyze in the time domain, the FFT can be used to transform the signal information into a frequency-based representation that makes the analysis easier and simpler. The FFT is a useful tool to move back and forth quickly between the time domain and the frequency domain in order to take advantage of the benefits each of these domains offer. In many cases, the mathematics in the frequency domain is simpler. The efficiency of a FFT allows a signal to be transformed to the frequency domain, processed, and moved back to the time domain without incurring a significant amount of overhead.

I would like to acknowledge Dennis Kertis for his contributions to this chapter.

5

DSP Architectures

General-purpose processors are designed to have broad functionality. These processors are designed to be used in a wide variety of applications. The performance goal of general-purpose processors is maximized performance over a broad range of applications. Specialized processors, on the other hand, are designed to take advantage of the limited functionality required by their applications to meet specific objectives. DSPs are a type of specialized processor. These processors are designed to focus on signal processing applications. Hence, there is little or no support to date for features such as virtual memory management, memory protection, and certain types of exceptions.

DSPs, as a type of specialized processor, have customized architectures to achieve high performance in signal processing applications. Traditionally, DSPs are designed to maximize performance for inner loops containing a product of sums. Thus, DSPs usually include hardware support for fast arithmetic, including single cycle multiply instructions (often both fixed-point and floating-point), large (extra wide) accumulators, and hardware support for highly pipelined, parallel computation and data movement. Support for parallel computation may include multiple pipelined execution units. To keep a steady flow of operands available, DSP systems usually rely on specialized, high bandwidth memory subsystems. Finally, DSPs feature specialized instructions and hardware for low overhead loop control and other specialized instructions and addressing modes that reduce the number of instructions necessary to describe the typical DSP algorithm.

Fast, Specialized Arithmetic

DSPs were designed to process certain mathematical operations very quickly. A seven tap FIR filter processing a 400 KHz signal, for example, must complete over 5.6 million multiplies and 5.6 million adds per second. DSP vendors recognized this bottleneck and have added specialized hardware to compute a multiply in only one cycle. This dramatically improves the throughput for these types of algorithms.

Another distinguishing feature of a DSP is its large accumulator register or registers. The accumulator is used to hold the summation of several multiplication operations without overflowing. The accumulator is several bits wider than a normal processor register which can help avoid overflow during the accumulation of many intermediate results. The extra bits are called *guard bits*. In DSPs the multiplier, adder and accumulator are almost always cascaded so that the "multiply and accumulate" operation so common to DSP algorithms can be specified in a single specialized instruction. This instruction (typically called a *MAC*) is one example of how DSPs are optimized to reduce the number of instruction fetches necessary for DSP algorithms. I'll describe others later in this chapter.

The MAC unit

At the core of the DSP CPU is the "MAC" unit (for multiply and accumulate), which gets its name from the fact that the unit consists of a multiplier that feeds an accumulator. To feed the multiplier most efficiently, a "data" bus is used for the "x" array, and a "coefficient" bus is used for the "a" array. This simultaneous use of two data buses is a fundamental characteristic of a DSP system. If the coefficients were constants, the user might want to store them in ROM. This is a common place to hold the program, and it allows the user to perform the MAC operations using coefficients stored in program memory and accessed via the program bus. Results of the MAC operation are built up in either of two accumulator registers. Both of these registers can be used equally. This allows the program to maintain two chains of calculations instead of one. The MAC unit is invoked with the MAC instruction. An example of a MAC instruction is shown in Figure 5.1. This one instruction is performing many operations in a single cycle; it reads two operands from memory, increments two pointers, multiplies the two operands together, and accumulates this product with the running total.

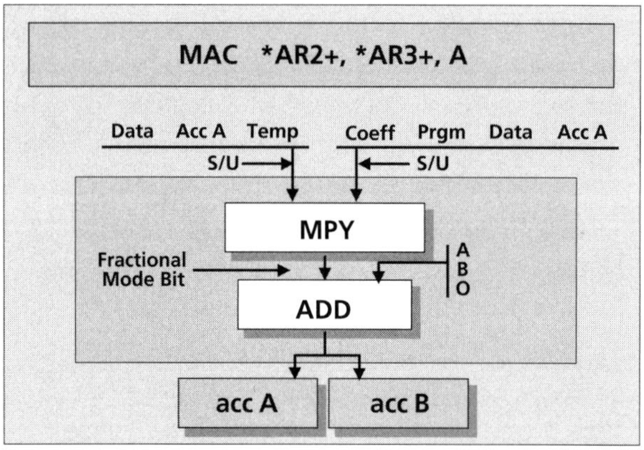

Figure 5.1 Multiply and accumulate functionality in a DSP (courtesy of Texas Instruments)

Parallel ALUs

Most DSP systems need to perform lots of general-purpose arithmetic (simple processes of addition and subtraction). For this type of process, a separate arithmetic logic unit (ALU) is often added to a DSP. The results of the ALU can be basic math or Boolean functions. These results are sent to either of the accumulators seen before. The ALU performs standard bit manipulation and binary operations. The ALU also has an adder. The adder assures efficient implementation of simple math operations without interfering with the specialized MAC unit.

Inputs to the ALU are, as with the MAC unit, very broad to allow the user maximum flexibility in obtaining efficient code implementation. The ALU can use either accumulator and the data and/or coefficient buses as single or dual inputs (Figure 5.2). The ALU generates the standard status bits including overflow flags for both A and B, zero flags for both A and B, and a carry bit for both A and B.

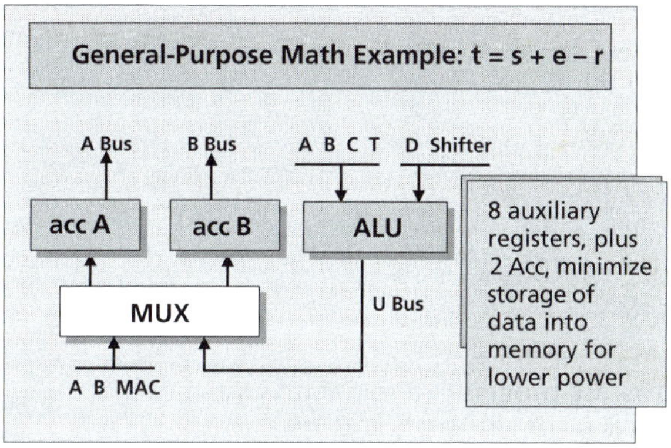

Figure 5.2 DSP with ALU (courtesy of Texas Instruments)

Numeric representation

While DSP units have traditionally favored fixed-point arithmetic, modern processors increasingly offer both fixed- and floating-point arithmetic. Floating-point numbers have many advantages for DSPs;

First, floating-point arithmetic simplifies programming by making it easier to use high level languages instead of assembly. With fixed-point devices, the programmer must keep track of where the implied binary point is. The programmer must also worry about performing the proper scaling throughout the computations to ensure the required accuracy. This becomes very error-prone and hard to debug as well as to integrate.

Floating-point numbers also offer greater dynamic range and precision than fixed-point. Dynamic range is the range of numbers that can be represented before an overflow or an underflow occurs. This range effectively indicates when a signal needs to be scaled. Scaling operations are expensive in terms of processor clocks and so scaling affects the performance of the application. Scaling data also causes errors due to truncation of data and rounding errors (also known as quantization errors). The dynamic range of a processor is determined by size of the exponent. So for an 8-bit exponent the range of magnitudes that can be represented would be:

1 * 2**(–127) .... 2 * 2**(127)

Floating-point numbers also offer greater precision. Precision measures the number of bits used to represent numbers. Precision can be used to estimate the impact of errors due to integer truncation and rounding. The precision of a floating-point number is determined by the mantissa. For a 32 bit floating-point DSP, the mantissa is generally 24 bits. So the precision offered by a 32 bit DSP with a mantissa of 24 bits is at least that of a 24 bit fixed-point device. The big difference is that the floating-point hardware automatically normalizes and scales the resultant data, maintaining 24 bit precision for all numbers large and small. In a fixed-point DSP, the programmer is responsible for performing this normalization and scaling operation.

Keep in mind that floating-point devices have some disadvantages as well:

- *Algorithmic issues* – Some algorithms, such as data compression, do not need floating-point precision and are better implemented on a fixed-point device.
- *More power* – Floating-point devices need more hardware to perform the floating-point operations and automatic normalization and scaling. This requires more die space for the DSP, which takes more power to operate.
- *Slower speed* – Because of the larger device size and more complex operations, the device runs slower than a comparable fixed-point device.
- *More expensive* – Because of the added complexity, a floating-point DSP is more expensive than fixed-point. A trade-off should be made regarding device cost and software programmer cost when programming these devices.

High Bandwidth Memory Architectures

There are several ways to increase memory bandwidth and provide faster access to data and instructions. The DSP architecture can be designed with independent memories allowing multiple operands to be fetched during the same cycle (a modified Harvard architecture). The number of I/O ports can be increased. This provides more overall bandwidth for data coming into the device and data leaving the device. A DMA controller can be used to transfer the data in and out of the processor, increasing the overall data bandwidth of the system. Access time to the memory can be decreased, allowing data to be accessed more quickly from memory. This usually requires adding more registers or on-chip memory (this will be discussed more later). Finally, independent memory banks can be used to reduce the memory access time. In many DSP devices, access time to memory is a fraction of the instruction cycle time (usually ½, meaning that two memory accesses can be performed in a single cycle). Access time, of course, is also a function of memory size and RAM technology.

Data and Instruction Memories

DSP-related devices require a more sophisticated memory architecture. The main reason is the memory bandwidth requirement for many DSP applications. It becomes imperative in these applications to keep the processor core fed with data. A single memory interface is not good enough. For the case of a simple FIR filter, each filter tap requires up to four memory accesses:

- Fetch the instruction.
- Read the sample data value.
- Read the corresponding coefficient value.
- Write the sample value to the next memory location (shift the data).

If this is implemented using a DSP MAC instruction, four accesses to memory are required for each instruction. Using a von Neumann architecture, these four accesses must all cross the same processor/memory interface, which can slow down the processor core to the point of causing it to wait for data. In this case, the processor is "starved" for data and performance suffers.

In order to overcome this performance bottleneck, DSP devices often use several buses to interface the CPU to memory: a data bus, a coefficient bus, and a program bus (Figure 5.3). All of these move data from memory to the DSP. An additional bus may be available to store results back out to memory. During each cycle, any number of buses may access the arrays of internal memory, and any one bus may talk to an external device. Internal control logic in a DSP prevents the possibility of bus collision if several buses try to access off-chip resources at the same time.

In order for the MAC instruction to execute efficiently, multiple buses are necessary to keep data moving to the CPU. The use of multiple buses makes the device faster[1]

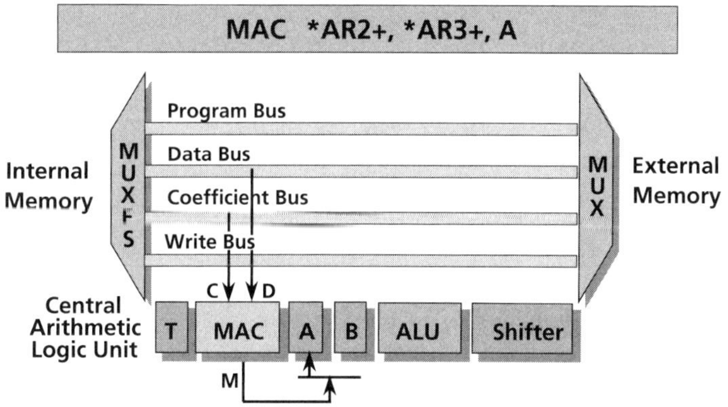

Figure 5.3 DSPs use multiple buses to read and write data (courtesy of Texas Instruments)

Memory Options

There are two different types of memories available on DSP architectures: ROM and RAM. Different DSP families and versions have different amounts of memory. Current DSPs contain up to 48K of ROM and up to 200K of RAM, depending on the specific device.

[1] If a bus is bidirectional there will be bus "turn-around" time built-in to the latency equation. A one-directional bus has fewer transistors, which will make data accesses and writes faster. The bottom line is that the simpler the bus, the faster it will operate.

RAM comes in two forms: single access (SARAM) and dual access (DARAM). For optimum performance, data should be stored in dual access memory and programs should be stored in SARAM (Figure 5.4). The main value of DARAM is that storing a result back to memory will not block a subsequent read from the same block of memory.

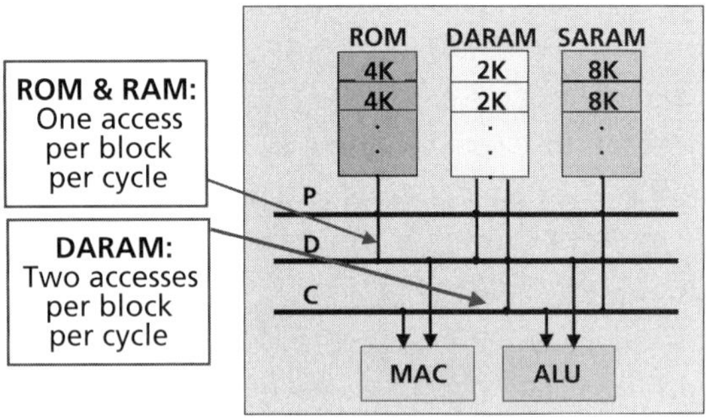

Figure 5.4 Memory options on a DSP (courtesy of Texas Instruments)

High Speed Registers

Many DSP processors include a bank of internal memory with fast, single cycle access times. This memory is used for storing temporary values for algorithms such as the FIR filter. The single cycle access time allows for the multiply and accumulate to complete in one cycle. In fact, a DSP system contains several levels of memory as shown in Figure 5.5. The register file residing on-chip is used for storing important program information and some of the most important variables for the application. For example, the delay operation required for FIR filtering implies information must be kept around for later use. It is inefficient to store these values in external memory because a penalty is incurred for each read or write from external memory. Therefore, these delay values are stored in the DSP registers or in the internal banks of memory for later recall by the application without incurring the wait time to access the data from external memory. The on-chip bank of memory can also be used for program and/or data and also has fast access time. External memory is the slowest to access, requiring several wait states from the DSP before the data arrives for processing. External memory is used to hold the results of processing.

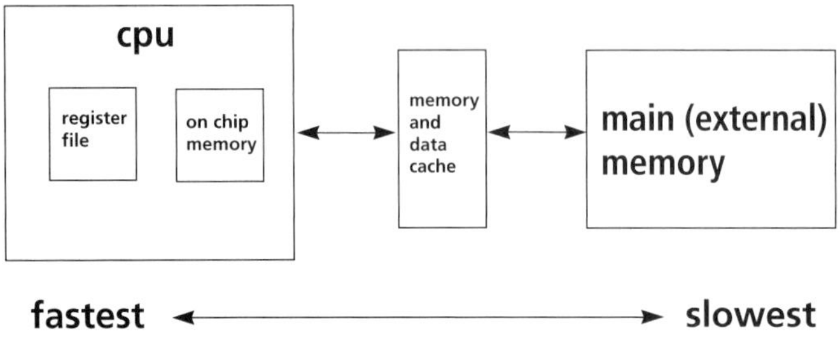

Figure 5.5 DSP memory system hierarchy

Memory Interleaving

DSP devices incorporate memory composed of separate banks with latches for separate addresses to these banks (Figure 5.6). The banks are interleaved sequentially. This allows multiple words to be fetched simultaneously as long as they are from a block whose length is equal to the number of banks. Bank organization can also be used to match bus bandwidths with memory access times. If separate addresses are provided, multiple words can be fetched as long as they are in different banks, a significant performance improvement in DSP applications.

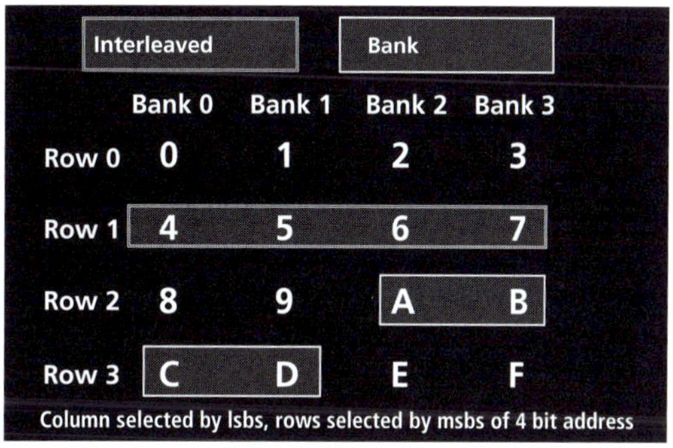

Figure 5.6 Interleaved memory banks improve overall performance

Interleaved bank organization is a special case of bank organization where all banks are addressed simultaneously and row aligned. This allows multiple accesses as long as they are from sequential addresses. This approach reduces the number of address latches required on the device.

Bank Switching

Modern DSPs also provide the capability for memory bank switching. Bank switching allows switching between external memory banks for devices requiring extra cycles to turn off after a read is performed. Using this capability, one external bus cycle can be automatically inserted when crossing memory bank boundaries inside program or data space. The DSP can also add one wait state between program and data accesses or add one wait state when crossing between user defined physical memory boundaries.

Memory Size

Not having enough memory on-chip requires that the designer supply off-chip memory, which is less power, size and cost efficient. However, a processor with far too much on-chip memory is not power, size or cost efficient either. Different applications have a wide range of memory requirements. Designers must analyze the requirements of the system and choose the DSP with the right mix of on-chip memory, performance, and cost to meet the application requirements. For example, the TI 5402 DSP is optimized for high performance at very low power and low cost. This particular device is designed to meet the needs of applications that do not require a large amount of on-chip memory. Some applications, such as the VOP gateway, however, require a significant amount of memory. In this case, another DSP with a larger on-chip memory would be required (Table 5.1).

System Example	Memory Required	On-Chip Memory
Portable Audio	21 Kwords	32 Kwords ('C5409)
Pay Phone	15 Kwords	16 Kwords ('C5402)
VoP Gateway (8 channels)	182 Kwords	200 Kwords ('C5420)

Table 5.1 On-chip memory requirements for various applications

Caches for DSPs

Cache systems often improve the average performance of a system, but not the worst case performance[2]. This results in less value for DSP systems which depend on predictable worst case performance. Some DSPs, however, contain cache systems. These are typically smaller and simpler than general-purpose microprocessor caches. They are often used to free memory bandwidth rather than reduce overall memory latency. These caches are typically instruction (program) caches.

DSPs that do not support data caches (or self-modifying code) have simple hardware because writes to cache (and consequently coherency) are not needed. DSP caches are often software controlled rather than hardware controlled. For example, internal on-chip memory in a DSP is often used as a software controlled data cache to reduce main memory latency. On-chip memory, however, is not classified as a cache because it has addresses in the memory map and no tag support. But on-chip memory may be used as a software cache.

Repeat buffers

DSPs often implement what is called a *repeat buffer*. A repeat buffer is effectively a one word instruction cache. This cache is used explicitly with a special "repeat" instruction in DSPs. The "repeat" instruction will repeat the execution of the next instruction "n"

[2] Performance improves significantly with cache up to 256KB. After that the performance benefit starts to flatten out. Once the cache reaches a certain size (around 512 KBytes) there is almost no additional benefit. This is primarily due to the fact that the larger the cache, the more difficult it is to keep track of the data in the cache and the longer it takes to determine the location and availability of data. Checking for and finding data in a small cache takes very little time. As the cache size grows, the time required to check for data increases dramatically while the likelihood of the data being found increases marginally.

times, where n can be an immediate value or a registered integer. This is a popular and powerful instruction to use on one-instruction inner loops[3].

As advanced form of the repeat buffer is the "*Repeat buffer with multiple entries.*" This is the same as the repeat buffer but with multiple instructions stored in a buffer. To use this instruction, the programmer must specify the block of instructions to be stored in the buffer. The instructions are copied into buffer on the first execution of the block. Once the instructions are in the block, all future passes through the loop execute out of the buffer[4].

Single-entry instruction cache

A similar form of instruction cache is called a *single entry instruction cache.* In this configuration, only a single block of contiguous code can be in the cache at once. This type of cache system allows multiple entry points into the sequential code rather than one single entry at the beginning. This type of cache permits indirection such as jumps, gotos, and returns, to access the buffer[5].

Associative instruction cache

An associative instruction cache allows multiple entries or blocks of code to reside in the cache. This is a fully associative system which avoids block alignment issues. Hardware controls which blocks are in cache. Systems of this kind often have special instructions for enabling, locking, and loading. A "manual" control capability also exists. This mode can be used to improve performance or ensure predictable response[6].

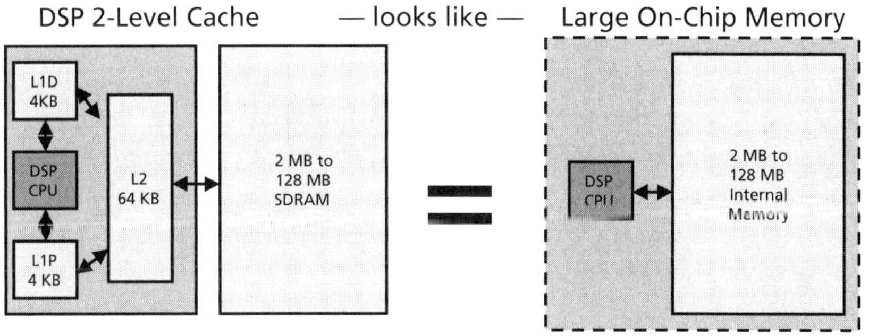

Figure 5.7 A two-level cache looks like a large on-chip memory (courtesy of Texas Instruments)

Dual cache

Finally, a dual cache memory system is one that allows on-chip memory to be configured as RAM (in the memory map) or cache. The programmer is allowed to configure the mode. The advantage of this approach is that it allows an initial program to be brought up quickly using the cache and then optimized for performance/predictability if needed.

[3] The TMS320C20 is an example of a DSP that uses this instruction.

[4] The Lucent DSP16xx is an example of a DSP that implements this type of instruction.

[5] The ZR3800x is an example of a DSP that implements this type of cache system.

[6] Example of this system include the TMS320C30 and the Motorola DSP96002.

Execution Time Predictability

Specialized processors with hard time constraints, such as DSPs, require designs to meet worst case scenarios. Predictability is very important so time responses can be calculated and predicted accurately[7]. Many modern general-purpose processors have complex architectures that make execution time predictability difficult (for example, superscalar architectures dynamically select instructions for parallel execution). There is little advantage in improving the average performance of a DSP processor if the improvement means that you can no longer predict the worst case performance. Thus, DSP processors have relatively straightforward architectures and are supported by development and analysis tools that help the programmer determine execution time accurately.

Most DSPs' memory systems limit the use of cache because conventional cache can only offer a probabilistic execution model. If the instructions the processor executes reside in the cache, they will execute at cache speeds. If the instruction and data are not in the cache, the processor must wait while the code and data are loaded into the caches.

The unpredictability of cache hits isn't the only problem associated with advanced architectures. Branch prediction can also cause potential execution time predictability problems. Modern CPUs have deep pipelines that are used to exploit the parallelism in the instruction stream. This has historically been one of the most effective ways for improving performance of general-purpose processors as well as DSPs.

Branch instructions can cause problems with pipelined machines. A branch instruction is the implementation of an if-then-else construct. If a condition is true then jump to some other location; if false then continue with the next instruction. This conditional forces a break in the flow of instructions through the pipeline and introduces irregularities instead of the natural, steady progression. The processor does not know which instruction comes next until it has finished executing the branch instruction. This behavior can force the processor to stall while the target jump location is resolved. The deeper the execution pipeline of the processor, the longer the processor will have to wait until it knows which instruction to feed into the pipeline next. This is one of the largest limiting factors in microprocessor instruction execution throughput and execution time predictability. These effects are referred to as *branch effects*. These branch effects are probably the single biggest performance inhibitor of modern processors.

Dynamic instruction scheduling can also lead to execution time predictability problems. Dynamic instruction scheduling means that the processor dynamically selects sequential instructions for parallel execution, depending on the available execution units and on dependencies between instructions. Instructions may be issued out of order when dependencies allow for it. Superscalar processor architectures use dynamic instruction scheduling. Many DSPs on the market today (although not all) are not supserscalar for this reason—the need to have high execution predictability.

---

[7] Execution predictability is a programmer's ability to predict the exact timing that will be associated with the execution of a specific segment of code. Execution predictability is important in DSP applications because of the need to predict performance accurately, to ensure real-time behavior, and to optimize code for maximum execution speed.

Direct Memory Access (DMA)

Direct memory access is a capability provided by DSP computer bus architectures that allows data to be sent directly from an attached device (such as a disk drive or external memory) to other memory locations in the DSP address space. The DSP is freed from involvement with the data transfer, thus speeding up overall computer operation.

DMA devices are partially dependent coprocessors which offload data transfers from the main CPU. DMA offers a high performance capability to DSPs. Modern DSP DMA controllers such as the TMS320C6x can transfer data at up to 800 Mbytes/sec (sustained). This DMA controller can read and write one 32-bit word every cycle. Offloading the work of transferring data allows the CPU to focus on computation. Typical types of transfers that a DMA may control are:

- Memory to memory transfers (internal to external).
- Transfers from I/O devices to memory.
- Transfers from memory to I/O devices.
- Transfers from/to communications ports and memory.
- Transfers from/to serial ports and memory.

DMA setup

Setup of a DMA operation requires the CPU to write to several DMA registers. Setup information includes information such as the starting address of the data to be transferred, the number of words to be transferred, transfer type and the addressing mode to be used, the direction of the transfers, and the destination address or peripheral for the data. The CPU signals to the DMA controller when the transaction request is complete.

The actual transfer of data involves the following steps:

1. The DMA requests the bus when it is ready to start the transfer.
2. The CPU completes any memory transactions currently in operation and grants the bus to the DMA controller.
3. The DMA transfers words until the transfer is complete or the CPU requests bus.
4. If the DMA grants the bus back before the transfer is complete, it must request it again to finish the transaction.
5. When the entire block is transferred, the DMA may optionally notify the CPU via an interrupt.

Managing conflicts and multiple requests

DMA controllers have built in features to reduce bus or resource conflict. Many DMA controllers synchronize transfers to the source or destination via interrupts. DMA controllers allow the programmer to move data off chip to on chip, or back and forth in off-chip and on-chip memory. Modern DSPs provide multiple memories and ports for

DMA use. DSPs implement independent data and address buses to avoid bus conflicts with the CPU. The Texas Instruments TMS320Cx, the Motorola DSP96002, and the Analog Devices ADSP-2106x families all implement some form of DMA capability.

Multiple channel DMA capability allows the DSP to manage multiple transaction requests at once with arbitration decisions for sub block transfers based on priority as well as destination and source availability (Figure 5.9). DSP vendors refer to this capability as multiple channels but in reality this is a transaction abstraction and not a physical hardware path. Each DMA channel has its own control registers to hold the transaction's information. The Texas Instruments TMS320C5x DSP, for example, has a six DMA channel capability (Figure 5.8).

The DMA available on some DSPs allows parallel and/or serial data streams to be up and/or down loaded to or from internal and/or external memory. This allows the CPU to focus on processing tasks instead of data movement overhead.

DMA Example

In the portable audio example of Figure 5.8, the DMA synchronizes the retrieval of compressed data from the mass storage as well as the flow of uncompressed data through the DAC into the stereo headphones. This frees the CPU to spend 100% of its cycles on what it does best—performing DSP functions efficiently.

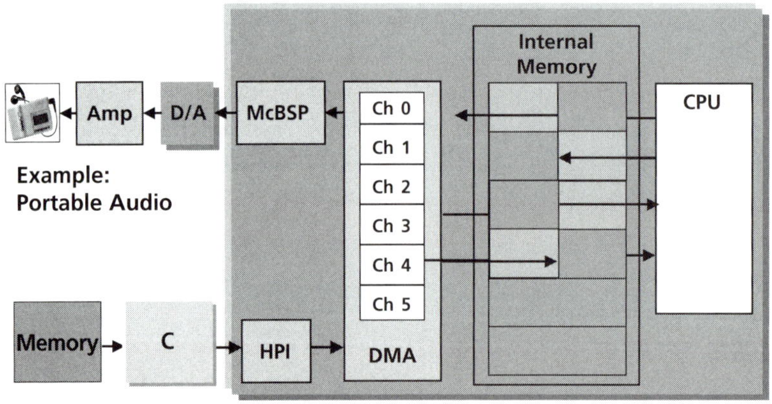

Figure 5.8 Six DMA channel capability in the Texas Instruments TMS320C5x DSP (courtesy of Texas Instruments)

DSP-specific features

DMA controllers are not just good at efficient data transfer. They often provide useful data manipulation functions as well, such as sorts and fills. Some of the more complex modes include:

- *Circular addressing* – This mode is very useful when the application is managing a circular buffer of some type.

- *Bit reversal* – This mode is efficient when moving the results of a FFT, which outputs data in bit-reversed order.
- *Dimensional transfers such as 1D to 2D* – This mode is very useful for processing multidimensional FFT and filters.
- *Striding though arrays* – This mode is useful when the application, for example, needs to access only the DC component of several arrays of data.

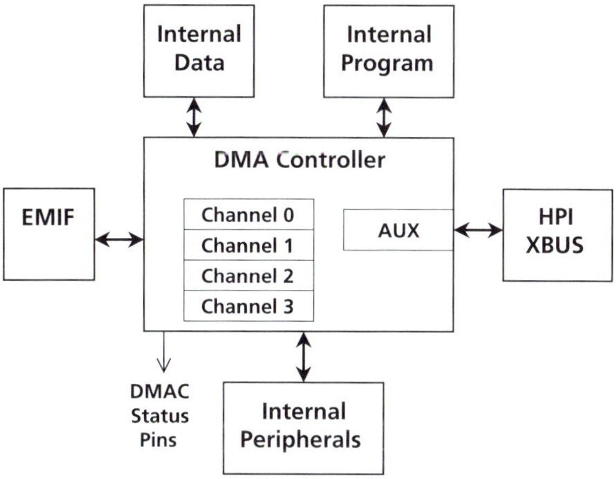

Figure 5.9 DMA controller for a DSP

Host Ports

Some DSP memory systems can also be accessed through a "host port." Host port interfaces (HPIs) provide for connection to a general-purpose processor. These host ports are specialized 8 or 16 bit bidirectional parallel ports. This type of port allows the host to perform data transfers to DSP memory space. Some host interfaces allow the processor to force instruction execution or interrupt setting on architectural state probing. Host ports can be used to configure a DSP as a partially dependent coprocessor. Host ports are also often used in development environments where users work on the host, which controls the DSP.

In general, the HPI is an asynchronous 8 or 16 bit link (depending on the device), intended for facilitating communication across a glueless interface to a wide array of external 'host' processors. Numerous control pins are available to allow this variety of connection modes and eliminate the need to clock the host with the DSP. The host port allows a DSP to look like a FIFO-like device to the host, greatly simplifying and cost reducing the effort of connecting two dissimilar processors. The HPI is very handy for booting a ROM-less DSP from a system host, or for exchanging input data and results between a DSP and host at run-time (Figure 5.10).

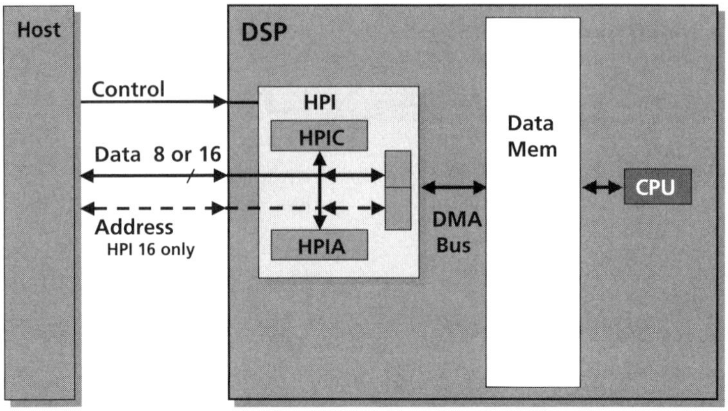

Figure 5.10 The host port interface can be used to transfer large amounts of data to and from a host (courtesy of Texas Instruments)

External Memory

In the solid-state portable audio application of Figure 5.11, the DMA reads the flash card memory using the EMIF (one of the memory options). The extended address reach allows the DMA to access the entire memory space of a flash card[8]. This is shown in abstract form in Figure 5.11. The 16 bit connection in Figure 5.10 is a data connection which transfers the 16 bit words to and from memory. The 23 bit connection is the address bus (8 MWords range). Different DSP devices have different extended program address reach (these devices may also have 16 bit I/O and data spaces, 64KWords apiece).

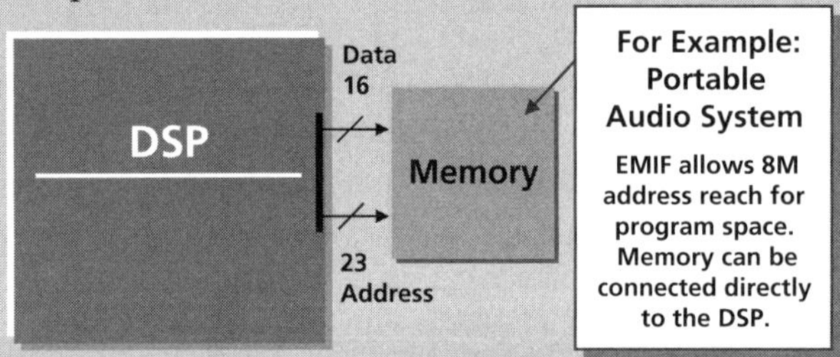

Figure 5.11 External memory interface provides flexibility for off-chip memory accesses (courtesy of Texas Instruments)

For off-chip memory accesses, the EMIF bus is active. This consumes power. By using internal memory, power consumption is reduced. Since the external buses are not used, the clock to these external buses can be stopped, which saves power. Since the internal memory operates at the core voltage of the DSP instead of the voltage of the external SRAM and EMIF buses, it consumes less power (Figure 5.12).

---

[8] CompactFlash cards are available in capacities from 4 MB to over 128 MB, SmartMedia cards are available in capacities of 2MB to over 8 MB. In this example a card no larger than 8 MB would be used.

Internal memory accesses use about half the power of external memory accesses in most cases. Another advantage of using internal memory instead of external memory is that it requires less board space.

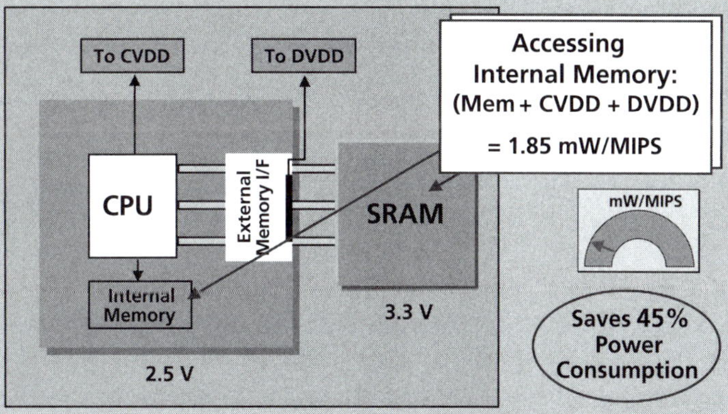

Figure 5.12 Internal memory on a DSP reduces power and increased performance (courtesy of Texas Instruments)

Pipelined Processing

In DSP processors, like most other processors, the slowest process defines the speed of the system. Processor architectures have undergone a transformation in recent years. It used to be that CPU operations (multiply, accumulate, shift with incrementing pointers, and so on) dominated the overall latency of the device and drove the execution time of the application. Memory accesses were trivial. Now it's the opposite. Geometries are so small that CPU functions can go very fast, but memory hasn't kept up. One solution to this problem is to break the instruction execution into smaller chunks. This way, while the processor is waiting for a memory access it can be performing at least part of the work it is trying to do.

When the various stages of different instances of an operation are overlapped in time, we say the operation has been pipelined. Instruction pipelines are a fundamental element of how a DSP achieves its high order of performance and significant operating speeds.

When *any* processor implements instructions, it must implement these standard sub-actions, or "phases" (Figure 5.13):

- *Prefetch* – Calculate the address of the instruction. Determine where to go for the next line of code, usually via the "PC" or program counter register found in most processors.
- *Fetch* – Collect the instruction. Retrieve the instruction found at the desired address in memory.
- *Decode* – Interpret the instruction. Determine what job it represents, and *plan* how to carry it out.

- *Access* – Collect the address of the operand. If an operand is required, determine *where* in data memory the value is located.
- *Read* – Collect the operand. Fetch desired operand from data memory.
- Execute; the instruction – or "do the job" desired in the first place.

In a DSP, each of these sub-phases can be performed quickly, allowing us to achieve the high MIP rates.

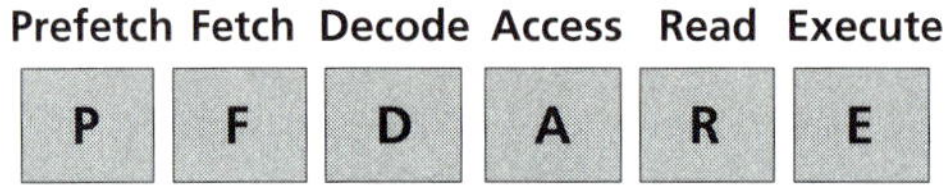

Figure 5.13 The sub-phases of a DSP instruction execution

All instructions must pass through these phases to be implemented. Each sub-phase takes one cycle. This sounds like a lot of overhead when speed is important. However, since each pipeline phase uses separate hardware resources from each other phase, it is possible for several instructions to be in process every cycle (Figure 5.14).

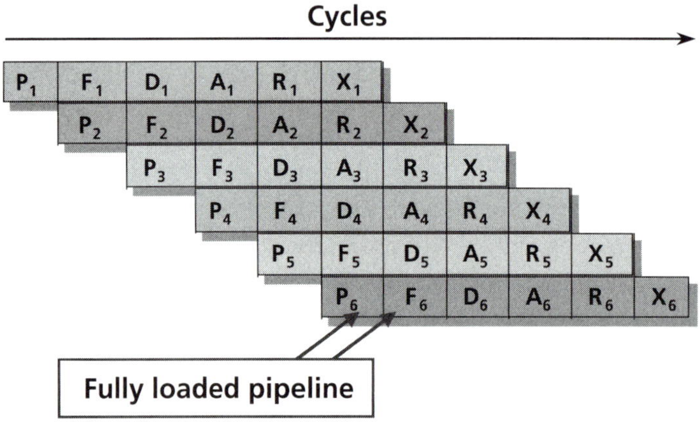

Figure 5.14 DSP instruction pipeline (courtesy of Texas Instruments)

As shown in Figure 5.14, instruction 1 (P1) first implements its prefetch phase. The instruction then moves to the fetch phase. The prefetch hardware is therefore free to begin working on instruction 2 (P2) while instruction 1 is using the fetch hardware. In the next cycle, instruction 1 moves on to the decode phase (D1), allowing instruction 2 to advance to the fetch phase (F2), and opening the prefetch hardware to start on instruction 3. This process continues until instruction 1 executes. Instruction 2 will now execute *one* cycle later, not six. This is what allows the high MIPs rate offered in pipelined DSPs. This process is performed automatically by the DSP, and requires no awareness on the part of the programmer.

Figure 5.15 shows the speed improvements to be gained by executing multiple instruction in parallel using the pipeline concept. By reducing the number of cycles required, pipelining also reduces the overall power consumed.

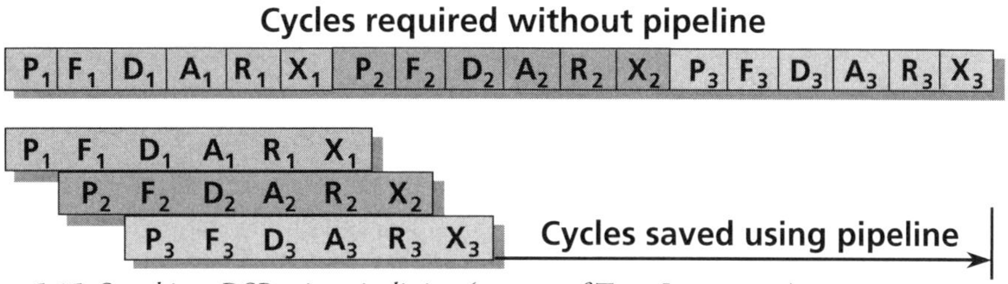

Figure 5.15 Speed in a DSP using pipelining (courtesy of Texas Instruments)

Adding pipeline stages increases the computational time available per stage, allowing the DSP to run at a slower clock rate. A slower clock rate minimizes data switching. A CMOS circuit consumes power when you switch. So requirements reducing the clock rate leads to lower power requirements[9]. The trade-off is that large pipelines consume more silicon area. In summary, pipelining, if managed effectively, can improve the overall power efficiency of the DSP.

Atomic Instructions

Atomic instructions assume a processor that operates in synch with a clock whose cycle time is Tc. The instructions are treated atomically; that is, one completes before the next one starts. An instruction consists of an integral number (N) of states (possibly 1) with a latency (L = N * Tc). Then program execution time is the sum of the latency time of each instruction in the dynamic stream executed by the processor (Figure 5.16).

One of the advantages of atomic instructions is that they reduce hardware cost to a minimum. Hardware resources can have maximum reuse within an instruction without conflicting with other instructions. For example, the ALU can be used for both address generation and data computation. State and value holding hardware can be optimized for single instruction processing without regard to partitioning for simultaneous instructions. Another advantage of atomic instructions is that they offer the simplest hardware and programming model.

State	Instruction Cycle							
Fetch	I1				I2			
Decode		I1				I2		
Read			I1				I2	
Execute				I1				I2

← Latency →

← Sample Period →

Figure 5.16 Atomic execution of instructions

[9] CMOS circuits consume power when they switch. By slowing the clock down to minimize switching, the power consumption in CMOS circuits goes down dramatically.

Atomic execution minimizes the hardware but is not necessarily hardware efficient. Atomic execution allows reuse between stages. The ALU, for example, can perform address computation and data computation at the same time. Cycle time and latency can also be optimized to minimize the hardware with regard to pipeline partitioning. Overall, pipeline hardware is used more efficiently than nonpipeline hardware. Hardware that is associated with each stage of computation is used every cycle instead of 1/N cycles for a N stage deep pipeline.

In an ideal pipeline, all instructions go through a linear sequence of states with the same order and the same number of states (N) and latency ($L = N * Tc$). An instruction starts after the previous instruction leaves its initial state. Therefore, the rate of instruction starts is 1/Tc (Figure 5.17). Program execution time is the number of dynamic instructions multiplied by the cycle time plus the latency of the last instruction.

State	Instruction Cycle							
Fetch	I1	I2	I3	I4	I5			
Decode		I1	I2	I3	I4	I5		
Read			I1	I2	I3	I4	I5	
Execute				I1	I2	I3	I4	I5

← Latency →

↔ Sample Period

Figure 5.17 Ideal pipeline

Pipeline hardware is used more efficiently in the ideal case. Hardware associated with each stage is used every cycle instead of 1/N cycles. (Some blocks within a stage may not be used).

Limitations

The depth of the pipeline is often increased to achieve higher clock rates. The less logic per stage, the smaller the clock period can be and the higher the throughput. Increasing MHz helps all instructions. There are limits to the throughput that can be gained by increasing pipeline depth, however. Given that each instruction is an atomic unit of logic, there is a loss of performance due to pipeline flushes during changes in program control flow. Another limitation to pipelines is the increased hardware and programmer complexity because of resource conflicts and data hazards. These limitations cause latency which has a significant impact on overall performance (on the order of MHz impact). Sources of latency increase, including register delays for the extra pipeline stages, register setup for the extra stages, and clock skew to the additional registers.

Common pipeline depths in traditional DSPs. Traditional DSPs have pipeline depths of 2–6. Several examples are listed in Table 5.2:

Table 5.2 Pipeline descriptions for various DSPs

Processor	Pipeline Description
Analog Devices 2101	Instruction Fetch/Decode Execute (including single cycle MAC)
Texas InstrumentsTMS320C10	Instruction Fetch/Decode Execute 1 Multiply Execute 2 Accumulate
Motorola DSP 5600's	Fetch Decode Execute
TMS320C30 (Floating-point)	Instruction Fetch Decode Operand Read/Write Execute
Texas Instruments TMS320C54x	Prefetch = calculate address of instruction Fetch = read instruction Decode = interpret instruction Access = calculate address of operand Read = read operand Execute = perform operation
Texas Instruments TMS320C6x (7 for common instructions)	PG = generate fetch address PS = send address to memory PW = wait for data PR = read opcode DP = decode packet DC = decode instruction E1 = execute common instructions (adds, subs, logic) Up to 12 worst case, depending on instructions E2—multiply E3 load E4 load E5 load E6—branch

The factors that prevent pipelines from achieving 100% overlaps include:

- *Resource Conflicts* – Two or more instructions in the pipeline need the same hardware during the cycle.
- *Data Hazards* – A subsequent instruction needs the result of a previous instruction before it has completed.

- *Branch Control Flow* – Change in flow due to a branch results in not knowing how to resolve the next instruction fetch until the branch has completed.
- *Interrupt Control Flow* – Change in flow due to a interrupt results in the any interrupt needing to be inserted atomically.

Resource Conflicts

As an example of a resource conflict, assume a machine with the following scenario:

- A Fetch, Decode, Read, Execute pipeline,
- A two-port memory, both ports of which can read or write,
- An ADD *R0, *R1, R2 instruction that reads the operands in the read stage,
- A STORE R4 *R3 instruction that reads the operand during the read stage, and stores to the memory in the execute stage.

This machine can execute a sequence of adds or stores in an ideal pipeline. However, if a store precedes the add then during the cycle where add is in the read stage and the store in execute stage, the memory with two ports will be asked to read two values and store a third (Figure 5.18).

State	Instruction Cycle							
Fetch	I1	ST	ADD	I3	I4	I5	I6	I7
Decode		I1	ST	ADD	I3	I4	I5	I6
Operands			I1	ST	ADD	I3	I4	I5
Execute				I1	ST	ADD	I3	I4

ADD reads two instructions in this cycle
and STORE writes one instruction

Figure 5.18 Pipeline diagram of a resource conflict

State	Instruction Cycle							
Fetch	I1	ST	ADD	I3	I4	I4	I5	I6
Decode		I1	ST	ADD	I3	I3	I4	I5
Operands			I1	ST	ADD	ADD	I3	I4
Execute				I1	ST	-	ADD	I4

Extra cycle added

Figure 5.19 Pipeline diagram of a resource conflict resolved

One solution to the resource conflict just described is for the programmer to insert a NOP between the store and the add instructions. This introduces a delay in the pipeline that removes the conflict, as shown in Figure 5.19. When the programmer must do this to prevent the conflict, the pipe is said to be unprotected or visible.

In some DSPs, the hardware detects the conflict and stalls the store instruction and all subsequent instructions for one cycle. This is said to be a protected or transparent pipeline. Regardless of whether the programmer or the hardware has to introduce the delay, the cycles per instruction performance of the CPU degrades due to the extra delay slot.

Resource conflicts are caused by various situations. Following are some examples of how resource conflicts are created:

- *Two-cycle write* – Some memories may use two cycles to write but one to read. Instructions writing to the memory can therefore overlap with instructions reading from the memory.
- *Multiple word instructions* – Some instructions use more than one word and therefore take two instruction fetch cycles. An example is an instruction with long intermediate data equal to the width of the word.
- *Memory bank conflict* – Memory supports multiple accesses only if locations are not in the same bank.
- *Cache misses or off-chip memory* – If the location is not in the cache or on-chip memory, the longer latency of going off-chip causes a pipeline stall.
- *Same memory conflict* – Processor supports multiple accesses only if memory locations are in different memories.

Data hazards

Data hazards are similar to resource hazards. An example of a data hazard is shown below:

If

ADD R1, R2, R3 (destination is R1)

is followed by

SUB R4, R1, R5

a data hazard exists if R1 is not written by the time SUB reads it. Data hazards can also occur in registers which are being used for address calculations, as shown in Figure 5.20.

State	Instruction Cycle							
Fetch	I1	ADD	SUB	I3	I4	I5	I6	I7
Decode		I1	ADD	SUB	I3	I4	I5	I6
Operands			I1	ADD	SUB	I3	I4	I5
Execute				I1	ADD	SUB	I3	I4

ADD writes at the beginning of this cycle
but SUB is trying to read at the beginning of it

Figure 5.20 Pipeline diagram of a data hazard

State	Instruction Cycle							
Fetch	I1	ADD	SUB	I3	I3	I4	I5	I6
Decode		I1	ADD	SUB	I3	I3	I4	I5
Operands			I1	ADD	SUB	SUB	I3	I4
Execute				I1	ADD	--	SUB	I3

Figure 5.21 Resolution of a data hazard

Extra cycle added

Bypassing (also called *forwarding*) is commonly used to reduce the data hazard problem. In addition to writing the result to the register file, the hardware detects the immediate use of the value and forwards it directly to a unit bypassing the write/read operation (Figure 5.21).

There are several types of data hazards. These include:

- *Read after Write* – This is where *j* tries to read a source before *i* writes it.
- *Write after Read* – This is where *j* writes a destination before it is read by *i*, so *i* incorrectly gets the new value. This type of data hazard can occur when an increment address pointer is read as an operand by the next instruction.
- *Write after Write* – This type of data hazard occurs when *j* tries to write an operand before it is written by *i*. The writes end up in the wrong order. This can only occur in pipelines that write in more than one pipe stage or execute instructions out of order.

Branching control flow

Branch conditions are detected in the decode stage of the pipeline. In a branch instruction, the target address is not known until the execute stage of the instruction. The subsequent instruction(s) have already been fetched, which can potentially cause additional problems.

This condition does not just happen in branches but also subroutine calls and returns. An occurrence of a branch instruction in the pipeline is shown in Figure 5.22.

State	Instruction Cycle							
Fetch	BR	I2	-	-	NOP1	NOP2	NOP3	NOP4
Decode		BR	I2	-	-	NOP1	NOP2	NOP3
Operands			BR	I2	-	-	NOP1	NOP2
Execute				BR	I2	-	-	NOP1

Figure 5.22 Occurrence of a branch in a DSP pipeline

The solution to branch effects is to "flush" or throw away every subsequent instruction currently in the pipeline. The pipeline is effectively stalled while the processor is busy fetching new instructions until the target address is known. Then the processor starts fetching the branch target.

This condition results in a "bubble" where the processor is doing nothing, effectively making the branch a multicycle instruction equal to the depth of the branch resolution in the pipeline. Therefore, the "deeper" the pipeline (the more stages of an instruction there are), the longer it takes to flush the pipeline and the longer the processor is stalled.

Another solution is called a *delayed* branch. This is essentially like a flush solution but has a programming model where the instructions after the branch will always be executed. The number of instructions is equal to the number of cycles that have to be flushed. The programmer fills the slots with instructions that do useful work if possible. Otherwise the programmer inserts NOPS. An example of a delayed branch with three delay slots is the following:

```
BRNCH   Addr.   ; Branch to new address
INSTR 1         ; Always executed
INSTR 2         ; Always executed
INSTR 3         ; Always executed
INSTR 4         ; Executed when branch not taken.
```

Another solution to the branch stall condition is the use of a "repeat buffer." For the special case of loops, the repeat buffer is a good solution. In this situation, special hardware exists at the top of the pipe to repeat the instruction "fetching." Local buffering of loop instructions occurs so fetch is immediate.

In a "conditional branch with annul" solution, processor interrupts are disabled. The succeeding instructions are then fetched. If the branch condition is not met then the execution proceeds as normal. If the condition is met, the processor will annul the instructions following the branch until the target instruction is fetched.

Some processors implement a branch prediction solution. In this approach, a paradigm is used to predict whether the branch will be taken or not. A cache of branch target locations is kept by the processor. The tag is the location of the branch, and the data is the location of the target of the branch. Control bits can indicate the history of the branch. When an instruction is fetched and if it is in the cache, it is a branch and a prediction is made. If the prediction is taken, the branch target is fetched next. Otherwise, the next sequential instruction is fetched. When resolved, if the proessor predicted correctly then execution proceeds without any stalls. If mis-predicted, the processor will flush the instructions past the branch instruction and re-fetch the correct instruction.

Branch prediction is another approach for handling branch instructions. This approach reduces the number of bubbles, depending on the accuracy of the prediction. In this approach the processor must not change the machine state until the branch is resolved. Because of the significant unpredictability in this approach, branch prediction is not used in DSP architectures.

Interrupt effects

Interrupts can be viewed as a branch to an interrupt service routine. Interrupts have a very similar effect to branches. The important issue in this situation is that the pipeline increases the processor interrupt response time.

In the TI TMS320C5x DSP, the interrupt processing is handled by inserting the interrupt at the decode stage in place of the current instruction (Figure 5.23). Inserting the entire bubble avoids all pipeline hazards and conflicts. Instructions are allowed to drain as they may have changed state such as address pointers. On return from the interrupt, I5 is the first instruction fetched.

State	Instruction Cycle							
Fetch	I4	I5	-	-	-	NOP1	NOP2	NOP3
Decode	I3	I4	INT	-	-		NOP1	NOP2
Operands	I2	I3	I4	INT	-	-	-	NOP1
Execute	I1	I2	I3	I4	INT	-	-	-

↑ Interrupt synchronized here

Figure 5.23 Interrupt processing in a pipelined processor

As another example of this type of interrupt processing, consider the JSR (jump to subroutine) instruction on the Motorola 5600x. JSR is stored as the first word of a two word interrupt vector. The second word is not fetched in this case (Figure 5.24).

State	Instruction Cycle							
Fetch	I3	I4	JSR	-	NOP1	NOP2	NOP3	
Decode	I2	I3	I4	JSR	-	NOP1	NOP2	
Execute	I1	I2	I3	I4	JSR	-	NOP1	

↑ Interrupt synchronized here

Figure 5.24 Interrupt processing in the Motorola 5600x

Pipeline Control

DSP programmers must also be knowledgeable of the processor pipeline control options. There are different types of processor pipeline control. The first is called *data stationary pipeline control.* In this approach, the program specifies what happens to one set of operands involving more than one operation at different points in time. This style of control tracks a single set of operands through a sequence of operations and is easier to read.

Time-stationary pipeline control is when the program specifies what happens at any one point in time to a pipeline involving more than one set of operands and operators. In this method the programmer specifies what the functional units (address generators, adders, multipliers) are doing at any point in time. This approach allows for more precise time control.

Interlocked or protected pipeline control allows the programmer to specify the instruction without control of the pipeline stages. The DSP hardware resolves resource conflicts and data hazards. This approach is much easier to program.

As an example of time stationary control, consider the Lucent DSP16xx MAC loop instruction shown below:

A0 = A0 + p   p = x * y   y= *R0++   x = *Pt ++

This instruction can be interpreted as follows: At the time the instruction executes,

- accumulate the value in p in A0;
- multiply the values in x and y and store in p;
- store the value of the memory location pointed to by R0 in y and increment the pointer;
- store the value of the memory location pointed to by Pt in x and increment the pointer.

where:

p = product register
x = data register
y = data register
A0 = accumulator register
R0 = pointer register
Pt = pointer register

As an example of data-stationary control, consider the Lucent DSP32xx MAC loop instruction shown below:

A0 = A0 + (*R5++ = *R4++) * *R3++

This instruction is interpreted as follows:

- Fetch the value of the memory location pointed to by R3 and store in y and increment the pointer.
- Fetch the value of the memory location pointed to by R4 and store in x and increment the pointer.
- Store the value fetched in the above step in the memory location pointed to by R5 and increment the pointer.
- Multiply the values stored above in x and y and accumulate in A0.

Specialized Instructions and Address Modes

As mentioned earlier, DSPs have specialized instructions to perform certain operations such as multiply and accumulate very quickly. In many DSPs, there is a dedicated hardware adder and hardware multiplier. The adder and multiplier are often designed to operate in parallel so that a multiply and an add can be executed in the same clock cycle. Special multiply and accumulate (MAC) instructions have been designed into DSPs for this purpose. Figure 5.25 shows the high level architecture of the multiply unit and adder unit in parallel.

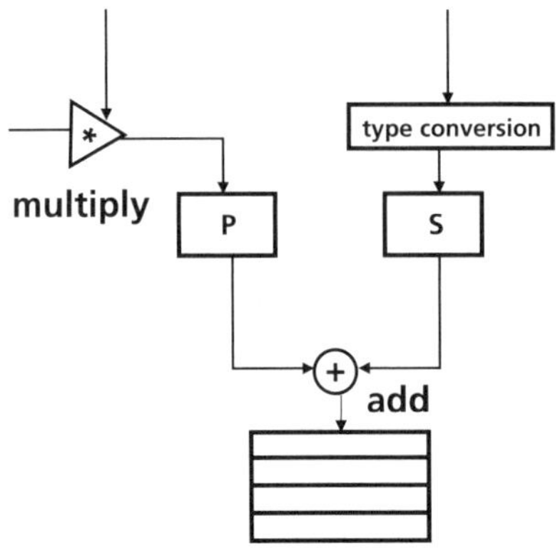

Figure 5.25 Adder and multiplier operating in parallel

DSPs can also perform multiple MAC operations using additional forms of parallelism. For example, the C5000 family of TI DSPs have two adders—one in the ALU and another within the MAC unit, as shown in Figure 5.26. This additional form of parallelism allows for the computation of two filter taps per cycle, effectively doubling performance for this type of operation. The RPTZ instruction is a zero overhead looping instruction that reduces overhead in this computation. This is another unique feature of DSPs that is not found in other more general-purpose microprocessors.

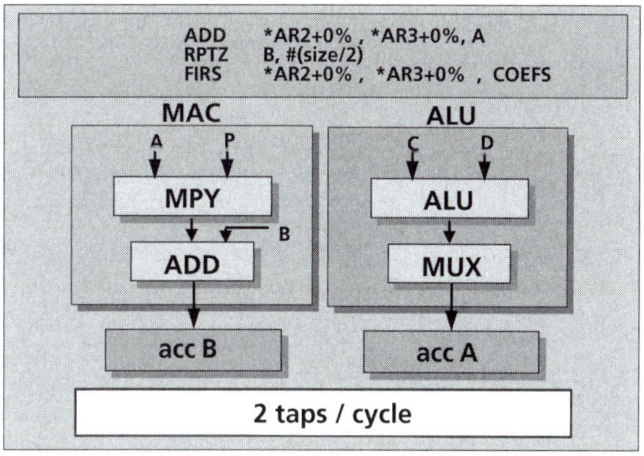

Figure 5.26 A DSP with two adders, one in the MAC unit and one in the ALU (courtesy of Texas Instruments)

The instructions in Figure 5.26 are part of a symmetrical FIR implementation. The symmetrical FIR first adds together the two data samples, which share a common coefficient. In Figure 5.27, the first instruction is a dual operand ADD and it performs that function using registers AR2 and AR3. Registers AR2 and AR3 point to the first and last data values, allowing the A accumulator to hold their sum in a single cycle. The pointers to these data values are automatically incremented (no extra cycles) to point to the next pair of data samples for the subsequent ADD. The repeat instruction (RPTZ) instructs the DSP to implement the next instruction "N/2" times. The FIRS instruction implements the rest of the filter at 1 cycle per each two taps. The FIRS instruction takes the data sum from the A accumulator, multiplies it with the common coefficient drawn from the program bus, and adds the running filter sum to the B accumulator. In parallel to the MAC unit performing a multiply and accumulation operation (MAC), the ALU is being fed the next pair of data values via the C and D buses, and summing them into the A accumulator. The combination of the multiple buses, multiple math hardware, and instructions that task them in efficient ways allows the N tap FIR filter to be implemented in N/2 cycles. This process uses *three* input buses *every cycle*. This results in a lot of overall throughput (at 10 nSec this amounts to 30 M words per sec—sustained, not "burst").

Another form of LMS instruction (another specialized instruction for DSPs), merges the LMS ADD with the FIR's MAC. This can reduce the load to 2N cycles for an N tap adaptive filter. As such, a 100th order system would run in about 200 cycles, vs. the expected 500. When selecting a DSP for use in your system, subtle performance issues like these can be seen to have a very significant effect on how many MIPS a given function will require.

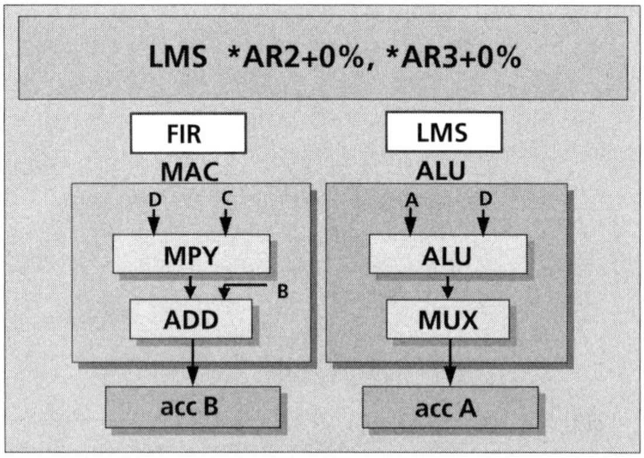

Figure 5.27 Operation of the LMS instruction (courtesy of Texas Instruments)

The LMS instruction on a TI 5000 family DSP works similar to a MAC (Figure 5.27). LMS uses two address registers (AR's) to point to the data and coefficient arrays. A normal MAC implements the FIR process in the B accumulator. And, the addition of the coefficient to the tap-specific weighted error is performed in parallel, with the result being sent to the A accumulator. The dual accumulators aid in this process by providing the additional resources. The FIRS and LMS are only a few of the special instructions present on modern DSPs. Other, even higher performance instructions have been created that give extra speed to algorithms such as code book search, polynomial evaluation, and viterbi decoder processes. Each offers savings of from 2:1 to to over 5:1 over their implementation using 'normal' instructions. Table 5.3 lists some of the other specialized instructions on digital signal processors that offer significant performance improvements over general-purpose instructions.

Instruction	Description	Use
.0.0.0.26.1 DADST		Channel decoder
DSADT	Subtract/add in 16-bit or double-precision mode	Channel decoder
CMPS	Compare upper word and lower word or accumulator and store larger value in memory	Channel decoder/Equalizer (Vererbi)
SRCCD	Store value of block repeat value	Code book search
SACCD	Read and update a value	Code book search
MPYA	Multiply and accumulate previous product	Code book search

Instruction	Description	Use
POLY	Shift and store in parallel with a multiply, add, round and store	Polynomial evaluation
LMS	Least mean square	LMS for adaptive filter
MACA	Multiple high part of accumulator	Lattice filter
FIRS	Perform a symmetric filter calculation	Symmetrical FIR
ST || MAC	Store value in memory in parallel with a multiple and accumulate operation	Lattice filter
ST || LD	Perform a store and a load in parallel	Polynomial evaluation

Table 5.3 Specialized DSP instructions for important DSP algorithms

Circular Addressing

In many signal processing applications, semi-infinite data streams are mapped into finite buffers. Many DSP systems that perform a data acquisition function require a robust way to store bursts or streams of data coming into the system and process that data in a well defined order. In these systems, the data moves through the buffer sequentially. This implies a first in first out (FIFO) model for processing data. A linear buffer (Figure 5.28) requires that the delay in the filter operation be implemented by manually moving data down the delay line. In this implementation, new data is written to the recently vacated spot at the top of the dedicated buffer. Many DSPs as well as data converters such as ADCs implement a FIFO queue as a circular buffer (Figure 5.29). Circular buffers realize the delay line in a filter operation, for example, by moving a pointer through the data instead of moving the data itself. The new data sample is written to a location that is one position above the previous sample[10]. In a circular buffer implementation, read and write pointers are used to manage the buffer. A write pointer is used to track where the buffer is empty and where the next write can occur. A read pointer is used to track where the data was last read from. In a linear buffer, when either pointer reaches the end of the buffer it must be reset to point to the start. In circular addressing modes, the resetting of these pointers is handled using modulo arithmetic.

Some DSP families have special memory mapped registers and dedicated circuitry to support one or more circular buffers. Auxiliary registers are used as pointers into the circular buffer. Two auxiliary registers need to be initialized with the start and end addresses of the buffer. Circular buffers are common in many signal processing applications such as telephone networks and control systems. Circular buffer implementations are used in these systems for filter implementations and other transfer functions. For filter operations, a new sample (x(n)) is read into the circular buffer, overwriting the oldest sample in the buffer. This new sample is stored in a memory location pointed to

[10] For a complete description, see application note SPRA292, Implementing Circular Buffers with Bit-Reversed Addressing, Henry Hendrix, Texas Instruments

by one of the DSP auxiliary registers, AR(i). The filter calculation is performed using x(n) and the pointer is then adjusted for the next input sample[11].

Depending on the application, the DSP programmer has the choice of overwriting old data if the buffer becomes filled, or waiting for the data in the circular buffer to be retrieved. In those applications where the buffer will be overwritten starting from the beginning of the buffer, the application simply needs to retrieve the data at any time before the buffer is filled (or ensure that the consumed rate of data is faster than the produced rate of data). Circular buffers can also be used in some applications as pre and post trigger mechanisms. Using circular buffers, it is possible for the programmer to retrieve data prior to a specific trigger occurring. If this failure of interest is tied to a triggering event, the software application can access the circular buffer to obtain information that led up to the failure occurring. This makes circular buffers good for certain types of fault isolation.

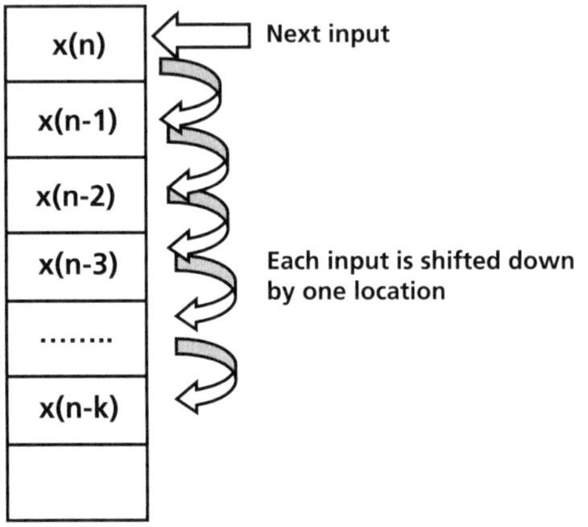

Figure 5.28 A linear buffer requires manual data movement

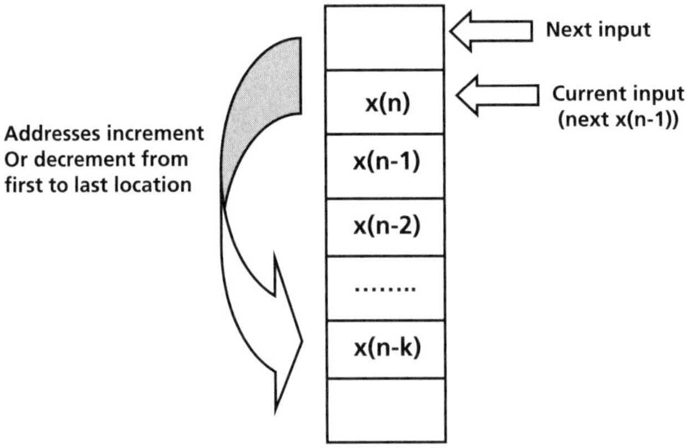

Figure 5.29 Circular buffer. Pointer moves through the data, the data itself is not moved

[11] For more, see Texas Instruments application note SPRA264, *Using the Circular Buffers on the TMS320C5x*, by Thomas Horner.

Bit-Reversed Addressing

Bit-reversed addressing is another special addressing mode used in many DSP applications. Bit-reversed addressing is used primarily in FFT computations. In these computations, either the inputs are supplied in bit-reversed order or the outputs are generated in bit-reversed order. DSPs are optimized to perform bit-reversed addressing. Calculations that use this addressing mode such as FFTs, are optimized for speed as well as decreased memory utilization. The savings come from the software not having to copy either the imput data or the output results back and forth to the standard addressing mode. This saves time and memory. Bit-reversed addressing is a mode in many DSPs, enabled by setting a bit in a mode status register. When this mode is enabled, all addresses using pre-defined index registers are bit-reversed on the output of the address. The logic uses indirect addressing with additional registers to hold a base address and an increment value (for FFT calculations, the increment value is one half the length of the FFT).

Examples of DSP Architectures

This section provides some examples of the various DSP architectures. DSP controllers and cores are discussed as well as high performance DSPs.

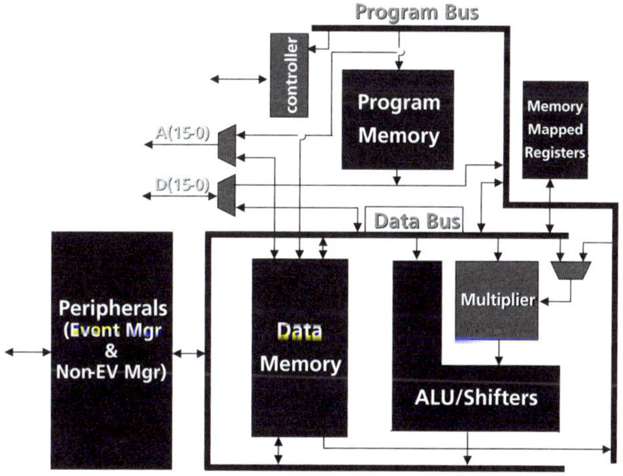

Figure 5.30 Architecture of a TI TMS320C24x DSP (courtesy of Texas Instruments)

Low-Cost Accumulator-Based Architecture

An example of a DSP microcontroller is the TMS320C24x (Figure 5.30). This DSP utilizes a modified Harvard architecture consisting of separate program and data buses and separate memory spaces for program, data and I/O. It is an accumulator-based architecture. The basic building blocks of this DSP include program memory, data memory, ALU and shifters, multipliers, memory mapped registers, peripherals and a controller. These low-cost DSPs are aimed at motor control and industrial systems, as well as solutions for industrial systems, multifunction motor control, and appliance and consumer applications. The low cost and relative simplicity of the architecture makes this DSP ideal for lower cost applications.

Low-Power DSP Architecture

The TI TMS320C55x DSP shown in Figure 5.31 is an example of a low power DSP architecture. This DSP design is a low power design based on advanced power management techniques. Examples of these power management techniques include:

- Selective enabling/disabling of six functional units (IDLE domains). This allows for greater power-down configurability and granularity (Figure 5.32).
- Automatic turn-on/off mechanism for peripherals and on-chip memory arrays. Only the memory array read from or written to, and registers that are in use, contribute to power consumption. This management functionality is completely transparent to the user.
- Memory accesses are minimized. A 32-bit program reduces the number of internal/external fetches. A cache system with burst fill minimizes off-chip memory accesses, which prevents the external bus from being turned on and off, thereby reducing power.
- Cycle count per task is minimized due to increased parallelism in the device. This DSP has two MAC units, four accumulators, four data registers, and five data buses. This DSP also has "soft-dual" instructions, for example a dual-write, double-write, and double-push in one cycle.
- The multiply unit is not used when an operation can be performed by a simpler arithmetic operation.

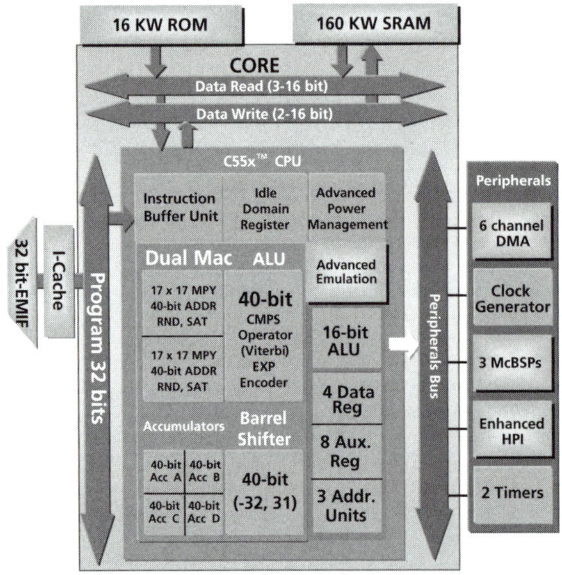

Figure 5.31 DSP architecture for the TMS320C5510

Event-Driven Loops Applications

Many DSP applications are event driven loops. The DSP waits for an event to occur, processes any incoming data, and goes back into the loop, waiting for the next event.

One way to program this into a DSP is to use NOPs. Once the event occurs, the DSP can start execution of the appropriate routine (the active region in Figure 5.32). At the end of the routine, the programmer has the option of sending the DSP back into an infinite loop type of a structure, or executing NOPs.

The other option is to use IDLE modes. These are power down modes available on the C5000. These modes power down different sections of the C5000 device to conserve power in areas that are not being used. As Figure 5.32 shows, idle modes use less power than NOPs.

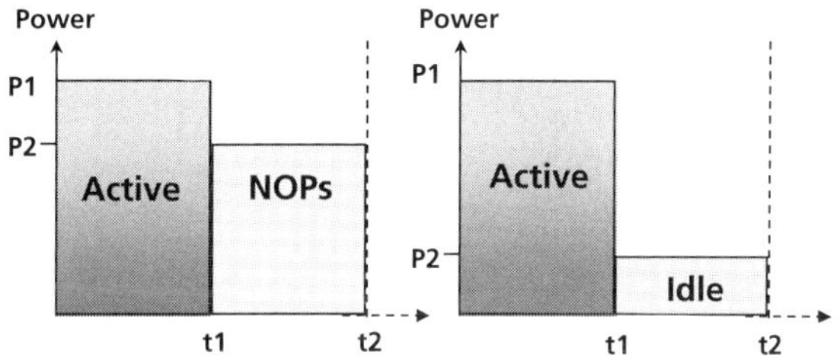

Figure 5.32 Replacing NOPs with idle modes saves power (courtesy of Texas Instruments)

A DSP With Idle Modes

The TI C5000 devices have three IDLE modes that can be initiated through software control. In the normal mode of operation, all the device components are clocked. In IDLE 1 mode, the CPU processor clock is disabled but the peripherals remain active. IDLE 1 can be used when waiting for input data to arrive from a peripheral device. IDLE 1 can also be used when the processor is receiving data but not processing it. Since the peripherals are active, they can be used to bring the processor out of this power down state.

High-Performance DSP

Figure 5.33 shows the architecture of a high performance TMS320C6x DSP. This device is a very long instruction word (VLIW) architecture. This is a highly parallel and deterministic architecture that emphasizes software-based flexibility and maximum code performance through compiler efficiency. This device has eight functional units on the DSP core. This includes two multipliers and six arithmetic units. These units are highly orthogonal, which provides the compiler with many execution resources to make the compilation process easier. Up to eight 32-bit RISC-like instructions are fetched by the CPU each cycle. The architecture of this device includes instruction packing, which allows the eight instructions to be executed in parallel, in serial, or parallel/serial combinations. This optimized scheme enables high performance from this device in terms of reduction of the number of program instructions fetched as well as significant power reductions.

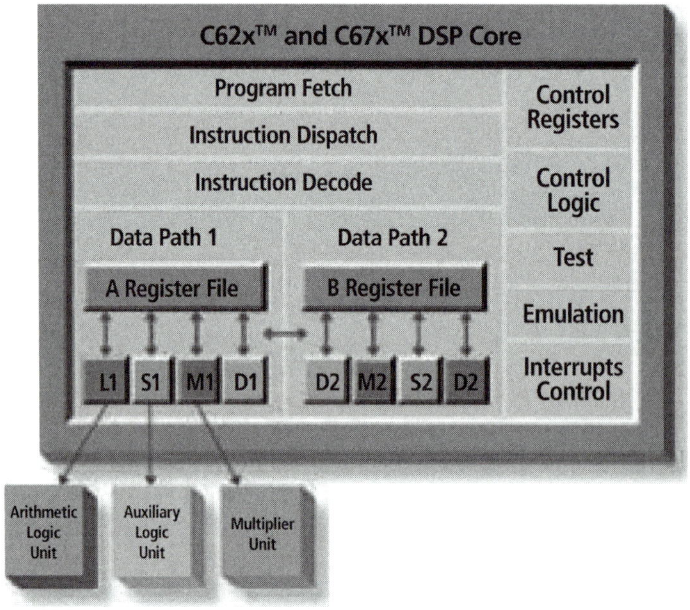

Figure 5.33 A high performance DSP—the TMS320C6x has dual data paths and orthogonal instruction units which boost overall performance (courtesy of Texas Instruments)

VLIW Load and Store DSP

Figure 5.33 is an example of a *load store* architecture for a VLIW DSP. This architecture contains a CPU, dual data paths, and eight independent orthogonal functional units. The VLIW architecture of this processor allows it to execute up to eight 32-bit instructions per cycle.

A detailed view of the register architecture is shown in Figure 5.34. Each of the eight individual execution units are orthogonal; they can independently execute instructions during each cycle. There are six arithmetic units and two multipliers. These orthogonal execution units are:

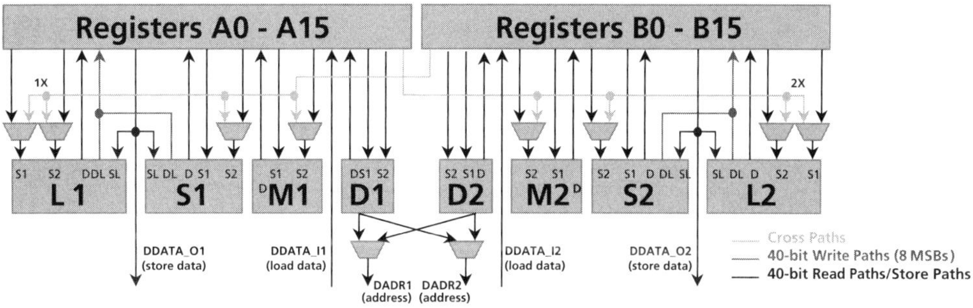

Figure 5.34 Detailed view of VLIW architecture (courtesy of Texas Instruments)

- L-Unit (L1, L2); this is a 40-bit Integer ALU used for comparisons, bit counting, and normalization.
- S-Unit (S1, S2); this is a 32-bit ALU as well as a 40-bit shifter. This unit is used for bitfield operations and branching.
- M-Unit (M1, M2); this is a 16 × 16 multiplier which produces a 32 bit result.
- D-Unit (D1, D2); this unit is used for 32-bit adds and subtracts. The unit also performs address calculations.

There is independent control and up to eight 32-bit instructions can be executed in parallel. There are two register files with 32 registers each and a cross path capability which allows instructions to be executed on each side of the register file.

The external memory interface (EMIF) is an important peripheral to the high performance DSP. The EMIF is an asynchronous parallel interface with separate strobes for program, data and I/O spaces. The primary advantages of the extended address range of the EMIF are prevention of bus contention and easier to interface to memory because no external logic is required. The EMIF is designed for direct interface to industry standard RAMs and ROMs.

Summary

There are many DSP processors on the market today. Each processor advertises its own set of unique functionality for improving DSP processing. However, as different as these processors sound from each other, there are a few fundamental characteristics that all DSP processors have. In particular, all DSP processors must:

- Perform high-speed arithmetic,
- Make multiple accesses to and from memory in a single cycle,
- Be able to process data to and from the real world.

To accomplish these goals, processor designers draw from a wide range of advanced architectural techniques. Thus, to create high performance DSP applications, programmers must master not only the relevant DSP algorithms, but also the idioms of the particular architecture. The next chapter will address some of these issues.

References

VLIW Architecture Emerges as Embedded Alternative, Alexander Wolf, Embedded Systems Programming, February 2001

DSP Building Blocks to Handle High Sample Rate Sensor Data, Chris Ciufo, COTS Journal, March/April 2000

DSP Processor Fundamentals, Architectures and Features, Phil Lapsley, Jeff Bier, Amit Shoham, and Edward A. Lee, BDTI, 1995

Programming DSPs using C: efficiency and portability trade-offs, by Gerard Vink, Embedded Systems, May 2000

Implementing Circular Buffers with Bit-reversed Addressing, Application Report SPRS292, Henry Hendrix, November 1997, Texas Instruments

6

Optimizing DSP Software

Introduction

Many of today's DSP applications are subject to real-time constraints. Many embedded DSP applications will eventually grow to a point where they are stressing the available CPU, memory or power resources. Understanding the workings of the DSP architecture, compiler and application algorithms can speed up applications, sometimes by an order of magnitude. This chapter will summarize some of the techniques that can improve the performance of your code in terms of cycle count, memory use and power consumption.

What Is Optimization?

Optimization is a procedure that seeks to maximize or minimize one or more performance indices. These indices include:

- Throughput (execution speed)
- Memory usage
- I/O bandwidth
- Power dissipation

Since many DSP systems are real-time systems, at least one (and probably more) of these indices must be optimized. It is difficult (and usually impossible) to optimize all these performance indices at the same time. For example, to make the application faster, the developer may require more memory to achieve the goal. The designer must weigh each of these indices and make the best trade-off.

The tricky part to optimizing DSP applications is understanding the trade-off between the various performance indices. For example, optimizing an application for speed often means a corresponding decrease in power consumption but an increase in memory usage. Optimizing for memory may also result in a decrease in power consumption due to fewer memory accesses but an offsetting decrease in code performance. The various trade-offs and system goals must be understood and considered before attempting any form of application optimization.

Determining which index or set of indices is important to optimize depends on the goals of the application developer. For example, optimizing for performance means that the developer can use a slow or less expensive DSP to do the same amount of work. In some embedded systems, cost savings like this can have a significant impact on the success of the product. The developer can alternatively choose to optimize the application to allow the addition of more functionality. This may be very important if the additional functionality improves the overall performance of the system, or if the developer can add more capability to the system such as an additional channel of a base station system. Optimizing for memory use can also lead to overall system cost reduction. Reducing the application size leads to a lower demand for memory, which reduces overall system cost. Finally, optimizing for power means that the application can run longer on the same amount of power. This is important for battery powered applications. This type of optimization also reduces the overall system cost with respect to power supply requirements and other cooling functionality required.

The Process

Generally, DSP optimization follows the 80/20 rule. This rule states that 20% of the software in a typical application uses 80% of the processing time. This is especially true for DSP applications that spend much of their time in tight inner loops of DSP algorithms. Thus, the real issue in optimization isn't how to optimize, but where to optimize. The first rule of optimization is "Don't!." Do not start the optimization process until you have a good understanding of where the execution cycles are being spent.

The best way to determine which parts of the code should be optimized is to profile the application. This will answer the question as to which modules take the longest to execute. These will become the best candidates for performance-based optimization. Similar questions can be asked about memory usage and power consumption.

DSP application optimization requires a disciplined approach to get the best results. To get the best results out of your DSP optimization effort, the following process should be used:

- *Do your homework* – Make certain you have a thorough understanding of the DSP architecture, the DSP compiler, and the application algorithms. Each target processor and compiler has different strengths and weaknesses and understanding them is critical to successful software optimization. Today's DSP optimizing compilers are advanced. Many allow the developer to use a higher order language such as C and very little, if any, assembly language. This allows for faster code development, easier debugging, and more reusable code. But the developer must understand the "hints" and guidelines to follow to enable the compiler to produce the most efficient code.
- *Know when to stop* – Performance analysis and optimization is a process of diminishing returns. Significant improvements can be found early in the process with

relatively little effort. This is the "low hanging fruit." Examples of this include accessing data from fast on-chip memory using the DMA and pipelining inner loops. However, as the optimization process continues, the effort expended will increase dramatically and further improvements and results will fall dramatically.

- *Change one parameter at a time* – Go forward one step at a time. Avoid making several optimization changes at the same time. This will make it difficult to determine what change led to which improvement percentage. Retest after each significant change in the code. Keep optimization changes down to one change per test in order to know exactly how that change affected the whole program. Document these results and keep a history of these changes and the resulting improvements. This will prove useful if you have to go back and understand how you got to where you are.
- *Use the right tools* – Given the complexity of modern DSP CPUs and the increasing sophistication of optimizing compilers, there is often little correlation between what a programmer thinks is optimized code and what actually performs well. One of the most useful tools to the DSP programmer is the profiler. This is a tool that allows the developer to run an application and get a "profile" of where cycles are being used throughput the program. This allows the developer to identify and focus on the core bottlenecks in the program quickly. Without a profiler, gross performance issues as well as minor code modifications can go unnoticed for long periods of time and make the entire code optimization process less disciplined.
- *Have a set of regression tests and use it after each iteration* – Optimization can be difficult. More difficult optimizations can result in subtle changes to the program behavior that lead to wrong answers. More complex code optimizations in the compiler can, at times, produce incorrect code (a compiler, after all, is a software program with its own bugs!). Develop a test plan that compares the expected results to the actual results of the software program. Run the test regression often enough to catch problems early. The programmer must verify that program optimizations have not broken the application. It is extremely difficult to backtrack optimized changes out of a program when a program breaks.

A general code optimization process (see Figure 6.1) consists of a series of iterations. In each iteration, the programmer should examine the compiler generated code and look for optimization opportunities. For example, the programmer may look for an abundance of NOPs or other inefficiencies in the code due to delays in accessing memory and/or another processor resource. These are the areas that become the focus of improvement. The programmer will apply techniques such as software pipelining, loop unrolling, DMA resource utilization, etc., to reduce the processor cycle count (we will talk more about these specific techniques later). As a last resort the programmer can consider hand-tuning the algorithms using assembly language.

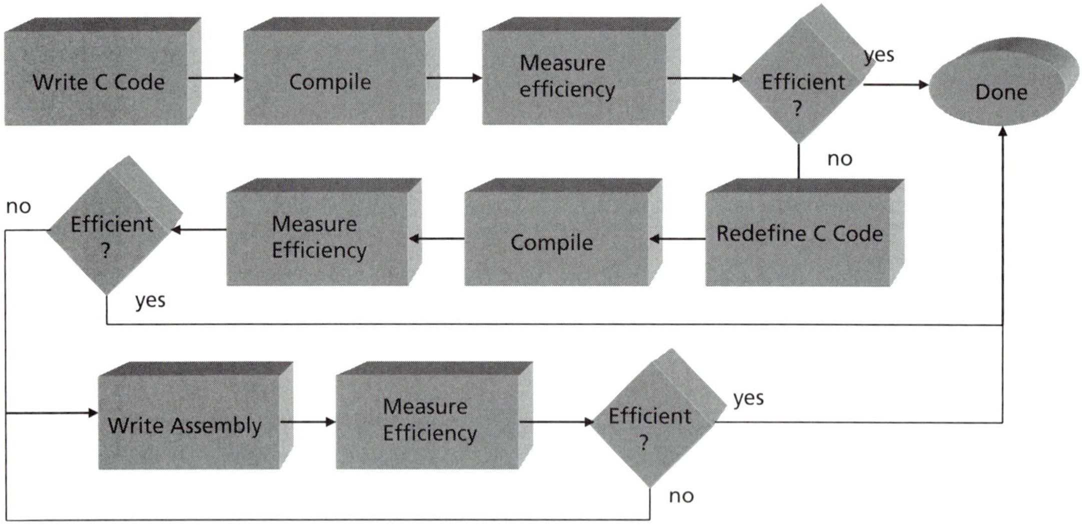

Figure 6.1 A general DSP code optimization process (courtesy of Texas Instruments)

Many times, the C code can be modified slightly to achieve the desired efficiency, but finding the right "tweak" for the optimal (or close to optimal) solution can take time and several iterations. Keep in mind that the software engineer/programmer must take responsibility for at least a portion of this optimization. There have been substantial improvements in production DSP compilers with respect to advanced optimization techniques. These optimizing compilers have grown to be quite complex due to the advanced algorithms used to identify optimization opportunities and make these code transformations. With this increased complexity comes the opportunity for errors in the compiler. You still need to understand the algorithms and the tools well enough that you can supply the necessary improvements when the compiler can't. In this chapter we will discuss how to optimize DSP software in the the context of this process.

Make The Common Case Fast

The fundamental rule in computer design as well as programming real-time DSP-based systems is "make the common case fast, and favor the frequent case." This is really just Amdahl's Law that says the performance improvement to be gained using some faster mode of execution is limited by how often you use that faster mode of execution. So don't spend time trying to optimize a piece of code that will hardly ever run. You won't get much out of it, no matter how innovative you are. Instead, if you can eliminate just one cycle from a loop that executes thousands of times, you will see a bigger impact on the bottom line. I will now discuss three different approaches to making the common case fast (by common case, I am referring to the areas in the code that consume the most resources in terms of cycles, memory, or power):

- Understand the DSP architecture.
- Understand the DSP algorithms.
- Understand the DSP compiler.

Make the Common Case Fast—DSP Architectures

DSP architectures are designed to make the common case fast. Many DSP applications are composed from a standard set of DSP building blocks such as filters, Fourier transforms, and convolutions. Table 6.1 contains a number of these common DSP algorithms. Notice the common structure of each of the algorithms:

- They all accumulate a number of computaions.
- They all sum over a number of elements.
- They all perform a series of multiplies and adds.

These algorithms all share some common characteristics; they perform multiplies and adds over and over again. This is generally referred to as the *sum of products (SOP)*.

Like we discussed earlier, a DSP is, in many ways, an application specific microprocessor. DSP designers have developed hardware architectures that allow the efficient execution of algorithms to take advantage of the algorithmic specialty in signal processing. For example, some of the specific architectural features of DSPs to accommodate the algorithmic structure of DSP algorithms include:

- Special instructions such as a single cycle multiple and accumulate (MAC). Many signal processing algorithms perform many of these operations in tight loops. Figure 6.2 shows the savings from computing a multiplication in hardware instead of microde in the DSP processor. A savings of four cycles is significant when multiplications are performed millions of times in signal processing applications.

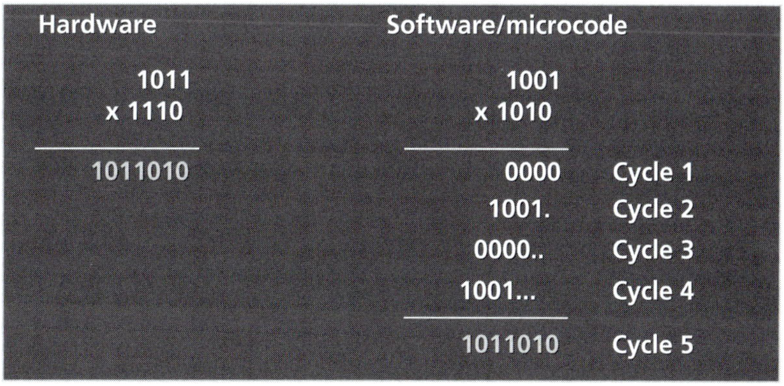

Figure 6.2 Special multiplication hardware speeds up DSP processing (courtesy of Texas Instruments)

- Large accumulators to allow for accumulating a large number of elements.

- Special hardware to assist in loop checking so this does not have to be performed in software, which is much slower.
- Access to two or more data elements in the same cycle. Many signal processing algorithms multiple two arrays of data and coefficients. Being able to access two operands at the same time makes these operations very efficient. The DSP Harvard architecture shown below allows for access of two or more data elements in the same cycle.

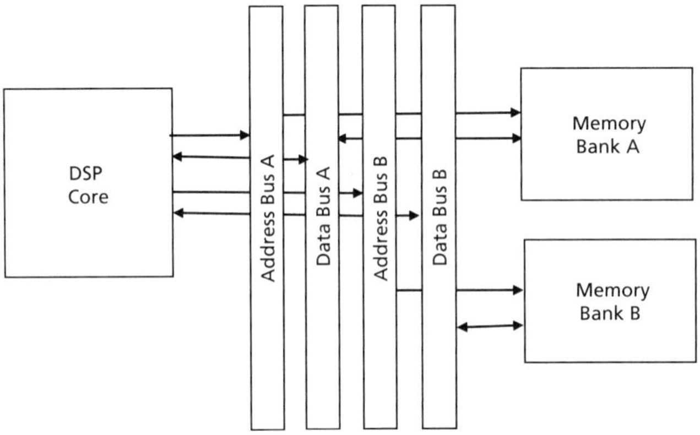

Figure 6.3 DSP Harvard architecture—multiple address and data busses accessing multiple banks of memory simultaneously

More detail on this topic have been discussed in the chapter on DSP architectures. The DSP developer must choose the right DSP architecture to accommodate the signal processing algorithms required for the application as well as the other selection factors such as cost, tools support, etc.

Equation
$y(n)=\sum_{k=0}^{M} a_k x(n-k)$
$y(n)=\sum_{k=0}^{M} a_k x(n-k)+\sum_{k=1}^{N} b_k y(n-k)$
$y(n)=\sum_{k=0}^{N} x(k)h(n-k)$
$X(k)=\sum_{n=0}^{N-1} x(n)\exp[-j(2\pi/N)nk]$
$F(u)=\sum_{x=0}^{N-1} c(u).f(x).\cos\left[\frac{\pi}{2N}u(2x+1)\right]$

Table 6.1 DSP algorithms share common characteristics

Make the Common Case Fast—DSP Algorithms

DSP algorithms can be made to run faster using techniques of algorithmic transformation. For example, a common algorithm used in DSP applications is the Fourier transform. The Fourier transform is a mathematical method of breaking a signal in the time domain into all of its individual frequency components[1]. The process of examining a time signal broken down into its individual frequency components is also called *spectral analysis* or *harmonic analysis.*

There are different ways to characterize a Fourier transforms:

- The Fourier transform (FT) is a mathematical formula using integrals:

$$F(u) = \int_{-\infty}^{\infty} f(x) e^{-x\pi i x u} d\omega$$

- The discrete Fourier transform (DFT) is a discrete numerical equivalent using sums instead of integrals which maps well to a digital processor like a DSP:

$$F(u) = \frac{1}{N} \sum_{x=0}^{N-1} f(x) e^{-x\pi i \omega u f N}{}_{j}$$

- The fast Fourier transform (FFT) is just a computationally fast way to calculate the DFT which reduces many of the redundant computations of the DFT.

How these are implemented on a DSP has a significant impact on overall performance of the algorithm. The FFT, for example, is a fast version of the DFT. The FFT makes use of periodicities in the sines that are multiplied to perform the transform. This significantly reduces the amount of calculations required. A DFT implementation requires N^2 operations to calculate a N point transform. For the same N point data set, using a FFT algorithm requires N * log2(N) operations. The FFT is therefore faster than the DFT by a factor of N/log2(n). The speedup for a FFT is more significant as N increases (Figure 6.4).

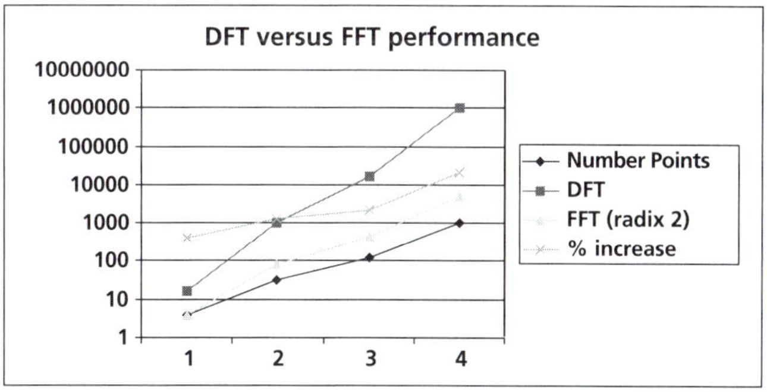

Figure 6.4 FFT vs. DFT for various sizes of transforms (logarithmic scale)

[1] Brigham, E. Oren, 1988, *The Fast Fourier Transform and Its Applications*, Englewood Cliffs, NJ: Prentice-Hall, Inc., p. 448.

Recognizing the significant impact that efficiently implemented algorithms have on overall system performance, DSP vendors and other providers have developed libraries of efficient DSP algorithms optimized for specific DSP architectures. Depending on the type of algorithm, these can downloaded from web sites (be careful of obtaining free software like this—the code may be buggy as there is no guarantee of quality) or bought from DSP solution providers.

Make the Common Case Fast—DSP Compilers

Just a few years ago, it was an unwritten rule that writing programs in assembly would usually result in better performance than writing in higher level languages like C or C++. The early "optimizing" compilers produced code that was not as good as what one could get by programming in assembly language, where an experienced programmer generally achieves better performance. Compilers have gotten much better and today there are very specific high performance optimizations performed that compete well with even the best assembly language programmers.

Optimizing compilers perform sophisticated program analysis including intraprocedural and interprocedural analysis. These compilers also perform data and control flow analysis as well as dependence analysis and often employ provably correct methods for modifying or transforming code. Much of this analysis is to prove that the transformation is correct in the general sense. Many optimization strategies used in DSP compilers are also strongly heuristic[2].

One effective code optimization strategy is to write DSP application code that can be *pipelined* efficiently by the compiler. *Software* pipelining is an optimization strategy to schedule loops and functional units efficiently. In modern DSPs there are multiple functional units that are orthogonal and can be used at the same time (Figure 6.5). The compiler is given the burden of figuring out how to schedule instructions so that these functional units can be used in parallel whenever possible. Sometimes this is a matter of a subtle change in the way the C code is structured that makes all the difference. In software pipelining, multiple iterations of a loop are scheduled to execute in parallel. The loop is reorganized in a way that each iteration in the pipelined code is made from instruction sequences selected from different iterations in the original loop. In the example in Figure 6.6, a five-stage loop with three iterations is shown. There is an initial period (cycles n and n+1), called the *prolog* when the pipes are being "primed" or initially loaded with operations. Cycles n+2 to n+4 is the actual pipelined section of the code. It is in this section that the processor is performing three different operations (C, B, and A) for three different loops (1, 2 and 3). There is a epilog section where the last remaining instructions are performed before exiting the loop. This is an example of a fully utilized set of pipelines that produces the fastest, most efficient code.

[2] Heuristics involves problem-solving by experimental and especially trial-and-error methods or relating to exploratory problem-solving techniques that utilize self-educating techniques (as the evaluation of feedback) to improve performance.

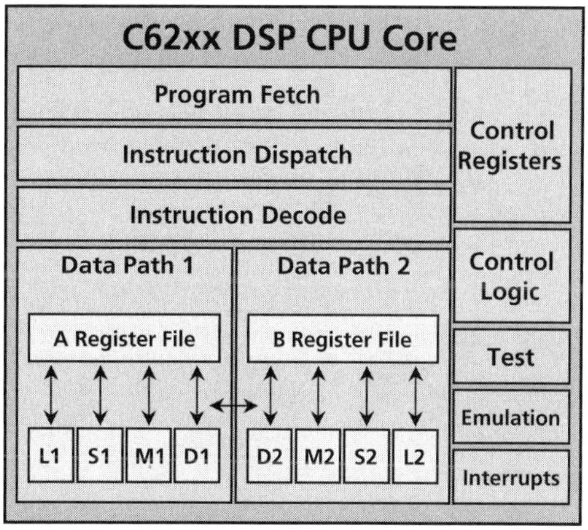

Figure 6.5 DSP architectures may have orthogonal execution units and data paths used to execute DSP algorithms more efficiently. In this figure, units L1, S1, M1, D1, and L2, S2, M2, and D2 are all orthogonal execution units that can have instructions scheduled for execution by the compiler in the same cycle if the conditions are right.

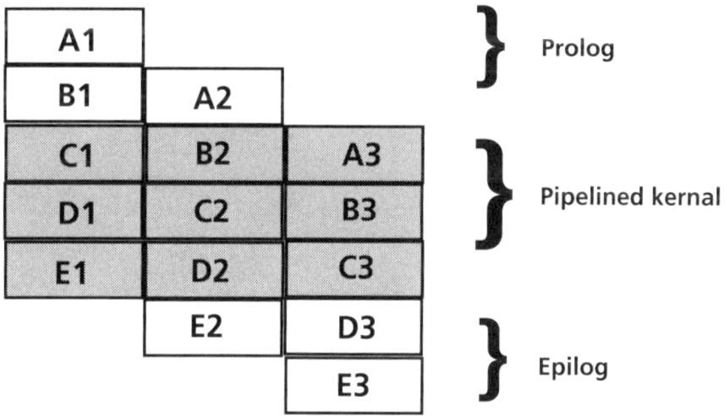

Figure 6.6 A five-stage instruction pipeline that is scheduled to be software pipelined by the compiler

Figure 6.7 shows a sample piece of C code and the corresponding assembly language output. In this case the compiler was asked to attempt to pipeline the code. This is evident by the piped loop prolog and piped loop kernel sections in the assembly language output. Keep in mind that the prolog and epilog sections of the code prime the pipe and flush the pipe, respectively, as shown in Figure 6.6. In this case, the pipelined code is not as good as it could be. You can spot inefficient code by looking for how many NOPs are in the piped loop kernel of the code. In this case the piped loop kernel has a total of five NOP cycles, two in line 16, and three in line 20. This loop takes a total of 10 cycles to execute. The NOPs are the first indication that a more efficient loop may be possible. But how short can this loop be? One way to estimate the minimum loop size is to determine what execution unit is being used the most. In this example, the D unit is used more than any other unit, a total of three times (lines 14, 15 and 21). There are two sides to a superscalar device, enabling each unit to be used twice (D1 and D2) per clock for a minimum two clock loop; two D operations in one clock and one D unit in the second clock. The compiler was smart enough to use the D units on both sides of the pipe (lines 14 and 15), enabling it to parallelize the instructions and only use one clock. It should be possible to perform other instructions while waiting for the loads to complete, instead of delaying with NOPs.

```
void example1(float *out, float *input1, float *input2)
{
  int i;

  for(i = 0; i < 100; i++)
   {
      out[i] = input1[i] * input2[i];
   }
}
```

```
1   _example1:
2   ;** ---------------------------------------------------------------*
3               MVK         .S2     0x64,B0
4
5               MVC         .S2     CSR,B6
6   ||          MV          .L1X        B4,A3
7   ||          MV          .L2X        A6,B5

8               AND         .L1X        -2,B6,A0
9               MVC           .S2X        A0,CSR
10  ;** ---------------------------------------------------------------*
11  L11:         ; PIPED LOOP PROLOG
12  ;** ---------------------------------------------------------------*
13  L12:         ; PIPED LOOP KERNEL

14              LDW         .D2         *B5++,B4        ;
15  ||          LDW         .D1         *A3++,A0        ;

16              NOP                        2
17  [ B0]       SUB         .L2         B0,1,B0         ;
18  [ B0]       B           .S2         L12             ;
19              MPYSP       .M1X        B4,A0,A0        ;
20              NOP                     3
21              STW         .D1         A0,*A4++        ;
22  ;** ---------------------------------------------------------------*
23              MVC         .S2         B6,CSR
24              B           .S2         B3
25              NOP                     5
26              ; BRANCH OCCURS
```

Figure 6.7 C example and the corresponding pipelined assembly language output

In the simple ***for*** loop, the two input arrays (array1 and array2) may or may not be dependent or overlap in memory. The same with the output array. In a language such as C/C++ this is something that is allowed and therefore the compiler must be able to handle this correctly. Compilers are generally pessimistic creatures. They will not attempt an optimization if there is a case where the resultant code will not execute properly. In these situations, the compiler takes a conservative approach and assumes the inputs can be dependent on the previous output each time through the loop. If it is known that the inputs are not dependent on the output, we can hint to the compiler by declaring input1 and input2 as "restrict," indicating that these fields will not change. In this example, "restrict" is a keyword in C that can be used for this purpose. This is also a trigger for enabling software pipelining which can improve throughput. This C code is shown in Figure 6.8 with the corresponding assembly language.

```
void example2(float *out, restrict float *input1, restrict float *input2)
{
  int i;
  for(i = 0; i < 100; i++)
   {
      out[i] = input1[i] * input2[i];
   }
}
```

```
_example2:
;** -----------------------------------------------------------------*
           MVK         .S2          0x64,B0
           MVC         .S2          CSR,B6
||         MV          .L1X         B4,A3
||         MV          .L2X         A6,B5
           AND         .L1X        -2,B6,A0
           MVC         .S2X        A0,CSR
||         SUB         .L2         B0,4,B0
;** ----------------------------------------------------------------*
L8:        ; PIPED LOOP PROLOG
           LDW         .D2         *B5++,B4      ;
||         LDW         .D1         *A3++,A0      ;
           NOP                     1
           LDW         .D2         *B5++,B4      ;@
||         LDW         .D1         *A3++,A0      ;@
   [ B0]   SUB         .L2         B0,1,B0       ;
   [ B0]   B           .S2         L9            ;
||         LDW         .D2         *B5++,B4      ;@@
```

```
||          LDW         .D1         *A3++,A0        ;@@

            MPYSP       .M1X        B4,A0,A5        ;
|| [ B0]    SUB         .L2         B0,1,B0         ;@

   [ B0]    B           .S2         L9              ;@
||          LDW         .D2         *B5++,B4        ;@@@
||          LDW         .D1         *A3++,A0        ;@@@

            MPYSP       .M1X        B4,A0,A5        ;@
|| [ B0]    SUB         .L2         B0,1,B0         ;@@

;** --------------------------------------------------------*
L9:         ; PIPED LOOP KERNEL

   [ B0]    B           .S2         L9              ;@@
||          LDW         .D2         *B5++,B4        ;@@@@
||          LDW         .D1         *A3++,A0        ;@@@@

            STW         .D1         A5,*A4++        ;
||          MPYSP       .M1X        B4,A0,A5        ;@@
|| [ B0]    SUB         .L2         B0,1,B0         ;@@@

;** --------------------------------------------------------*
L10:         ; PIPED LOOP EPILOG
            NOP                     1

            STW         .D1         A5,*A4++        ;@
||          MPYSP       .M1X        B4,A0,A5        ;@@@

            NOP                     1

            STW         .D1         A5,*A4++        ;@@
||          MPYSP       .M1X        B4,A0,A5        ;@@@@

            NOP                     1
            STW         .D1         A5,*A4++        ;@@@
            NOP                     1
            STW         .D1         A5,*A4++        ;@@@@
;** --------------------------------------------------------*
             MVC        .S2         B6,CSR
             B          .S2         B3
             NOP                    5
            ; BRANCH OCCURS
```

Figure 6.8 Corresponding pipelined assembly language output

There are a few things to notice in looking at this assembly language. First, the piped loop kernel has become smaller. In fact, the loop is now only two cycles long. Lines 44–47 are all executed in one cycle (the parallel instructions are indicated by the || symbol) and lines 48–50 are executed in the second cycle of the loop. The compiler, with the additional dependency information we supplied it with the "restrict" declaration, has been able to take advantage of the parallelism in the execution units to schedule the inner part of the loop very efficiently. The prolog and epilog portions of the code are much larger now. Tighter piped kernels will require more priming operations to coordinate all of the execution based on the various instruction and branching delays. But once primed, the kernel loop executes extremely fast, performing operations on various iterations of the loop. The goal of software pipelining is, like we mentioned earlier, to make the common case fast. The kernel is the common case in this example, and we have made it very fast. Pipelined code may not be worth doing for loops with a small loop count. But for loops with a large loop count, executing thousands of times, software pipelining produces significant savings in performance while also increasing the size of the code.

In the two cycles the piped kernel takes to execute, there are a lot of things going on. The right hand column in the assembly listing indicates what iteration is being performed by each instruction (Each "@" symbol is a iteration count. So, in this kernel, line 44 is performing a branch for iteration n+2, lines 45 and 46 are performing loads for iteration n+4, line 48 is storing a result for iteration n, line 49 is performing a multiply for iteration n+2, and line 50 is performing a subtraction for iteration n+3, all in two cycles!). The epilog is completing the operations once the piped kernel stops executing. The compiler was able to make the loop two cycles long, which is what we predicted by looking at the inefficient version of the code.

The code size for a pipelined function becomes larger, as is obvious by looking at the code produced. This is one of the trade-offs for speed that the programmer must make.

Software pipelining does not happen without careful analysis and structuring of the code. Small loops that do not have many iterations may not be pipelined because the benefits are not realized. Loops that are large in the sense that there are many instructions per iteration that must be performed may not be pipelined because there are not enough processors resources (primarily registers) to hold the key data during the pipeline operation. If the compiler has to "spill" data to the stack, precious time will be wasted having to fetch this information from the stack during the execution of the loop.

An In-Depth Discussion of DSP Optimization

While DSP processors offer tremendous potential throughput, your application won't achieve that potential unless you understand certain important implementation techniques. We will now discuss key techniques and strategies that greatly reduce the overall number of DSP CPU cycles required by your application. For the most part, the main object of these techniques is to fully exploit the potential parallelism in the processor and in the memory subsystem. The specific techniques covered include:

- Direct memory access;
- Loop unrolling; and
- More on software pipelining.

Direct Memory Access

Modern DSPs are extremely fast; so fast that the processor can often compute results faster than the memory system can supply new operands—a situation known as "data starvation." In other words, the bottleneck for these systems becomes keeping the unit fed with data fast enough to prevent the DSP from sitting idle waiting for data. Direct memory access is one technique for addressing this problem.

Direct memory access (DMA) is a mechanism for accessing memory without the intervention of the CPU. A peripheral device (the DMA controller) is used to write data directly to and from memory, taking the burden off the CPU. The DMA controller is just another type of CPU whose only function is moving data around very quickly. In a DMA capable machine, the CPU can issue a few instructions to the DMA controller, describing what data is to be moved (using a data structure called a *transfer control block* (TCB)), and then go back to what it was doing, creating another opportunity for parallelism. The DMA controller moves the data in parallel with the CPU operation (Figure 6.9), and notifies the CPU when the transfer is complete.

DMA is most useful for copying larger blocks of data. Smaller blocks of data do not have the payoff because the setup and overhead time for the DMA makes it worthwhile just to use the CPU. But when used smartly, the DMA can result in huge time savings. For example, using the DMA to stage data on- and off-chip allows the CPU to access the staged data in a single cycle instead of waiting multiple cycles while data is fetched from slower external memory.

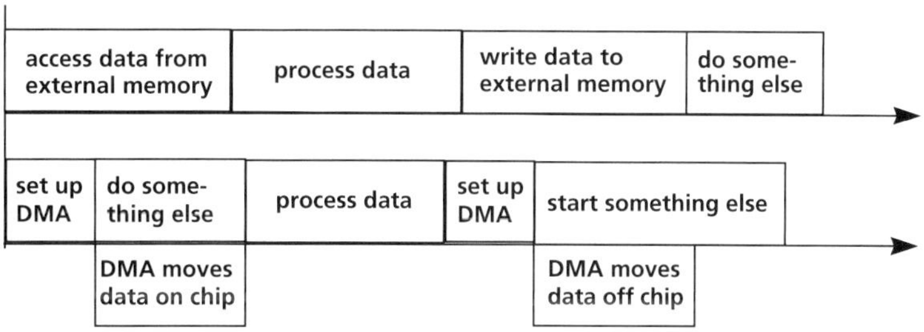

Figure 6.9 Using DMA instead of the CPU can offer big performance improvements because the DMA handles the movement of the data while the CPU is busy performing meaningful operations on the data

Using DMA

Because of the large penalty associated with accessing external memory, and the cost of getting the CPU involved, the DMA should be used wherever possible. The code for this is not too overwhelming. The DMA requires a data structure to describe the data it is going to access (where it is, where its going, how much, etc.). A good portion of this structure can be built ahead of time. Then it is simply a matter of writing to a memory-mapped DMA enable register to start the operation (Figure 6.10). It is best to start the DMA operation well ahead of when the data is actually needed. This gives the CPU something to do in the meantime and does not force the application to wait for the data to be moved. Then, when the data is actually needed, it is already there. The application should check to verify the operation was successful and this requires checking a register. If the operation was done ahead of time, this should be a one time poll of the register, and not a spin on the register, chewing up valuable processing time.

```
--------------------------------------------------------------------------------------------
/* Addresses of some of the important DMA registers */
#define DMA_CONTROL_REG   (*(volatile unsigned*)0x40000404)
#define DMA_STATUS_REG    (*(volatile unsigned*)0x40000408)
#define DMA_CHAIN_REG            (*(volatile unsigned*)0x40000414)

/* macro to wait for the DMA to complete and signal the status register */
#define DMA_WAIT                 while(DMA_STATUS_REG&1) {}

/* pre-built tcb structure */
typedef struct {

     tcb setup fields

} DMA_TCB;

--------------------------------------------------------------------------------------------
extern DMA_TCB tcb;

/* set up the remaining fields of the tcb structure -
    where you want the data to go, and how much you want to send */
tcb.destination_address = dest_address;
tcb.word_count = word_count;

/* writing to the chain register kicks off the DMA operation */
DMA_CHAIN_REG = (unsigned)&tcb;

Allow the CPU to do other meaningful work....

/* wait for the DMA operation to complete */
DMA_WAIT;
```

Figure 6.10 Code to set up and enable a DMA operation is pretty simple. The main operations include setting up a data structure (called a TCB in the example above) and performing a few memory mapped operations to initialize and check the results of the operation

Staging Data

The CPU can access on-chip memory much faster than off-chip or external memory. Having as much data as possible on chip is the best way to improve performance. Unfortunately, because of cost and space considerations most DSPs do not have a lot of on-chip memory. This requires the programmer to coordinate the algorithms in such a way to efficiently use the available on-chip memory. With limited on-chip memory, data must be staged on- and off-chip using the DMA. All of the data transfers can be happening in the background, while the CPU is actually crunching the data. Once the data is in internal memory, the CPU can access the data in on-chip memory very quickly (Figure 6.12).

Smart layout and utilization of on-chip memory, and judicious use of the DMA can eliminate most of the penalty associated with accessing off-chip memory. In general, the rule is to stage the data in and out of on-chip memory using the DMA and generate the results on chip. Figure 6.11 shows a template describing how to use the DMA to stage blocks of data on and off chip. This technique uses a double-buffering mechanism to stage the data. This way the CPU can be processing one buffer while the DMA is staging the other buffer. Speed improvements over 90% are possible using this technique.

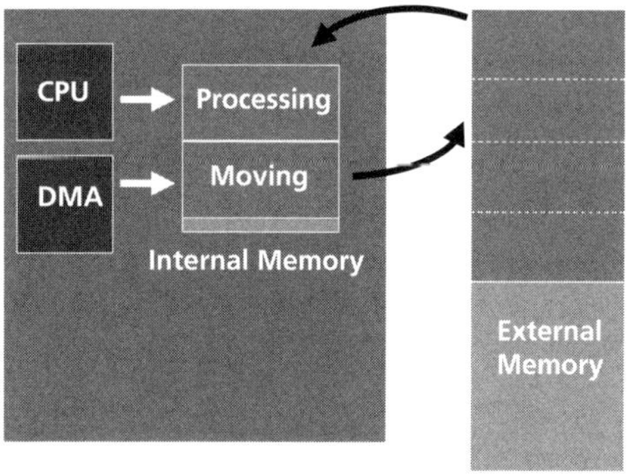

Figure 6.11 Template for using the DMA to stage data on and off chip

```
INITIALIZE TCBS

DMA SOURCE DATA 0 INTO SOURCE BUFFER 0
WAIT FOR DMA TO COMPLETE

DMA SOURCE DATA 1 INTO SOURCE BUFFER 1
PERFORM CALCULATION AND STORE IN RESULT BUFFER

FOR LOOP_COUNT =1 TO N-1
   WAIT FOR DMA TO COMPLETE
   DMA SOURCE DATA I+1 INTO SOURCE BUFFER [(I+1)%2]
   DMA RESULT BUFFER[(I-1)%2] TO DESTINATION DATA
   PERFORM CALCULATION AND STORE IN RESULT BUFFER
END FOR

WAIT FOR DMA TO COMPLETE
DMA RESULT BUFFER[(I-1)%2] TO DESTINATION DATA
PERFORM CALCULATION AND STORE IN RESULT BUFFER

WAIT FOR DMA TO COMPLETE
DMA LAST RESULT BUFFER TO DESTIMATION DATA
```

Figure 6.12 With limited on-chip memory, data can be staged in and out of on-chip memory using the DMA and leaving the CPU to perform other processing

Writing DSP code to use the DMA does have some cost penalties. Code size will increase, depending on how much of the application uses the DMA. Using the DMA also adds increased complexity and synchronization to the application. Code portability is reduced when you add processor specific DMA operations. Using the DMA should only be done in areas requiring high throughput.

An example

As an example of this technique, consider the code in Figure 6.13. This code snippet sums a data field and computes a simple percentage before returning. The code in Figure 6.13 consists of 5 executable lines of code. In this example, the "processed_data" field is assumed to be in external memory of the DSP. Each access of a processed_data element in the loop will cause an external memory access to fetch the data value.

```
int i;
float sum;

/*
**sum data field
*/
sum = 0.0f;
for(i=0; i<num_data_points; i++;)
{
            sum += processed_data[i];
}

/*
**Compute percentage and return
*/
return(MTH_divide(sum,num_data_points));
} /* end */
```

Figure 6.13 A simple function consisting of five executable lines of code

The code shown in Figure 6.14 is the same function shown in Figure 6.12 but implemented to use the DMA to transfer blocks of data from external memory to internal or on-chip memory. This code consists of 36 executable lines of code, but runs much faster than the code in Figure 6.12. The overhead associated with setting up the DMA, building the transfer packets, initiating the DMA transfers, and checking for completion are relatively small compared to the fast memory accesses of the data from on-chip memory. This code snippet was instrumented following the guidelines in the template of Figure 6.13. This code also performs a loop unrolling operation when summing the data at the end of the computation (we'll talk more about loop unrolling later). This also adds to the speedup of this code. This code snippet uses semaphores to protect the on-chip memory and DMA resources. Semaphores and other operating system functions are discussed in a separate chapter on real-time operating systems.

```
#define NUM_BLOCKS   2
#define BLOCK_SIZE(DATA_SIZE/NUM_BLOCKS)

int i, block;
float sum = 0;
float sum0, sum1, sum2, sum3, sum4, sum5,
      sum6, sum7;

/* array contains pointers to internal memory buffers*/
float *processed_data[2];
unsigned index = 0;

/* wait for on chip memory semaphore */
SEM_pend(g_aos_onchip_avail_sem, SYS_FOREVER);

MDMA_build_1d (tcb,              /*tcb pointer */
          0,                     /*prev tcb in chain */
(void*)processed_data[0]         /* destination address */
(void*)p_processed_data,         /* source addr */
BLOCK_SIZE);                     /*  num words to tran */

MDMA_update (tcb,                /*tcb pointer */
   MDMA_ADD,BLOCK_SIZE /*  src  update mode */
   MDMA_TOGGLE,BLOCK_SIZE /*  dst update mode*/

MDMA_initiate_chain(1_tcb);

for(block=0, block<NUM_BLOCKS; block++)
{
  /* point to current buffers */
  internal_processed_data = processed_data[index];

  /* swap buffers */
  index ^= 1;


  if(block < (NUM_BLOCKS - 1))
  {
    MDMA_initiate_chain(1-tcb);
  }
  else
  {
    MDMA_wait();
  }
```

```
/*
**sum data fields to compute percentages
*/
   sum0 = 0.0;
   sum1 = 0.0;
   sum2 = 0.0;
   sum3 = 0.0;
   sum4 = 0.0;
   sum5 = 0.0;
   sum6 = 0.0;
   sum7 = 0.0;

for (i=0; i<BLOCK_SIZE; i+=8)
{
  sum0 += internal_processed_data[i    ];
  sum1 += internal_processed_data[i + 1];
  sum2 += internal_processed_data[i + 2];
  sum3 += internal_processed_data[i + 3];
  sum4 += internal_processed_data[i + 4];
  sum5 += internal_processed_data[i + 5];
  sum6 += internal_processed_data[i + 6];
  sum7 += internal_processed_data[i + 7];
}
   sum += sum0 + sum1 + sum2 + sum3 + sum4 +
            sum5 + sum6 + sum7;

} /* block loop */

/* release on chip memory semaphore */
SEM_post(g_ aos_onchip_avail_sem);
```

Figure 6.14. The same function enhanced to use the DMA. This function is 36 executable lines of code but runs much faster than the code in Figure 6.8.

The code in Figure 6.14 runs much faster than the code in Figure 6.13. The penalty is an increased number of lines of code, which takes up memory space. This may or may not be a problem, depending on the memory constraints of the system. Another drawback to the code in Figure 6.14 is that it is a bit less understandable and portable than the code in Figure 6.13. Implementing the code to use the DMA requires the programmer to make the code less readable, which could possibly lead to maintainability problems. The code is now also tuned for a specific DSP. Porting the code to another DSP family may require the programmer to re-write this code to use the different resources on the new DSP.

Pending vs. Polling

Pending vs. polling

The DMA can be considered a resource, just like memory and the CPU. When a DMA operation is in progress, the application can either wait for the DMA transfer to complete or continue processing another part of the application until the data transfer is complete. There are advantages and disadvantages to each approach. If the application waits for the DMA transfer to complete, it must poll the DMA hardware status register until the completion bit is set. This requires the CPU to check the DMA status register in a looping operation that wastes valuable CPU cycles. If the transfer is short enough, this may only require a few cycles to do and may be appropriate. For longer data transfers, the application engineer may want to use a synchronization mechanism like a semaphore to signal when the transfer is complete. In this case, the application will *pend* on a semaphore through the operating system while the transfer is taking place. The application will be swapped with another application that is ready to run. This swapping of tasks incurs overhead as well and should not be performed unless the overhead associated with swapping tasks is less than the overhead associated with simply polling on the DMA completion. The wait time depends on the amount of data being transferred.

Figure 6.16 shows some code that checks for the transfer length and performs either a DMA polling operation (if there are only a few words to transfer), or a semaphore pend operation (for larger data transfer sizes). The "break even" length for data size is dependent on the processor and the interface structure and should be prototyped to determine the optimal size.

The code for a pend operation is shown in Figure 6.16. In this case, the application will perform a SEM_pend operation to wait for the DMA transfer to complete. This allows the application to perform other meaningful work by temporarily suspending the currently executing task and switching to another task to perform other processing. The details of this operating system model are discussed in the chapter on real time operating systems. When the operating system suspends one task and begins executing another task, a certain amount of overhead is incurred. The amount of this overhead is dependent on the DSP and operating system.

Figure 6.18 shows the code for the polling operation. In this example, the application will continue polling the DMA completion status register for the operation to complete. This requires the use of the CPU to perform the polling operation. Doing this prevents the CPU from doing other meaningful work. If the transfer is short enough and the CPU only has to poll the status register for a short period of time, this approach may be more efficient. The decision is based on how much data is being transferred and how many cycles the CPU must spend polling. If the poll takes less time than the overhead to go through the operating system to swap out one task and begin executing another, this approach may be more efficient. In that context, the code snippet below checks for the transfer length and, if the length is less than the breakeven transfer length, a function will be called to poll the DMA transfer completion status. If the length is greater than the pre-determined cutoff transfer length, a function will be called to set-up the DMA to interrupt on completion of the transfer. This ensures the most efficient processing of the completion of each DMA operation.

```
if (transfer_length < LARGE_TRANSFER)
            IO_DRIVER();
else
            IO_LARGE_DRIVER();
endif
```

Figure 6.15 A code snippet that checks for the transfer length and calls a driver function that will either poll the DMA completion bit in the DSP status register or pend on an operating system semaphore

```
/* wait for port to become available */
while(g_io_channel_status[ dir] & ACTIVE_MASK)
{
            /* poll */
}

/* submittcb */
*(g_io_chain_queue_a[ dir]) = (unsigned int)tcb;

/* wait for transfer to complete */
sem_status = SEM_ pend(handle, SYS_FOREVER);
```

Figure 6.16 A code snippet that pends on a Semaphore for DMA completion

```
/* wait for port to become available */
while (g_io_channel_status[ dir] & ACTIVE_MASK)
{
            /* poll */
}

/* submittcb */
*(g_io_chain_queue_a[ dir]) = (unsigned int)tcb;


/* wait for transfer to complete by polling the
   DMA status register */

status = *((vol_uint)*)g_io_channel_status[ dir];
while  ((status & DMA_ACTIVE_MASK) ==
            DMA_CHANNEL_ACTIVE_MASK)
{
            status = *((vol_uint*)g_io_channel_status[ dir];
}
.
.
.
```

Figure 6.17 A code snippet that polls for DMA completion

Managing Internal Memory

One of the most important resources for a DSP is its on-chip or internal memory. This is the area where data for most computations should reside because access to this memory is so much faster than off-chip or external memory. Since many DSPs do not have a data cache because of determinism unpredictability, software designers should think of a DSP internal memory as a sort of programmer managed cache. Instead of the hardware on the processor caching data for performance improvements with no control by the programmer, the DSP internal data memory is under full control of the DSP programmer. Using the DMA, data can be cycled in and out of the internal memory in the background, with little or no intervention by the DSP CPU. If managed correctly and efficiently, this internal memory can be a very valuable resource.

It is important to map out the use of internal memory and manage where data is going in the internal memory at all times. Given the limited amount of internal memory for many applications, not all the program's data can reside in internal memory for the duration of the application timeline. Over time, data will be moved to internal memory, processed, perhaps used again, and moved to external memory when it is no longer needed. Figure 6.19 shows an example of how a memory map of internal DSP memory might look during the timeline of the application. During the execution of the application, different data structures will be moved to on-chip memory, processed to form additional structures on chip, and eventually be moved off chip to external memory to be saved, or overwritten in internal memory when the data is no longer needed.

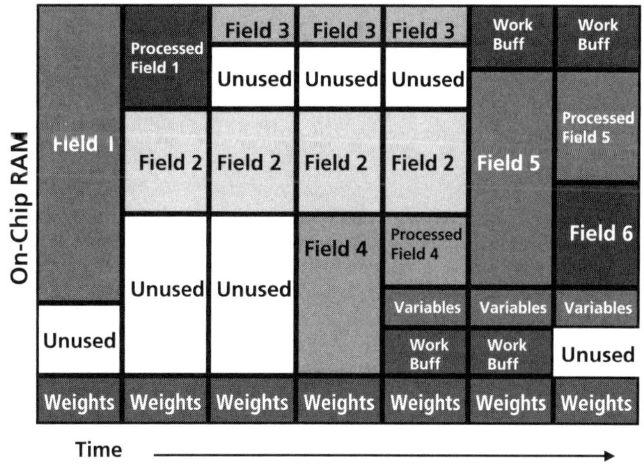

Figure 6.18 Internal memory of a DSP must be managed by the programmer

Loop Unrolling

The standard rule when programming superscalar and VLIW devices is "Keep the pipelines full!" A full pipe means efficient code. In order to determine how full the pipelines are, you need to spend some time inspecting the assembly language code generated by the compiler.

To demonstrate the advantages of parallelism in VLIW-based machines, let's start with a simple looping program shown in Figure 6.19. If we were to write a serial assembly language implementation of this, the code would similar to that in Figure 6.20. This loop uses one of the two available sides of the superscalar machine. By counting up the instructions and the NOPs, it takes 26 cycles to execute each iteration of the loop. We should be able to do much better.

Filling the Execution Units

There are two things to notice in this example. Many of the execution units are not being used and are sitting idle. This is a waste of processor hardware. Second, there are many delay slots in this piece of assembly (20 to be exact) where the CPU is stalled waiting for data to be loaded or stored. When the CPU is stalled, nothing is happening. This is the worst thing you can do to a processor when trying to crunch large amounts of data.

There are ways to keep the CPU busy while it is waiting for data to arrive. We can be doing other operations that are not dependent on the data we are waiting for. We can also use both sides of the VLIW architecture to help us load and store other data values. The code in Figure 6.21 shows an implementation designed for a CPU with multiple execution units. While the assembly looks very much like conventional serial assembly, the run-time execution is very unconventional. Instead of each line representing a single instruction that is completed before the next instruction is begun, each line in Figure 6.21 represents an individual operation, an operation that might be scheduled to execute in parallel with some other operation. The assembly format has been extended to allow the programmer to specify which execution unit should perform a particular operation and which operations may be scheduled concurrently. The DSP compiler automatically determines which execution unit to use for an operation and indicates this by naming the target unit in the extra column that precedes the operand fields (the column containing D1, D2, etc.). in the assembly listing. To indicate that two or more operations may proceed in parallel, the lines describing the individual operations are "joined" with a parallel bar (as in lines 4 and 5 of Figure 6.21). The parallelism rules are also determined by the compiler. Keep in mind that if the programmer decides to program the application using assembly language, the responsibility for scheduling the instructions on each of the available execution units as well as determining the parallelism rules falls on the programmer. This is a difficult task and should only be done when the compiler generated assembly does not have the required performance.

The code in Figure 6.21 is an improvement over the serial version. We have reduced the number of NOPs from 20 to 5. We are also performing some steps in parallel. Lines 4 and 5 are executing two loads at the same time into each of the two load units (D1 and D2) of the device. This code is also performing the branch operation earlier in the loop and then taking advantage of the delays associated with that operation to complete operations on the current cycle.

The code in Figure 6.21 shows an implementation designed for a CPU with multiple execution units. While the assembly looks very much like conventional serial assembly, the run-time execution is very unconventional. Instead of each line representing a single instruction that is completed before the next instruction is begun, each line in Figure 6.21 represents an individual *operation*, an operation that might be scheduled to execute in parallel with some other operation. The assembly format has been extended to allow the programmer to specify which execution unit should perform a particular operation and which operations may be scheduled concurrently. The programmer specifies which execution unit to use for an operation by naming the target unit in the extra column that precedes the operand fields (the column containing D1, D2, etc). To indicate that two or more operations may proceed in parallel, the lines describing the individual operations are "joined" with a parallel bar (as in lines 4 and 5 of Figure 6.21).

```
void example1(float *out, float *input1, float *input2)
{
  int i;

  for(i = 0; i < 100; i++)
   {
      out[i] = input1[i] * input2[i];
   }
}
```

Figure 6.19 Simple for loop in C

```
;
;    serial implementation of loop (26 cycles per iteration)
;
L1:          LDW        *B++,B5     ; load B[i] into B5
             NOP        4           ; wait for load to complete

             LDW        *A++,A4     ; load A[i] into A4
             NOP        4           ; wait for load to complete

             MPYSP      B5,A4,A4    ; A4 = A4 * B5
             NOP        3           ; wait for mult to complete

             STW        A4,*C++     ; store A4 in C[i]
             NOP        4           ; wait for store to complete

             SUB        i,1,i       ; decrement i
        [i]  B          L1          ; if i != 0, goto L1
             NOP        5           ; delay for branch
```

Figure 6.20 Serial assembly language implementation of C loop

```
;  using delay slots and duplicate execution units of the device
;  10 cycles per iteration

L1:          LDW       .D2    *B++,B5     ; load B[i] into B5
||           LDW       .D1    *A++,A4     ; load A[i] into A4

             NOP              2           ; wait load to complete
             SUB       .L2    i,1,i       ; decrement i
        [i]  B         .S1    L1          ; if i != 0, goto L1

             MPYSP     .M1X   B5,A4,A4    ; A4 = A4 * B5
             NOP       3                  ; wait mpy to complete

             STW       .D1    A4,*C++     ; store A4 into C[i]
```

Figure 6.21 A more parallel implementation of the C loop

Reducing Loop Overhead

Loop unrolling is a technique used to increase the number of instructions executed between executions of the loop branch logic. This reduces the number of times the loop branch logic is executed. Since the loop branch logic is overhead, reducing the number of times this has to execute reduces the overhead and makes the loop body, the important part of the structure, run faster. A loop can be unrolled by replicating the loop body a number of times and then changing the termination logic to comprehend the multiple iterations of the loop body (Figure 6.22). The loops in Figures 6.22a and 6.22b each take four cycles to execute, but the loop in Figure 6.22b is doing four times as much work! This is illustrated in Figure 6.23. The assembly language kernel of this loop is shown in Figure 6.23a. The mapping of variables from the loop to the processor is shown in Figure 6.23b. The compiler is able to structure this loop such that all required resources are stored in the register file, and the work is spread across several of the execution units. The work done by cycle for each of these units is shown in Figure 6.23c.

```
for (i = 0;  i < 128;  i ++)
{
        sum1 + = const[i] * input[128 - i];
}
```

a.

```
for (i = 0;  i < 32;  i ++)
{
        sum1 += const[i] * input[128 - i];
        sum2 += const[2*i] * input[128 - (2*i)];
        sum3 += const[3*i] * input[128 - (3*i)];
        sum4 += const[4*i] * input[128 - (4*i)];
}
```

b.

Figure 6.22 Loop unrolling a) a simple loop b) the same loop unrolled 4 times

```
        NOP         1
        SUB     .L2     B0,1,B0

        B       .S2     L3
||      LDW     .D2     *B5--,B4
||      LDW     .D1     *A3++,A0

        ADDSP   .L1     A5,A4,A4
||      MPYSP   .M1X    B4,A0,A5
```

a. the assembly language kernal

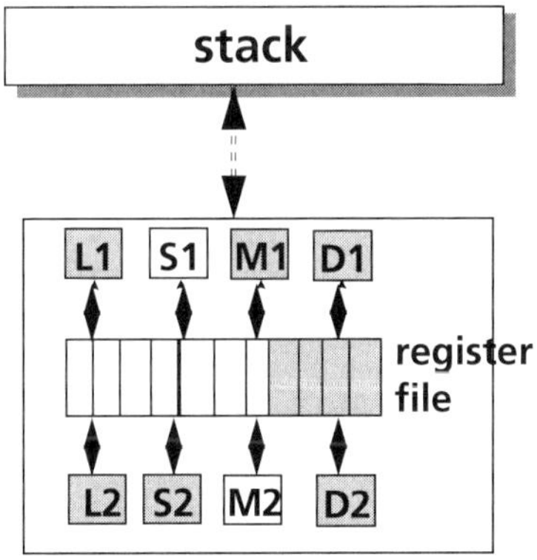

b. the resources used (shaded) on the processor

1 POINT								
CYCLE	D1	S1	L1	M1	D2	S2	L2	M2
1								
2							SUB	
3	LOAD				LOAD	BRANCH		
4			ADD	MPY				

c. resources use by cycle

Figure 6.23 Implementation of a simple loop

Now look at the implementation of the loop unrolled four times in Figure 6.24. Again, only the assembly language for the loop kernel is shown. Notice that more of the register file is being used to store the variables needed in the larger loop kernel. An additional execution unit is also being used, as well as a several bytes from the stack in external memory (Figure 6.24b). Also, the execution unit utilization shown in Figure 6.24c indicates the execution units are being used more efficiently while still maintaining a four cycle latency to complete the loop. This is an example of using all the available resources of the device to gain significant speed improvements. Although the code size looks bigger, it actually runs faster than the loop in Figure 6.22a.

Fitting the Loop to Register Space

Unrolling too much can cause performance problems. In Figure 6.25, the loop is unrolled eight times. At this point, the compiler cannot find enough registers to map all the required variables. When this happens, variables start getting stored on the stack, which is usually in external memory somewhere. This is expensive because instead of a single cycle read, it can now take many cycles to read each of the variables each time it is needed. This causes things to break down, as shown in Figure 6.25. The obvious problems are the number of bytes that are now being stored in external memory (88 vs. 8 before) and the lack of parallelism in the assembly language loop kernel. The actual kernel assembly language was very long and inefficient. A small part of it is shown in Figure 6.25b. Notice the lack of "||" instructions and the new "NOP" instructions. These are, effectively, stalls to the CPU when nothing else can happen. The CPU is waiting for data to arrive from external memory.

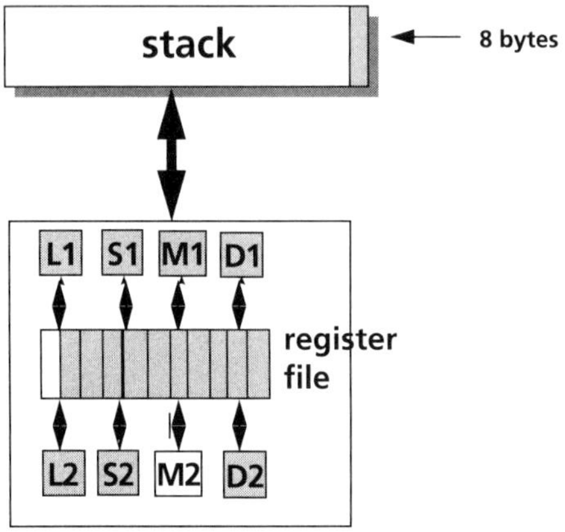

a. the register files and execution unit being used

```
         LDW      .D2      *B2++(32),B0
||       LDW      .D1      *A3--(48),A5

         ADDSP    .L2      B9,B8,B8
||       SUB      .L1      A1,1,A1
||       MPYSP    .M2X     A8,B0,B9
||       LDW      .D2      *B6++(48),B9
||       LDW      .D1      *A6++(64),A8

         ADDSP    .L2      B0,B3,B3
||       B        .S1      L3
||       MPYSP    .M2X     A5,B9,B0
||       LDW      .D1      *A4--(16),A5
||       LDW      .D2      *B4--(64),B0

         ADDSP    .L2      B9,B1,B1
||       ADDSP    .L1      A9,A7,A7
||       MPYSP    .M2X     A5,B7,B9
||       MPYSP    .M1X     B0,A8,A9
||       LDW      .D2      *B5++(16),B7
||       LDW      .D1      *A0--(32),A8
```

b. the assembly language kernal

4 POINT								
CYCLE	D1	S1	L1	M1	D2	S2	L2	M2
1	LOAD				LOAD			
2	LOAD		SUB		LOAD		ADD	MPY
3	LOAD	BRANCH			LOAD		ADD	MPY
4	LOAD		ADD	MPY	LOAD		ADD	MPY

c. utilization of the execution units

Figure 6.24 Implementation of a loop unrolled four times

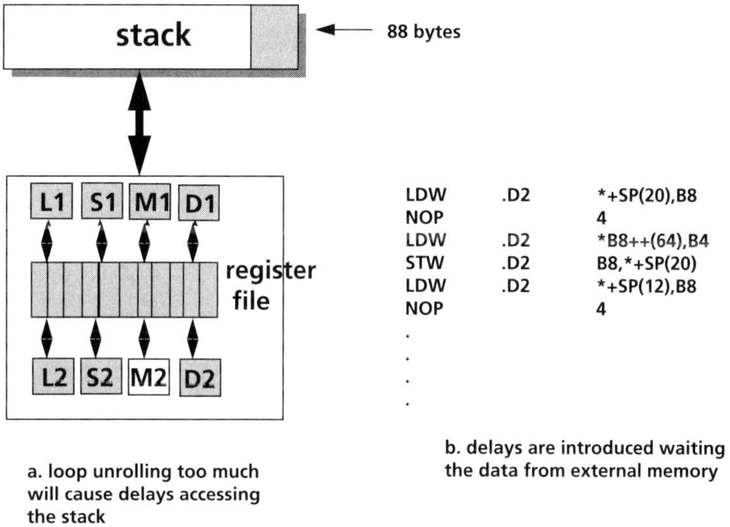

Figure 6.25 Loop unrolled eight times. Too much of a good thing!

Trade-offs

The drawback to loop unrolling is that it uses more registers in the register file as well as execution units. Different registers need to be used for each iteration. Once the available registers are used, the processor starts going to the stack to store required data. Going to the off-chip stack is expensive and may wipe out the gains achieved by unrolling the loop in the first place. Loop unrolling should only be used when the operations in a single iteration of the loop do not use all of the available resources of the processor architecture. Check the assembly language output if you are not sure of this. Another drawback is the code size increase. As you can see in Figure 6.24, the unrolled loop, albeit faster, requires more instructions and, therefore, more memory.

Software Pipelining

One of the best performance strategies for the DSP programmer is writing code that can be pipelined efficiently by the compiler. Software pipelining is an optimization strategy to schedule loops and functional units efficiently. In software pipelining, operations from different iterations of a software loop are performed in parallel. In each iteration, intermediate results generated by the previous iteration are used. Each iteration will also perform operations whose intermediate results will be used in the next iteration. This technique produces highly optimized code through maximum use of the processor functional units. The advantage to this approach is that most of the scheduling associated with software pipelining is performed by the compiler and not by the programmer (unless the programmer is writing code at the assembly language level). There are certain conditions which must be satisfied for this to work properly and we will talk about that shortly.

DSPs may have multiple functional units available for use while a piece of code is executing. In the case of the TMS320C6X family of VLIW DSPs, there are eight functional units that can be used at the same time, if the compiler can determine how to utilize all of them efficiently. Sometimes, subtle changes in the way the C code is structured can make all the difference. In software pipelining, multiple iterations of a loop are scheduled to execute in parallel. The loop is reorganized so that each iteration in the pipelined code is made from instruction sequences selected from different iterations in the original loop. In the example in Figure 6.26, a five-stage loop with three iterations is shown. As we discussed earlier, the initial period (cycles n and n+1), called the *prolog*, is when the pipes are being "primed" or initially loaded with operations. Cycles n+2 to n+4 are the actual pipelined section of the code. It is in this section that the processor is performing three different operations (C, B, and A) for three different loops (1, 2 and 3). The epilog section is where the last remaining instructions are performed before exiting the loop. This is an example of a fully utilized set of pipelines that produces the fastest, most efficient code.

We saw earlier how loop unrolling offers speed improvements over simple loops. Software pipelining can be faster than loop unrolling for certain sections of code because, with loop unrolling, the prolog and epilog are only performed once (Figure 6.27).

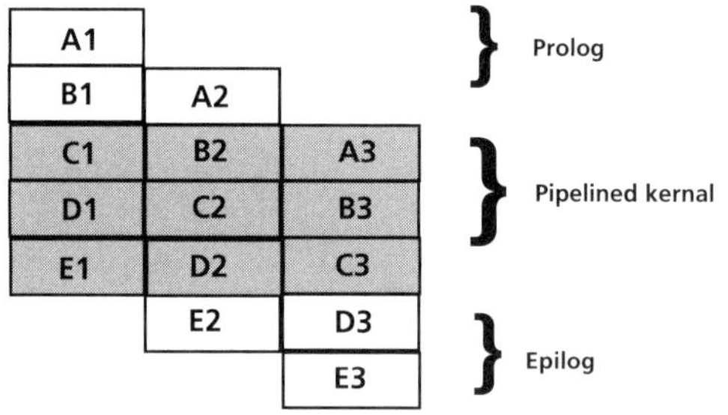

Figure 6.26 A five stage pipe that is software pipelined

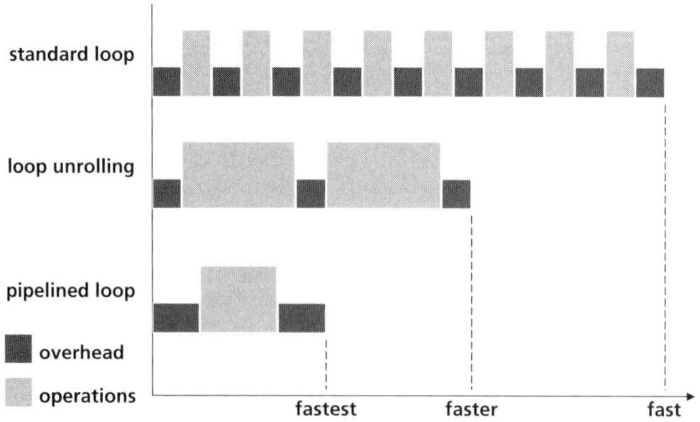

Figure 6.27 Standard loop overhead vs. loop unrolling and software pipelining. The standard loop uses a loop check each iteration through the loop. This is considered overhead. Loop unrolling still checks the loop count but less often. Software pipelining substitutes the loop check with prolog and epilog operations that prime and empty a pipelined loop operation which performs many iterations of the loop in parallel.

An Example

To demonstrate this technique, lets look at an example. In Figure 6.28, a simple loop is implemented in C. This loop simply multiplies elements from two arrays and stores the result in another array.

```
void example1(float *out, float *input1, float *input2)
{
 int i;

 for(i = 0; i < 100; i++)
 {
   out[i] = input1[i] * input2[];
 }
}
```

Figure 6.28 A simple loop in C

```
1    ;
2    ;  serial implementation of loop (26 cycles per iteration)
3    ;
4    L1:        LDW      *B++,B5      ;load B[i] into B5
5               NOP      4            ; wait for load to complete
6
7               LDW      *A++,A4      ; load A[i] into A4
8               NOP      4            ; wait for load to complete
9
10              MPYSP    B5,A4,A4     ; A4 = A4 * B5
11              NOP      3            ; wait formult to complete
12
13              STW      A4,*C++      ; store A4 in C[i]
14              NOP      4            ; wait got store to complete
15
16              SUB      i,1,i        ; decrement i
17        [i]   B        L1           ; if i != 0, goto L1
18              NOP      5            ; delay for branch
```

Figure 6.29 Assembly language output of the simple loop

A serial implementation

The assembly language output for this simple loop is shown in Figure 6.29. This is a serial assembly language implementation of the C loop; serial in the sense that there is no real parallelism (use of the other processor resources) taking place in the code. This is easily to see by the abundance of NOP operations in the code. These NOPs are instructions that do nothing but wait and burn CPU cycles. These are required and inserted in the code when the CPU is waiting for memory fetches or writes to complete. Since a load operation takes five CPU cycles to complete, the load word operation in line 4 (LDW) is followed by a NOP with a length of 4 indicating that the CPU must now wait four additional cycles for the data to arrive in the appropriate register before it can be used.

These NOPs are extremely inefficient and the programmer should endeavor to remove as many of these delay slots as possible to improve performance of the system. One way of doing this is to take advantage of the other DSP functional resources.

A minimally parallel implementation

A more parallel implementation of the same C loop is shown in Figure 6.30. In this implementation, the compiler is able to use more of the functional units on the DSP. Line 4 shows a load operation, similar to the previous example. However, this load is explicitly loaded into the D2 functional unit on the DSP. This instruction is followed by another load into the D1 unit of the DSP. What is going on here? While the first load is taking place, moving data into the D2 unit, another load is initiated that loads data into the D1 register. These operations can be performed during the same clock cycle because the destination of the data is different. These values can be preloaded for use later and we do not have to waste as many clock cycles waiting for data to move. As you can see in this code listing, there are now only two NOP cycles required instead of four. This is a step in the right direction. The "||" symbol means that the two loads are performed during the same clock cycle.

```
1       ; using delay slots and duplicate execution units of the device
2       ; 10 cycles per iteration
3
4       L1:             LDW       .D2     *B++,B5      ;load B[i] into B5
5       ||              LDW       .D1     *A++,A4      ;load A[i] into A4
6
7                       NOP               2            ; wait load to complete
8                       SUB       .L2     i,1,i        ;decrement i
9             [i]       B         .S1     L1           ; if i != 0, goto L1
10
11                      MPYSP   .M1X      B5,A4,A4     ; A4 = A4 * B5
12                      NOP       3                    ; wait mpy to complete
13
14                      STW       .D1     A4,*C++      ;store A4 into C[i]
```

Figure 6.30 Assembly language output of the simple loop exploiting the parallel orthogonal execution units of the DSP

Compiler-generated pipeline

Figure 6.31 shows the same sample piece of C code and the corresponding assembly language output. In this case, the compiler was asked (via a compile switch) to attempt to pipeline the code. This is evident by the piped loop prolog and piped loop kernel sections in the assembly language output. In this case, the pipelined code is not as good as it could be. Inefficient code can be located by looking for how many NOPs there are in the piped loop kernel of the code. In this case the piped loop kernel has a total of five NOP cycles, two in line 16, and three in line 20. This loop takes a total of ten cycles to execute. The NOPs are the first indication that a more efficient loop may be possible. But how short can this loop be? One way to estimate the minimum loop size is to determine what execution unit is being used the most. In this example, the D unit is used more than any other unit, a total of three times (lines 14, 15 and 21). There are two sides to this VLIW device, enabling each unit to be used twice (D1 and D2) per clock for a minimum two clock loop; two D operations in one clock and one D unit in the second clock. The compiler was smart enough to use the D units on both sides of the pipe (lines 14 and 15), enabling it to parallelize the instructions and only use one clock. It should be possible to perform other instructions while waiting for the loads to complete, instead of delaying with NOPs.

```
void example1(float *out, float *input1, float *input2)
{
  int i;

  for(i = 0; i < 100; i++)
   {
      out[i] = input1[i] * input2[i];
   }
}
```

```
1    _example1:
2    ;** ----------------------------------------------------------*
3                MVK       .S2        0x64,B0
4
5                MVC       .S2        CSR,B6
6    ||          MV        .L1X       B4,A3
7    ||          MV        .L2X       A6,B5

8                AND       .L1X       -2,B6,A0
9                MVC       .S2X       A0,CSR
10   ;** ----------------------------------------------------------*
11   L11:          ; PIPED LOOP PROLOG
12   ;** ----------------------------------------------------------*
13   L12:          ; PIPED LOOP KERNEL

14               LDW       .D2        *B5++,B4      ;
15   ||          LDW       .D1        *A3++,A0      ;

16               NOP                    2
17   [ B0]       SUB       .L2        B0,1,B0       ;
18   [ B0]       B         .S2        L12           ;
19               MPYSP     .M1X       B4,A0,A0      ;
20               NOP                  3
21               STW       .D1        A0,*A4++      ;
22   ;** ---------------------------------------------------------*
23               MVC       .S2        B6,CSR
24               B         .S2        B3
25               NOP                  5
26               ; BRANCH OCCURS
```

Figure 6.31 C example and the corresponding pipelined assembly language output

An implementation with restrict keyword

In the simple "for" loop, it is apparent that the inputs are not dependent on the output. In other words, there are no dependencies. But the compiler does not know that. Compilers are generally pessimistic creatures. They will not optimize something if the situation is not totally understood. The compiler takes the conservative approach and assumes the inputs can be dependent on the previous output each time through the loop. If it is known that the inputs are not dependent on the output, we can hint to the compiler by declaring the input1 and input2 as "restrict," indicating that these fields will not change. This is a trigger for enabling software pipelining and saving throughput. This C code is shown in Figure 6.32 with the corresponding assembly language.

```
void example2(float *out, restrict float *input1, restrict float *input2)
{
  int i;

  for(i = 0; i < 100; i++)
   {
      out[i] = input1[i] * input2[i];
   }
}
```

```
_example2:
;** ---------------------------------------------------------------*
            MVK         .S2         0x64,B0

            MVC         .S2         CSR,B6
||          MV          .L1X        B4,A3
||          MV          .L2X        A6,B5

            AND         .L1X        -2,B6,A0

            MVC         .S2X        A0,CSR
||          SUB         .L2         B0,4,B0

;** ---------------------------------------------------------------*
L8:         ; PIPED LOOP PROLOG

            LDW         .D2         *B5++,B4      ;
||          LDW         .D1         *A3++,A0      ;

            NOP                     1

            LDW         .D2         *B5++,B4      ;@
||          LDW         .D1         *A3++,A0      ;@

    [ B0]   SUB         .L2         B0,1,B0       ;
```

```

   [ B0]    B          .S2        L9          ;
||          LDW        .D2        *B5++,B4    ;@@
||          LDW        .D1        *A3++,A0    ;@@

            MPYSP      .M1X       B4,A0,A5    ;
|| [ B0]    SUB        .L2        B0,1,B0     ;@

   [ B0]    B          .S2        L9          ;@
||          LDW        .D2        *B5++,B4    ;@@@
||          LDW        .D1        *A3++,A0    ;@@@

            MPYSP      .M1X       B4,A0,A5    ;@
|| [ B0]    SUB        .L2        B0,1,B0     ;@@

;** ------------------------------------------------------------*
L9:         ; PIPED LOOP KERNEL

   [ B0]    B          .S2        L9          ;@@
||          LDW        .D2        *B5++,B4    ;@@@@
||          LDW        .D1        *A3++,A0    ;@@@@

            STW        .D1        A5,*A4++    ;
||          MPYSP      .M1X       B4,A0,A5    ;@@
|| [ B0]    SUB        .L2        B0,1,B0     ;@@@

;** ------------------------------------------------------------*
L10:         ; PIPED LOOP EPILOG
            NOP                   1

            STW        .D1        A5,*A4++    ;@
||          MPYSP      .M1X       B4,A0,A5    ;@@@

            NOP                   1

            STW        .D1        A5,*A4++    ;@@
||          MPYSP      .M1X       B4,A0,A5    ;@@@@

            NOP                  1
            STW        .D1       A5,*A4++     ;@@@
            NOP                  1
            STW        .D1       A5,*A4++     ;@@@@
;** ------------------------------------------------------------*
             MVC       .S2        B6,CSR
             B         .S2        B3
             NOP                  5
            ; BRANCH OCCURS
```

Figure 6.32 Corresponding pipelined assembly language output

There are a few things to notice in looking at this assembly language. First, the piped loop kernel has become smaller. In fact, the loop is now only two cycles long. Lines 44–47 are all executed in one cycle (the parallel instructions are indicated by the || symbol) and lines 48–50 are executed in the second cycle of the loop. The compiler, with the additional dependency information we supplied with the "restrict" declaration, has been able to take advantage of the parallelism in the execution units to schedule the inner part of the loop very efficiently. But this comes at a price. The prolog and epilog portions of the code are much larger now. Tighter piped kernels will require more priming operations to coordinate all of the execution based on the various instruction and branching delays. But once primed, the kernel loop executes extremely fast, performing operations on various iterations of the loop. The goal of software pipelining is, like we mentioned earlier, to make the common case fast. The kernel is the common case in this example, and we have made it very fast. Pipelined code may not be worth doing for loops with a small loop count. But for loops with a large loop count, executing thousands of times, software pipelining is the only way to go.

In the two cycles the piped kernel takes to execute, there are a lot of things going on. The right hand column in the assembly listing indicates what iteration is being performed by each instruction. Each "@" symbol is a iteration count. So, in this kernel, line 44 is performing a branch for iteration n+2, lines 45 and 46 are performing loads for iteration n+4, line 48 is storing a result for iteration n, line 49 is performing a multiply for iteration n+2, and line 50 is performing a subtraction for iteration n+3, all in two cycles! The epilog is completing the operations once the piped kernel stops executing. The compiler was able to make the loop two cycles long, which is what we predicted by looking at the inefficient version of the code.

The code size for a pipelined function becomes larger, as is obvious by looking at the code produced. This is one of the trade-offs for speed that the programmer must make.

In summary, when processing arrays of data (which is common in many DSP applications) the programmer must inform the compiler when arrays are not dependent on each other. The compiler must assume that the data arrays can be anywhere in memory, even overlapping each other. Unless informed of array independence, the compiler will assume the next load operation requires the previous store operation to complete (as to not load stale data). Independent data structures allows the compiler to structure the code to load from the input array before storing the last output. Basically, if two arrays are not pointing to the same place in memory, using the "restrict" keyword to indicate this independence will improve performance. Another term for this technique is *memory disambiguation.*

Enabling Software Pipelining

The compiler must decide what variables to put on the stack (which take longer to access) and which variables to put in the fast on-chip registers. This is part of the register allocator of a compiler. If a loop contains too many operations to make efficient use of the processor registers, the compiler may decide to not pipeline the loop. In cases like that, it may make sense to break up the loop into smaller loops that will enable the compiler to pipeline each of the smaller loops (Figure 6.33).

Instead of:

```
for (expression)
{
        Do A
        Do B
        Do C
        Do D
}
```

Try:

```
for (expression)
        Do A

for (expression)
        Do B

for (expression)
        Do C

for (expression)
        Do D
```

Figure 6.33 Breaking up larger loops into smaller loops may enable each loop to be pipelined more efficiently

The compiler will not attempt to software pipeline a loop when there are not enough resources (execution units, registers, etc) to allow it, or if the compiler determines that it is not worth the effort to pipeline a loop because the benefit does not outweigh the gain (for example, the amount of cycles required to produce the prolog and epilog far outweighs the amount of cycles saved in the loop kernel). But the programmer can intervene in some cases to improve the situation. With careful analysis and structuring of the code, the programmer can make the necessary modification at the high language level to allow certain loops to pipeline. For example, some loops have so many processing requirements inside the loop that the compiler cannot find enough registers and execution units to map all the required data and instructions. When this happens, the compiler will not attempt to pipeline the loop. Also, function calls within a loop will not be pipelined because the compiler has a hard time resolving the function call. Instead, if you want a pipelined loop, replace the function call with an inline expansion of the function.

Interrupts and Pipelined Code

Because an interrupt in the middle of a fully primed pipe destroys the synergy in instruction execution, the compiler may protect a software pipelining operation by disabling interrupts before entering the pipelined section and enabling interrupts on the way out (Figure 6.34). Lines 11 and 69 of Figure 6.32 show interrupts being disabled prior to the prolog and enabled again just after completing the epilog. This means that the price of the efficiency in software pipelining is paid for in a nonpreemptible section of code. The programmer must be able to determine the impact of sections of nonpreemptible code on real time performance. This is not a problem for single task applications. But it may have an impact on systems built using a tasking architecture. Each of the software pipelined sections must be considered a blocking term in the overall tasking equation (see the chapter on real-time operating systems for a discussion on schedulability analysis).

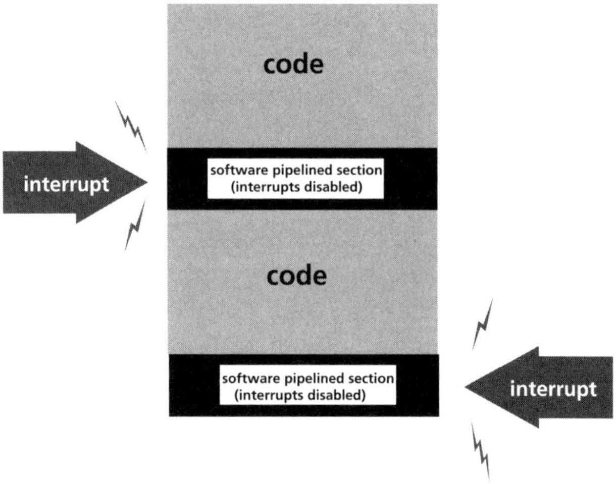

Figure 6.34 Interrupts may be disabled during a software pipelined section of code

More on DSP Compilers and Optimization

Not too long ago, it was an unwritten rule that writing programs in assembly would usually result in better performance than writing in higher level languages like C or C++. The early "optimizing" compilers solved the problem of "optimization" at too general and simplistic a level. The results were not as good as a good assembly language programmer. Compilers have gotten much better and today there are very specific high performance optimizations that compete well with even the best assembly language programmers.

Optimizing compilers perform sophisticated program analysis including intra-procedural and interprocedural analysis. These compilers also perform data flow and control flow analysis as well as dependence analysis and often require provably correct methods for modifying or transforming code. Much of this analysis is to prove that the transformation is correct in the general sense. Many optimization strategies used in DSP compilers are strongly heuristic[3].

Compiler Architecture and Flow

The general architecture of a modern compiler is shown in Figure 6.35. The front end of the compiler reads in the DSP source code, determines whether the input is legal, detects and reports errors, obtains the meaning of the input, and creates an intermediate representation of the source code. The intermediate stage of the compiler is called the *optimizer*. The optimizer performs a set of optimization techniques on the code including:

- Control flow optimizations.
- Local optimizations.
- Global optimizations.

The back end of the compiler generates the target code from the intermediate code, performs the optimizations on the code for the specific target machine, and performs instruction selection, instruction scheduling and register allocation to minimize memory bandwidth, and finally outputs object code to be run on the target.

[3] Heuristics involves problem-solving by experimental and especially trial-and-error methods or relating to exploratory problem-solving techniques that utilize self-educating techniques (as the evaluation of feedback) to improve performance (*Webster's English Language Dictionary*).

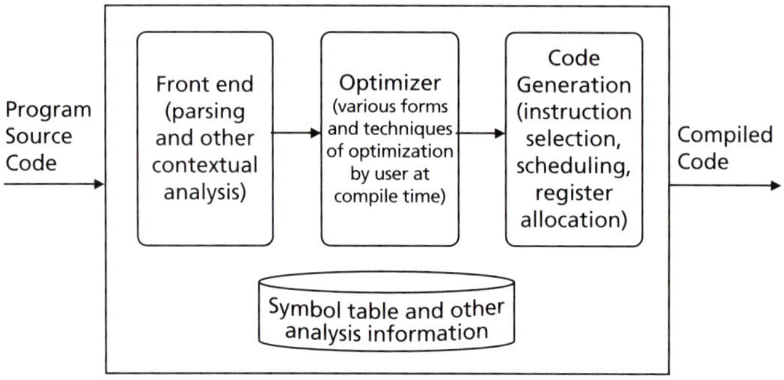

Figure 6.35 General architecture of a compiler

Compiler Optimizations

Compilers perform what are called *machine independent* and *machine dependent* optimizations. Machine independent optimizations are those that are not dependent on the architecture of the device. Examples of this are:

- *Branch optimization* – This is a technique that rearranges the program code to minimize branching logic and to combine physically separate blocks of code.
- *Loop invariant code motion* – If variables used in a computation within a loop are not altered within the loop, the calculation can be performed outside of the loop and the results used within the loop (Figure 6.2).

```
do i = 1,100                    j = 100
        j = 10                  do i = 1,100
        x(i) = x(i) + j                 x(i) = x(i) + j
enddo                           enddo
.......                         .......
........                        ........
```

Figure 6.36 Example of code motion optimization

- *Loop unrolling* – In this technique, the compiler replicates the loop's body and adjusts the logic that controls the number of iterations performed. Now the code effectively performs the same useful work with less comparisons and branches. The compiler may or may not know the loop count. This approach reduces the total number of operations but also increases code size. This may or may not be an issue. If the resultant code size is too large for the device cache (if one is being used), then the resulting cache miss penalty can overcome any benefit in loop overhead. An example of loop unrolling was discussed earlier in the chapter.
- *Common subexpression elimination* – In common expressions, the same value is recalculated in a subsequent expression. The duplicate expression can be eliminated by using the previous value. The goal is to eliminate redundant or multiple com-

putations. The compiler will compute the value once and store it in a temporary variable for subsequent reuse.

- *Constant propagation* – In this technique, constants used in an expression are combined, and new constants are generated. Some implicit conversions between integers and floating-point types may also be done. The goal is to save memory by the removal of these equivalent variables.
- *Dead code elimination* – This approach attempts to eliminate code that cannot be reached or where the results are not subsequently used.
- *Dead store elimination* – This optimization technique will try to eliminate stores to memory when the value stored will never be referenced in the future. An example of this approach is code that performs two stores to the same location without having an intervening load. The compiler will remove the first store because it is unnecessary.
- *Global register allocation* – This optimization technique allocates variables and expressions to available hardware registers using a "graph coloring" algorithm[4].
- *Inlining* – Inlining replaces function calls with actual program code (Figure 6.3). This can speed up the execution of the software by not having to perform function calls with the associated overhead. The disadvantage to inlining is that the program size will increase.

```
do i = 1,n
        j = k(i)
        call subroutine(a(i),j)
         call subroutine(b(i),j)
         call subroutine(c(i),j)
........

subroutine INL(x,y)
        temp1 = x * y
        temp2 = x / y
        temp3 = temp1 + temp2
end
```

```
do i = 1,n
        j = k(i)
        temp1 = a(i) * y
        temp2 = a(i) / y
        temp3 = temp1 + temp2
        temp1 = b(i) * y
        temp2 = b(i) / y
        temp3 = temp1 + temp2
        temp1 = c(i) * y
        temp2 = c(i) / y
        temp3 = temp1 + temp2
```

Figure 6.37 Inlining replaces function calls with actual code which increases performance but may increase program size

[4] The problem of assigning data values to registers is a key challenge for compiler engineers. This problem can be converted into one of graph-coloring. In this approach, attempting to color a graph with N colors is equivalent to attempting to allocate data into N registers. Graph coloring is then the partitioning of the vertices of a graph into a minimum number of independent sets.

- *Strength reduction* – The basic approach with this form of optimization is to use cheaper operations instead of more expensive ones. A simple example of this is to use a compound assignment operator instead of an expanded one, since fewer instructions are needed:

  Instead of:

  ```
  for ( i=0; i < array_length; i++)
          a[i] = a[i] + constant;
  ```

  Use:

  ```
  for ( i=0; i < array_length; i++)
          a[i] += constant;
  ```

Another example of a strength reduction optimization is using shifts instead of multiplication by powers of two.

- *Alias disambiguation* – Aliasing occurs if two or more symbols, pointer references, or structure references refer to the same memory location. This situation can prevent the compiler from retaining values in registers because it cannot be certain that the register and memory continue to hold the same values over time. Alias disambiguation is a compiler technique that determines when two pointer expressions cannot point to the same location. This allows the compiler to freely optimize these expressions.
- *Inline expansion of runtime-support library functions* – This optimization technique replaces calls to small functions with inline code. This saves the overhead associated with a function call and provides increased opportunities to apply other optimizations.

The programmer has control over the various optimization approaches in a compiler, from aggressive to none at all. Some specific controls are discussed in the next section.

Machine dependent optimizations are those that require some knowledge of the target machine in order to perform the optimization. Examples of this type of optimization include:

- *Implementing special features* – This includes instruction selection techniques that produce efficient code, selecting an efficient combination of machine dependent instructions that implement the intermediate representation in the compiler.
- *Latency* – This involves selecting the right instruction schedules to implement the selected instructions for the target machine. There are a large number of different schedules that can be chosen and the compiler must select one that gives an efficient overall schedule for the code.

- *Resources* – This involves register allocation techniques which includes analysis of program variables and selecting the right combination of registers to hold the variables for the optimum amount of time such that the optimal memory bandwidth goals can be met. This technique mainly determines which variables should be in which registers at each point in the program.

Instruction Selection

Instruction selection is important in generating code for a target machine for a number of reasons. As an example, there may be some instructions on the processor that the C compiler cannot implement efficiently. Saturation is a good example. Many DSP applications perform saturation checks on video and image processing applications. To manually write code to saturate requires a lot of code (check sign bits, determine proper limit, etc). Some DSPs, for example, can do a similar operation in one cycle or as part of another instruction (i.e., replace a multiply instruction, MPY with a saturate multiply, SMUL=1). But a compiler is often unable to use these algorithm-specific instuctions which the DSP provides. So the programmer often has to force their use. To get the C compiler to use specific assembly language instructions like this, one approach is to use what are called *intrinsics*. Intrinsics are implemented with assembly language instructions on the target processor. Some examples of DSP intrinsics include:

- short _abs(short src); absolute value
- long _labs(long src); long absolute value
- short _norm(short src); normalization
- long _rnd(long src); rounding
- short _sadd(short src1, short src2); saturated add

One benefit to using intrinsics like this is that they are automatically inlined. Since we want to run a processor instruction directly, we would not want to waste the overhead of doing a call. Since intrinsics also require things like the saturation flag to be set, they may be longer than one processor instruction. Intrinsics are better than using assembly language function calls since the compiler is ignorant of the contents of these assembly language functions and may not be able to make some needed optimizations.

Figure 6.38 is an example of using C code to produce a saturated add function. The resulting assembly language is also shown for a C5x DSP. Notice the amount of C code and assembly code required to implement this basic function.

C Code:

```
int sadd(int a, int b)
{
  int result;
  result = a + b;
  if (((a^b) & 0x8000) ==
  0)
  {
    if ((result ^ a) &
  0x8000)
        result = ( a < 0)
  ? 0x8000 : 0x7FFF;
  }
  return result;
}
```

Compiler Output:

```
_sadd:
   MOV T1, AR1
   XOR T0, T1
   BTST @#15, T1, TC1
   ADD T0, AR1
   BCC L2,TC1
   MOV T0, AR2
   XOR AR1, AR2
   BTST @#15, AR2, TC1
   BCC L2,!TC1
   BCC L1,T0 < #0
   MOV #32767, T0
   B L3
L1:    MOV #-32768, AR1
L2:    MOV AR1, T0
L3:    return
```

Figure 6.38 Code for a saturated add funtion

Now look at the same function in Figure 6.39 implemented with intrinsics. Notice the significant code size reduction using algorithm specific special instructions. The DSP designer should carefully analyze the algorithms required for the application and determine whether a specific DSP or family of DSPs supports the class of algorithm with these special instructions. The use of these special instructions in key areas of the application can have a significant impact on the overall performance of the system.

C Code

```
int sadd(int a, int
  b)
{
  return
  _sadd(a,b);
}
```

Compiler Output

```
_sadd:
  BSET ST3_SATA
  ADD T1, T0
  BCLR ST3_SATA
  return
```

Figure 6.39 Saturated add using DSP Intrinsics

Latency and instruction scheduling

The order in which operations are executed on a DSP has a significant impact on length of time to execute a specific function. Different operations take different lengths of time due to differences in memory access times, and differences in the functional unit of the processor (different functional units in a DSP, for example, may require different lengths of time to complete a specific operation). If the conditions are not right, the processor may delay or stall. The compiler may be able to predict these unfavorable conditions and reorder some of the instructions to get a better schedule. In the worst case, the compiler may have to force the processor to wait for a period of time by inserting delays (sometimes called *NOP* for "no operation") into the instruction stream to force the processor to wait for a cycle or more for something to complete such as a memory transfer.

Optimizing compilers have instruction schedulers to perform one major function; to reorder operations in the compiled code in an attempt to decrease its running time. DSP compilers have sophisticated schedulers that search for the optimal schedule (within reason; the compiler has to eventually terminate and produce something for the programmer!). The main goals of the scheduler are to preserve the meaning of the code (it can't "break" anything), minimize the overall execution time (by avoiding extra register spills to main memory for example), and operate as efficiently as possible from a usability standpoint.

From a DSP standpoint, loops are critical in many embedded DSP applications. Much of the signal processing performed by DSPs is in loops. Optimizing compilers for DSP often contain specialized loop schedulers to optimize this code. One of the most common examples of this is the function of software pipelining.

An example of software pipelining was given earlier in the chapter. Software pipelining is the execution of operations from different iterations of a software loop in parallel. In each iteration, intermediate results generated by the previous iteration are used and operations are also performed whose intermediate results will be used in the next iteration. This produces highly optimized code and makes maximum use of the processor functional units. Software pipelining is implemented by the compiler if the code structure is suited to making these transformations. In other words, the programmer must produce the right code structure to the compiler such that it can recognize the conditions are right to pipeline the loop. For example, when multiplying two arrays inside a loop, the programmer must inform the compiler when the two arrays do not point to the same space in memory. (Compilers must assume arrays can be anywhere in memory, even overlapping one another. Unless informed of array independence, they will assume the next load requires the previous load to complete. By informing the compiler of this independent structure (something as simple as using a keyword in the C code) allows the compiler to load from the input array before storing last output, as shown in the code snippet below where the "restrict" keyword is used to show this independence.

```
void example (float *out, restrict float *input1, restrict float *input2)
{
 int i;
 for (i=0; i<100; i++)
{
out[ i ] = input1[ i ] * input2[ i ];
}
}
```

The primary goal of instruction scheduling is to improve running time of generated code. But be careful how this is measured. For example, measuring the quality of the produced code using a simple measure such as "Instructions per second" is misleading. Although this is a common metric in many advertisements, it may not be indicative of the quality of code produced for the specific application running on the DSP. That is why developers should spend time measuring the time to complete a fixed representative task for the application in question. Using industry benchmarks to measure overall system performance is not a good idea because the information is too specific to be used in a broad sense. In reality there is no single metric that can accurately measure quality of code produced by the compiler. We will discuss this more later.

Register allocation

On-chip DSP registers are the fastest locations in the memory hierarchy (see Figure 6.40). The primary responsibility of the register allocator is to make efficient use of the target registers. The register allocator works with the scheduled instructions generated by the instruction scheduler and finds an optimal arrangement for data and variables in the processors registers to avoid "spilling" data into main memory where it is more expensive to access (performance). By minimizing register spills the compiler will generate higher performing code by eliminating expensive reads and writes to main memory.

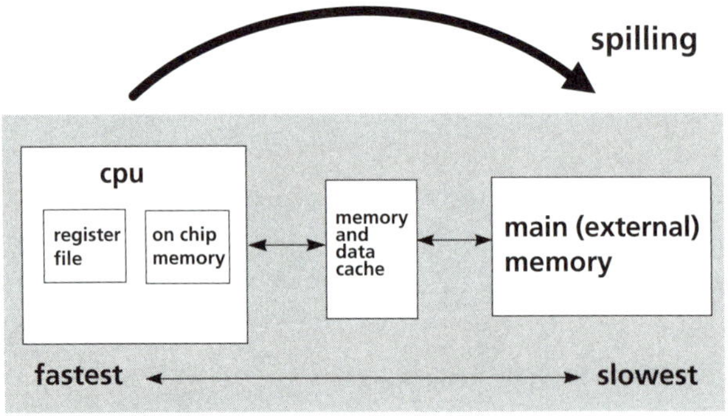

Figure 6.40 Processor memory hierarchy. On-chip registers and memory are the fastest way to access data, typically one cycle per access. Cache systems are used to increase the performance when requiring data from off chip. Main external memory is the slowest.

Sometimes the code structure forces the register allocator to use external memory. For example, the C code in Figure 6.41 (which does not do anything useful) shows what can happen when too many variables are required to perform a specific calculation. This function requires a number of different variables, x0…x9. When the compiler attempts to map these into registers, not all variables are accommodated and the compiler must spill some of the variables to the stack. This is shown in the following code.

```
int foo(int a, int b, int c, int d)
{
   int x0, x1, x2, x3, x4, x5, x6, x7, x8, x9;

   x0 = (a&0xa);
   x1 = (b&0xa) + x0;

   x2 = (c&0xb) + x1;
   x3 = (d&0xb) + x2;

   x4 = (a&0xc) + x3;
   x5 = (b&0xc) + x4;

   x6 = (c&0xd) + x5;
   x7 = (d&0xd) + x6;

   x8 = (a&0xe);
   x9 = (b&0xe);

   return (x0&x1&x2&x3)|(x4&x5)|(x6&x7&x8+x9);
}
```

Figure 6.41 C code snippet with a number of variables

```
;****************************************************************************
;* FUNCTION NAME: foo                                                       *
;*                                                                          *
;*   Regs Modified     : A1,A2,A3,A4,V1,V2,V3,V4,V9,SP,LR,SR                *
;*   Regs Used         : A1,A2,A3,A4,V1,V2,V3,V4,V9,SP,LR,SR                *
;*   Local Frame Size  : 0 Args + 4 Auto + 20 Save = 24 byte                *
;****************************************************************************
;****************************************************************************
        .compiler_opts --abi=ti_arm9_abi --code_state=16
        .state16
;       opt470 rob.if rob.opt

        .sect    ".text"
        .clink
        .thumbfunc _foo
        .state16
        .global  _foo

;****************************************************************************
;* FUNCTION NAME: foo                                                       *
```

```
;*                                                                  *
;*  Regs Modified     : A1,A2,A3,A4,V1,V2,V3,V4,V9,SP,LR,SR         *
;*  Regs Used         : A1,A2,A3,A4,V1,V2,V3,V4,V9,SP,LR,SR         *
;*  Local Frame Size  : 0 Args + 4 Auto + 20 Save = 24 byte         *
;*****************************************************************************
_foo:
;* ------------------------------------------------------------------------*
     PUSH     {A4, V1, V2, V3, V4, LR}
;** 21 -----------------------   C$1 = a&(C$12 = 10);
     MOV    V1, #10          ; |21|
     MOV    V2, A1
     AND    V2, V1
     MOV    V9, V2
;** 21 -----------------------   C$2 = (b&C$12)+C$1;
     AND    V1, A2
     ADD    V3, V2, V1       ; |21|
;** 21 -----------------------   C$3 = (c&(C$11 = 11))+C$2;
     MOV    V1, #11          ; |21|
     MOV    V2, A3
     AND    V2, V1
     ADD    V2, V3, V2       ; |21|
     STR    V2, [SP, #0]  **** this is an example of register spilling. SPindicates the
stack pointer
;** 21 -----------------------   C$4 = (d&C$11)+C$3;
     AND    V1, A4
     ADD    V1, V2, V1       ; |21|
     MOV    LR, V1
;** 21 -----------------------   C$5 = (a&(C$10 = 12))+C$4;
     MOV    V1, #12          ; |21|
     MOV    V4, A1
     AND    V4, V1
     MOV    V2, LR
     ADD    V2, V2, V4       ; |21|
;** 21 -----------------------   C$6 = (b&C$10)+C$5;
     AND    V1, A2
     ADD    V1, V2, V1       ; |21|
;** 21 -----------------------   C$8 = (c&(C$9 = 13))+C$6;
     MOV    V4, #13          ; |21|
     AND    A3, V4
     ADD    A3, V1, A3       ; |21|
;** 21 -----------------------   C$7 = 14;
;** 21 -----------------------   return C$1&C$2&C$3&C$4|C$5&C$6|(a&C$7)+(b
&C$7)&(d&C$9)+C$8&C$8;
```

```
    MOV   V4, V9
    AND   V4, V3
    LDR   V3, [SP, #0]   **** this is an example of register spilling. SP indicates
the stack pointer
    AND   V4, V3         ; |21|
    MOV   V3, LR
    AND   V4, V3         ; |21|
    AND   V2, V1
    ORR   V2, V4         ; |21|
    MOV   V1, #14        ; |21|
    AND   A1, V1
    AND   A2, V1
    ADD   A2, A2, A1     ; |21|
    MOV   A1, #13        ; |21|
    AND   A4, A1
    ADD   A1, A3, A4     ; |21|
    AND   A1, A2         ; |21|
    AND   A1, A3         ; |21|
    ORR   A1, V2
    POP   {A4, V1, V2, V3, V4}
    POP   {A3}
    BX    A3
```

Figure 6.42 Register spilling caused by lack of register resources

One way to reduce or eliminate register spilling is to break larger loops into smaller loops, if the correctness of the code can be maintained. This will enable the register allocator to treat each loop independently, thereby increasing the possibility of finding a suitable register allocation for each of the sub-functions. Figure 6.43 below is a simple example of breaking larger loops into smaller loops to enable this improvement.

```
Instead of:                          Try:

for (expression)                     for (expression) {
{                                            Do A    }
        Do A                         for (expression) {
        Do B                                 Do B    }
        Do C                         for (expression) {
        Do D                                 Do C    }
}                                    for (expression) {
                                             Do D    }
```

Figure 6.43 Some loops are too large for the compiler to pipeline. Reducing the computational load within a loop may allow the compiler to pipeline the smaller loops!

Eliminating embedded loops as shown below can also free up registers and allow for more efficient register allocation in DSP code.

```
for  x
    for  y                           pull out and make
                            -->      a separate function
            do some work             (more registers for
    end                              pipelining)
end
```

Compile Time Options

Many DSP optimizing compilers offer several options for code size vs. performance. Each option allows the programmer to achieve a different level of performance vs. code size. These options allow more and more aggressive code size reduction from the compiler. DSP compilers support different levels of code performance. Each option allows the compiler to perform different DSP optimization techniques, for example:

1. *First level of optimization – Register level optimizations.* This level of optimization may include techniques such as:
   - Simplification of control flow.
   - Allocation of variables to registers.
   - Elimination of unused code.
   - Simplification of expressions and statements.
   - Expand calls to inline functions.
2. *Second level of optimization – Local optimization.* This level of optimization may include techniques such as:
   - Local copy/constant propagation.
   - Removal of unused assignments.
   - Elimination of local common expressions.
3. *Third level of optimization – Global optimization.* This level of optimization may include techniques such as:
   - Loop optimizations.
   - Elimination of global common sub-expressions.
   - Elimination of global unused assignments.
   - Loop unrolling.
4. *Highest level of optimization – File optimizations.* This level of optimization may include techniques such as:
   - Removal of functions that are never called.
   - Simplification of functions with return values that are never used.
   - Inlines calls to small functions (regardless of declaration).
   - Reordering of functions so that attributes of called function are known when caller is optimized.
   - Identification of file-level variable characteristics.

The different levels of optimization and the specific techniques used at each level obviously vary by vendor and compiler. The DSP programmer should study the compiler manual to understand what each level does and experiment to see how the code is modified at each level.

Understanding what the compiler is thinking

There will always be situations where the DSP programmer will need to get pieces of information from the compiler to help understand why an optimization was or was not made. The compiler will normally generate information on each function in the generated assembly code. Information regarding register usage, stack usage, frame size, and how memory is used is usually listed. The programmer can usually get information concerning optimizer decisions (such as to inline or not) by examining information written out by the compiler usually stored in some form of information file (you may have to explicitly ask the compiler to produce this information for you using a compiler command which should be documented in the users manual). This output file contains information as to how optimizations were done or not done (check your compiler—in some situations, asking for this information can sometimes reduce the amount of optimization actually performed).

Programmer Helping Out the Compiler

Part of the job of an optimizing compiler is figuring out what the programmer is trying to do and then helping the programmer achieve that goal as efficiently as possible. That's why well structured code is better for a compiler—its easier to determine what the programmer is trying to do. This process can be aided by certain "hints" the programmer can provide to the compiler. The proper "hints" allow the compiler to be more aggressive in the optimizations it makes to the source code. Using the standard compiler options can only get you so far towards achieving optimal performance. To get even more optimization, the DSP programmer needs to provide helpful information to the compiler and optimizer, using mechanisms called *pragmas*, *intrinsics*, and *keywords*. We already discussed intrinsics as a method of informing the compiler about special instructions to use in the optimization of certain algorithms. These are special function names that map directly to assembly instructions. Pragmas provide extra information about functions and variables to the preprocessor part of the compiler. Helpful keywords are usually type modifiers that give the compiler information about how a variable is used. *Inline* is a special keyword that causes a function to be expanded in place instead of being called.

Pragmas

Pragmas are special instructions to the compiler that tell it how to treat functions and variables. These pragmas must be listed before the function is declared or referenced. Examples of some common pragmas the the TI DSP include:

Pragma	Description
CODE_SECTION(symbol, "section name") [;]	This pragma allocates space for a function in a given memory segment
DATA_SECTION(symbol, "section name") [;]	This pragma allocates space for a data variable in a given memory segment
MUST_ITERATE(min, max, multiple) [;]	This pragma gives the optimizer part of the compiler information on the number of times a loop will repeat
UNROLL (n) [;]	This pragma when specified to the optimizer, tells the compiler how many times to unroll a loop.

An example of a pragma to specify the loop count is shown in Figure 6.44. The first code snippet does not have the pragma inserted and is less efficient than the second snippet which has the pragma for loop count inserted just before the loop in the source code.

Inefficient loop code

C Code	Compiler output
int sum(const short *a, int n)	_sum
{	MOV #0, AR1
int sum = 0;	BCC L2,To<=#0
int i;	SUB #1, To, AR2
for (i=0; i<n; i++)	MOV AR2,CSR
{	RPT CSR
sum += a[i];	ADD *AR0+, AR1,AR1
}	MOV AR1,T0
return sum;	return
}	

Efficient loop code

C Code	Compiler output
int sum(const short *a, int n)	_sum
{	SUB #1, T0, AR2
int sum = 0;	MOV AR2, CSR
int i;	MOV #0, AR1

```
        #pragma MUST_ITERATE(1)
        for (i=0; i<n; i++)
        {
        sum += a[i];
        }
        return sum;
}
```

```
RPT CSR
ADD *AR0+,AR1,AR1
MOV AR1,T0
return
```

```
voidfirFilter(short *x, int f, short *y, int N, int M, QScale)
{ int i, j, sum;
   #pragmaUNROLL(2)                          Unroll outer loop
   for (j = 0; j < M; j++) {
      sum = 0;
      #pragmaUNROLL(2)                       Unroll inner loop
      for (i = 0; i < N; i++)
            sum += x[i + j] *filterCoeff[f][i];
      y[j] = sum >>QScale;
      y[j] &= 0xfffe;
   }}
```

Figure 6.44 Example of using pragmas to improve the efficiency of the code

Intrinsics

Modern optimizing compilers have special functions called *intrinsics* that map directly to inlined DSP instructions. Intrinsics help to optimize code quickly. They are called in the same way as a function call. Usually intrinsics are specified with some leading indicator such as the underscore (_).

As an example, if a developer were to write a routine to perform saturated addition[5] in a higher level language such as C, it would look similar to the code in Figure 6.45. The result assembly language for this routine is shown in Figure 6.46. This is quite messy and inefficient. As an alternative, the developer could write a simple routine calling a built in saturated add routine (Figure 6.47) which is much easier and produces cleaner and more efficient assembly code (Figure 6.48). Figure 6.49 shows some of the available intrinsics for the TMS320C55 DSP. Many modern DSPs support intrinsic libraries of this type.

[5] Saturated add is a process by which two operands are added together and, if the result is an overflow, the result is set to the maximum positive value. This is useful in certain multimedia applications where it is more desirable to have a result that is max positive instead of an overflow which effectively becomes a negative number which looks undesirable in an image, for example.

```
Int saturated_add(int a, int b)
{
   int result;

   result = a + b;

// check to see of a and b have the same sign

   if (((a^b) & 0x8000) == 0)
   {
     // if a and b have the same sign, check for underflow or overflow
     if ((result ^ a) & 0x8000)
     {
        // if the result has a different sign than a then underflow or overflow has
        // occurred. If a is negative, set result to max negative
        // If a is positive, set result to max positive
        result = ( a < 0) ? 0x8000 : 0x7FFF;
     }
   }
return result;
```

Figure 6.45. C code to perform saturated add

```
Saturated_add:
        SP = SP - #1
                                ; End Prolog Code
        AR1 = T1                ; |5|
        AR1 = AR1 + T0          ; |5|
        T1 = T1 ^ T0            ; |7|
        AR2 = T1 & #0x8000      ; |7|
        if (AR2!=#0) goto L2    ; |7|
                                ; branch occurs ; |7|
        AR2 = T0                ; |7|
        AR2 = AR2 ^ AR1         ; |7|
        AR2 = AR2 & #0x8000     ; |7|
        if (AR2==#0) goto L2    ; |7|
                                ; branch occurs ; |7|
        if (T0<#0) goto L1      ; |11|
                                ; branch occurs ; |11|
        T0 = #32767             ; |11|
        goto L3                 ; |11|
                          ; branch occurs ; |11|
L1:
        AR1 = #-32768       ; |11|
L2:
        T0 = AR1            ; |14|
L3:
                          ; Begin Epilog Code
```

```
        SP = SP + #1        ; |14|
return                      ; |14|
                            ; return occurs ; |14|
```

Figure 6.46 TMS320C55 DSP Assembly code for the saturated add routine

```
int sadd(int a, int b)
{
return _sadd(a,b);
}
```

Figure 6.47 TMS320C55 DSP code for the saturated add routine using a single call to an intrinsic

```
Saturated_add:
        SP = SP - #1
                                    ; End Prolog Code
        bit(ST3, #ST3_SATA) = #1
        T0 = T0 + T1                ; |3|
                                    ; Begin Epilog Code
        SP = SP + #1                ; |3|
        bit(ST3, #ST3_SATA) = #0
        return                      ; |3|
                                    ; return occurs ; |3|
```

Figure 6.48 TMS320C55 DSP assembly code for the saturated add routine using a single call to an intrinsic

Intrinsic	**Description**
int _sadd(int src1, int src2);	Adds two 16–bit integers, with SATA set, producing a saturated 16–bit result.
int _smpy(int src1, int src2);	Multiplies src1 and src2, and shifts the result left by 1. Produces a saturated 16–bit result. (SATD and FRCT set).
int _abss(int src);	Creates a saturated 16–bit absolute value. _abss(0x8000) => 0x7FFF (SATA set)
int _smpyr(int src1, int src2);	Multiplies src1 and src2, shifts the result left by 1, and rounds by adding 2 15 to the result. (SATD and FRCT set)
int _norm(int src);	Produces the number of left shifts needed to normalize src.
int _sshl(int src1, int src2);	Shifts src1 left by src2 and produces a 16-bit result. The result is saturated if src2 is less than or equal to 8. (SATD set)
long _lshrs(long src1, int src2);	Shifts src1 right by src2 and produces a 32-bit result. Produces a saturated 32–bit result. (SATD set)
long _laddc(long src1, int src2);	Adds src1, src2, and Carry bit and produces a 32-bit result.

Intrinsic	Description
long _lsubc(long src1, int src2);	Optimizing C Code Subtracts src2 and logical inverse of sign bit from src1, and produces a 32-bit result.

Figure 6.49 Some Intrinsics for the TMS320C55 DSP (courtesy of Texas Instruments)

Keywords

Keywords are type modifiers that give the compiler information about how a variable is used. These can be very helpful in helping the optimizer part of the compiler make optimization decisions. Some common keywords in DSP compilers are:

- *Const* – This keyword defines a variable or pointer as having a constant value. The compiler can allocate the variable or pointer into a special data section which can be placed in ROM. This keyword will also provide information to the compiler that allows it to make more aggressive optimization decisions.
- *Interrupt* – This keyword will force the compiler to save and restore context and enable interrupts on exit from a particular pipelined loop or function.
- *Ioport* – This defines a variable as being in I/O space (this keyword is only used with global or static variables).
- *On-chip* – Using this keyword with a variable or structure will guarantee that that memory location is on-chip.
- *Restrict* – This keyword tells the compiler that only this pointer will access the memory location it points to (i.e., no aliasing of this location). This allows the compiler to perform optimization techniques such as software pipelining.
- *Volatile* – This keyword tells the compiler that this memory location may be changed without compiler's knowledge. Therefore the memory location should not be stored in a temporary register and, instead, be read from memory before each use.

Inlining

For small infrequently called functions it may make sense to paste them directly into code. This eliminates overhead associated with register storage and parameter passing. Inlining uses more program space, but speeds up execution, sometimes significantly. When functions are inlined, the optimizer can optimize the function and the surrounding code in new context. There are two types of inlining; static inlining and normal inlining. With static inlining the function being inlined is only placed in the code where it will be used. Normal inlining also has a function definition which allows the function to be called. The compiler, if specified, will automatically inline functions if the size is small enough. Inlining can also be definition controlled, where the programmer specifies which functions to inline.

Reducing Stack Access Time

When using a real-time operating system (RTOS) for task driven systems, there is overhead to consider that increases with the number of tasks in the system. The overhead in a task switch (or mailbox pend or post, semaphore operation, and so forth) can vary based on where the operating system structures are located. If the structures are in off-chip memory, the access time to perform the operation can be much longer than if the structure was in on-chip memory. The same holds true for the task stack space. If this is in off-chip memory, the performance suffers proportionally to the number of times the stack has to be accessed.

One solution is to allocate the stack in on-chip memory. If the stack is small enough, this may be a viable thing to do. But if there are many tasks in the system, there will not be enough on-chip memory to store all of the task stacks. However, special code can be written to move the stack on chip when it is needed the most. Before the task (or function) is complete, the stack can be moved back off chip. Figure 6.52 shows the code to do this. Figure 6.53 is a diagrammatic explanation of the steps to perform this operation. The steps are as follows:

1. Compile the C code to get a .asm file.
2. Modify the .asm file with the code in the example.
3. Assemble the new .asm file.
4. Link the system.

You need to be careful when doing this. This type of modification should not be done in a function that calls other functions. Also, interrupts should be disabled when performing this type of operation. Finally, the programmer needs to ensure that the secondary stack in on-chip memory does not grow too large, overwriting other data in on-chip memory!

```
SSP      .set       0x80001000
SSP2     .set       0x80000FFC

         MVK        SSP, A0
||       MVK        SSP2,B0

         MVKH               SSP,A0
||       MVKH              SSP2,B0

         STW.D1      SP,*A0
||       MV         .L2      B0,SP
```

Figure 6.52 Modifying the stack pointer to point to on-chip memory

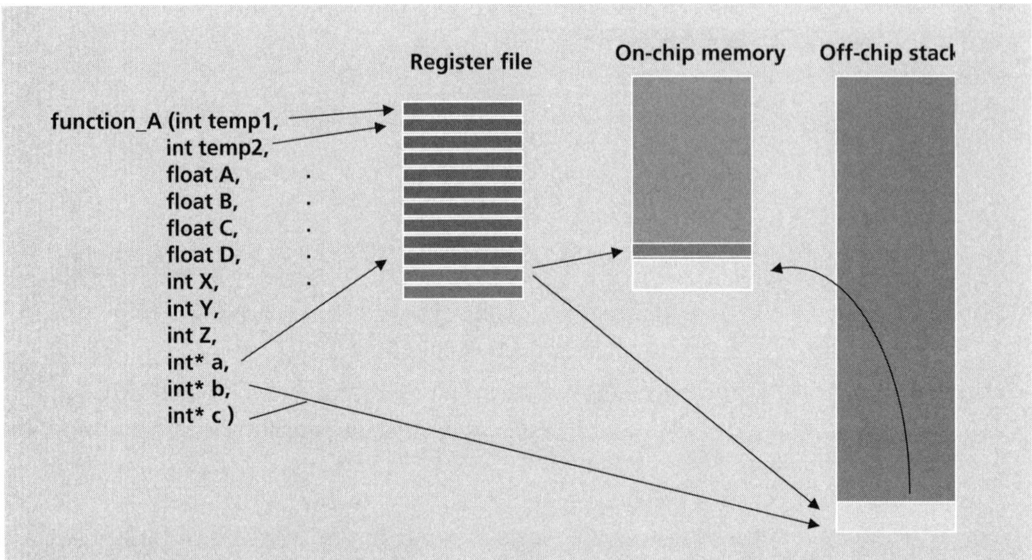

Figure 6.53 Moving the stack pointer on chip to increase performance in a DSP application

Compilers Helping Out the Programmer

Compilers do their best to optimize applications based on the wishes of their programmers. Compilers also produce output that documents the decisions they were able to make or not make based on the specific source code provided to them by the programmer. (See Figure 6.54) By analyzing this output, the programmer can understand the specific constraints and decisions and make appropriate adjustments in the source code to improve the performance of the compiler. In other words, if the programmer understands the thought process of the compilation process, they are then able to re-orient the application to be more consistent with that thought process. The information below is an example of output generated by the compiler that can be analyzed by the programmer. This output can come in various forms, a simple text output file or a more fancy user interface to guide the process of analyzing the compiled output. A more detailed discussion of compiler tuning tools to help this analysis process is given in Appendix B.

```
;*----------------------------------------------------------------------------*
;*   SOFTWARE PIPELINE INFORMATION
;*
;*      Known Minimum Trip Count         : 1
;*      Known Max Trip Count Factor      : 1
;*      Loop Carried Dependency Bound(^) : 0
;*      Unpartitioned Resource Bound     : 1
;*      Partitioned Resource Bound(*)    : 1
;*      Resource Partition:
;*                                A-side   B-side
;*      .L units                    0        0
;*      .S units                    0        1*
;*      .D units                    1*       1*
;*      .M units                    1*       0
;*      .X cross paths              1*       0
;*      .T address paths            1*       1*
;*      Long read paths             0        0
;*      Long write paths            0        0
;*      Logical  ops (.LS)          0        0     (.L or .S unit)
;*      Addition ops (.LSD)         1        1     (.L or .S or .D unit)
;*      Bound(.L .S .LS)            0        1*
;*      Bound(.L .S .D .LS .LSD)    1*       1*
;*
;*      Searching for software pipeline schedule at ...
;*         ii = 1  Schedule found with 8 iterations in parallel
;*      done
;*
;*      Collapsed epilog stages     : 7
;*      Prolog not entirely removed
;*      Collapsed prolog stages     : 2
;*
;*      Minimum required memory pad : 14 bytes
;*
;*      Minimum safe trip count     : 1
;*----------------------------------------------------------------------------*
```

Key Information for Loop

ii = 1 (iteration interval = 1 cycle)
Means: Single Cycle Inner Loop

B side .M unit not used
Means: Only one MPY per cycle

Figure 6.54 Compiler output can be used to diagnose compiler optimization results

Summary of Coding Guidelines

Here is a summary of guidelines that the DSP programmer can use to produce the most highly optimized code for the DSP application. Many of these recommendations are general to all DSP compilers. The recommendation is to develop a list like the one below for whatever DSP device and compiler is being used. It will provide a useful reference for the DSP programming team during the software development phase of a DSP project.

General programming guidelines

1. Avoid removing registers for C compiler usage. Otherwise valuable compiler resources are being thrown away. There are some cases when it makes sense to use these resources. For example, it is acceptable to preserve a register for an interrupt routine.
2. To optimize functions selectively, place in separate files. This lets the programmer adjust the level of optimization on a file-specific basis.
3. Use the least possible number of *volatile* variables. The compiler cannot allocate registers for these, and also can't inline when variables are declared with the volatile keyword.

Variable declaration

1. Local variables/pointers are preferred instead of globals. The compiler uses stack-relative addressing for globals, which is not as efficient. If the programmer will frequently be using a global variable in a function, it is better to assign the global to a local variable and then use it.
2. Declare globals in file where they are used the most.
3. Allocate most often used elements of a structure, array or bit-field in the first element, lowest address or LSB; this eliminates the need for extra bytes to specify the address offset.
4. Use unsigned variables instead of int wherever possible; this provides a larger dynamic range and gives extra information to the compiler for optimization.

Variable declaration (data types)

1. Pay attention to data type significance; The better the information provided to the compiler, the better the efficiency of the resulting code.
2. Only use casting if absolutely necessary, casting can use extra cycles, and can invoke wrong RTS functions if done wrong.
3. Avoid common mistakes in data type assumptions; Avoid code that assumes *int* and *long* are the same type. Also, use *int* for fixed-point arithmetic, since *long* requires a call to a library which is less efficient. Also avoid code that assumes *char* is 8 bits or *long long* is 64 bits for the same reasons.
4. May be more convenient to define your own data types. *Int16* for 16-bit integer (*int*) and *Int32* for 32-bit integer (*long*). Experiment and see what is best for your DSP device and application.

Initialization of variables

1. Initialize global variables with constants at load time. This eliminates the need to have code copy values over at run-time.
2. When assigning the same values to global variables, rearrange code if it makes sense to do so. For example use a=b=c=3; instead of a=3; b-3;c=3). The first uses a register to store the same value to all, the second produces 3 separate long stores;

MOV #3, AR1	MOV #3, *(#_a)
MOV AR1, *(#_a)	MOV #3, *(#_b)
MOV AR1, *(#_b)	MOV #3, *(#_c)
MOV AR1, *(#_c)	
(17 bytes)	(18 bytes)

3. Memory alignment requirements and stack management
4. Group all like data declarations together. The compiler will usually align a 32-bit data on even boundary, so it will pad an extra 16-bit word in if needed.

5. Use the .align linker directive to guarantee stack alignment on even address. Since the compiler needs 32-bit data aligned on an even boundary, it starts the stack on an even boundary.

Loops

1. Split up loops comprised of two unrelated operations.
2. Avoid function calls and control statements inside loops; the compiler needs to preserve loop context in case of a call to a function. By taking function calls and control statements outside a loop if possible, the compiler can take advantage of the special looping optimizations in the hardware (for example the localrepeat and blockrepeat in the TI DSP) to further optimize the loop (Figure 6.53).

```
for (expression)
{
        Do A
        Call X   <----   All function calls in inner loops must be
        Do C             inlinedinto the calling function!!
}
```

Figure 6.53 Do not call functions in inner loops of performance critical code

3. Keep loop code small to enable compiler use of local repeat optimizations.
4. Avoid deeply nested loops; more deeply nested loops use less efficient types of looping.
5. Use an *int* or *unsigned int* instead of *long* for loop counter. DSP hardware generally uses a 16-bit register for a counter.
6. Use pragmas (if available) to give the compiler better information about loop counts.

Control code

1. The DSP compiler may generate similar code for nested if-then-else and switch-case statements if the number of cases is less than eight. If greater than eight, the compiler will generate a .switch label section.
2. For highly dense compare code, use switch statements instead of if-then-else.
3. Place the most common case at the start, since the compiler checks in order.
4. For single conditionals, it is always best to test against 0 instead of !0; For example, 'if (a==0)' produces more efficient code than 'if (a!=1)'.

Functions

1. When a function is only called by other functions in same file, make it a *static* function. This will allow the compiler to inline functions better.
2. When a global variable is only used by functions in the same file, make it a *static* variable.
3. Group minor functions in a single file with functions that use them. This makes file-level optimization better.

4. Too many parameters in function calls become inefficient; Once DSP registers are used up, the rest of the parameters go on the stack. Accessing variables from the stack is very inefficient.
5. Parameters that are used frequently in the subroutine should be passed in registers.

Intrinsics

1. There are some instructions on a DSP that the C compiler cannot implement efficiently. For example, the saturation function is hard to implement using standard instructions on many DSPs. To saturate manually requires a lot of code (check sign bits, determine proper limit, etc). DSPs that support specific instrinsics like saturate will allow the DSP to execute the function much more efficiently.

When developing an application, it is very easy (and sometimes even required) to use generic routines to do various computations. Many time the application developer does not realize how much overhead can be involved in using these generic routines. Often times a more generalized version of an algorithm or function is used because of simple availability instead of creating a more specialized version that better fits the specific need. Creating large numbers of specialized routines is generally a bad programming style as well as a maintenance headache. But strategic use of specialized routines can greatly improve performance in high performance code segments.

Use libraries

2. Some optimizations are more macro or global level optimizations. These optimizations are performed at the algorithm level. This is somewhat unique to DSP where there are many common routines such as FFT, FIR filters, IIR filters, and so on. Eventually, just about every DSP developer will be required to use one of these functions in the development of a DSP application. For common functions used in DSP, vendors have developed highly efficient implementations of these algorithms that can be easily reused. Many of these algorithms are implemented in C and are tested against standards and well documented. Examples include:
   - FFT
   - Filtering and convolution
   - Adaptive filtering
   - Correlation
   - Trigonometric (i.e., sine)
   - Math (, max, log, div)
   - Matrix computations

Although many of these algorithms are very common routine in DSP, they can be complex to implement. Writing one from scratch would require an in-depth knowledge of how the algorithm works (for example an FFT), in-depth knowledge of the DSP architecture in order to optimize the algorithm, possibly expertise at assembly coding, which is hard to find, and time to get everything working right and optimized.

Profile-Based Compilation

Because there is a trade-off between code size and higher performance, it is often desirable to compile some functions in a DSP application for performance and others for code size. In fact, the ideal code size and performance for your application needs may be some combination of the different levels of optimization across all of the application functions. The challenge is in determining which functions need which options. In an application with 100 or more functions, each with five possible options, the number of option combinations starts to explode exponentially. Because of this, manual experimentation can take weeks of effort, and the DSP developer may rarely arrive at a solution close to optimal for the particular application needs. Profile-based compilation is one available technique which helps to solve this challenge by automating the entire process.

Profile-based compilation will automatically build and profile multiple compiler option sets. For example, this technique can build the entire application with the highest level of optimization and then profile each function to obtain its resulting code size and cycle count. This process is then repeated using the other compiler options at the remaining code-size reduction levels. The result is a set of different code size and cycle count data points for each function in the application. That data can then be plotted to show the most interesting combinations of functions and compiler options (Figure 6.54). The ideal location for the application is always at the origin of the graph in the lower left hand corner where cycle count and code size are both minimized.

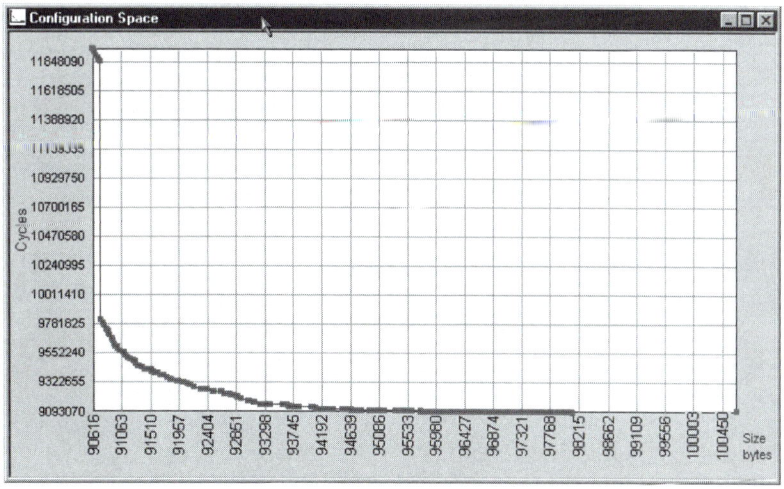

Figure 6.54 Profile-based compilation shows the various trade-offs between code size and performance

Advantages

An advantage of a profile-based environment is the ability for the application developer to select a profile by selecting the appropriate point on the curve, depending on the overall system needs. This automatic approach saves many hours of experimenting manually. The ability to display profiling information (cycle count by module or function for example) allows the developer to see the specifics of each piece of the application and work on that section independently, if desired (Figure 6.55).

Solution for Size = 86816 Cycles = 9375193

Function	Cycles	Size	OptionSet	Options
inflateInit2_	126	464	Minimum Size	-o3 -oi0 -ms3
inflate_blocks_new	71	260	Minimum Size	-o3 -oi0 -ms3
deflateInit2_	276	1032	Minimum Size	-o3 -oi0 -ms3
putShortMSB	45	52	Minimum Size	-o3 -oi0 -ms3
inflate	1123	1792	Minimum Size	-o3 -oi0 -ms3
inflate_blocks_reset	264	312	Speed	-o3 -oi0 -ms1
do_compress	665	804	Minimum Size	-o3 -oi0 -ms3
deflate	2011	1224	Minimum Size	-o3 -oi0 -ms3
deflateEnd	142	356	Speed	-o3 -oi0 -ms1
inflateReset	47	136	Speed	-o3 -oi0 -ms1
inflate_trees_dynamic	497	428	Minimum Size	-o3 -oi0 -ms3
inflate_trees_bits	287	168	Minimum Size	-o3 -oi0 -ms3
do_decompress	559	792	Minimum Size	-o3 -oi0 -ms3
bi_flush	235	208	Minimum Size	-o3 -oi0 -ms3
inflate_trees_fixed	2668	460	Minimum Size	-o3 -oi0 -ms3
_tr_align	973	1176	Minimum Size	-o3 -oi0 -ms3
_tr_init	68	260	Minimum Size	-o3 -oi0 -ms3
flush_pending	551	232	Minimum Size	-o3 -oi0 -ms3

Figure 6.55 Profiling information for DSP functions

Issues with Debugging Optimized Code

One word of caution: do not optimize programs that you intend to debug with a symbolic debugger. The compiler optimizer rearranges assembler-language instructions which makes it difficult to map individual instructions to a line of source code. If compiling with optimization options, be aware that this rearrangement may give the appearance that the source-level statements are executed in the wrong order when using a symbolic debugger.

The DSP programmer can ask the compiler to generate symbolic debugging information for use during a debug session. Most DSP compilers have an option for doing this. DSP compiler have directives that will generate symbolic debugging directives used by C source-level debugger. The downside to this is that it forces the compiler to disable many optimizations. The compiler will turn on the maximum optimization compatible with debugging. The best solution for debugging DSP code however, is first to verify the program's correctness and then start turning on optimizations to improve performance.

Summary of the Code Optimization Process

Given that this book is about managing the DSP software development process, we must now expand our definition of the code optimization process first discussed at the beginning of the chapter. Figure 6.56 shows an expanded software development process for code optimization.

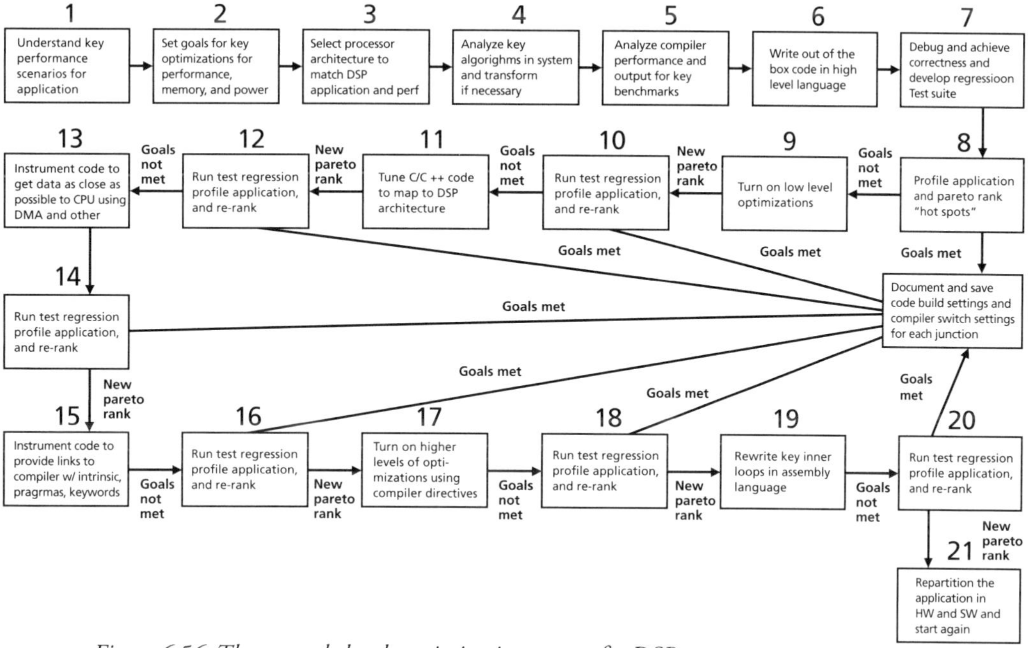

Figure 6.56 The expanded code optimization process for DSP

Although this process may vary based on the application, this process is a general flow for all DSP applications. There are 21 steps in this process that will be summarized below.

- *Step 1* – This step involves understanding the key performance scenarios for the application. A performance scenario is a path through the DSP application that will stress the available resources in the DSP. This could be performance, memory, and/or power. Once these key scenarios are understood, the optimization process can focus on these "worst case" performance paths. If the developer can reach performance goals in these conditions, all other scenarios should meet their goals as well.
- *Step 2* – Once the key performance scenarios are understood, the developer then selects the key goals for each resource being optimized. One goal may be to consume no more than 75% of processor throughput or memory, for example. Once these goals are established, there is something to measure progress towards as well as stopping criteria. Most DSP developers optimize until they reach a certain goal and then stop.

- *Step 3* – Once these goals are selected, the developer, if not done already, selects the DSP to meet the goals of the application. At this point, no code is run, but the processor is analyzed to determine whether it cam meet the goals through various modeling approaches.
- *Step 4* – This step involves analyzing key algorithms in the system and making any algorithm transformations necessary to improve the efficiency of the algorithm. This may be in terms of performance, memory, or power. An example of this is selecting a fast Fourier transform instead of a slower discrete Fourier transform.
- *Step 5* – This step involves doing a detailed analysis of the key algorithms in the system. These are the algorithms that will run most frequently in the system or otherwise consume the most resources. These algorithms should be benchmarked in detail, sometimes even to the point of writing these key algorithms in the target language and measuring the efficiency. Given that most of the application cycles may be consumed here, the developer must have detailed data for these key benchmarks. Alternatively, the developer can use industry benchmark data if there are algorithms that are very similar to the ones being used in the application. Examples of these industry benchmarks include the Embedded Processor Consortium at eembc.org and Berkeley Design Technology at bdti.com.
- *Step 6* – This step involves writing "out of the box" C/C++ code which is simply code with no architecture specific transformations done to "tune" the code to the target. This is the simplest and most portable structure for the code. This is the desired format if possible, and code should only be modified if the performance goals are not met. The developer should not undergo a thought process of "make it as fast as possible." Symptoms of this thought process include excessive optimization and premature optimization. Excessive optimization is when a developer keeps optimizing code even after the performance goals for the application have been met. Premature optimization is when the developer begins optimizing the application before understanding the key areas that should be optimized (following the 80/20 rule). Excessive and premature optimizaiton are dangerous because these consume project resources, delay releases, and compromise good software designs without directly improving performance.
- *Step 7* – This is the "make it work right before you make it work fast" approach. Before starting to optimize the application which could potentially break a working application because of the complexity involved, the developer must make the application work correctly. This is most easily done by turning off optimizations and debugging the application using a standard debugger until the application is working. To get a working application, the developer must also create a test regression that is run to confirm that the application is indeed working. Part of this regression should be used for all future optimizations to ensure that the application is still working correctly after each successive round of optimization.

- *Step 8* – This step involves running the application and collecting profiling information for each function of the application. This profiling information could be cycles consumed for each function, memory used, or power consumed. This data can then be pareto ranked to determine the biggest contributors to the performance bottlenecks. The developer can then focus on these key parts of the application for further optimization. If the goals are met, then no optimizations are needed. If the goals are not met the developer moves on to step 9.
- *Step 9* – In this step, the developer turns on basic compiler optimizations. These include many of the machine independent optimizations that compilers are good at finding. The developer can select options to reduce cycles (increase performance) or to reduce memory size. Power can also be considered during this phase.
- *Steps 10, 12, 14, 16, and 18* – These steps involve re-running the test regression for the application and measuring performance against goals. If the goals are met then the developer is through. If not, the developer re-profiles the application and establishes a new pareto rank of top performance, memory, or power bottlenecks.
- *Step 13* – This step involves re-structuring or tuning the C/C++ code to map the code more efficiently to the DSP architecture. Examples of this were discussed earlier in the chapter. Re-structuring C or C++ code can lead to significant performance improvements but makes the code potentially less portable and readable. The developer should first attempt this where there are the most significant performance bottlenecks in the system.
- *Step 15* – In this step, the C/C++ code is instrumented with very specific information to help the compiler make more aggressive optimizations based on these "hints" from the developer. Three common forms of this instrumentation include using special instructions with intrinsics, pragmas, and keywords.
- *Step 17* – If the compiler supports multiple levels of optimization, the developer should proceed to turn on higher levels of optimization to allow the compiler to be more aggressive in searching for code transformations to yield more performance gains. These higher levels of optimization are advanced and will cause the compiler to run longer searching for these optimizations. Also, there is the chance that a more aggressive optimization will do something to break the code or otherwise cause it to behave incorrectly. This is where a regression test suite that is run periodically is so important.
- *Step 19* – As final resort, the developer rewrites the key performance bottlenecks in assembly language if result can yield performance improvements above what the compiler can produce. This is a final step, since writing assembly language reduces portability, increases maintainability, decreases readability, and it has other side effects. For advanced architectures, assembly language may mean writing highly complex and parallel code that runs on multiple independent execution units. This can be very difficult to learn to do well and generally there are not many DSP developers that can become expert assembly language programmers without a significant ramp time.

- *Step 21* – Each of the optimization steps described above are iterative. The developer may iterate through these phases multiple times before moving on to the next phase. If the performance goals are not being met by this phase, the developer needs to consider re-partitioning the hardware and software for the system. These decisions are painful and costly but they must be considered when goals are not being met by the end of this optimization process.

As shown in the following figure, when moving through the code optimization process, improvements become harder and harder to achieve. When optimimzing generic "out of the box" C or C++ code, just a few optimizations can lead to significant performance improvements. Once the code is tuned to the DSP architecture and the right hints are given to the compiler in the form of intrinsics, pragmas, and keywords, additional performance gains are difficult to achieve. The developer must know where to look for these additional improvements. That's why profiling and measuring are so important. Even assembly language programming cannot always yield full entitlement from the DSP device. The developer ultimately has to decide how far down the curve in Figure 6.57 the effort should go, as the cost/benefit ratio gets less justifiable given the time required to perform the optimizations at these lower levels of the curve.

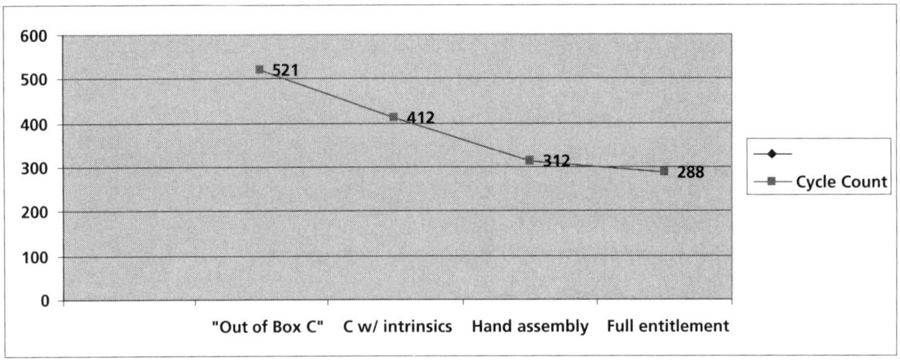

Figure 6.57 Achieving full entitlement. Most of the significant performance improvements are achieved early in the optimization process.

Summary

Coding for speed requires the programmer to match the expression of the application's algorithm to the particular resources and capabilities of the processor. Key among these fit issues is how data is staged for the various operations. The iterative nature of DSP algorithms also makes loop efficiency critically important. A full understanding of when to unroll loops and when and how to pipeline loops will be essential if you are to write high-performance DSP code—even if you rely on the compiler to draft most of such code.

Whereas hand-tuned assembly code was once common for DSP programmers, modern optimizing DSP compilers, with some help, can produce very high performance code.

Embedded real-time applications are an exercise in optimization. There are three main optimization strategies that the embedded DSP developer needs to consider:

- *DSP architecture optimization*: DSPs are optimized microprocessors that perform signal processing functions very efficiently by providing hardware support for common DSP functions;
- *DSP algorithm optimization*: Choosing the right implementation technique for standard and often used DSP algorithms can have a significant impact on system performance, and
- *DSP compiler optimization*: DSP compilers are tools that help the embedded programmer exploit the DSP architecture by mapping code onto the resources in such a way as to utilize as much of the processing resources as possible, gaining the highest level of architecture entitlement as possible.

References

TMS320C62XX Programmers Guide, Texas Instruments, 1997

Computer Architecture, A Quantitative Approach, by John L Hennesey and David A Patterson, copyright 1990 by Morgan Kaufmann Publishers, Inc., Palo Alto, CA

The Practice of Programming, by Brian W. Kernighan and Rob Pike, Addison Wesley, 1999

TMS320C55x_DSP_Programmer's_Guide

TMS320C55xx Optimizing C/C++ Compiler User's Guide (SPRU281C)TMS320C55x DSP Programmer's Guide (SPRU376A)

Generating Efficient Code with TMS320 DSPs: Style Guidelines (SPRA366)

How to Write Multiplies Correctly in C Code (SPRA683)

TMS320C55x DSP Library Programmer's Reference (SPRU422)

7

Power Optimization Techniques Using DSP

Introduction

Despite the importance of power consumption and memory use, relatively little emphasis has been placed on optimizing power and memory for DSP applications. This chapter will provide some guidelines on optimizing DSP applications for power.

The requirements for low power design come from several areas including mobility concerns for portable applications using batteries, ergonomics related areas that drive packaging and other thermal cooling constraints, and finally to overall system value that comes from requirements that support channel density and other electricity cost factors (Figure 7.1).

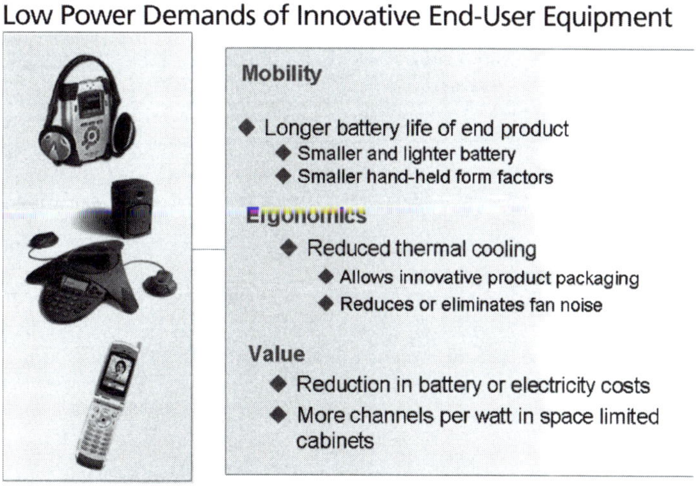

Figure 7.1 Low power demands of innovative end user equipment (courtesy of Texas Instruments)

Each milliwatt of power in a mobile device should be considered in terms of battery life. Even larger systems such as broadband and DSL (digital subscriber line) systems are sensitive to power even though these systems are effectively plugged into the wall. Heat generated by these devices requires more cooling equipment and limits the amount of channels per area (or channels per device) in these systems. In this sense, power reduction has become a key design goal for embedded systems programmers including DSP designers. The key challenge is to achieve an acceptable performance/ power level. DSPs are used primarily to achieve performance and computational goals because of the aforementioned support for these operations. As these complex algorithms become more ubiquitous in portable applications, this performance/power trade-off is becoming very important. In many cases processing power must be offset with higher power requirements unless a disciplined process is followed throughout the life cycle to manage power and performance.

Just as code size and speed impact cost, power consumption also affects cost. The more power consumed by an embedded application, the larger the battery required to drive it. For a portable application, this can make the product more expensive, unwieldy, and undesirable. To reduce power, you need to make the application run in as few cycles as possible, considering that each cycle consumes a measurable amount of energy. In this sense, it would seem that performance and power optimization are similar—consume the fewest number of cycles to get both performance and power optimization goals. Performance and power optimization strategies share similar goals but have subtle differences, as will be shown shortly.

The key challenge for embedded DSP developers is to develop an application solution that meets the right performance, price and power targets. Achieving all three of these goals is a challenge. In order to achieve this, some level of power optimization, in hardware as well as software, may be required (Figure 7.2). The achievement of these simultaneous goals requires life cycle management, tools and techniques as well as the right DSP device.

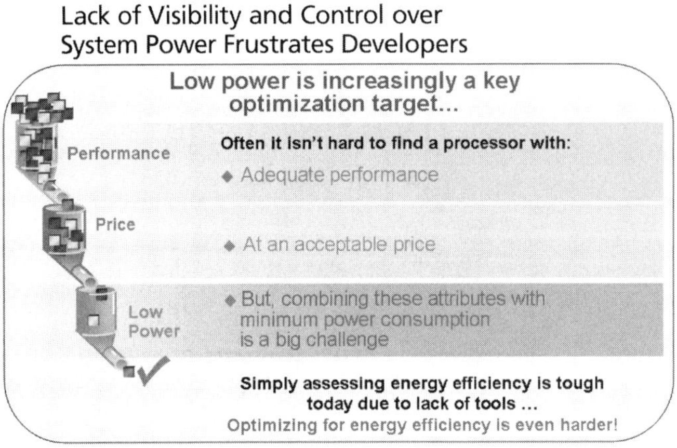

Figure 7.2 Lack of visibility and control over system power frustrates developers (courtesy of Texas Instruments)

Most of the dynamic power consumed in embedded applications comes not from the CPU but from the processes used to get data from memory to the CPU. Each time the CPU accesses external memory, buses are turned on, and other functional units must be powered on and utilized to get the data to the CPU. This is where the majority of power is consumed. If the programmer can design DSP applications to minimize the use of external memory, efficiently move data into and out of the CPU, and make efficient use of peripherals and cache to prevent cache thrashing, cycling of peripherals on and off, and so on, the overall power consumption of the application will be reduced significantly. Figure 7.3 shows the two main power contributors in embedded DSP applications. The compute block includes the CPU and this is where the algorithmic functions are performed. The other is the memory transfer block and this is where the memory subsystems are utilized by the application. The memory transfer block is where the majority of the power is consumed by a DSP application.

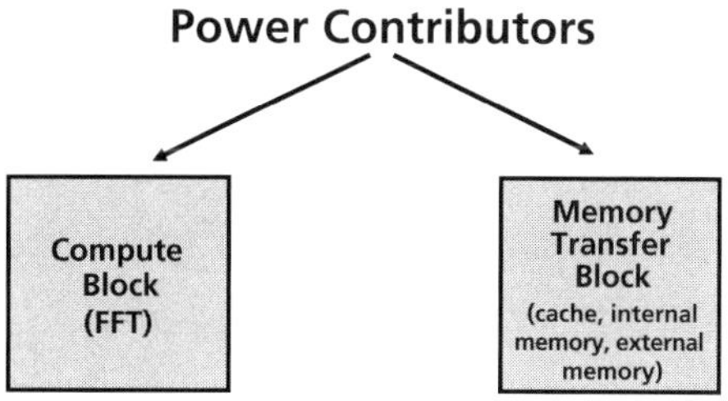

Figure 7.3 The main power contributors for a DSP application are in the memory transfer functions, not in the compute block.

The process of power optimization is a key DSP application management component because it's a life cycle process that should be planned and managed accordingly. The DSP team lead must understand the entire system, hardware as well as software in order to manage this effort effectively. It's not entirely a software related issue.

For example, some DSPs are designed for low-power operation. These DSPs are built with specific manufacturing processes that are tailored for low power consumption. These architectures use special hardware capability to manage power consumption on the device more efficiently while at the same time maintaining the architectural features necessary to provide the performance expected from a DSP.

On the software side, application optimization starts with the operating system, and builds up through the application. Each of these approaches will be discussed in this chapter.

As with most optimization techniques, including those that deal with performance and memory, power optimization is governed in many ways with the compound 80/20 rule for net power in mobile applications (Figure 7.4). In the pie on the left, 80% of the time, approximately, in a mobile application, the product is in some sort of standby mode. That means that only 20% of the time is spent in the operating mode. In that 20% of the time slice that the application is operating, approximately 80% of that operating time is actually using a subset of the peripherals or a subset functions that can be delivered by that new device or system. Therefore, you need to be operating in a managed way where you only want to use functions that are being developed. Consequently, 20% of the time of that 20% is when you are actually using full functionality of the device. Take 20% × 20% and you see that full operating power is only really needed approximately 4% of the time. So the question is, what you do with the rest of the time to manage power more efficiently? This is where we need to spend the time discussing these techniques.

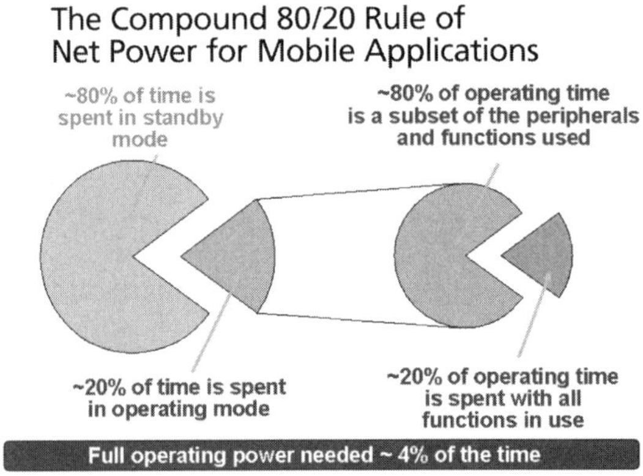

Figure 7.4 The compound 80/20 rule of net power for mobile applications (courtesy of Texas Instruments)

Standby power is an increasingly important parameter in optimizing for energy efficiency. In Figure 7.5, the left chart shows the operating power pretty much close to full throttle over the time discrete elements of this slide. We see from the percentage of power used over time that quite a bit of power is being burned in the left-hand chart. If you go to the right chart, we see one can optimize for very low standby power most of the time and utilize very efficient power by being in standby mode for a significant amount of time. Also, when you turn on, you can turn on the full operating power as highlighted in the second column or you can turn on some subset of full operating power with some peripherals on, some peripherals off, as highlighted by the fifth column in this time slice. As you can see, much less energy is consumed and we can get to a much lower minimum net operating power by using standby much more efficiently as well as only turning on the functions that you need to turn on.

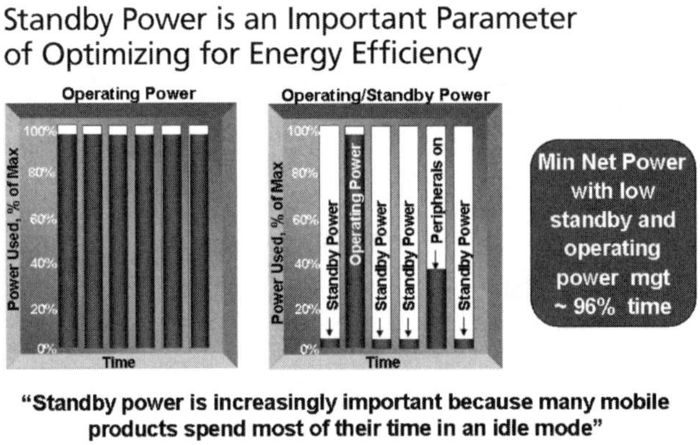

Figure 7.5 Standby power is an important parameter of optimizing for energy efficiency (courtesy of Texas Instruments)

The management of embedded DSP applications in the realm of power optimization should not be concerned with primarily a comparison of milliwatts per megahertz or other similar measure. As Figure 7.6 shows, this is really just the tip of the power iceberg. There is a more complicated process to manage than simply comparing these types of measurements. The process also includes an analysis of what are the on-chip memory vs. off-chip memory accesses of the application. Often, the penalty for accessing external memory can cause a significant amount and sometimes even a dominating amount of the system power consumption. Minimizing off-chip I/O accesses can also save power. With a peripheral on-chip such as USB, the DSP engineer can save a significant amount of power vs. having to access it off-chip. It also makes sense to be turning on and using power only on peripherals that need to be in utilization so a good granularity and flexibility of turning peripherals off and on saves power. In addition to that, minimizing off-chip I/O accesses can save power and CPU voltage and CPU frequency, keeping the voltage as low as possible and scaling the frequency down can also have a significant impact on CPU or core power consumption. Going even further down to the base of the concern is standby power. When items are turned off they can still leak power through leaky transistors. Therefore the application can suffer significant consumption of power even though it is not doing much functional work. Therefore, milliwatts (mW) of standby power is an important element to the more global power story. The intelligent use of idle and standby states and the granularity of being able to get into and out of these states as well as the speed of being able to switch states can have a significant impact on the amount of power your system burns in the application.

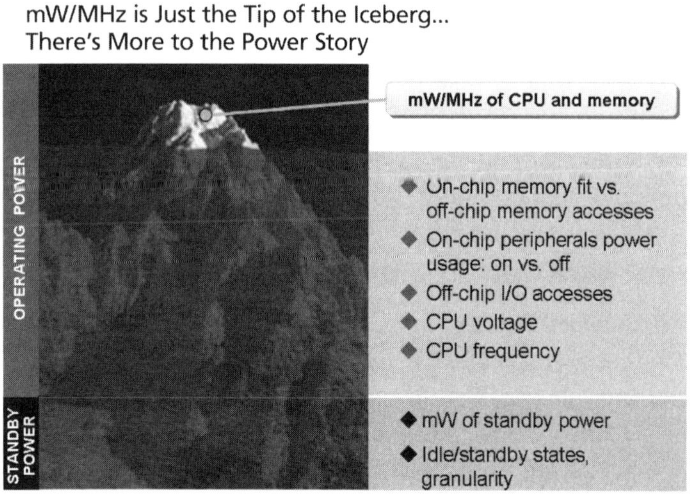

Figure 7.6 mW/MHz is only part of the power story for embedded systems

The main power saving strategies used in embedded applications include:

- Reducing the power supply voltage if possible.
- Running at a lower clock frequency.
- Disabling functional units with control signals when not in use.
- Disconnecting parts of the device from the power supply when not in use.

There are two main power management styles:

- *Static power management* – This approach does not depend on CPU activity. An example of this is a user-activated power-down mode.
- *Dynamic power management* – This approach is based on CPU activity. An example of this is disabling inactive function units.

We will discuss both of these approaches in this chapter.

Some system design considerations to understand as the engineer begins the power optimization phase include:

- *Internal CPU activity* – Instruction complexity (the number of parallel operations being performed by the instruction), the utilization of internal buses including data patterns on the buses, repeat instructions, and so on.
- *System clock and switching rates* – For example, if the clock speed doubles, the current will double.
- *On-chip vs. off-chip memory accesses* – On-chip requires less power because the external memory interface is not driven during internal accesses.
- *ROM vs. RAM* – Code execution from ROM requires about 10% less CPU current than the same code executing from SRAM.
- Capacitive loading of outputs and how to control it.
- *Address visibility* – Addresses passed to an external address bus even during internal memory accesses—useful for debugging but should be turned off when debugging complete.
- *Power down modes* – IDLE modes to conserve power.

Power Optimization Techniques in DSP Devices

Semiconductor process technology improvements drive much of the reduction in DSP power consumption. Smaller-geometry transistors in the core's logic require decreased operating voltages. The effects of lower voltages can be dramatic: decreasing DSP core voltages from 1.6 volts in available products to 1 volt in future releases may reduce active power consumption by as much as 80 percent. Additionally, processes are being tuned for low power consumption through the use of low-leakage transistors with a high transition threshold voltage (V_T). These transistors minimize quiescent current (I_Q) during periods when the circuitry is powered on but inactive, such as during system standby modes. DSP devices now have standby power consumption of only 0.12 mW, less than one percent of the standby power required by devices just a few years ago.

During the early phases of a DSP application life cycle, the engineer must analyze the application to determine the appropriate balance between active and standby power consumption in the application. This is an important design issue that affects the choice of the DSP. If the application is a large, heat-sensitive system, active power consumption is the driving factor. In portable battery-operated systems, standby power is the dominating factor. Cell phones, for instance, can drain a battery just as readily during waiting times as during talk times. MP3 players tend to be either on or off, so standby power consumption is less important. Some handheld instruments may be on and draining power even when they are not in active use. The DSP designers do not have any direct control over whether the process technology used in a DSP lowers quiescent as well as active current. However, the engineer must be able to weigh the balance of active and standby power needed by the application. Only then can they consider the process capabilities among other factors in selecting a DSP.

DSP hardware has incorporated a number of power saving and optimization techniques into the architectures of these devices. Aside from the natural savings coming from continued improvement in semiconductor processor technology, architecture advances have also been made to modern DSP devices. These include:

- Increased parallelism and dedicated instructions for a class of DSP operations including the idle instruction which forces the program to wait until an interrupt or a reset occurs. Power down modes during the idle depend on a configuration register.
- Extensive use of clock gating peripherals, memory units, clock tree structure and logic fanout.
- Memory access—clock gated memory access, reduction in access requirements—for example, internal buffer queues (IBQ or loop buffers).
- Custom designed data paths for both address and data.
- Designed using low leakage process technology.

Many DSP devices are developed using complementary metal oxide semiconductor (CMOS) technology. CMOS transistors drain only minimal current (leakage) when they are open or closed. They consume power only while switching. The power dissipated depends on the voltage swing, load capacitance, operating frequency, and switching activity (the fraction of wires that actually toggle).

Power dissipation depends linearly on frequency and quadratically on voltage. Nothing can be done in software about capacitive load; software techniques that reduce switching power address both frequency and voltage, although at different levels of granularity.

The current that flows through a CMOS gate when not switching is called *leakage*. This used to be relatively small (1%) but the percentage is now growing with deep submicron geometries (30% at 90nm technology nodes and below). This also effects standby operation of the device. To reduce leakage dissipation, software techniques can selectively turn off parts of the device (hardware is also effective at doing this).

There is a significant difference between active and static power dissipation:

Active – This is proportional to capacitance, frequency of transistor switching, voltage and number of bits (transistors switching).

Active power dissipation is in linear proportion to the device frequency (the power will drop if the frequency is cut in half). But it is not possible just to scale voltage without also watching the frequency. A significant drop in voltage means you must drop frequency to keep device in the active range. For example you cannot run a DSP at 1 gigahertz and then drop voltage by half without a corresponding reduction in frequency. If you cut frequency in half, your code will take longer to execute (about twice as long).

Static power consumption (Pstatic below) involves dropping voltage in standby to get lower standby power dissipation. This hasn't been a big issue in the past but it is starting to become an issue at increased technology nodes.

It is also possible to turn off voltage completely (this is the concept of power domains in DSP device architectures). Scaling back voltage as well as controlling power domains can be done under software control.

In summary:

- Power consumption in CMOS circuits:

  P total = Pactive + Pstatic

- Active (Dynamic) Power = when doing work: minimize

  Pactive ~ C x F x V**2 x Nsw

where,

C = capacitance

F = frequency

V = voltage

N = number of bits switching

Turn units, peripherals OFF when not doing work

- Standby Static Power = leakage when not switching = LOW

  Pstatic ~ V × Iq

  Low leakage process technology, transistors

As an example, consider a cell phone application. Active power consumption would be considered the talk time and standby power consumption is considered standby mode in a cell phone. Another example is an audio jukebox. The active power consumption is the music play time and the static power consumption is the "backpack" time.

An example of architectural features to support power optimization is shown in Figure 7.7. In this figure additional MAC units (multiply accumulate) and ALUs (arithmetic logic units) as well as additional registers have been added in order to perform more operations in parallel. Parallel operations are one way to gain power advantage at the DSP architecture level.

Within the core itself, a dual-multiply-accumulate (MAC) data path doubles performance on repetitive arithmetic tasks, allowing the core to return to a low-power state more quickly. In most tasks, array values in both MACs can be multiplied by the same coefficient. So with specific coding the coefficient can be fetched only once, then loaded into both MAC units, enabling operations to take place with only three fetches per array multiplication cycle instead of four.

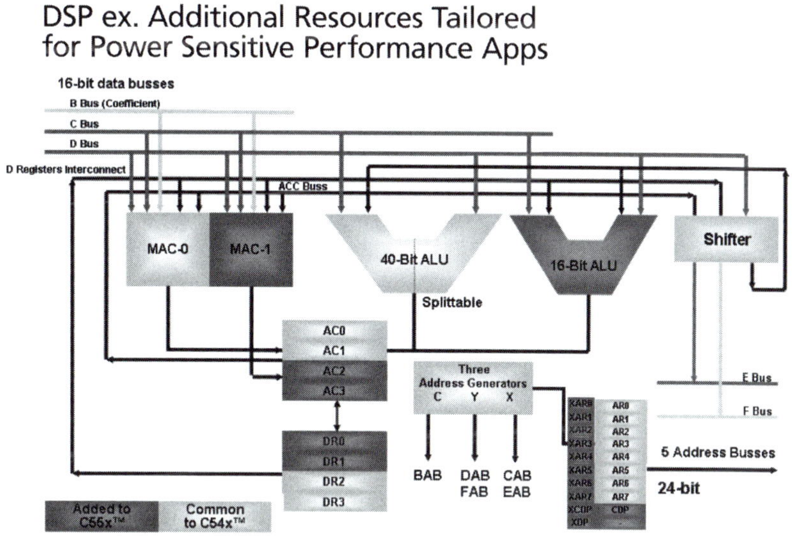

Figure 7.7 Additional resources tailored for power sensitive performance applications (courtesy of Texas Instruments)

The following steps are required to calculate overall device power consumption:

1. *Algorithm partitioning* – The algorithm under consideration should be broken into sections of unique device activity and the power requirements for these sections should be calculated separately. These sections can then be time averaged to determine the overall device current requirements.
2. *CPU activity* – The current contribution to CPU activity can be determined by examining code and determining the time-averaged current for each algorithm partition.
3. *Memory usage* – Scale the current in step 2 based on memory usage. Use of on-chip memory requires less current than off-chip (because of additional current due to external interface). Running from ROM vs. RAM is also less efficient.
4. *Peripherals* – Consider the additional current required by the timer, standard serial port, host port interface and so on.
5. *Current due to outputs* – Consider the current required by the algorithm to operate the external address and data buses.
6. *Calculation of average current* – If the power supply is observed over the full duration of device activity, different segments of activity will exhibit different current levels for different lengths of time.

7. *Effects of temperature and supply voltage on device operating current* – Include the effects of these factors after the total device current has been calculated.

Another example of a DSP architecture is shown in Figure 7.8. This is a model of the TI 6x architecture which has a relatively clean and orthogonal clustered VLIW instruction set that represents a good target for advanced ILP compilers. This helps in system level power consumption reduction. Because of the focus on performance and power, many DSPs have exposed pipeline and control to the programming interface to allow more control of instruction selection and scheduling. Figure 7.8 represents a power logical model of this device. The peripherals are represented as switches that can be turned on and off. DSPs provide this feature to DSP engineers. The ability to switch off peripherals, the large power consumption components of a DSP, when not in use is significant in overall system level power optimization and reduction. The sliding scales on the left represent the ability to scale power and voltage in these devices. The external memory interface (EMIF), cache subsystem, and direct memory access (DMA) also have the capability to be switched off when not in use. The memory can also be configured or scaled depending on the application to provide optimal memory configurations which can provide power savings.

Processor Architecture and Process Technology
Designed for Low Power

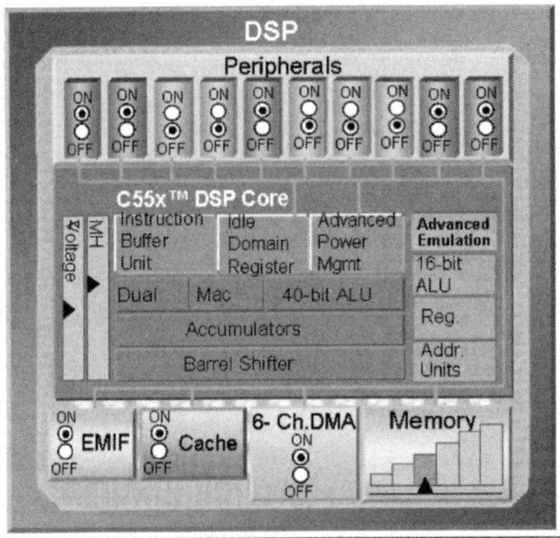

	Core + Memory (active)	Idle 3 Mode (standby)
DSP	0.46 mA/MHz	0.1 mA

Comprehensive Chip Level Low Power Management

- **Multiple standby modes**
- **Low standby power**
  - **Low leakage transistor = ultra low standby current = 0.012 mA**
- **Low/efficient operating power**
  - **mW/MHz core and memory**
  - **Ability to turn on and off individual peripherals**
  - **I/O driver power**
  - **SRAM size options from 32 KB to 320 KB for efficient code fit**
  - **Voltage scaling 1.2V, 1.6V; frequency scaling up to 300 MHz**
- **Core attributes**
  - **Designed for granular power management**

Figure 7.8 DSP processor architectures designed for low power consumption (courtesty of Texas Instruments)

Many DSP devices have integrated cache and ROM that may total hundreds of kilobytes. This is far too much memory to be active all the time without considerable power drain. It is a common characteristic of many DSP algorithms that most memory fetches hit within a small cache area a vast majority of the time. Power-saving DSP architectures take advantage of this tendency by dividing memory into fine-grained blocks, then enabling only the block that is in use. In practice, a distributed addressing scheme gates a narrowing funnel of row, column and enable lines, as well as the clock tree structure, so that only a single block of, say, 2K words becomes active for each access.

Memory subsystems are designed to provide power savings by monitoring and regulating power to only those parts of the memory that are being accessed and disabling those parts that are not being accessed (Figure 7.9). By disabling clocks to those parts of the memory that are not being used, significant power savings are realized.

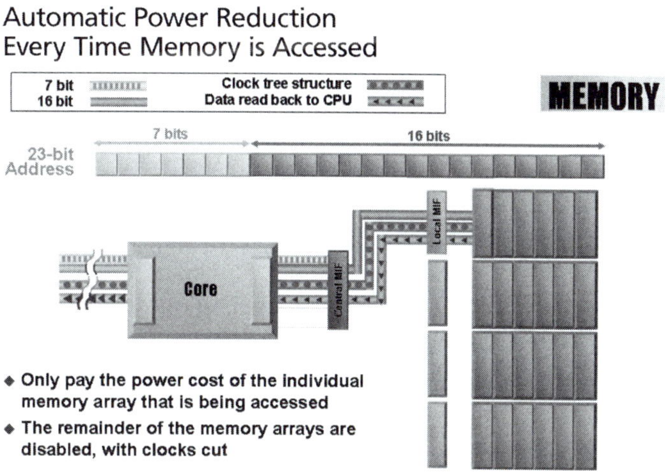

Figure 7.9 Automatic power reduction using memory (courtesy of Texas Instruments)

DSP peripherals are also architected to provide power savings when not in use. As an example consider Figure 7.10. In this scenario, data is detected from a sensor using the multichannel buffered serial port (McBSP) and sent to the direct memory access (DMA) which is then sent to the memory subsystem. When no data is being received from the DMA, the memory subsystem is disabled, lowering overall power consumption. Likewise, when there is no data being sent from the McBSP to the DMA, the DMA is automatically disabled. Finally, when there is no input data being detected from the sensor in the McBSP, this unit is also disabled. This intelligent peripheral management leads to significant power reduction for DSP applications.

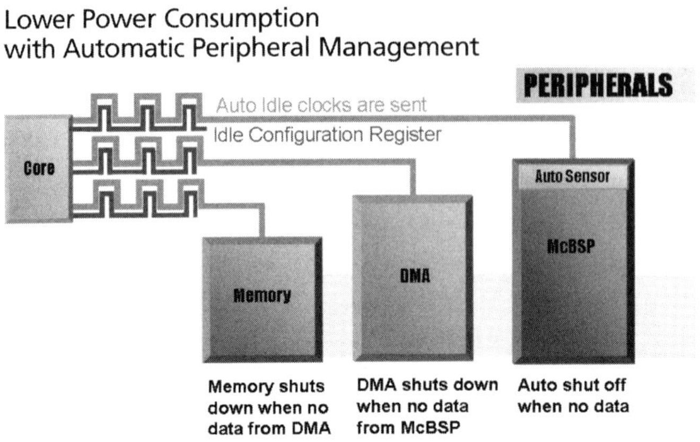

Figure 7.10 Low power consumption and automatic peripheral management (courtesy of Texas Instruments)

In order to get the most accurate model of the power consumption in a DSP application, the engineer must follow a disciplined approach to obtaining power consumption estimates. A simple Excel worksheet can help, although other tooling and support is also available. Before starting the design, the engineer must enter in the respective values for the proposed application including the percentage of loading in the CPU as well as whether the peripherals are being used or not. The engineer must get accurate estimates of overall device consumption by summing the usage of each of these peripherals as well as the CPU (see Figure 7.11). This process must start before the engineer actually obtains the hardware—a simple data sheet with general and worst case estimates is not good enough. Going through the detailed analysis will also give the engineer characterization points, and answer key questions such as "what will the overall change in power consumption be if I use this I/O peripheral?" This analysis also provides much more detail than a first order approximation of power using rough consumption numbers based on switching rates. A second order approximation will lead to a much more accurate overall system estimate.

Specific Chip Function Lever Power Calculation Worksheet that You Drive with Your Loads/Uses

(a)

	Frequency	Idle Status	%Utilization	%Writes	Bits	%Switch	Trace (In.)	Cap. (pF)	Other
CLKGEN	72	Active							
CPU	72	Active	75						
CLKOUT	72	Disabled					1	10	
EMIF	72	Idle	0	100	16	75	2	6	Sync
HPI	72	Idle	0	100	16	0	1	10	
DMA	72	Active							
Ch. 0		Enabled	0.06		16	100			
Ch. 1		Enabled	0.005		16	100			
Ch. 2		Disabled	0		16	0			
Ch. 3		Disabled	0		16	0			
Ch. 4		Disabled	0		16	0			
Ch. 5		Disabled	0		16	0			
McBSP0	1.41	Active	100		32	100	1	10	External
McBSP1	36	Idle	0		32	100	1	10	External
McBSP2	36	Idle	0		32	100	1	10	Internal
MMC1	36	Active	5.3	0		100	2	3	
SD1	9	Idle	0	0		100	1	10	
MMC2	9	Idle	0	0		100	1	10	
SD2	9	Idle	0	0		0	1	10	
I2C	12	Active	0			100	1	10	
Timer 0	5	Idle	100				1	10	TOUT enabled
Timer 1	20	Idle	100						
WDT									
ADC		Idle	0						
USB		Idle	25			100	6 feet		
RTC									
GPIO/XF	0		0			0	1	10	0

(b)

(c)

Figure 7.11 Example of a power calculation worksheet that is used to map application use cased for more accurate power estimated (courtesy of Texas Instruments)

Figures 7.12 a, b, and c show an example of the overall power reductions that can be achieved by disabling various peripherals. In Figure 7.12a, the EMIF and the DMA have been turned off resulting in a noticeable power savings. In Figure 7.12b, additional peripherals including the McBSPs, timer, USB, and GPIO have been turned off resulting in more power savings. In Figure 7.12c, the entire device has been powered down, resulting in no power dissipation.

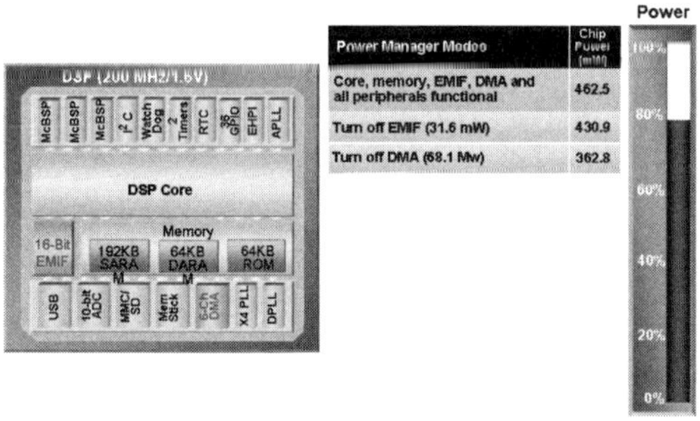

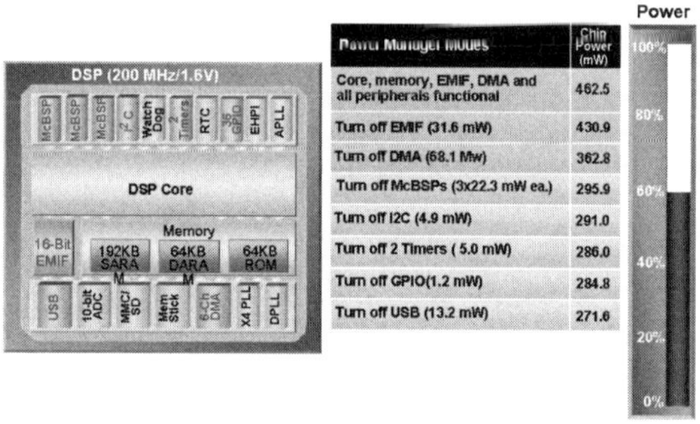

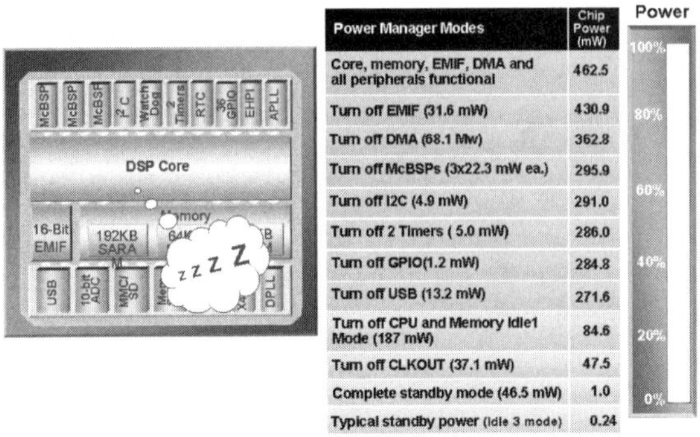

Power Manager Modes	Chip Power (mW)
Core, memory, EMIF, DMA and all peripherals functional	462.5
Turn off EMIF (31.6 mW)	430.9
Turn off DMA (68.1 Mw)	362.8
Turn off McBSPs (3x22.3 mW ea.)	295.9
Turn off I2C (4.9 mW)	291.0
Turn off 2 Timers (5.0 mW)	286.0
Turn off GPIO(1.2 mW)	284.8
Turn off USB (13.2 mW)	271.6
Turn off CPU and Memory Idle1 Mode (187 mW)	84.6
Turn off CLKOUT (37.1 mW)	47.5
Complete standby mode (46.5 mW)	1.0
Typical standby power (Idle 3 mode)	0.24

Figure 7.12 Power reduction example of turning peripherals off (courtesy of Texas Instruments)

Power Optimization for DSP Operating Systems

As real-time operating systems are becoming more popular in DSP systems, the DSP engineer effectively writes software not to the underlying hardware but to the operating system. It's the RTOS that manages the hardware resources as well as the synchronization between the various software components. The RTOS also manages the scheduling of the various tasks in the system and meeting these deadlines is a function of how much work there is to do and what the deadline is. Clock frequency is a dependent parameter in this equation. Changing the clock frequency changes the time it takes to complete the work. For low power applications, the engineer must be cautious when reducing the clock frequency so that the application still completes its work in the required timeframe (Figure 7.13).

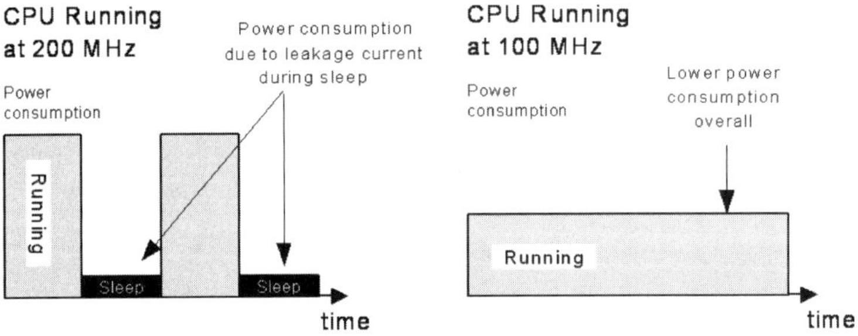

Figure 7.13 Example of power consumption scaling (courtesy of Texas Instruments)

Power and frequency scaling are best handled at the OS level, because the RTOS has the ability to control not only the software but the hardware as well, including peripherals as well as CPU. Power and frequency are monitored and scaled dynamically as the application is running. The RTOS must implement the desired capability for power control within a DSP application. This capability must provide the following:

- Manage all power-related functions in the RTOS application, both statically, configured by the application developer, and dynamically, called at runtime by the application.

- Selectively idle clock domains.
- Specify a power-saving function to be called at boot time to turn off unnecessary resources.
- Dynamically change voltage and frequency at runtime.
- Activate chip-specific and custom sleep modes and provide central registration and notification of power events to system functions and applications.

An example of this would be a user pressing a PDA button; the entire DSP clock tree may be gated off (sleep mode) and then woken up when user presses the PDA button to turn the system back on.

The RTOS power solution must also contain device-specific settings for viable voltage and frequency combinations for the device. This data is actually used to change the voltage and frequency settings for the device. There can be several "legal" combinations of V/F but the system engineer is ultimately responsible for producing these valid combinations.

Many of these capabilities can be provided as an application programming interface (API) to the application (Figure 7.14).

The power manager software interfaces directly to the DSP hardware by writing to and reading from a clock idle configuration register, using data from the platform-specific power-scaling library (PSL) that controls the core clock rate and voltage-regulation circuitry. This power management system also supports callbacks to application code before and after scaling to allow for preparation before scaling has occurred as well as cleanup work after the scaling has occurred. The function also provides the ability to make queries to determine the present voltage and frequency, supported frequencies and scaling latencies.

Some of the user characteristics of this system include:

- The user must understand the timing and scheduling issues of the system, know when frequency can be decreased, and when it must be increased.
- The user must understand the effects that frequency scaling may have on the peripherals, and how to reprogram the peripherals accordingly.

Some of the other constraints include:

- Voltage scaling cannot be done unless a voltage regulator is present. However, frequency scaling can still be done without voltage scaling.
- Must have code generation tools to rebuild configuration data.
- The power management software does not control the peripheral clocks, only the main CPU clocks.

Following is a simple example of a boot-time power reduction module that is called during reset and immediately turns off power to unneeded peripherals and clock domains.

```
Void bootPowerReduction()
{
    ...
    // Call chip support library to turn OFF the DSP's CLKOUT signal
    CHIP_FSET(ST3_55, CLKOFF, 1);

    // Turn off timer 1 whenever Peripherals clock domain is idle
    hTimer = TIMER_open(TIMER_DEV1, TIMER_OPEN_RESET);
    TIMER_FSETH(hTimer, TCR, IDLEEN, 1);

    // Turn off unused clock domains.
    status = PWRM_idleClocks(PWRM_IDLECACHE, &idleStatus);

    // Turn off LEDs
    DSK5510_LED_init();
    for (i = 0; i < NUM_LEDS; i++) {
        DSK5510_LED_off(i);
    }

    // Turn off other resources not needed...
}
```

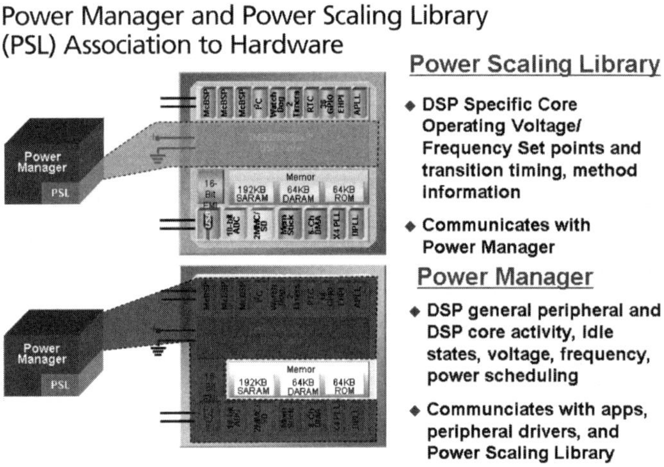

Figure 7.14 Power manager and power scaling library components (courtesy of Texas Instruments)

An example of a power scaling library (PSL) function to reduce the voltage and frequency of a device is shown in Figure 7.15. The basic sequence of events is as follows:

1. User code calls into PSL to reduce frequency.
2. PSL lowers frequency and may also lower voltage (depending on data structure).
3. PSL waits for the new frequency to be reached but not the voltage.
4. Time spent in PSL is around 70us, the time for the PLL to lock to the new frequency.
5. User code calls into PSL to increase frequency.
6. PSL does not return until both frequency and voltage have been scaled.

7. Must wait for the voltage to increase because it must be increased before the frequency is increased.
8. Longer time spent in PSL (time for voltage increase + time for PLL locking).

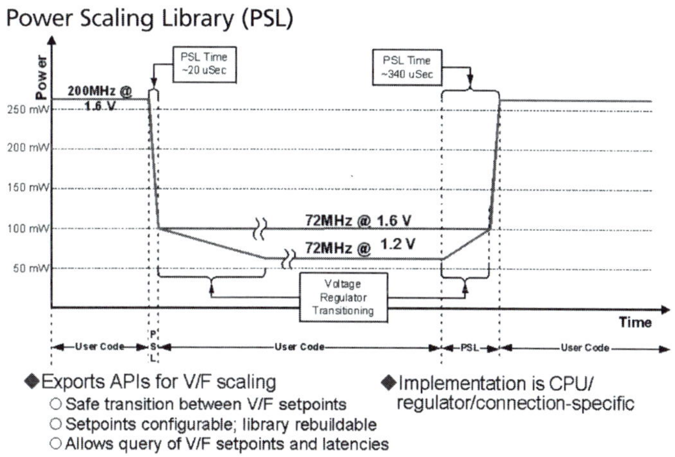

Figure 7.15 Example of the power scaling library function (courtesy of Texas Instruments)

For the specific example in Figure 7.15, the operations are as follows:

1. The frequency is backed off from 200 MHz to 72 MHz.
2. Modify voltage, first change frequency then change voltage from 1.6 volts to 1.2 volts, the latency is determined by how fast voltage is bled off.
3. The application can be doing work during this ramp down phase.
4. Must ramp up voltage, then the frequency is boosted back up.
5. At this point the application can't do anything during the ramp up, the power manager hangs on to processor during this time.
6. If the latency is OK, then the voltage can be dropped, if not then the power manager just needs to do the frequency scaling. Long latencies to ramp cause responsiveness issues so it's better to just stick with frequency scaling.

The architecture of an embedded application using the Power Manager software component is shown in Figure 7.16. The power manager software component contains the following:

- Loosely coupled architecture to RTOS kernel.
- A set of library routines; this executes in the client's context.
- Configurable by the application programmer.
- Uses platform-specific adaptation library for V/F scaling.
- Application, drivers, CLK register for notifications.
- Application triggers actions.

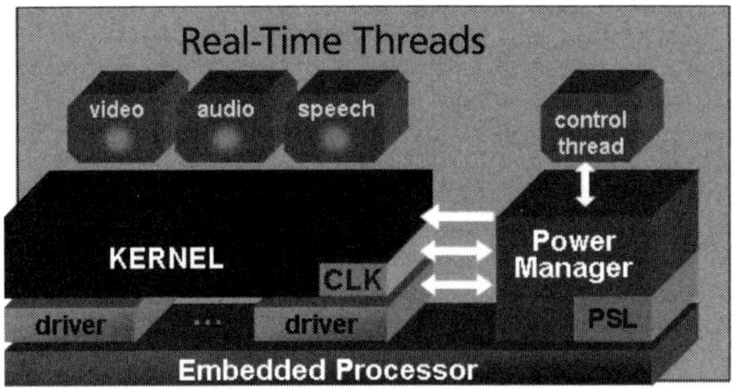

Figure 7.16 Software architecture of embedded application using the power manager (courtesy of Texas Instruments)

The role of the power manager software component is the following:

- Coordinates power management actions across the RTOS and entire application.
- Export APIs to enable application-driven power management. This includes:
  - Gating clocks and activating sleep modes.
  - Safely manage changes to V/F set points.
- Implements boot-time power savings.
- Idles clocks automatically when threads blocked.
- Adapts RTOS clock services upon frequency scaling.
- Provides registry for power event notifications.
  - Clients register/un-register for events at runtime.
  - Notify clients of events.
  - Support multiple instances of a client.
  - Allow clients to have delayed completion.

The power manager operates following the sequence shown in Figure 7.17. In this example, clients register and are notified about frequency-voltage events. The steps correspond to the numbers in Figure 7.17, with 1–3 as the registration sequence. Steps 4–7 represent the scaling sequence.

1. The application code registers to be notified of changes in frequency-voltage set points. The application developer is responsible for determining which parts of the application need to be notified.
2. A driver uses DMA to transfer data to and from external memory registers to be notified.
3. Packaged binary code registers to be notified.
4. The application decides to change the set point and calls the power API to initiate the set point change.
5. The power manager checks if the requested new set point is allowed for all registered clients, based on parameters they passed at registration, then notifies them of the impending set point change.

6. The power manager calls the F/V scaling library to change the set point. The F/V scaling library writes to the clock generation and voltage regulation hardware as appropriate to change the set point safely.
7. Following the change, the power manager notifies the appropriate clients.

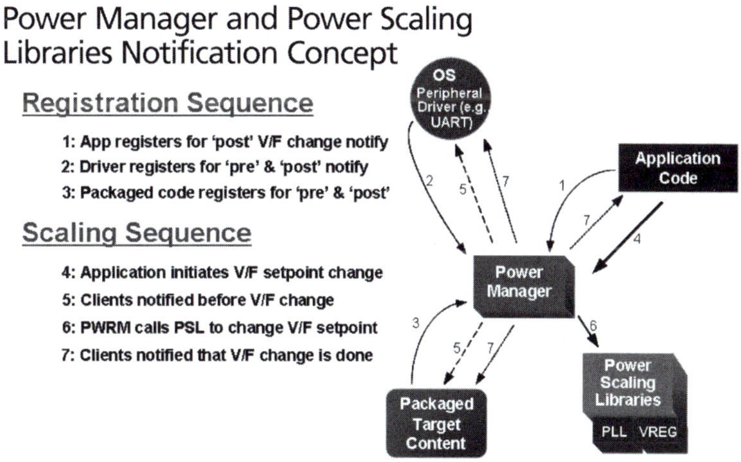

Figure 7.17. Power manager and power scaling library notification concept (courtesy of Texas Instruments)

Users may need to perform peripheral modifications that may be required as a result of the upcoming scaling operation. For example the user may need to stop a timer prior to changing the clock frequency. Users may also need to perform peripheral modifications that may be required as a result of the just completed scaling operation, like reprogramming and restarting a timer after changing the clock frequency.

Power Optimization Techniques for DSP Applications

There are several power management techniques available in application software development. The DSP engineer should consider each of these as they are architecting the software design and implementation for low power applications:

1. *Saving power at boot time* – The DSP engineer should specify a power-saving function to be called automatically at boot time. In many devices, including DSP, the device comes up after reset with everything turned on and clocks fed to every hardware module, effectively maximum power mode. The DSP engineer should design a low power module that is called as soon as possible after boot mode and immediately turns off those parts of the device and those peripherals that are not used or will not be used in the immediate future. This starts saving power immediately!
2. *Scaling voltage and frequency* – Dynamically change the operating voltage and frequency of the CPU based on the techniques discussed earlier.

3. *Using sleep modes and idling clock domains* – The DSP engineer should set custom sleep modes to save power during inactivity, including idling specific clock domains and peripherals to reduce active power consumption.
4. *Coordinating sleep and scaling* – Coordinate sleep modes and Voltage/Frequency scaling using registration and notification mechanisms as necessary. If the device is dormant you can be more aggressive at saving power. In these cases reduce voltage when necessary.
5. *Develop event driven vs. polling applications* – Applications should be written to be event driven instead of polling. In the case where a poll must be used, a time based poll should be used. In simple audio applications this can save 40% of the total power.

Code modeling is another effective technique to address power optimization. Each instruction a DSP executes consumes a certain amount of power (the cheapest instruction is the one you do not execute!). Several of the lessons learned from code performance optimization can be directly applied to the power optimization problem:

- The fewer instructions that are in your program, the less power it will consume.
- Fewer instructions mean a smaller memory footprint—the smaller the memory footprint, the fewer memory chips are required to keep powered up.
- Reducing the number of fetches from memory reduces the amount of power required to make the fetch from memory.
- The smaller the application, the more likely that significant parts of the application can be cached which reduces power required to execute the application.

The general approach for optimizing power using code modeling starts with profiling the application. Run a cycle profiler on the code to identify the "hot spots"—those areas in the code where most of the cycles are being spent. DSP applications spend a disproportionate amount of time in a small number of critical loops. These critical loops where the application spends most of the cycles is also most likely the place where the application is consuming the most power. Focusing on this small number of loops can provide most, if not all, of the power reduction needed for the application. As discussed earlier, many DSPs are designed to execute critical loops very efficiently using hardware support such as zero-overhead looping mechanisms. Programmers can use special instructions such as repeat block instructions (RPTB in the C5x DSP, for example) that eliminate the branch latency overhead for every iteration of the loop. Loop buffers provide another mechanism to help the programmer reduce power. If a critical application loop is small enough to fit within this special cache, the DSP will fetch instructions from this very low power, high-speed buffer rather than from slower, more power hungry off-chip memory. The DSP programmer should examine the code produced by the compiler to determine if loops are candidates for the loop buffer. Loops that are slightly larger than what would fit in the loop buffer should be "tweaked" if possible such that they fit within this buffer. It is possible to have a loop which apparently has more instructions but lower code size than an equivalent loop but runs faster and consumes less power because it fits in the loop buffer.

Keep in mind that fetches from memory consume power. Memory buses must be switched on and off and this is a power consumer. Modern optimizing DSP compilers will attempt to avoid fetching the same value repeatably from memory. However, if the code has complex pointer manipulations, particularly multiple pointer instructions, the compiler might not be able to determine that a referenced memory location always has the same value. Therefore, the DSP programmer should avoid using complicated pointer expressions where possible. Use array structures as an alternative. Write your application in a straightforward manner. This is a situation where becoming more clever with code production will not lead to the most optimal solution. Many modern optimizing compilers can make many of these transformations for you automatically—give the compiler a chance before making these code transformation decisions.

The link command file can also be used to direct the placement of critical sections of the application in on-chip memory, which is not only faster than off-chip memory, but consumes less power as well. DSP programmers should attempt to place as much of the critical code and data segments as possible in on-chip memory. Allocating each critical section its own section gives the linker more flexibility and freedom to pack the structures as it sees fit. The DSP programmer should take advantage of existing power-saving library features and use them in the application where needed. Keep in mind that the problems of size, speed, and power are not independent. What might be the optimal solution for speed might not necessarily be the optimal power solution. Caution should be used when trying to optimize the last few cycles of an application when the trade-off could be a disproportionate penalty in power, for example. The same rule can be stated again for clarity; be familiar with the application as well as the device in order to gain the most out of the optimization phase.

The DSP programmer must also address the need for accurate power measurement and estimation. This is important to DSP system designers who, in many cases, can only rely on device specifications that contain worst case power estimations. This worst case data forces the system designer to design to a worst case scenario for power. While at times this may be the correct approach, there will be situations where the DSP system designer will need to verify power consumption in the system, or part of the system. In these instances, having a robust set of power estimation and measurement tools helps the designer to make cost saving trade-offs very quickly. DSP vendors are beginning to produce such power estimation and measurement tools as the power requirements for embedded and portable products continue to look to the processing power of DSP while maintaining low power consumption for their portable applications.

Using Idle Modes

Modern DSPs contain special instructions that help reduce power. For example, the "idle" instruction on the C5x DSP will perform the following:

- Scales down the clock frequency.
- Turns off peripherals (peripherals can actually consume more power than the CPU).
- Turns off the cache.

This effectively puts part of the DSP to sleep, which dramatically reduces the power required for the device. A special interrupt is required to take the device out of this "sleep" mode.

Consider the following code snippet:

```
% cl55 -o power.c

#include <stdio.h>

char *reverse(char *str)
{
    int i;
    char *beg = str;
    char *end = str + strlen(str) - 1;

    for (i=0; i < strlen(str)/2; i++)
    {
       char t = *end;
       *end-- = *beg;
       *beg++ = t;
    }

    return str;
}
```

This function contains a call in the loop predicate (strlen(str)/2) which causes the length of the loop to exceed the loop buffer size and prevents the loop from being moved into the loop buffer. This prevents this loop from taking advantage of the power savings offered by the loop buffer mechanism. A small change to this loop such as shown below will now qualify this loop for movement into the loop buffer where the appropriate optimizations can be realized.

```
char *reverse_rptblocal(char *str)
{
    int i;
    char *beg = str;
    char *end = str + strlen(str) - 1;
    int len = strlen(str);

    for (i=0; i < len/2; i++)
    {
       char t = *end;
       *end-- = *beg;
       *beg++ = t;
    }

    return str;
}
```

Top Ten Power Optimizations

Many optimization techniques have been discussed in the last couple of chapters and techniques for power optimization have both similarities and differences to the techniques already discussed. Below is a list of the top ten power optimization techniques for the DSP engineer to use in guiding the application-level process of power optimization. A majority of these techniques are independent in the sense that they can be used in isolation or conjunction with other techniques.

1. Architect software to have natural "idle" points (including a low power boot). By intelligent architecting of the software, stopping or idle points in the application can be used as triggering points for turning off parts of the device or peripherals including the DMA. This makes the global management of the power efficiency of an application easier to manage.
2. Use interrupt-driven programming (instead of polling, and use the DSP operating system to block the application as necessary).
3. Code and data placement close to processor to minimize off-chip accesses (and overlays from nonvolatile to fast memory). Minimize off-chip memory accesses that require turning on and off buses and other peripherals. Use hierarchical memory models as much as possible, including the cache). Close to the CPU means less access time and less hardware used to get to the data.
4. Perform necessary code and data size optimizations to reduce the application footprint, memory and corresponding leakage caused by a larger required memory for the application.
5. Optimize for speed. There is, in many ways, a direct relationship between performance and power optimization. By optimizing the application for performance, the DSP engineer can build up a larger time base for more CPU idle modes or reduced CPU frequency opportunities (the engineer must benchmark and experiment to perform the correct trade-offs!).
6. Don't over calculate, use minimum data widths if at all possible. This leads to reduced bus activity which saves power as well as using smaller multipliers which will also save power.
7. Use the DMA for efficient transfer of data (instead of the CPU). This optimization technique has already been discussed.
8. Use co-processors to efficiently handle/accelerate frequent/specialized processing. Co-processors are optimized to perform certain computationally intensive tasks very efficiently, which also saves power.
9. Use more buffering and batch processing to allow more computation at once and more time in low power modes.
10. Use the DSP operating system to scale the Voltage/Frequency and analyze and benchmark to determine these points. (The engineer should ensure that the application is running correctly before doing this as well as other optimizations. This is the "make it work right, then make it work fast at low power" principle.)

Power Optimization Life Cycle

As mentioned earlier in this chapter, power optimization, like other optimization techniques, is a life cycle activity; even more so with power since it affects the purchase of the device as well as what you do after it. In this section we review the basic life cycle steps to power optimization.

A typical development flow will include application design, code and build, debug, and an analyze and tune phase, as shown in Figure 7.18.

In the application design phase, the engineer will be choosing the device to use in the application as well as performing some high level application design steps. During this phase, the engineer will be comparing power numbers at the device level. This can be done by looking at device specification sheets or a more detailed analysis of the application as it relates to the device. The goal is to find or otherwise develop detailed device-specific spreadsheets. These should provide detailed chip-specific internal function power as well as peripheral function power and other detailed data of operating power and standby power. With information at this level of detail, the engineer can plan the application net power consumption according to the utilization functions on the device. A detailed power spreadsheet provides for very easy trial configurations and allows the engineer to configure it however they feel they will configure their device as well as apply loads, however they wish to load their application to get a reasonably close proxy of what the power consumption will be for their design.

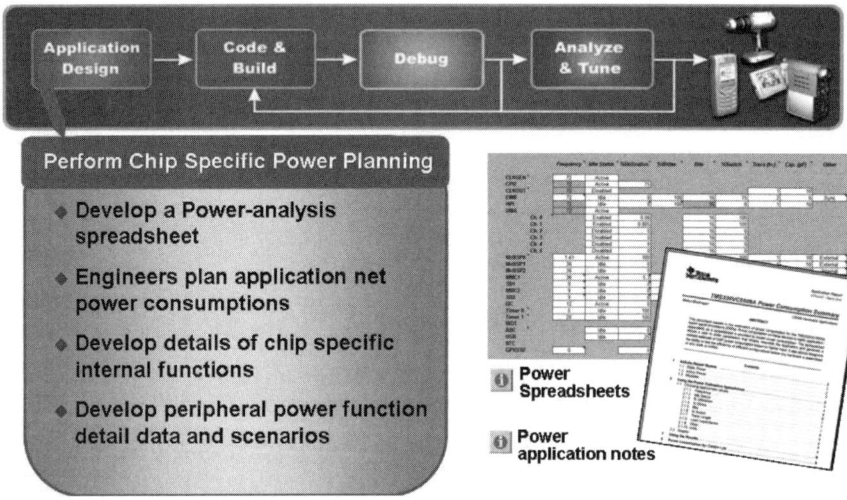

Figure 7.18 Power planning during the application design phase (courtesy of Texas Instruments)

The next phase of the power optimization life cycle is *code* and *build* (Figure 7.19). At this phase of the application development life cycle, the engineer is making a decision on the operating system. This is the level of abstraction the engineering team will be programming to. At this phase, in order to utilize the appropriate level of power optimization techniques, the engineer should look for OS support for power and frequency tuning. The combination of software at this phase that supports the coding and building of the application are chip specific power scaling libraries and a power aware OS manager. The power scaling libraries, which can be developed, purchased, or otherwise obtained, allow for supporting scaling of the frequency as well scaling of the voltage on these devices and provide query options for frequency voltage, scaling latencies, and other kinds of things that relate to the actual operating mode and the transitioning between your modes for power-efficient performance and callbacks, before and after scaling operations that accommodate these types of applications.

A *power manager* in the operating system is the software component that has the ability to activate at boot time the power savings that relate to low power modes that the engineer may want to be in, as opposed to full ON, which is difficult for boot products. It does allow managing clock domains and the application sleep and idle states, supervises and controls the power scaling and the voltage scaling as well as controls the I/O and the peripherals being turned on and off and it supports a central registry for a power event notification for your application to discuss with the power manager. It works in conjunction with the power scaling library of the particular device to provide a combined power management software capability.

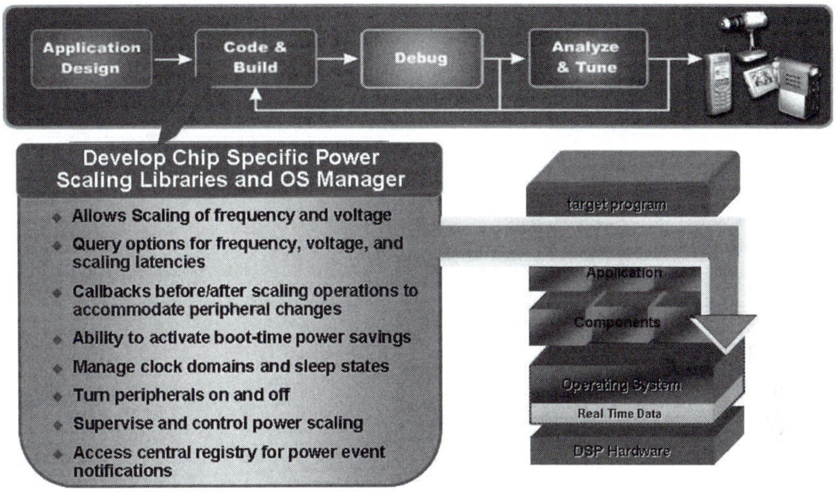

Figure 7.19 Code and build phase of the power optimization life cycle (courtesy of Texas Instruments)

At a minimum, the combined capability of a power scaling library and a power manager software component provides the DSP engineer the ability to:

1. *Idle clock domains* – The engineer can idle specific clock domains to reduce active power consumption.
2. *Saving power at boot time* – This allows the engineer to specify a power-saving function to be called automatically at boot time. This function can idle power-using peripherals as desired.
3. *Scaling voltage and frequency* – The engineer can dynamically change the operating voltage and frequency of the CPU. This is called *V/F scaling*. Since power usage is linearly proportional to the frequency and quadratically proportional to the voltage, using the PWRM module can result in significant power savings.

4. *Using sleep modes* – The engineer can set custom sleep modes to save power during inactivity. These can be set statically or at run-time.
5. *Coordinating sleep and scaling* – The engineer can coordinate sleep modes and V/F scaling using registration and notification mechanisms provided by the PWRM module.

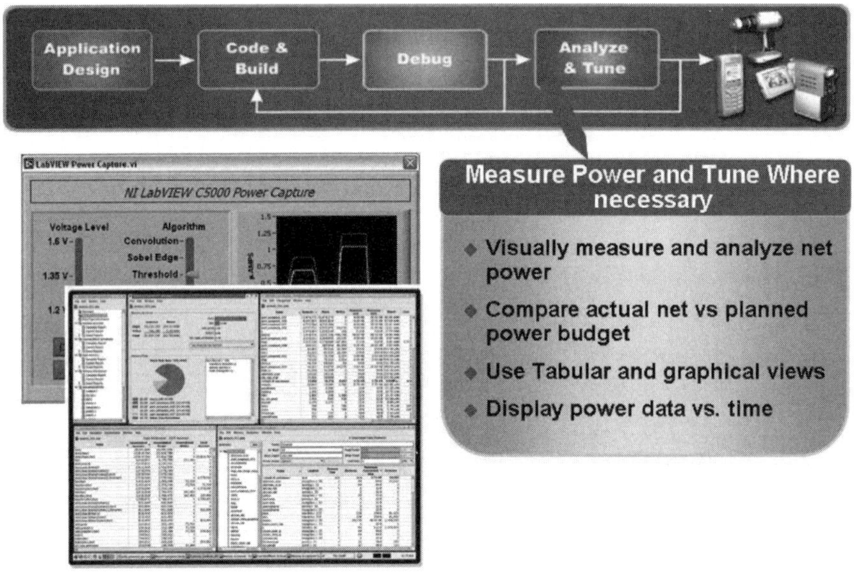

Figure 7.20 Analyze and tune phase of the power optimization life cycle (courtesy of Texas Instruments)

Once the DSP engineer has captured the plan and the power consumed, and implemented the code and the power scaling library as well as the OS power manager and power-efficient software with the application, the next phase is to measure results (Figure 7.20). In the analyze and tune phase of the life cycle the engineer wants to see what they have and possibly go back and tune it according to feedback. During this phase, the engineer should use the appropriate power measurement techniques and tools measure and analyze power at the core system level. The engineer should measure from the CPU core area as well as from the peripheral I/O area. If possible, this measurement should be viewed graphically in a data vs. time perspective. There are several tools that allow the engineer to both model and measure power in an application. One such tool is PowerEscape (www.powerescape.com), which allows algorithm developers to tune their C code for data efficiency in an architecture independent way, long before the memory architecture is defined. As we now know, data-efficient algorithms always perform faster and consume less energy. Intimate knowledge of the application domain is required to make algorithmic changes that significantly improve data efficiency. Such modifications therefore have to be made by the people who invent these algorithms and during the design of data-intensive applications. The developer can use a graphical user interface (Figure 7.21) during the power optimization process. In this figure, the user can get quick access to power profiles of the application, estimates of memory accesses, power consumption, cache misses, etc. This power optimization process fits within a more general power optimization framework as shown in Figure 7.22. There are also post-implementation power measurement tools available from companies such as National Instruments (www.ni.com).

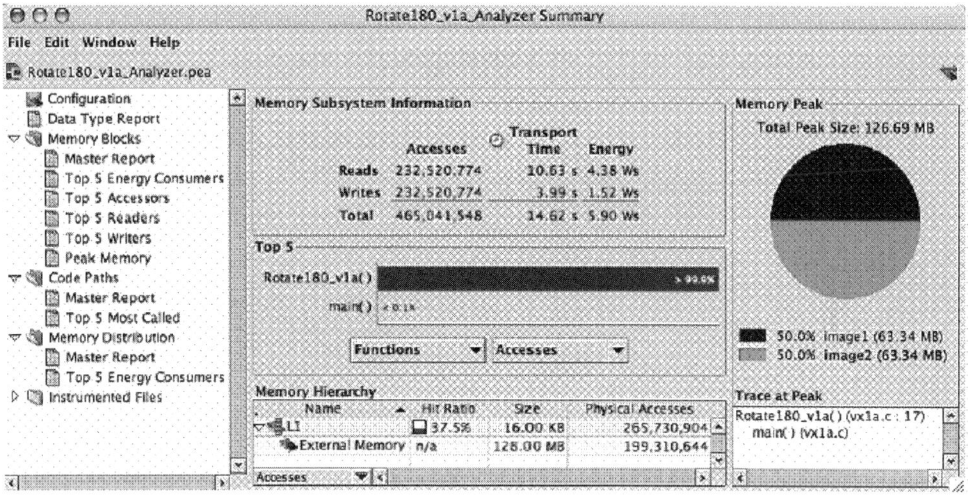

Figure 7.21 Power optimization tools such as PowerEscape provide the developer with aids to optimize application software early in the life cycle (figure courtesy of PowerEscape, www.powerescape.com)

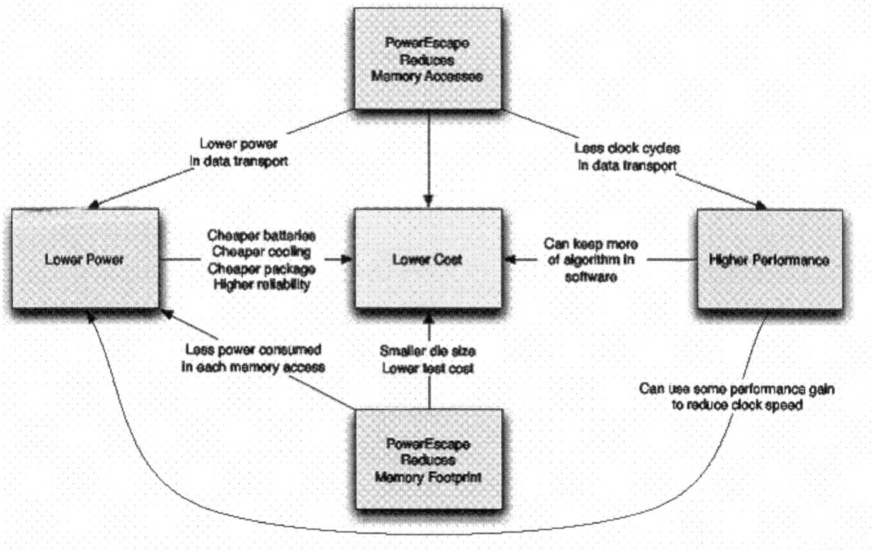

Figure 7.22 Power optimization framework as defined by PowerEscape (figure courtesy of PowerEscape, www.powerescape.com)

Power Optimization Summary

In summary, keep the following points in mind when attempting to optimize an application for power:

- Do less:
  - Execute fewer instructions.
  - Access memory less (the compiler can help here by performing automatic optimizations such as common sub-expression elimination).
  - Keep in mind that many performance optimizations also save power.
- Use low power resources:
  - Use the lowest power functional units for an operation when possible.
- Identify critical code and data:
  - Put important code on chip.
  - Put important data on chip.
  - Restructure the code to get maximum utilization of caches.
  - Keep the program "close" to the CPU (less bus and pin traffic).

As Figure 7.23 describes, the whole power story includes operating power as well as standby power. Standby power is addressed at the device level and the DSP engineer performs the necessary steps to select the right device for the application. The operating power is addressed with power scaling libraries, power aware operating system managers, and also application level optimization techniques.

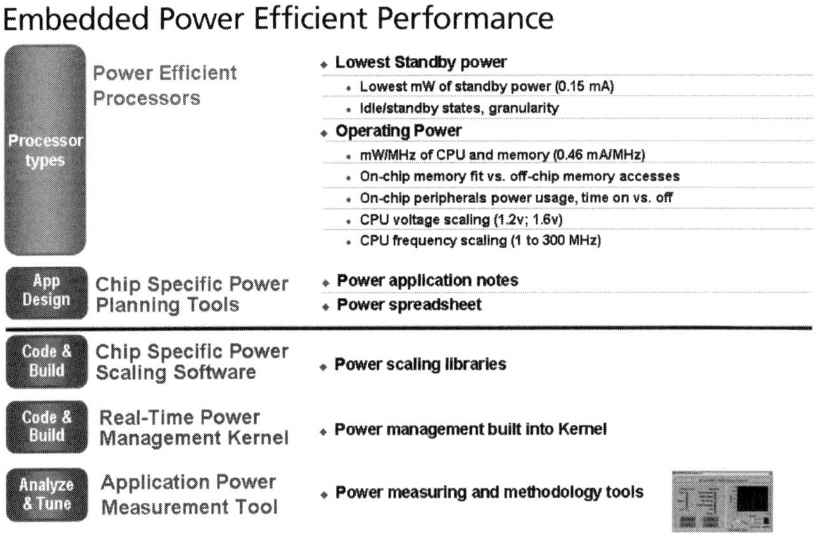

Figure 7.23 Components of the entire power spectrum for DSP application development (courtesy of Texas Instruments)

DSP vendors are designing devices that allow for power efficient operation. Architectural features such as clock idling, power rail gating, and dynamic voltage and frequency scaling can all help DSP system designers reduce power consumption in their systems.

In order for the DSP engineer to use these techniques effectively however, they must obtain sufficient information about power consumption during application operation. DSP power information and control must focus not just on typical core and memory consumption, but also operating modes, peripherals and I/O load.

The engineer must use the available tools and information from the vendor to obtain greater visibility into power consumption by different functions on the chip during real-time operation. To design a complete power optimization strategy, the engineer must select the right device at the right process technology and feature set, work to obtain detailed power usage information for the device and the application, and exploit the power-saving techniques discussed in this chapter to design an application tuned for power just as carefully as they design for performance.

Below is a summary of the up-front design decisions and runtime power management techniques for DSP power optimization.

Decision	Description
Choose a low-power silicon process	Choosing a power-efficient process (such as low-leakage CMOS) is the most important up-front decision, and directly drives power efficiency.
Choose low-power components	Components that have been designed from the ground up for power efficiency (for example, a processor with multiple clock and voltage domains) will provide dramatic savings in overall system power.
Partition separate voltage and clock domains	By partitioning separate domains, different components can be wired to the appropriate power rail and clock line, eliminating the need for all circuitry to operate at the maximum required by any specific module.
Enable scaling of voltage and frequency	Designing in programmable clock generators and programmable voltage sources will enable dynamic voltage and frequency scaling by the application or OS. Also, designing the hardware to minimize scaling latencies will enable broader usage of the scaling techniques.
Enable gating of different voltages to modules	Some circuit modules (e.g., static RAMs) require less voltage in retention mode vs. normal operation mode. By designing in voltage gating circuitry, power consumption can be reduced during inactivity, while still retaining state.
Utilize interrupts to alleviate polling by software	Often software is required to poll an interface periodically to detect events. For example, a keypad interface routine might need to spin or periodically wake to detect and resolve a keypad input. Designing the interface to generate an interrupt on keypad input will not only simplify the software, but it will also enable event-driven processing and activation of processor idle and sleep modes while waiting for interrupts.
Use hierarchical memory model	Leveraging caches and instruction buffers can drastically reduce off-chip memory accesses, and subsequent power draw.

Decision	Description
Reduce loading of outputs	Decreasing capacitive and DC loading on output pins will reduce total power consumption.
Minimize number of active PLLs	Using shared clocks can reduce the number of active clock generators, and their corresponding power draw. For example, a processor's on-board PLL might be bypassed in favor of an external clock signal.
Use clock dividers for fast selection of an alternate frequency	A common barrier to highly-dynamic frequency scaling is the latency of re-locking a PLL on a frequency change. Adding a clock divider circuit at the output of the PLL will allow instantaneous selection of a different clock frequency.

Table 7.1 Up-front design decisions

Technique	Description
Gate clocks off when not needed	By turning off clocks that are not needed, unnecessary active power consumption is eliminated.
On boot turn off unnecessary power consumers	Processors typically boot up fully powered, at a maximum clock rate. There will inevitably be resources powered that are not needed yet, or that may never be used in the course of the application. At boot time, the application or OS can traverse the system, turning off or idling unnecessary power consumers.
Gate power to subsystems only as needed	A system may include a power-hungry module that need not be powered at all times. For example, a mobile device may have a radio subsystem that only needs to be ON when in range of the device it is to communicate with. By gating power on-demand, unnecessary power dissipation can be avoided.
Activate peripheral low-power modes	Some peripherals have built-in low power modes that can be activated when the peripheral is not immediately needed. For example, a device driver managing a codec over a serial port may command the codec to a low power mode when there is no audio to be played, or if the whole system is being transitioned to a low-power sleep mode.
Leverage peripheral activity detectors	Some peripherals (e.g., disk drives) have built-in activity detectors that can be programmed to power down the peripheral after a period of inactivity.
Utilize auto-refresh modes	Dynamic memories and displays will typically have a self or auto-refresh mode where the device will efficiently manage the refresh operation on its own.
Benchmark application to find minimum required frequency and voltage	Typically, systems are designed with excess processing capacity built in, either for safety purposes, or for future extensibility and upgrades. For the latter case, a common development technique is to fully exercise and benchmark the application to determine excess capacity, and then 'dial-down' the operating frequency and voltage to that which enables the application to fully meet its requirements, but minimizes excess capacity. Frequency and voltage are usually not changed at runtime, but are set at boot time, based upon the benchmarking activity.

Technique	Description
Adjust CPU frequency and voltage based upon gross activity	Another technique for addressing excess processing capacity is to periodically sample CPU utilization at runtime, and then dynamically adjust the frequency and voltage based upon the empirical utilization of the processor. This "interval-based scheduling" technique improves on the power-savings of the previous static benchmarking technique because it takes advantage of the dynamic variability of the application's processing needs.
Dynamically schedule CPU frequency and voltage to match predicted work load	The "interval-based scheduling" technique enables dynamic adjustments to processing capacity based upon history data, but typically does not do well at anticipating the future needs of the application, and is therefore not acceptable for systems with hard real-time deadlines. An alternate technique is to dynamically vary the CPU frequency and voltage based upon predicted workload. For example, using dynamic, fine-grained comparison of work completed vs. the worst-case execution time (WCET), and deadline of the next task, the CPU frequency and voltage can be dynamically tuned to the minimum required. This technique is most applicable to specialized systems with data-dependent processing requirements that can be accurately characterized. Inability to fully characterize an application usually limits the general applicability of this technique.
Optimize execution speed of code	Developers often optimize their code for execution speed. In many situations the speed may be good enough, and further optimizations are not considered. When considering power consumption, faster code will typically mean more time for leveraging idle or sleep modes, or a greater reduction in the CPU frequency requirements. Note that in some situations speed optimizations may actually increase power consumption (for example, more parallelism and subsequent circuit activity), but in others there may be power savings.
Use low-power code sequences and data patterns	Different processor instructions exercise different functional units and data paths, resulting in different power requirements. Additionally, because of data bus line capacitances and the inter-signal capacitances between bus lines, the amount of power required is affected by the data patterns that are transferred over the bus. Analyzing the effects of individual instructions and data patterns is an extreme technique that is sometimes used to maximize power efficiency.
Use code overlays to reduce fast memory requirements	For some applications, dynamically overlaying code from nonvolatile to fast memory will reduce both the cost and power consumption of additional fast memory.
Enter a reduced capability mode on a power change	When there is a change in the capabilities of the power source, for instance when a laptop goes from AC to battery power, a common technique is to enter a reduced capability mode with more aggressive runtime power management. A similar technique can be employed in battery-only systems, where a battery monitor detects reduced capacity, and activates more aggressive power management, such as slowing down the CPU, not enabling viewing via a power-hungry LCD display, and so on.

Technique	Description
Trade-off accuracy vs. power consumption	It may be the case that an application is over-calculating results (such as using long integers when shorts would suffice). Accepting less accuracy in some calculations can drastically reduce processing requirements. For example, certain signal processing applications may be able to tolerate more noise in the results, which enables reduced processing, and reduced power consumption.

Table 7.2 Runtime power management techniques

References

"Optimize performance and power consumption with DSP hardware, software," Leon Adams, DSP Strategist and Raj Agrawala, C5000 Product Manager, Texas Instruments, Courtesy of *Power Management Design Line*, June 6, 2005 (13:04 PM).

URL: http://www.embedded.com/showArticle.jhtm

http://www.embedded.com/showArticle.jhtml?articleID=164300796

For more info, see SPRAA19 http://focus.ti.com/docs/apps/catalog/resources/appnoteabstract.jhtml?abstractName=spraa19

"Power-Optimizing Embedded Applications," *Proceedings of the Embedded Systems Conference*, Scott Gary, March 2004

http://www.esconline.com/archive/04sfpapers.htm

***Using the Power Scaling Library*, SPRA848, Texas Instruments.**

http://focus.ti.com/docs/apps/catalog/resources/appnoteabstract.jhtml?abstractName=spra848a

TMS320 DSP/BIOS User's Guide, SPRU423, Texas Instruments.

TMS320C5000 DSP/BIOS Application Programming Interface (API) Reference Guide, SPRA404, Texas Instruments.

I would like to acknowledge and give special thanks to Leon Adams and Scott Gary for their significant contributions to this chapter.

8

Real-Time Operating Systems for DSP

In the early days of DSP, much of the software written was low level assembly language that ran in a loop performing a relatively small set of functions . There are several potential problems with this approach:

- The algorithms could be running at different rates. This makes scheduling the system difficult using a "polling" or "super loop" approach.
- Some algorithms could overshadow other algorithms, effectively starving them. With no resource management, some algorithms could never run.
- There are no guarantees of meeting real-time deadlines. Polling in the fashion described is nondeterministic. The time it takes to go through the loop may be different each time because the demands may change dynamically.
- Nondeterministic timing.
- No or difficult interrupt preemption.
- Unmanaged interrupt context switch.
- "Super loop" approach is difficult to understand and maintain.

As application complexity has grown, the DSP is now required to perform very complex concurrent processing tasks at various rates. A simple polling loop to respond to these rates has become obsolete. Modern DSP applications must respond quickly to many external events, be able to prioritize processing, and perform many tasks at once. These complex applications are also changing rapidly over time, responding to ever changing market conditions. Time to market has become more important than ever. DSP developers, like many other software developers, must now be able to develop applications that are maintainable, portable, reusable, and scalable.

Modern DSP systems are managed by real-time operating systems that manage multiple tasks, service events from the environment based on an interrupt structure, and effectively manage the system resources.

What Makes an OS an RTOS?

An operating system is a computer program that is initially loaded into a processor by a boot program. It then manages all the other programs in the processor. The other programs are called *applications* or *tasks*. (I'll describe tasks in more detail later.)

The operating system is part of the system software layer (Figure 8.1). The function of the system software is to manage the resources for the application. Examples of system resources that must be managed are peripherals like a DMA, HPI, or on-chip memory. The DSP is a processing resource to be managed and scheduled like other resources.

The system software provides the infrastructure and hardware abstraction for the application software. As application complexity grows, a real-time kernel can simplify the task of managing the DSP MIPS efficiently using a multitasking design model. The developer also has access to a standard set of interfaces for performing I/O as well as handling hardware interrupts.

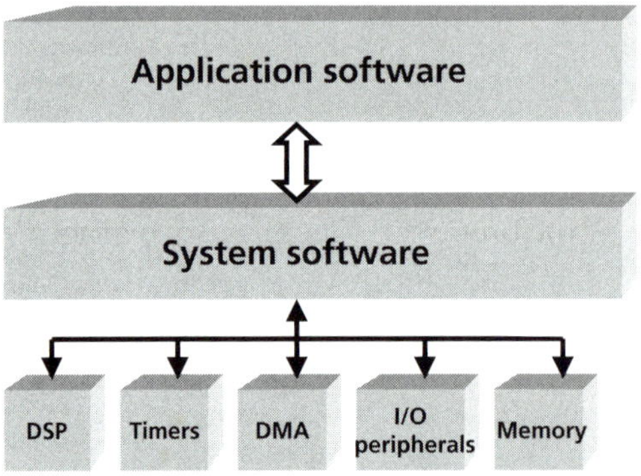

Figure 8.1 Embedded DSP software components

Migration to next generation processors is also easier because of the hardware abstraction provided by a real-time kernel and associated peripheral support software.

The applications make use of the operating system by making requests for services through a defined application program interface (API).

A general-purpose operating system performs these services for applications:

- In a multitasking operating system where multiple programs can be running at the same time, the operating system determines which applications should run in what order and how much time should be allowed for each application before giving another application a turn.
- It manages the sharing of internal memory among multiple applications.
- It handles input and output to and from attached hardware devices, such as hard disks, printers, and dial-up ports.

- It sends messages to each application or interactive user about the status of operation and any errors that may have occurred.
- It can offload the management of what are called *batch jobs* (for example, printing) so that the initiating application is freed from this work.
- On computers that can provide parallel processing, an operating system can manage how to divide the program so that it runs on more than one processor at a time.

In some respects, almost any general-purpose operating system (Microsoft's Windows NT for example) can be evaluated for its real-time operating system qualities. But, here we will define a real-time operating system (RTOS) to be a specialized type of operating system that guarantees a certain capability within a specified time constraint.

An operating system must have certain properties to qualify it as a real-time operating system. Most importantly, a RTOS must be multithreaded and pre-emptible. The RTOS must also support task priorities. Because of the predictability and determinism requirements of real-time systems, the RTOS must support predictable task synchronization mechanisms. A system of priority inheritance must exist to limit any priority inversion conditions. Finally, the RTOS behavior should be known to allow the application developer to accurately predict performance of the system[1].

The purpose of a RTOS is to manage and arbitrate access to global resources such as the CPU, memory, and peripherals. The RTOS scheduler manages MIPS and real-time aspects of the processor. The memory manager allocates, frees, and protects code and memory. The RTOS drivers manage I/O devices, timers, DMAs, and so on. Aside from that, real-time operating systems do very little explicit resource management to provide acceptable predictability of task completion times.

Selecting an RTOS

A good RTOS is more than a fast kernel! Execution efficiency is only part of the problem. You also need to consider cost of ownership factors. Having the best interrupt latency or context switch time won't seem so important if you find the lack of tool or driver support puts your project months behind schedule.

Here are some points to consider:

Is it specialized for DSP? Some real-time operating systems are created for a special applications such as DSP or even a cell phone. Others are more general-purpose operating systems. Some of the existing general-purpose operating systems claim to be real-time operating systems.

How good is the documentation and support?

[1] An example is the interrupt latency (that is time from interrupt to task run). This has to be compatible with the application requirements and has to be predictable. This value depends on the number of simultaneous pending interrupts. For every system call, this value would be the maximum time to process the interrupt. The interrupt latency should be predictable and independent from the number of objects in the system. Another example of an important measure is the maximum time the RTOS and drivers mask the interrupts.

How good is the tool support? The RTOS should also be delivered with good tools to develop and tune the application.

How good is the device support? Are there drivers for the devices you plan to use? For those you are most likely to add in the future?

Can you make it fit? More than general-purpose operating systems, RTOS should be modular and extensible. In embedded systems the kernel must be small because it is often in ROM and RAM space may be limited. That is why many RTOSs consist of a micro kernel that provides only essential services:

- scheduling
- synchronization
- interrupt handling
- multitasking

Is it certifiable? Some systems are safety critical and require certification, including the operating system

DSP Specializations

Choosing a DSP-oriented RTOS can have many benefits. A typical embedded DSP application will consist of two general components, the application software and the system software.

DSP RTOSs have very low interrupt latency. Because many DSP systems interface with the external environment, they are event or interrupt driven. Low overhead in handling interrupts is very important for DSP systems. For many of the same reasons, DSP RTOSs also ensure that the amount of time interrupts are disabled is as short as possible. When interrupts are disabled (context switching, etc) the DSP cannot respond to the environment.

A DSP RTOS also has very high performance device independent I/O. This involves basic I/O capability for interacting with devices and other threads. This I/O should also be asynchronous, have low overhead, and be deterministic in the sense that the completion time for an I/O transfer should not be dependent on the data size.

A DSP RTOS must also have specialized memory management. A DSP RTOS should provide the capability to define and configure system memory efficiently.

Capability to align memory allocations and multiple heaps with very low space overhead is important. The RTOS will also have the capability to interface to the different types of memory that may be found in a DSP system, including SRAM, SDRAM, fast on chip memory, etc.

A DSP RTOS is more likely to include an appropriate Chip Support Library (CSL) for your device. I'll discuss CSLs in more detail later in this chapter.

Concepts of RTOS

Real-time operating systems require a set of functionality to effectively perform their function, which is to be able to execute all of their tasks without violating specified timing constraints. This section describes the major functions that real-time operating systems perform.

Task-Based

Functions—task-based

A task is a basic unit of programming that an operating system controls. Each different operating system may define a task slightly differently. A task implements a computation job and is the basic unit of work handled by the scheduler. The kernel creates the task, allocates memory space to the task and brings the code to be executed by the task into memory. A structure called a *task control block* (TCB) is created and used to manage the schedule of the task. A task is placeholder information associated with a single use of a program that can handle multiple concurrent users. From the program's point-of-view, a task is the information needed to serve one individual user or a particular service request. .

Types of real-time tasks:

Periodic – Executes regularly according to a precomputed schedule.

Sporadic – Triggered by external events or conditions, with a predetermined maximum execution frequency.

Spontaneous – Optional real-time tasks that are executed in response to external events (or opportunities), if resources are available.

Ongoing tasks – Fair-share threads.

Multitasking

Today's microprocessors can only execute one program instruction at a time. But because they operate so fast, they can be made to appear to run many programs and serve many users simultaneously. This illusion is created by rapidly multiplexing the processor over the set of active tasks. The processor operates at speeds that make it seem as though all of the user's tasks are being performed at the same time.

While all multitasking depends on some time of processor multiplexing, there are many strategies for "scheduling" the processor. In systems with priority-based scheduling, each task is assigned a priority depending on its relative importance, the amount of resources it is consuming, and other factors. The operating system preempts tasks having a lower priority value so that a higher priority task is given a chance to run. I'll have much more to say about scheduling later in this chapter.

Forms of Multitasking

There are three important multitasking algorithms:

- *Preemptive* – With this algorithm, if a high-priority task becomes ready for execution it can immediately pre-empt the execution of a lower-priority task and acquire the processor without having to wait for the next regular rescheduling. In this context, "immediately" means after the scheduling latency period. This latency is one of the most important characteristics of a real-time kernel and largely defines the system responsiveness to external stimuli.
- *Cooperative* – With this algorithm, if a task is not ready to execute it voluntarily relinquishes the control of the processor so that other tasks can run. This algorithm does not require much scheduling and generally is not suited for real-time applications.
- *Time-sharing* – A pure time sharing algorithm has obvious low responsiveness (limited by the length of scheduling interval). Nevertheless, a time sharing algorithm is always implemented in real-time operating systems. The reason for this is that there is almost always more than one nonreal-time task in the real-time system (interaction with user, logging of accounting information, various other bookkeeping tasks). These tasks have low priority and are scheduled with a time-sharing policy in the time when no tasks of higher priority are ready for execution.

Rapid Response to Interrupts

An interrupt is a signal from a device attached to a computer or from a program within the computer that causes the RTOS to stop and figure out what to do next. Almost all DSPs and general-purpose processors are interrupt-driven. The processor will begin executing a list of computer instructions in one program and keep executing the instructions until either the task is complete or cannot go any further (waiting on a system resource for example) or an interrupt signal is sensed. After the interrupt signal is sensed, the processor either resumes running the program it was running or begins running another program.

An RTOS has code that is called *an interrupt handler*. The interrupt handler prioritizes the interrupts and saves them in a queue if more than one is waiting to be handled. The scheduler program in the RTOS then determines which program to give control to next.

An interrupt request (IRQ) will have a value associated with it that identifies it as a particular device. The IRQ value is an assigned location where the processor can expect a particular device to interrupt it when the device sends the processor signals about its operation. The signal momentarily interrupts the processor so that it can decide what processing to do next. Since multiple signals to the processor on the same interrupt line might not be understood by the processor, a unique value must be specified for each device and its path to the processor.

In many ways, interrupts provide the "energy" for embedded real-time systems. The energy is consumed by the tasks executing in the system. Typically, in DSP systems, interrupts are generated on buffer boundaries by the DMA or other equivalent hardware. In this way, every interrupt occurrence will ready a DSP RTOS task that iterated on full/empty buffers. Interrupts come from many sources (Figure 8.2) and the DSP RTOS must effectively manage multiple interruptions from both inside and outside the DSP system.

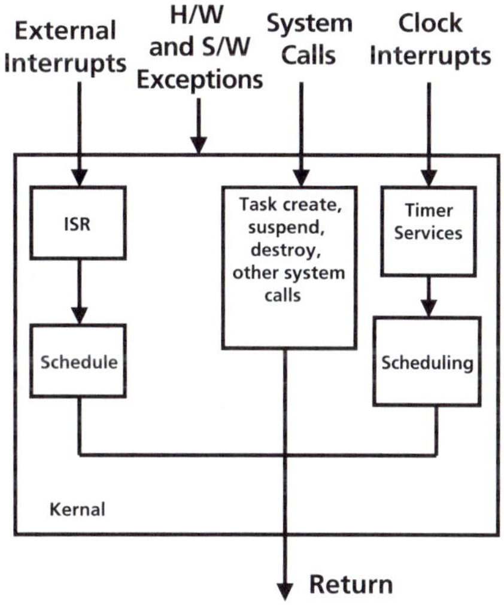

Figure 8.2 An RTOS must respond to and manage multiple interruptions from inside and outside the application

RTOS Scheduling

The RTOS scheduling policy is one of the most important features of the RTOS to consider when assessing its usefulness in a real-time application. The scheduler decides which tasks are eligible to run at a given instant, and which task should actually be granted the processor. The scheduler runs on the same CPU as the user tasks and this is already the penalty in itself for using its services. There are a multitude of scheduling algorithms and scheduler characteristics. Not all of these are important for a real-time system.

A RTOS for DSP requires a specific set of attributes to be effective. The task scheduling should be priority based. A task scheduler for DSP RTOS has multiple levels of interrupt priorities where the higher priority tasks run first. The task scheduler for DSP RTOS is also preemptive. If a higher priority task becomes ready to run, it will immediately pre-empt a lower priority running task. This is required for real-time applications. Finally, the DSP RTOS is event driven. The RTOS has the capability to respond to external events such as interrupts from the environment. DSP RTOS can also respond to internal events if required.

Scheduler Jargon

Multitasking jargon is often confusing to newcomers because there are so many different strategies and techniques for multiplexing the processor. The key differences among the various strategies revolve around how a task loses control and how it gains control of the processor. While these are separate design decisions (for the RTOS designer), they are often treated as being implicitly linked in marketing literature and casual conversation.

Scheduling regimens viewed by how tasks lose control:

- Only by voluntary surrender. This style is called *cooperative multitasking*. For a task to lose control of the processor, the task must voluntarily call the RTOS. These systems continue to multitask only so long as all the tasks continue to share graciously.
- Only after they've finished their work. Called *run to completion*. To be practical, this style requires that all tasks be relatively short duration.
- Whenever the scheduler says so. Called *preemptive*. In this style the RTOS scheduler can interrupt a task at any time. Preemptive schedulers are generally more able to meet specific timing requirements than others. (Notice that if the scheduler "says so" at regular fixed intervals, then this style is called *time slicing*.

Scheduling regimens viewed by how tasks gain control:

- By being next in line. A simple FIFO task queue. Sometimes mistakenly called *round-robin*. Very uncommon.
- By waiting for it's turn in a fixed-rotation. A slop cyclic scheduler. The "super loop" mentioned at the beginning of this chapter is this form of cyclic scheduler.If the cycle is only allowed to restart at specific fixed intervals, it's called a *rate cyclic* scheduler.
- By waiting a specific amount of time. A very literal form of multiplexing in which each ready to execute task is given the processor for a fixed-quantum of time. If the tasks are processed in FIFO order, this style is called a *round-robin* scheduler. If the tasks are selected using some other scheme it's considered a time-slicing scheduler.
- By having a higher priority than any other task wanting the processor. A priority-based or "prioritzed" scheduler.

Not all of these combinations make sense, but even so, it's important to understand that task interruption and task selection are separate mechanisms. Certain combinations are so common (e.g., premptive prioritized), that one trait (e.g., prioritized) is often misconstrued to imply the other (e.g., preemptive). In fact, it is perfectly reasonable to have a prioritized, nonpreemptive (e.g. run to completion) scheduler. For technical reasons, prioritized-preemptive schedulers are the most frequently used in RTOSs.

The scheduler is a central part of the kernel. It executes periodically and whenever the state of a thread changes. A single-task system does not really need a scheduler since there is no competition for the processor. Multitasking implies scheduling because there are multiple tasks competing for the processor . A scheduler must run often enough to monitor the usage of the CPU by the tasks. In most real-time systems, the scheduler is invoked at regular intervals. This invocation is usually the result of a periodic timer interrupt. The period in which this interrupt is invoked is called the *tick* size or the system "heartbeat." At each clock interrupt the RTOS updates the state of the system by analyzing task execution budgets and making decisions as to which task should have access to the system CPU.

Task states

Further refinement comes when we allow tasks to have states other than ready for execution. The major states of a typical RTOS include:

- Sleeping – The thread is put into a sleeping state immediately after it is created and initialized. The thread is released and leaves this state upon the occurrence of an event of the specified type(s). Upon the completion of a thread that is to execute again it is reinitialized and put in the sleeping state.
- Ready – The thread enters the ready state after it is released or when it is pre-empted. A thread in this state is in the ready queue and eligible for execution. (The RTOS may also keep a number of ready queues. In fixed-priority scheduling, there will be a queue for each priority. The RTOS scheduler rather than simply admitting the tasks to the CPU, makes a decision based on the task state and priority.)
- Executing – A thread is in the executing state when it executes
- Suspended (Blocked) – A thread that has been released and is yet to complete enters the suspended or blocked state when its execution cannot proceed for some reason. The kernel puts a suspended thread in the suspended queue. There are various reasons for a multitasking DSP to suspend. A task may be blocked due to resource access control. A task may be waiting to synchronize execution with some other task(s). The task may be held waiting for some reason (I/O completion and jitter control). There may be no budget or aperiodic[2] job to execute (this is a form of bandwidth control). The RTOS maintains different queues for tasks suspended or blocked for different reasons (for example, a queue for tasks waiting for each resource).
- Terminated – A thread that will not execute again enters the terminated state when it completes. A terminated thread may be destroyed.

[2] Tasks are not always periodic. Tasks that are not periodic are called *aperiodic*. Examples of aperiodic tasks include operator requests, emergency message arrivals, threshold crossing notifications, keyboard presses, mouse movements, and detection of incoming objects.

Different RTOS's will have slightly different states. Figure 8.3 shows the state model for a DSP RTOS, DSP/BIOS, called *DSP/BIOS* from Texas Instruments.

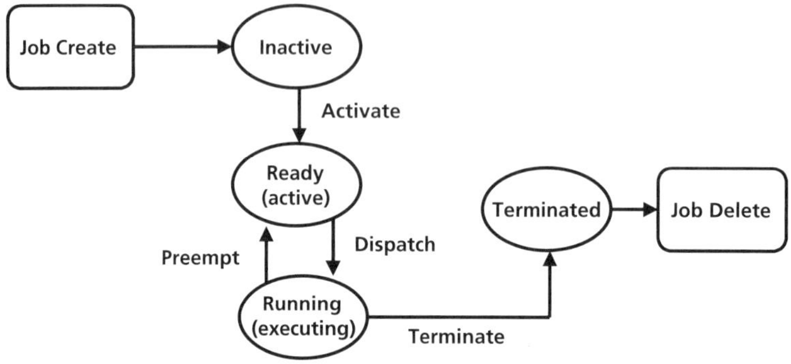

Figure 8.3 State model (with preemption) for the Texas Instruments DSP BIOS RTOS

Most RTOSs provide the user the choice of round-robin or FIFO scheduling of threads of equal priority. A *time slice* (execution budget) is used for round-robin scheduling. At each clock interrupt the scheduler reduces the budget of the thread by the tick size. Tasks will be pre-empted if necessary (A FIFO scheduling algorithm uses an infinite time slice.) At each clock interrupt, the RTOS scheduler updates the ready queue and either returns control to the current task or preempts the current task if it is no longer the highest priority ready task. Because of the previous actions, some threads may become ready (released upon timer expiration) and the thread executing may need to be pre-empted. The scheduler updates the ready queue accordingly and then gives control to the thread at the head of the highest priority queue.

Cyclic Executives

Cyclic executives (also called *interrupt control loops*) were until recently a popular method for designing simple DSP based real-time systems. The "cyclic" name comes from the way the control loop "cycles" through a predetermined path in the code in a very predictable manner. It is a very simple architecture because there is minimal "executive" code in the system which leads to minimal system overhead. These types of systems require only a single regular (periodic) interrupt to manage the control loop.

Steps in cyclic executive system model

An example of a cyclic executive is shown in Figure 8.4. Cyclic executive systems will generally perform one-time start-up processing (including enabling the system interrupt). The system then enters an infinite loop. Each cycle through the loop (after the first cycle) is initiated by a periodic interrupt. During each cycle, the system will first accept any required input. Application processing is then performed on the data read into the system. The resulting outputs are then generated. At the end of the cycle, the software enters an idle loop waiting for the interrupt to initiate the next cycle. During this idle time, the system could perform some low level built-in-test processing.

Disadvantages to cyclic executive system model

This basic cyclic executive system model is effective for simple systems where all processing is done at the rate established by the periodic interrupt. It may be unnecessary, or even undesirable, to run all "tasks" at this same rate. This is one of the drawbacks of this model – inefficiency.

This has an effect on the timing of the data output. If each task in the system has a different execution time, the timing of the data output, in relation to the regular period of the process, varies. This effect is called *jitter* and will occur on alternate cycles through the system. Depending on the amount of this jitter and the sensitivity of the system this effect may be totally unsatisfactory for a real-time system. One solution is to pad the shorter running task(s) with worthless computations to add time until it the shorter running task(s) take the same amount of time to complete as the other task(s). This is another inelegant approach to solving this synchronization problem. If a change is made to one of the tasks, the timing of the system may have to be adjusted to reestablish the output timing stability.

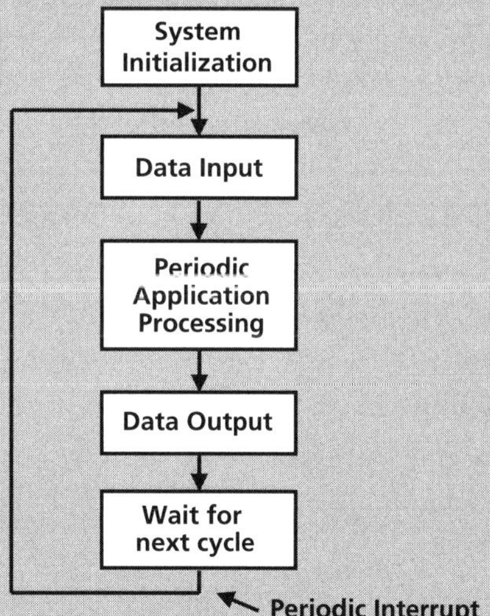

Figure 8.4 Simple model of a cyclic executive

Advantages to cyclic executive system model

The Cyclic executive approach has some advantages. The model is conceptually simple and potentially very efficient, requiring relatively little memory and processing overhead.

Disadvantages to cyclic executive system model

Despite the advantages to this approach, cyclic executives have some serious disadvantages. It is very difficult to develop systems of any complexity with this model because the cyclic executive model does not support arbitrary task processing rates. All of the tasks in the system must run at a rate that is a harmonic of the periodic interrupt. Cyclic executives also require even simple control flow relationships to be hard coded into the program flowing from lower task rates to higher task rates. Because cyclic executives are nonpreemptive, the system developer must divide the lower rate processing into sometimes unnatural "tasks" that can be run each cycle. Cyclic executives do not support aperiodic processing required by many DSP-based real-time systems. It is very difficult and error-prone to make changes to the system once it is working. Even relatively small changes to the system may result in extensive repartitioning of activities, especially if the processor loading is high, to achieve system stability again. Any changes in processing time in the system can affect the relative timing of input or output operations. This will require revalidating the behavior of any control law parameters in the system. Any changes to the application requires that a timing analysis must be performed to prevent any undesirable side effects on the overall system operation.

The RTOS Kernel

A kernel can be defined as the essential center of an operating system. The kernel is the core of an operating system that provides a set of basic services for the other parts of the operating system. A kernel consists of a basic set of functions:

- An interrupt handler to handle the requests that compete for the kernel's services.
- A scheduler to determine which tasks share the kernel's processing time and in which order they shall use this time.
- A manager of resources that manage system resources usage such as memory or storage and how they shall be shared among the various tasks in the system.

There may be other components of operating systems such as file managers, but these are considered outside the essential requirements of the kernel needed for core processing capability.

A RTOS consists of a kernel that provides the basic operating system functions. There are three reasons for the kernel to take control from the executing thread and execute itself:

- To respond to a system call.
- To perform scheduling and service timers.
- To handle external interrupts.

The general priority of an RTOS scheduling service is as follows (Figure 8.5):

Highest priority: Hardware interrupts – Hardware interrupts have the highest priority. One hardware interrupt can interrupt another.

Next Highest Priority: Software interrupts – The RTOS may also support a number of software interrupts. Software interrupts have a lower priority than hardware interrupts and are prioritized within themselves. A higher software interrupt can pre-empt a lower priority software interrupt. All hardware interrupts can pre-empt software interrupts. A software interrupt is similar to a hardware interrupt and is very useful for real-time multitasking applications. These interrupts are driven by an internal RTOS clock module. Software interrupts run to completion and do not block and cannot be suspended.

Next Highest Priority: Tasks – Tasks have lower priority than software interrupts. Tasks can also be pre-empted by higher priority tasks. Tasks will run to completion or yield if blocked waiting for a resource or voluntarily to support certain scheduling algorithms such as round robin.

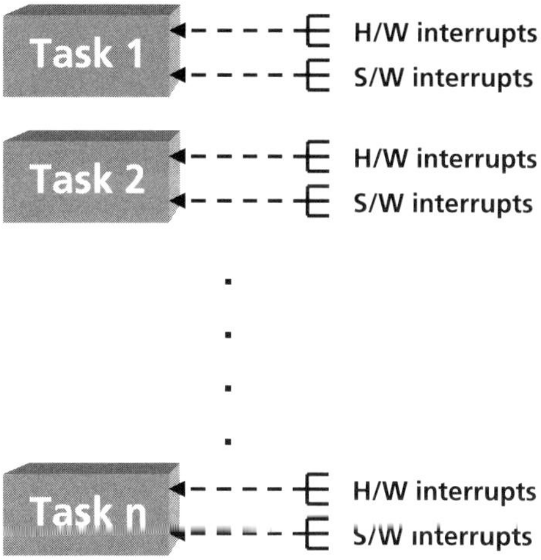

Figure 8.5 The relative priority of activity in a DSP RTOS; hardware interrupts, software interrupts, and tasks

System Calls

The application can access the kernel code and data via application programming interface (API) functions. An API is the specific method prescribed by a computer operating system or by an application program by which a programmer writing an application program can make requests of the operating system or another application. A *system call* is a call to one of the API functions. When this happens, the kernel saves context of calling task, switches from user mode to kernel mode (to ensure memory protection), executes the function on behalf of the calling task, and returns to user mode. Table 8.1 describes some common APIs for the DSP/BIOS RTOS.

API Name	Function
ATM_cleari, ATM_clearu	Atomically clear memory location and return the previous value
ATM_seti, ATM_setu	Atomically set memory and return previous value
CLK_countspms	Number of hardware timer counts per millisecond
CLK_gethtime, CLK_getltime	Get high and low resolution time
HWI_denable, HWI_disable	Globally enable and disable hardware interrupts
LCK_create, LCK_delete	Create and delete a resource lock
LCK_pend, LCK_post	Aquire and relinquish ownership of a resource lock
MBX_create, MBX_delete, MBX_pend, MBX_post	Create, delete, wait for a message, and post a message to a mailbox
MEM_alloc, MEM_free	Allocate from a memory segment, free a block of memory
PIP_get, PIP_put	Get and put a frame into a pipe
PRD_getticks, PRD_tick	Set the current tick counter, Advance the tick counter and dispatch periodic functions
QUE_Create, QUE_delete, QUE_dequeue, QUE_empty	Create and delete an empty queue, remove from front of queue and test for an empty queue
SEM_pend, SEM_post	Wait for and signal a semaphore
SIO_issue, SIO_put	Send and put a buffer to a stream
SWI_enable, SWI_getattrs	Enable software interrupts, Get attributes of a software interrupt
SYS_abort, SYS_error	Abort program execution, flag error condition
TSK_create, TSK_delete	Create a task for execution, delate a task
TSK_disable, TSK_enable	Disable and enable the RTOS scheduler
TSK_getenr, TSK_setpri	Get the task environment, set a task priority

Table 8.1 Example APIs for DSP/BIOS RTOS

Dynamic Memory Allocation

Dynamic memory allocation allows the allocation of memory at variable addresses. The actual address of the memory is usually not a concern to the application. One of the disadvantages of dynamic memory allocation is the potential for memory fragmentation. This occurs when there is enough memory in total but an allocation of a memory block fails due to lack of a single large enough block of memory. RTOS must manage the heap effectively to allow maximum use of the heap.

Chip Support Software for DSP RTOS

A part of the DSP RTOS support infrastructure is software to support the access of the various chip peripherals. This Chip Support Library (CSL) is a runtime library used to configure and control all of the DSP on-chip peripherals such as the cache, DMA, external memory interface (EMIF), multichannel buffered serial port (MCBSP), the timers, high speed parallel interfaces (HPI), etc. The CSL provides the peripheral initialization as well as the runtime control. This software fits well into a DSP RTOS package to provide the necessary abstraction from the hardware that the DSP programmer requires to be productive and effective. DSP vendors can provide this software written (mostly) in C and already optimized for code size and speed. Figure 8.6 shows how the CSL fits into the DSP RTOS infrastructure.

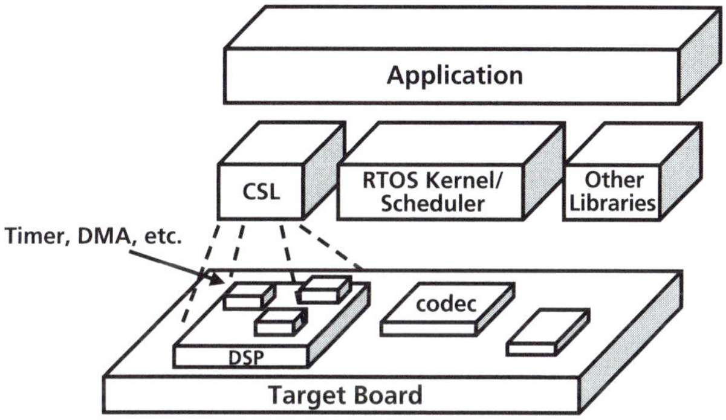

Figure 8.6 Chip Support Library to support a DSP RTOS infrastructure (courtesy of Texas Instruments)

Benefits for the CSL

Because different DSP chips and families of chips support different peripherals (a TMS320C2x DSP device may support specific peripherals to support motor control such as a CAN device, while a DSP such as the TMS320C5x device, which is optimized for low power and portable applications, may support a different set of peripherals including a Multichannel buffered serial port for optimized I/O), a different chip support library is required for each device. The hardware abstraction provided by these libraries adds very little overhead to the system and provides several benefits to the DSP developer:

- *Shortened development time* – The DSP developer will not be burdened with understanding a large number of memory mapped registers for the DSP device. A hardware abstraction layer in the CSL has already defined this (Figure 8.7).
- *Standardization* – A properly developed CSL will have a uniform methodology for developing the API definitions to be used by the DSP developer. The level of abstraction and the coverage from a macro level will be consistent.
- *Tool support* – DSP tools support is often available to make the job of generating customized configuration structures and initializing registers easier.

- Resource management – Basic resource management for peripherals that have multiple channels, ports, devices, and so on is much easier.

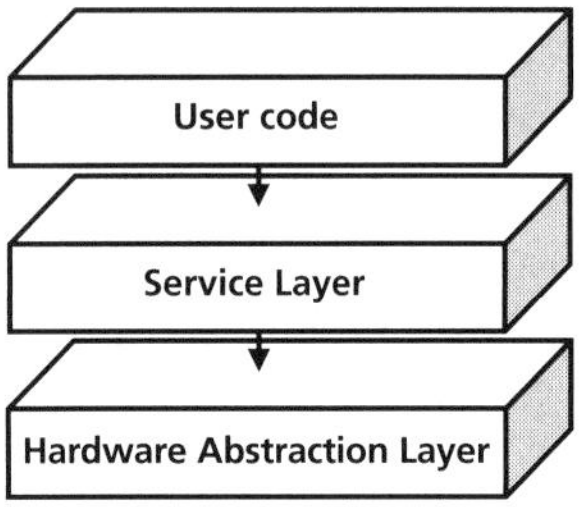

Figure 8.7 Hardware abstraction makes managing DSP peripherals easier

Example

Figure 8.8 is an example of a simple CSL function. This example opens an EDMA channel, configures it, and closes it when finished.

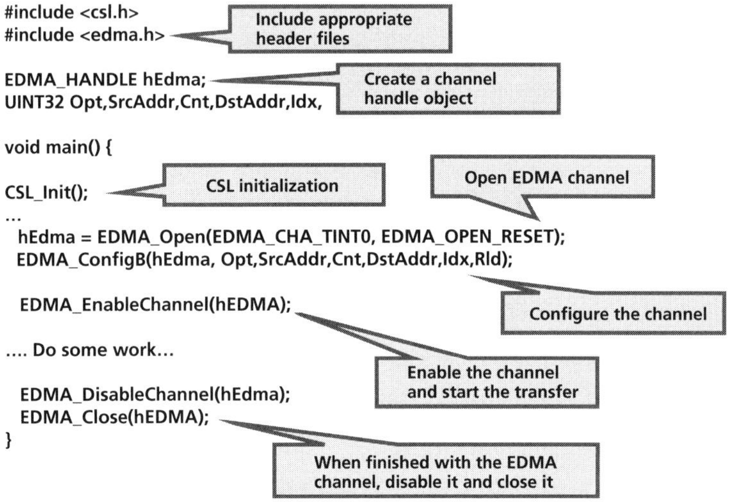

Figure 8.8 A simple CSL example for an EDMA DSP peripheral (courtesy of Texas Instruments)

DSP RTOS Application Example

As an example of the techniques and concepts that have been discussed in this chapter, a simple motor control example will be discussed. In this example, a single DSP will be used to control a motor (Figure 8.9). The DSP will also be responsible for interfacing to an operator using a keypad, updating a simple display device, and sending data out one of the DSP ports. The operator uses the keypad to control the system. The motor speed must be sampled at a 1 KHz rate. A timer will be used to interrupt processing at this rate to allow the DSP to execute some simple motor control algorithms. At each interrupt, the DSP will read the motor speed, run some simple calculations, and adjust the speed of the motor accordingly. Diagnostic data is transmitted out the RS232 port when selected by the operator using the keypad. The summary of these motor control requirements are shown in Table 2.

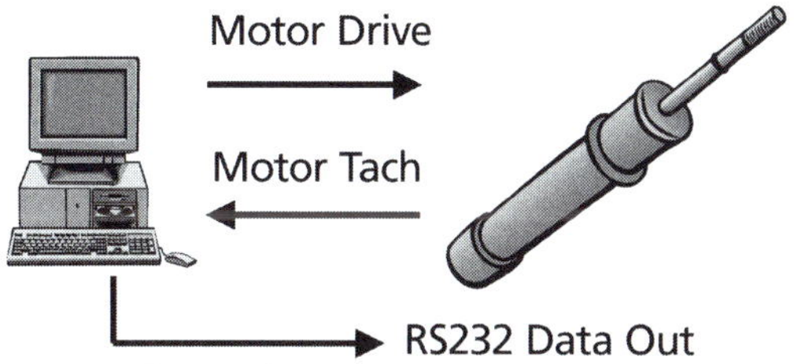

Figure 8.9 A simple motor control example

Defining the Threads

The first step in developing a multitasking system is to architect the application into isolated independent execution threads. There are tools available to help the system designer during this phase. This architecture definition will produce data flow diagrams, system block diagrams, or finite state machines. An example of a set of independent execution threads is shown in Figure 8.10. There are four independent threads in this design: the main motor control algorithm which is a periodic task, running at a 1 KHz rate, a keypad control thread which is an aperiodic task controlled by the operator, a display update thread which is a periodic task executing at a 2 Hz rate, and a data output thread which runs as a background task and outputting data when there is no other processing required.

Requirements
Control motor speed (1 KHz sampling rate – dV/dT)
Accept keyboard commands to control the motor, change the display, or send data out the RS232 port
Drive a simple display and refresh 2 times per second
Send data out the RS232 port when there is nothing else to do

Table 8.2 Summary of motor control requirements

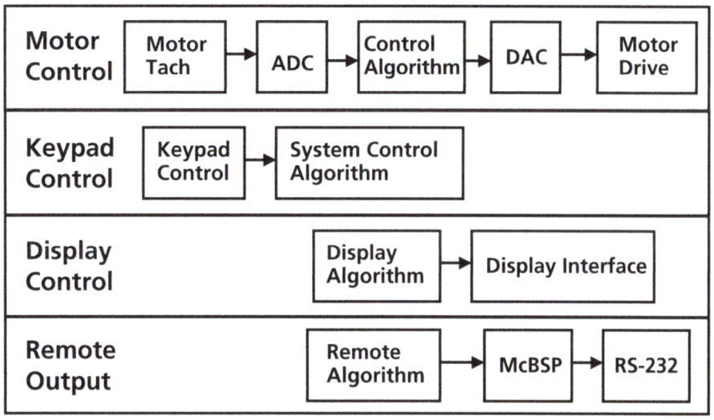

Figure 8.10 Motor control application divided into independent execution threads

Determining Relative Priority of Threads

Once the major threads of the application have been defined, the relative priority of the threads must be determined. Since this motor control example is a real-time system (there are hard real-time deadlines for critical operations to complete), there must be a priority assigned to the thread execution. In this example, there is one hard real-time thread, the motor control algorithm that must execute at a 1 KHz rate. There are some soft real-time tasks in the system as well. The display update at a two hertz rate is a soft-real time task (this is a soft real-time task because although the display update is a requirement, the system will not fail if the display is not updated precisely twice per second.). The keypad control is also a soft real-time task but since it is the primary control input, it should have a higher priority than the display update. The remote output thread is a background task that will run when no other task is running.

Using Hardware Interrupts

The motor control system will be designed to use a hardware interrupt to control the motor control thread. Interrupts have fast context switching times (faster than that of thread context switch) and can be generated from the timer on the DSP. The priority of the tasks in the motor control example are shown in Table 3. This is an example of a rate monotonic priority assignment; the shorter the period of the thread, the higher the priority.

Along with the priority, the activation mechanism is described. The highest priority motor control thread will use a hardware interrupt to trigger execution. Hardware interrupts are the highest priority scheduling mechanism in most real time operating systems (Figure 8.11). The keypad control function is an interface to the environment (operator console) and will use a hardware interrupt to signal the keypad control actions. This priority will be at a lower priority than the motor control interrupt. The display control thread will use a software interrupt to schedule a display update at a 2 hertz rate. Software interrupts operate similar to hardware interrupts but at a lower priority than hardware interrupts (but at a higher priority than threads). The lowest priority task, the remote output task, will be executed as an idle loop. This idle loop will runs continuously in the background while there are no other higher priority threads to run.

Thread Period

The periodicity of each thread is also described in Table 8.3. The highest priority thread, the motor control thread, is a periodic thread. Like many DSP applications this thread processes data periodically, in this case at a 1 KHz rate. This motor control example is actually a multirate system. This means there are multiple periodic operations in the system (motor control and display control, for example). These threads operate at different rates. DSP RTOSs allow for multiple periodic threads to run . Table 3 shows how the two periodic threads in this system are architected within the framework of the RTOS. For each of these threads, the DSP system designer must determine the period the specific operation must run and the time required to complete the operation. The DSP developer will then program the DSP timer functions in such a way to produce an interrupt or other scheduling approach to enable the thread to run at the desired rate. Most DSP RTOSs have a standard clock manager and API function to perform this setup operation.

Task	Rate	Priority	Periodic or aperiodic	Activation mechanism
Motor control	1 KHz	1	Periodic	Hardware Interrupt
Keypad control	5 hertz	2	Aperiodic	Hardware Interrupt
Display control	2 hertz	3	Periodic	Software interrupt
Remote output I	Background	4	Aperiodic	Idle loop (runs continuously in the background)

Table 8.3 Assignment of priorities to motor control threads and the activation mechanisms (courtesy of Texas Instruments)

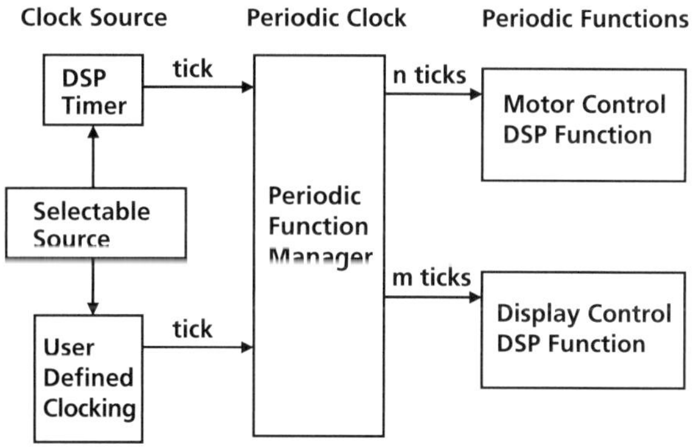

Figure 8.11 Periodic threads in the motor control example (courtesy of Texas Instruments)

Summary

The complexity of real-time systems continues to increase. As DSP developers migrate from a model of "programming in the small," where the DSP application performs a single task, to one of "programming in the large," where the DSP application performs a number of complicated tasks, scheduling the system resources optimally is becoming more and more difficult. An RTOS provides the needed abstraction from this complexity while providing the performance required of real-time systems. An RTOS can be configured by the DSP developer to schedule the various tasks in the system based on a set of rules, handle external events quickly and efficiently, and allocate and manage system resources.

Commercial DSP RTOSs provide a good value to the DSP developer. They come already debugged (having been used by many commercial users in the past) and with good documentation and tooling support. DSP RTOSs have also been optimized to provide the smallest footprint possible and with the best performance possible for key functions such as context switching and interrupt latency processing.

The most common DSP RTOS implement prioritized-preemptive scheduling policies. Systems based on this scheduling approach are popular in real-time systems because they are the most readily analyzed of the more sophisticated scheduling mechanisms. We will now turn our attention to the analysis of these scheduling policies.

Multiprocessing pitfalls

Multiprocessing, although required for complex real-time systems, suffers from potential dangers that must be carefully considered to prevent unexpected system performance and correctness problems. Multiprocessing problems can be very subtle and difficult to diagnose and fix. Some problems cannot be reliably removed through "debugging," so careful design and extreme discipline are critical. A major role of the RTOS is to encapsulate as much of these problems as practical and to establish the "metaphor" used to structure the discipline.

This chapter will outline the major issues with implementing multiprocessing systems. We will first discuss the danger of deadlock. We will also discuss the ways in which multiprocessing systems share common resources. Designers of DSP-based systems and embedded systems in general must constantly manage the allocation of scarce resources; these may include on-chip memory, the DMA controller, a peripheral, an input or output buffer, etc. This chapter will introduce ways to synchronization on shared resources as well as other forms of task synchronization.

Deadlock

One common danger is that of "deadlock," also known as *deadly embrace*. A deadlock is a situation in which two computer programs sharing the same resource are effectively preventing each other from accessing the resource, resulting in both programs ceasing to function. The earliest computer operating systems ran only one program at a time. In these situations, all of the resources of the system were always available to the single running program, and thus deadlock wasn't an issue. As operating systems evolved, they began to support timesharing and to interleave the execution of multiple programs. In early timesharing systems, programs were required to specify in advance what resources they needed so that they could avoid conflicts with other programs running at the same time. Still, deadlocks weren't a problem. But then, as timesharing systems evolved, some began to offer dynamic allocation of resources. These systems allowed a program or task to request further allocation of resources after it had begun running. Suddenly deadlock was a problem—in some cases a problem that would freeze the entire system several times a day.

The following scenario illustrates how deadlock develops and how it is related to dynamic (or partial) allocation of resources:

- Program 1 requests resource A and receives it.
- Program 2 requests resource B and receives it.
- Program 1 requests resource B and is queued up, pending the release of B.
- Program 2 requests resource A and is queued up, pending the release of A.
- Now neither program can proceed until the other program releases a resource. The OS cannot know what action to take.

At this point the only alternative is to abort (stop) one of the programs.

Deadlock Prerequisites

As our short history of operating systems illustrates, several conditions must be met before a deadlock can occur:

- *Mutual exclusion* – Only one task can use a resource at once.
- *Multiple pend and wait* – There must exist tasks which are holding resources while waiting for others.
- *Circular pend* – Circular chain of tasks hold resources that are needed by others in the chain (cyclic processing).
- *Preemption* – A resource can only be released voluntarily by a task.

Unfortunately, these prerequites are quite naturally met by almost any dynamic, multitasking system, including the average real-time embedded system.

"Indefinite Postponement" is another dangerous system condition in which a task that wishes to gain access to a resource is never allowed to do so because there are always other tasks gaining access before it. This condition is also called starvation or lockout.

Handling Deadlock

There are two basic strategies for handling deadlock: make the programmer responsible or make the operating system responsible. Large, general-purpose operating systems usually make it their responsibility to avoid deadlock. Small footprint RTOSs, on the other hand, usually expect the programmer to follow a discipline that avoids deadlock.

Deadlock prevention

Programmers can prevent deadlock by disciplined observance of some policy that guarantees that at least one of the prerequisites of deadlock (mutual exclusion, hold and wait, no preemption of the resource, and circular wait) can never occur in the system. The mutual exclusion condition can be removed, for example, by making resources sharable. This is difficult, if not impossible, to do in many cases. Another approach is to remove the circular pend condition. This can be solved by setting a particular order in which resources are requested (for example, all processes must request resource A before B). A third approach is to remove the multiple pend and wait conditions by requiring each task to always acquire all resources that will be used together as one "atomic" acquisition. This increases the risk of starvation and must be used carefully. Finally, you can allow preemption, in other words, if a task can't acquire a particular resource within a certain time, it must surrender *all* resources it holds and try again.

Deadlock detection and recovery

Some operating systems are able to manage deadlock for the programmer, either by detecting deadlocks and recovering or by refusing to grant (avoiding) requests that might lead to deadlock. This approach is difficult because the deadlock is generally hard to detect. Resource allocation graphs can be used but this can also be a very difficult task to perform in a large real-time system. Recovery is, in most cases, not easy. One extreme approach is to reset the system Another approach is to perform a rollback to a pre-deadlock state. In general, deadlock detection is more appropriate for general-purpose operating systems than RTOSs.

Perhaps the crudest form of deadlock detection and recovery is the watchdog timer. In event of a deadlock, the timer will elapse (deadlock detection), and reboot the system (recovery). As a last ditch safety net for very infrequent or as yet undiscovered deadlocks this is acceptable.

Deadlock avoidance

Here the RTOS dynamically detects whether honoring a resource request could cause it to enter an unsafe state. To implement a deadlock avoidance approach, the RTOS must implement an algorithm which allows all four deadlock preconditions to occur but which will also ensure that the system never enters a deadlock state. On resource allocation, the RTOS needs to examine dynamically the resource allocation state and take action based on the *possibility* of deadlock occurring. The resource allocation state is the number of resources available, the number of resources allocated, and maximum demand on the resources of each process.

Avoiding Unsafe States

A deadlock avoidance algorithm must do more than just avoid the state of deadlock—it must avoid all unsafe states. A "safe" state is a state from which the system can allocate resources to each task (up to the maximum they require) in some order and still avoid deadlock. The following example may help clarify the difference.

Assume their are 12 blocks of memory that may be allocated. The following state is "unsafe":

Process	Has	May need
Task 1	5	10
Task 2	2	4
Task 3	3	9

There are still two blocks left but Task1 and Task 3 can deadlock (cannot get all the resources they may need) even if task 2 completes and frees up two more blocks.

As an example of a safe state, assume there are twelve blocks of memory that may be allocated.

Process	Has	May need
Task 1	5	10
Task 2	2	4
Task 3	2	9

There are still three blocks left and task 2 could complete, and then task1 with the two resources freed by task 2 and finally task 3 could complete with the resources freed up from task 1.

Shared Resource Integrity

A second hazard of concurrent programming is resource corruption. When two (or more) interleaved threads of execution both try to update the same variable (or otherwise modify some shared resource), the two threads can interfere with each other, corrupting the state of the variable or shared resource.

As an example, consider a DSP system with the following characteristics:

- Shared memory (between tasks);
- Load-store architecture (like many of the DSP architectures today);
- Preemptive RTOS;
- Two tasks running in the system, T1 and T2 (T2 has higher priority);
- Both tasks needing to increment a shared variable X by one.

There are two possible interleavings for these tasks on the system, as shown below:

Interleaving 1

```
t1: ........
t1: ld  A,x
t1: add A,1,A
t1: st A, x
                <t2 : preempts
t2: .......
t2: ld  A,x
t2: add A,1,A
t2: st  A, x
t2: ........
                <t2 :completes
t1:  ......
result: x+2
```

Interleaving 2

```
t1: ........
t1:ld  A,x
                <t2 : preempts
t2: .......
t2:ld  A,x
t2: add A,1,A
t2:st A, x
t2: ........
                <t2 :completes
t1: add A,1,A
t1:st A, x
t1:  ......
result: x+1
```

Solution—Mutual Exclusion

Interleaving 1 does not work correctly because the preemption occurred in the middle of a read/modify/write cycle. The variable X was shared between the tasks. For correct operation, any read/modify/write operation on a shared resource must be executed *atomically*; that is, in such a way that it cannot be interrupted or pre-empted until complete. Sections of code that must execute atomically are *critical sections*. A critical section is a sequence of statements that must appear to be executed indivisibly.

Another way of viewing this problem is to say that while the shared variable can be shared, it can only be manipulated or modified by one thread at a time; that is, any time one thread is modifying the variable, all other threads are excluded from access. This access discipline (or type of synchronization) is called *mutual exclusion*.

Synchronizing Tasks for Mutual Exclusion

There are a number of potential solutions to the synchronization requirement. Some approaches are dependent on the DSP hardware. These approaches can be very efficient but they are not always useable.

One approach is to disable interrupts. If you simply disable the DSP processor's interrupts, then it becomes impossible to have a context switch occur since all interrupts (except for perhaps a nonmaskable interrupts, which is usually reserved for when something serious has to occur or has occurred—for example resetting the DSP) are disabled. Note that this only works if you have a single processor—multiprocessor DSP systems get more complicated and will not be discussed here.

Another approach that is hardware-related is to use an atomic instruction to record the old state of a bit, and then change its state. Some DSPs have specialized instructions referred to as *test and set bit*. The DSP hardware implements an instruction that does atomically what are normally a pair of operations. This instruction will examine a specified bit in memory, set a condition code depending on what that memory location contains, and then set the bit to a value of 1. This instruction, in affect, will "atomic-ize" several memory transactions in such a way that an interrupt cannot occur between the two memory transactions thereby avoiding shared data problems in multiprocessing systems.

With mutual exclusion, only one of the application processes can be in the critical section at a time (meaning having the program counter pointing to the critical section code). If no processes are in the critical section, and a process wants to enter the critical section, that process can participate in a decision making process to enter the critical section.

Other software approaches involve asking a resource manager function in an RTOS for access to the critical section. RTOSes use mechanisms called *semaphores* to control access. A semaphore is a synchronization variable that takes on positive integer values.

We will now discuss these approaches and the advantages and trade-offs of each.

Mutual Exclusion via Interrupt Control

One approach to mutual exclusion is proactively to take steps to prevent the task that is accessing a shared resource or otherwise executing in a critical section from being interrupted. Disabling interrupts is one approach. The programmer can disable interrupts prior to entering the critical section, and then enable interrupts after leaving the critical section:

```
.......
disable_interrupts( );
critical_section( );
enable_interrupts( );
.......
```

This approach is useful because no clock interrupts occur and the process has exclusive access to the CPU and cannot be interrupted by any other process. The approach is simple and efficient.

There are some significant drawbacks to the approach. Disabling interrupts effectively turns off the preemptive scheduling ability of your system. Doing this may adversely affect system performance. The more time spent in a critical section with interrupts turned off, the greater the degradation in system latency. If the programmer is allowed to disable interrupts at will, he or she may easily forget to enable interrupts after a critical section. This leads to system hangs that may or may not be obvious. This is a very error prone process and, without the right programmer discipline, could lead to significant integration and debug problems. Finally, this approach works for single processor systems but does not work for multiprocessor systems.

RTOSs provide primitives to help a programmer manage critical sections without having to resort to messing with the system interrupts.

Mutual Exclusion with a Simple Semaphore

In its simplest form, a semaphore is just a thread safe flag that is used to signal that a resource is "locked." Most modern instruction sets include instructions that are especially designed to so that they can be used as a simple semaphore. The two most common suitable instructions are test and set instructions and swap instructions. The critical feature is that the flag variable must be both tested and modified as a single, atomic operation.

Note that "semaphore" is often used to mean something more complex than a simple thread-safe flag. That's because the "semaphore" service offered by many operating systems is a special, more complex (to implement at least) form of semaphore (technically a "blocking semaphore") which automatically suspends the waiting process if the requested resource is not busy. I'll discuss how to use blocking semaphores in the next section.

The most primitive techique for synchronizing on a critical section is called *busy waiting*. In this approach, a task sets and checks a shared variable (a simple semaphore) that acts as a flag to provide synchronization. This approach is also called *spinning* with the flag variables called *spin locks*. To signal a resource (use that resource), a task sets the value of a flag and proceeds with execution. To wait for a resource, a task checks the value of the flag. If the flag is set, the task may proceed with using the resource. If the flag is not set, the task will check again.

The disadvantages of busy waits are that the tasks spend processor cycles checking conditions without doing useful work. This approach can also lead to excessive memory traffic. Finally, it can become very difficult to impose queuing disciplines if there is more than one task waiting on a condition. To work properly, the busy wait must test the flag *and set it* if the resource is available, as a single atomic operation. Modern architectures typically include an instruction designed with this operation in mind.

Test and set instruction

The test and set instruction is atomic by hardware design. This instruction allows a task to access a memory location in the following way:

- If the bit accessed is a 0 then set it to 1 and return 0.
- If the bit is 1 return 1.

The pseudocode for this operation is as follows:

```
boolean
testset int i
{
  if ( i == 0)
  {
      i = 1;
      return true;
```

```
        }
        else
            return false;
    }
```

The task must test the location for 0 before executing a wait. If the value is 0 then the task may proceed. If it is not a 0, then the task must wait. After completing the operation, the task must reset the location to 0.

Swap instruction

The swap instruction is similar to test and set. In this case, a task wishing to execute a semaphore operation will swap a 1 with the lock bit. If the task gets back a 0 (zero) from the lock, it can proceed. If the task gets back a 1 then some other task is active with the semaphore and it must retest.

The following example uses a test and set instruction in a busy wait loop to control entry to a critical section.

```
int var = 0 or 1;
{
    while ( testset(var) );
    < critical section >
     var = 0;
}
```

When There's No Atomic Update

If the operating system or the language provide no synchronization primitives, or if disabling and enabling interrupts is unacceptable from a system response standpoint (it is easy to abuse the disable/enable interrupts approach and easy to make mistakes if overused or not managed correctly), a suitable busy wait (for two competing processes) can be constructed using two variables. There are difficulties in doing this, however. Protocols that use busy loops are difficult to design, understand, and prove correct. It gets worse when considering more than 2 tasks. Testing of programs may not examine rare interleavings that break mutual exclusion or lead to livelock. These approaches are inefficient and can be corruptible by a single rogue task.

The following code works for two tasks. In this approach, "turn" must be initialized to 1 or 2. Also, livelock can still occur if a task dies in the critical section.

```
task t1;
....
flag1 := up;
turn := 2;
while flag2 = up and turn = 2 do nop;
```

```
... execute critical code ...
flag1 := down;
......
end task1;
task t1;
....
flag2 := up;
turn := 1;
while flag1 = up and turn = 1 do nop;
... execute critical code ...
flag2 = down;
......
end task1;
```

Mutual Exclusion via Blocking Semaphores

A blocking semaphore is a nonnegative integer variable that can be used with operating system services Wait() and Signal() to implement several different kinds of synchonization. The Wait() and Signal() services behave as follows:

- Wait(S): If the value of the semaphore, S, is greater than zero then decrement its value by one; otherwise delay the task until S is greater than zero (and then decrement its value).
- Signal(S): Increment the value of the semaphore, S, by one.

The operating system guarantees that the actions of wait and signal are atomic. Two tasks, both executing wait or signal operations on the same semaphore, cannot interfere with each other. A task cannot fail during the execution of a semaphore operation (which requires support from the underlying operating system).

When a task must wait on a blocking semaphore, the following run time support actions are performed:

- The task executes a wait (semaphore pend).
- The RTOS is invoked.
- The task is removed from the CPU.
- The task is placed in a queue of suspended tasks for that semaphore.
- The RTOS runs another task.
- Eventually a signal executes on that semaphore.
- The RTOS will pick out one of the suspended tasks awaiting a signal on that semaphore and make it executable. (The semaphore is only incremented if no tasks are suspended.)

Blocking semaphores are indivisible in the sense that once a task has started a wait or signal, it must continue until it is complete. The RTOS scheduler will not pre-empt the task when a signal or wait is executing. The RTOS may also disable interrupts so no external event interrupts it[3]. In multiprocessor systems, an atomic memory swap or test and set instructions are required to lock the semaphore.

Mutual exclusion via blocking semaphores removes the need for busy-wait loops. An example of this type of operation is shown below:

```
var mutex : semaphore; (*initially zero *)

task t1; (* waiting task *)
....
wait(mutex);
... critical section
signal(mutex)
....
end t1;

task t2; (* waiting task *)
....
wait(mutex);
... critical section
signal(mutex)
....
end t2;
```

This example allows only one task at a time to have access to a resource such as on-chip memory or the DMA controller. In Figure 8.12, the initial count for the mutual exclusion semaphore is set to one. Task 1, when accessing a critical resource, will call Pend upon entry to the critical region and call the Post operation on exit from the resource.

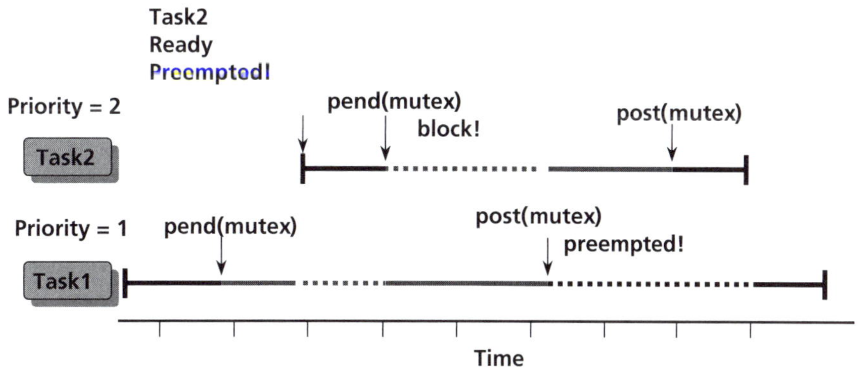

Figure 8.12 Using a mutual exclusion semaphore

In summary, semaphores are software solutions—they are not provided by hardware. Semaphores have several useful properties:

[3] Disabling interrupts on multiprocessors can be time-consuming, as the message is passed to all processors.

- *Machine independence* – Semaphores are implemented in software and are one of the standard synchronization mechanisms in DSP RTOSs.
- *Simple* – Semaphore use is relatively simple, requiring a request and a relinquish operation which are usually API calls into the RTOS.
- *Correctness is easy to determine* – For complex real-time systems, this is very important.
- *Works with many processes* – Semaphore use is not limited by the complexity of the system.
- *Different critical sections with different semaphores* – Different system resources can be protected using different semaphores. This significantly reduces the system responsiveness. If all important sections of code and system resources were protected by a single semaphore, the "waiting line" to use the semaphore would be unacceptable with respect to overall system response time.
- *Can acquire multiple resources simultaneously* – If you have two or more identical system resources that need to be protected, a "counting" semaphore can be used to manage the "pool" of resources.

Mutual Exclusion Through Sharable Resources

All of the synchronization techniques we've reviewed so far have a common disadvantage: they require the programmer to follow a particular coding discipline each time they use the resource. A more reliable and maintainable solution to the mutual exclusion problem is to make the resources sharable.

Many operating systems offer fairly high-level, sharable data structures, including queues, buffers, and data pipes that are guaranteed thread safe because they can only be accessed through synchronized operating system services.

Circular Queue or Ring Buffer

Often, information can be communicated between tasks without requiring explicit synchronization if the information is transmitted via a circular queue or ring buffer. This type of buffer is viewed as a ring of items. The data is loaded at the tail of the ring buffer. Data is read from the head of the buffer. This type of buffer mechanism is used to implement FIFO request queues, data queues, circulating coefficients, and data buffers. An example of the logic for this type of buffer is shown below.

Producer

in = pointer to tail

```
repeat
      produce an item in nextp
      while counter=n do nop;
      buffer[in]:= nextp;
```

```
        in := in +1 mod n;
        counter := counter+1;
  until false.
```

Consumer
out = pointer to head

```
repeat
        while counter=0 do nop;
        nextc:= buffer[out];
        out := out +1 mod n;
        counter := counter-1;
     consume the item in nextc
  until false.
```

Pseudocode for a Ring Buffer

The advantage of this structure is that if pointer updates are atomic and the queue is shared between only a single reader and writer, then no explicit synchronization is necessary. As with any finite (bounded) buffer, the readers and writers must be careful to observe two critical correctness conditions:

- The producer task must not attempt to deposit data into the buffer if the buffer is full.
- The consumer task cannot be allowed to extract objects from the buffer if the buffer is empty.

If multiple readers or writers are allowed, explicit mutual exclusion must be ensured so that the pointers are not corrupted.

Double Buffering

Double buffering (often referred to as ping pong buffering) is the use of two equal buffers where the buffer being written to is switched with the buffer being read from and alternately back. The switch between buffers can be performed in hardware or software.

This type of buffering scheme can be used when time-relative data needs to be transferred between cycles of different rates, or when a full set of data is needed by one task but can only be supplied slowly by another task. Double buffering is used in many telemetry systems, disk controllers, graphical interfaces, navigational equipment, robot controls and other applications. Many graphics displays, for example, use two buffers where the buffer being written to is switched with the buffer being read from and alternately back. The switch can be performed in hardware or software.

Basic Buffer Size Calculation

A buffer is a set of memory locations that provide temporary storage for data that are being passed from one task to another, when the receiving task is consuming the data slower than they are being sent for some bounded period of time. For example, data from the environment may be burst into a DSP at a high rate for a finite period of time. After the data has been input to the DSP, a task is scheduled to process the data in the buffer.

The data is typically sent at a greater rate for some finite time, called a *burst duration*. If the data is produced at a rate of P(t) and can be consumed at a rate of C(t) (where C(t) <P(t)) for a burst duration T, the DSP application engineer must determine the size of the buffer in order to prevent the data from being lost (often systems can be analyzed in terms of a single burst duration that re-occurs periodically, such as the burst of input data coming from a sensor at a predefined rate).

If the production rate is constant (P(t) = P) and the consumption rate is constant (C(t) = C) over the burst duration, the buffer size is calculated to be;

B = (P – C)T

This assumes P>C for some burst duration. If not, a buffer is not needed. The buffer can be emptied prior to the next burst. If this is not the case, then the system is unstable.

Example

Assume a device is providing real-time data to a DSP via DMA at 9600 byes/second in bursts of 2 seconds. The computer is capable of processing the data at 800 bytes/second. The minimum buffer size is;

B =(9600-800)*2=17600 bytes

In order to determine how often bursts can occur before the system becomes unstable:

800*Tp = 9600*T

Tp =9600 * 2 / 800 = 24 sec

- In the constant rate analysis, the excess data was at the maximum at the end of the burst period because P>C for the entire burst period. In the case where P(t) and C(t) are variable but P(t) > C(t) for the entire burst duration then the maximum size still occurs at the end of the burst duration and P-C can just be replaced with a integral of P(t)-C(t) over the burst duration. In this case the burst duration should be defined as the time that P(t)>C(t) is another solution to this problem. The solution in this case is to assign tasks to the same priority that need to be mutual exclusive. All resources are released when a task fails to acquire a resource. The RTOS could allow stealing of resources if a task tries to allocate a resource that is already allocated. This requires a resource context save and restoration. This approach is called the *priority ceiling protocol.*

Data Pipes

As an example of support for data transfer between tasks in a real-time application, DSPBIOS provides I/O building blocks between threads (and interrupts) using a concept called *pipes*.

A pipe is an object that has 2 sides, a "tail," which is the producer side and a "head," which is the consumer side (Figure 8.13). There are also "notify functions" that are called when the system is ready for I/O. The data buffer used in this protocol can be divided into a fixed number of frames, referred to as "nframes" as well as a frame size, "framesize."

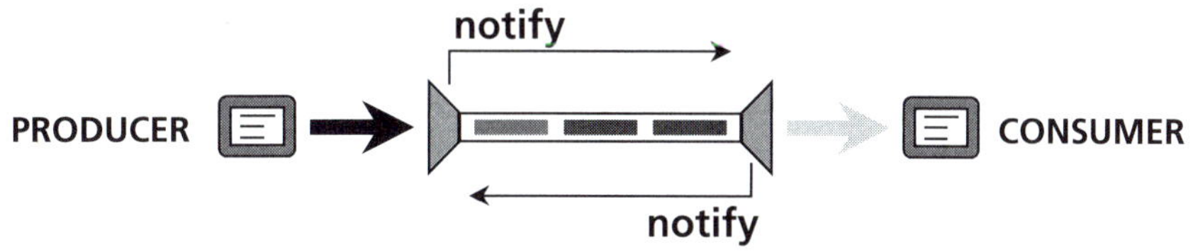

Figure 8.13 Data pipes in DSP/BIOS (courtesy of Texas Instruments)

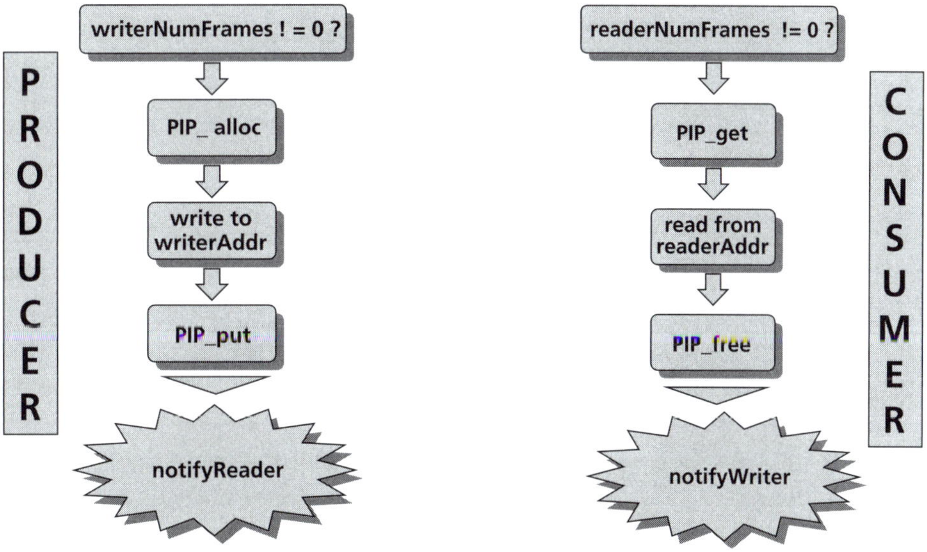

Figure 8.14 Steps required to implement pipes in a DSP application (courtesy of Texas Instruments)

A separate pipe should be used for each data transfer thread (Figure 8.14). A pipe should have a single reader and a single writer (point to point protocol). The notify function can perform operations on the software interrupt mailbox. During program startup, a notifywriter function is called for each pipe.

Writing to a pipe

To write to a pipe, a function first checks the writerNumFrames in the pipe object to make sure at least one empty frame is available. The function calls PIP_alloc to get an empty frame. If more empty frames are available, PIP_alloc will run the notifywriter function for each frame. The function writes the data to the address pointed to by writerAddr up to writerSize. Finally, the function should call PIP_put to put the frame in the pipe. Calling PIP_put will call the notifyReader function to synchronize with the consumer. (Figure 8.14).

Reading from a pipe

To read from a pipe, a function first checks the readerNumFrames in the pipe object to ensure at least one full frame is available. The function calls PIP_get to get a full frame from the pipe. If more full frames are available PIP_alloc will run the notifyReader function for each frame. The function reads the data pointed to by the readerAddr up to readerSize. Finally, the function calls PIP_free to recycle the frame for reuse by PIP_alloc. Calling PIP_free will call the notifyWriter function to synchronize with the producer (Figure 8.14).

Example

As an example, consider part of a telecommunications system that transforms an 8 KHz input (voice) stream into a 1 KHz output (data) stream using an 8:1 compression algorithm operating on 64 point data frames. There are several tasks that can be envisioned for this system:

1. There must be an input port that is posted by a periodic interrupt at an 8 KHz rate. (Period = 0.125 msec). An input device driver will service this port and pack the frames with 64 data points. This is task 1.
2. There is an output port that is updated by a posted periodic interrupt at a 1 KHz rate. (Period = 1 msec). An output device driver will service this port and unpack the 8 point output frames. This is task 2.
3. Task 3 is the compression algorithm at the heart of this example that executes once every 8 ms. (64 * 0.125 and 8 *1). It needs to be posted when both a full frame and an empty frame are available. Its execution time needs to be considerably less than 8 msec to allow for other loading on the DSP.

The tasks are connected by two pipes, one from task 1 to task 3 and one from task 3 to task 1. Pipe one will need two frames of 64 data words. Pipe two will need two frames of 8 data words.

When the input device driver has filled a frame, it notifies the pipe by doing a PIP_put call to pipe 1. Pipe 1 signals to task 3 using a SIG_andn call to clear a bit to indicate a full input frame. Task 3 will post if the bit indicating an empty frame is already zero.

When the output device driver has emptied a frame, it notifies the pipe by doing a PIP_free call to pipe 1. Pipe 1 posts to task 3 using a SIG_andn call to clear a bit to indicate an empty output frame. Task 3 will post if the bit indicating a full frame is available is already zero. The pseudocode for these tasks is shown below.

Code for task1

```
inputInterrupt(){
service hardware
if ('current frame is full') {
HWI_enter;
PIP_put(inputPipe);   /* enqueue current full framc */
   PIP_alloc(inputPipe); /* dequeue next empty frame */
   HWI_exit;              /* dispatch pending signals */
}
 return;
     }
```

Code for task 2

```
outputInterrupt(){
service hardware
if ('current frame is empty') {
HWI_enter;
PIP_free(outputPipe); /* enqueue current empty frame */
   PIP_get(outputPipe);  /* dequeue next full frame */
   HWI_exit;              /* dispatch pending signals */
]
 return;
     }
```

Code for task 3

```
compressionSignal(inputPipe, outputPipe){
PIP_get(inputPipe);          /* dequeue full frame */
PIP_alloc(outputPipe);       /* dequeue empty frame */
do compression algorithm   /* read/write data frame */
PIP_free(inputPipe);          /* recycle input frame */
   PIP_out(outputPipe):         /* output data */
return;
}
```

Pseudocode for Telecommunication System Tasks

Note that the communication and synchronization is independent of the nature of the input drivers, the compression routine, or even whether the input/output interrupts are really every data point (other than frame sizes).

Other Kinds of Synchronization

While shared resources are the primary reason tasks need to synchronize their activity, there are other situations where some form of synchronization is necessary. Sometimes one task needs to wait for others to complete before it can proceed (rendezvous). Sometimes a task should proceed only if certain conditions are met (conditional). The following sections explain how to achieve these kinds of synchronization.

Task Synchronization (Rendezvous)

Semaphores are often used for task synchronization. Consider the case of two tasks executing in a DSP. Both tasks have the same priority. If task A becomes ready before task B, task A will post a semaphore and task B will pend on the synchronization semaphore, as shown in Figure 8.15. If task B has a higher priority than task A and task B is ready before task A, then task A may be pre-empted after releasing a semaphore using the Post operation (Figure 8.16).

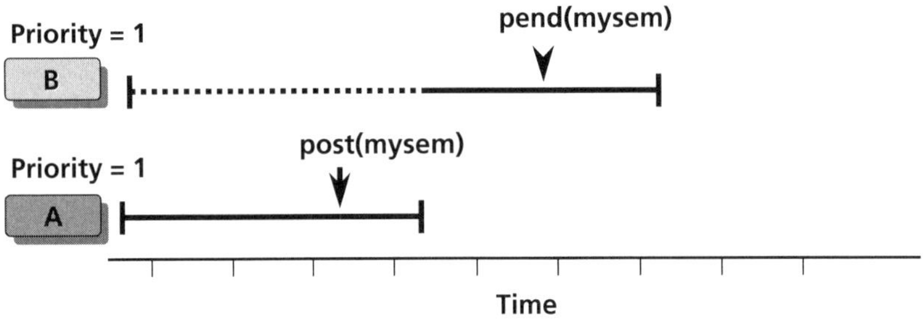

Figure 8.15 Synchronization semaphore; Tasks A and B are the same priority and A is ready before B (courtesy of Texas Instruments)

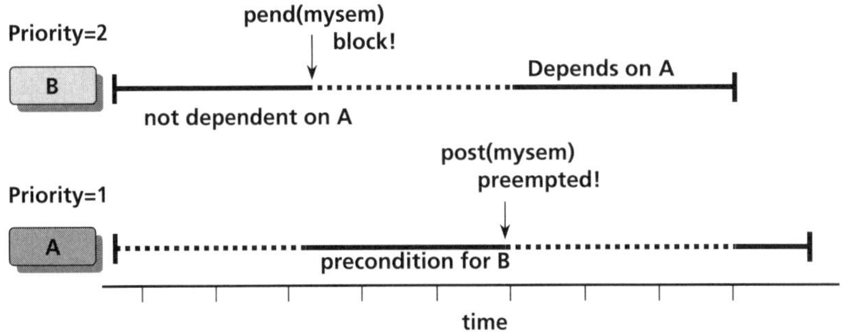

Figure 8.16 Synchronization semaphore; Task B is higher priority than task A and B is ready before A (courtesy of Texas Instruments)

Conditional Synchronization

Conditional synchronization overview

Conditional synchronization is a type of synchronization that is needed when a task wishes to perform an operation that can only logically, or safely, be performed if another task has itself taken some action or is some defined state. An example of this type of synchronization is a buffer. A buffer is used to link (decouple) two tasks, one of which writes data that the other which reads data. The use of a buffer link is known as a producer-consumer system. A buffer can be used for efficient handling of fluctuations of speed between the tasks, different data sizes between consumer and producer, preserving time relativeness, and re-ordering of data.

An example of conditional synchronization using semaphores is shown below;

```
var semi : semaphore; (*initially zero *)

task t1; (* waiting task *)
....
wait(semi);
....
end t1;

task t2; (* waiting task *)
....
signal(semi);
....
end t2;
```

Summary

Current execution (multiprocessing) provides many benefits to the development of complex real-time systems. However, one of the significant problems with multiprocessing is that concurrent processes often need to share data (maintained in data structures in shared memory) as well as resources such as on-chip memory, or an I/O port or peripheral. There must be controlled access to this shared data or resource or some processes may obtain an inconsistent view of the data or incorrectly use the hardware resource. The actions performed by the concurrent processes will ultimately depend on the order in which their execution is interleaved. This leads to nondeterminism in their execution, which is catastrophic for real-time systems and may be extremely hard to diagnose and correct.

Concurrency errors can be very subtle and intermittent. It is very important to identify and properly synchronize everything shared among threads or tasks. One of the advantages of an RTOS is to make this problem more manageable.

The DSP RTOS manages mutual-exclusion in multiprocessing systems by guaranteeing that critical sections will not be executed by more than one process simultaneously. These sections of code usually access shared variables in a common store or access shared hardware like on-chip DSP memory or the DMA controller. What actually happens inside the critical section is immaterial to the problem of mutual-exclusion. For real-time systems there is usually an assumption that the section is fairly short in order to reduce overall system responsiveness.

Semaphores are used for two main purposes in real-time DSP systems: task synchronization and resource management. A DSP RTOS implements sempahore mechanisms. The application programmer is responsible for the proper management of the resources, but the semaphore can be used as an important tool in this management.

Schedulability and Response Times

So far we've said that determinism was important for analysis of real-time systems. Now we're going to show you the analysis. In real-time systems, the process of verifying whether a schedule of task execution meets the imposed timing constraints is referred to as schedulability analysis. In this chapter, we will first review the two main categories of scheduling algorithms, static and dynamic. Then we will look at techniques to actually perform the analysis of systems to determine schedulability. Finally, we will describe some methods to minimize some of the common scheduling problems when using common scheduling algorithms.

Scheduling Policies in Real-Time Systems

There are several approaches to scheduling tasks in real-time systems. These fall into two general categories, fixed or static priority scheduling policies and dynamic priority scheduling policies.

Many commercial RTOSs today support fixed-priority scheduling policies. Fixed-priority scheduling algorithms do not modify a job's priority while the task is running. The task itself is allowed to modify its own priority for reasons that will become apparent later in the chapter. This approach requires very little support code in the scheduler to implement this functionality. The scheduler is fast and predictable with this approach. The scheduling is mostly done offline (before the system runs). This requires the DSP system designer to know the task set a-priori (ahead of time) and is not suitable for tasks that are created dynamically during run time. The priority of the task set must be determined beforehand and cannot change when the system runs unless the task itself changes its own priority.

Dynamic scheduling algorithms allow a scheduler to modify a job's priority based on one of several scheduling algorithms or policies. This is a more complicated approach and requires more code in the scheduler to implement. This leads to more overhead in managing a task set in a DSP system because the scheduler must now spend more time dynamically sorting through the system task set and prioritizing tasks for execution based on the scheduling policy. This leads to nondeterminism, which is not favorable, especially for hard real-time systems. Dynamic scheduling algorithms are online scheduling algorithms. The scheduling policy is applied to the task set during the execution of the system. The active task set changes dynamically as the system runs. The priority of the tasks can also change dynamically.

Static Scheduling Policies

Examples of static scheduling policies are rate monotonic scheduling and deadline monotonic scheduling. Examples of dynamic scheduling policies are earliest deadline first and least slack scheduling.

Rate-monotonic scheduling

Rate monotonic scheduling is an optimal fixed-priority policy where the higher the frequency (1/period) of a task, the higher is its priority. This approach can be implemented in any operating system supporting the fixed-priority preemptive scheme, such as DSP/BIOS and VxWorks. Rate monotonic scheduling assumes that the deadline of a periodic task is the same as its period. Rate monotonic scheduling approaches are not new, being used by NASA, for example, on the Apollo space missions.

Deadline-monotonic scheduling

Deadline monotonic scheduling is a generalization of the Rate-Monotonic scheduling policy. In this approach, the deadline of a task is a fixed (relative) point in time from the beginning of the period. The shorter this (fixed) deadline, the higher the priority.

Dynamic Scheduling Policies

Dynamic scheduling algorithms can be broken into two main classes of algorithms. The first is referred to as a "dynamic planning based approach." This approach is very useful for systems that must dynamically accept new tasks into the system; for example a wireless base station that must accept new calls into the system at a some dynamic rate. This approach combines some of the flexibility of a dynamic approach and some of the predictability of a more static approach. After a task arrives, but before its execution begins, a check is made to determine whether a schedule can be created that can handle the new task as well as the currently executing tasks. Another approach, called the *dynamic best effort apprach*, uses the task deadlines to set the priorities. With this approach, a task could be pre-empted at any time during its execution. So, until the deadline arrives or the task finishes execution, we do not have a guarantee that a timing constraint can be met. Examples of dynamic best effort algorithms are Earliest Deadline First and Least Slack scheduling.

Earliest deadline first scheduling

Earliest deadline first scheduling is a dynamic priority preemptive policy. With this approach, the deadline of a task instance is the absolute point in time by which the instance must complete. The task deadline is computed when the instance is created. The operating system scheduler picks the task with the earliest deadline to run. A task with an earlier deadline preempts a task with a later deadline.

Least slack scheduling

Least slack scheduling is also a dynamic priority preemptive policy. The slack of a task instance is the absolute deadline minus the remaining execution time for the instance to complete. The OS scheduler picks the task with the shortest slack to run first. A task with a smaller slack preempts a task with a larger slack. This approach maximizes the minimum lateness of tasks.

Dynamic priority preemptive scheduling

In a dynamic scheduling approach such as dynamic priority preemptive scheduling, the priority of a task can change from instance to instance or within the execution of an instance. In this approach a higher priority task preempts a lower priority task. Very few commercial RTOS support such policies because this approach leads to systems that are hard to analyze for real-time and determinism properties. This section will focus instead on static scheduling policies.

Scheduling Periodic Tasks

Many DSP systems are multirate systems. This means there are multiple tasks in the DSP system running at different periodic rates. Multirate DSP systems can be managed using nonpreemptive as well as preemptive scheduling techniques. Nonpreemptive techniques include using state machines as well as cyclic executives.

Analyzing Scheduling Behavior in Preemptive Systems

Preemptive approaches include using scheduling algorithms similar to the ones just described (rate monotonic scheduling, deadline monotonic scheduling, etc).This section will overview techniques to analyze a number of tasks statically in a DSP system in order to schedule them in the most optimal way for system execution. In the sections that follow, we'll look at three different ways of determining whether a particular set of tasks will meet it's deadlines: rate monotonic analysis, completion time analysis, and response time analysis.

Rate Monotonic Analysis

Some of the scheduling strategies presented earlier presented a means of scheduling but did not give any information on whether the deadlines would actually be met.

Rate Monotonic analysis addresses how to determine whether a group of tasks, whose individual CPU utilization is known, will meet their deadlines. This approach assumes a priority preemption scheduling algorithm. The approach also assumes independent tasks. (no communication or synchronization).[4]

In this discussion, each task discussed has the following properties:

- Each task is a periodic task which has a period T, which is the frequency with which it executes.
- An execution time C, which is the CPU time required during the period.
- A utilization U, which is the ratio C/T.

A task is schedulable if all its deadlines are met (i.e., the task completes its execution before its period elapses.) A group of tasks is considered to be schedulable if each task can meet its deadlines.

Rate monotonic analysis (RMA) is a mathematical solution to the scheduling problem for periodic tasks with known cost. The assumption with the RMA approach is that the total utilization must always be less than or equal to 100%.[5]

For a set of independent periodic tasks, the rate monotonic algorithm assigns each task a fixed-priority based on its period, such that the shorter the period of a task, the higher the priority. For three tasks T1, T2, and T3 with periods of 5, 15 and 40 msec respectively the highest priority is given to the task T1, as it has the shortest period, the medium priority to task T2, and the lowest priority to task T3. Note the priority assignment is independent of the application's "priority," i.e., how important meeting this deadline is to the functioning of the system or user concerns.

Calculating Worst-Case Execution Time

In order to determine task worst case execution time, accurate modeling or measuring is required. In order to determine worst case performance statically (by code observation), the following steps can be used:

- Decompose task code into a directed graph of basic blocks. (a basic block is section of straight line code).
- Estimate processor performance on basic blocks (caches, pipelines, memory conflicts, memory wait states can make this a challenge without an accurate simulator).
- Collapse the basic blocks (for instance if the execution path can go through either of two basic blocks choose the longest one). Loops can be collapsed using knowledge of maximum bounds.

---

[4] The restriction of no communication or synchronization may appear to be unrealistic, but there are techniques for dealing with this that will be discussed later.

[5] Any more and you are exceeding the capacity of the CPU. Are you asking for more computing power than you have? If so, forget it!

To model task behavior accurately, loops and recursion must be bounded. An alternative approach is to set up worst case conditions from an execution standpoint and simulate tasks counting cycles. Simulation and processor simulators will be discussed more later.

If all tasks in the system under observation are worst cased to meet deadlines, often very poor utilization is obtained. If hard and soft deadlines are present, the following guideline can be used:

- All hard real-time tasks should be schedulable using worst-case execution times and worst-case arrival rates.
- All tasks should be schedulable using average execution times and average arrival rates.

Utilization bound theorem

The RMA approach to scheduling tasks uses a utilization bound theorem to determine schedulability. Using this theorem, a set of n independent periodic tasks scheduled by the rate monotonic algorithm will always meet its deadlines, for all task phasings, if the total utilization of the task set is lower than the bound given in Table 8.4. This table shows the utilization bound for several examples of task numbers.

Number of Tasks	Utilization Bound
1	1.0
2	.828
3	.779
4	.756
5	.743
6	.734
7	.728
8	.720
9	.720
infinity	.69

Table 8.4 Utilization bound for different task sets

This theory is a worst case approximation. For a randomly chosen group of tasks, it has been shown that the likely upper bound is 88%. The special case where all periods are harmonic has an upper bound of 100%. The algorithm is stable in conditions where there is a transient overload. In this case, there is a subset of the total number of tasks, namely those with the highest priorities that will still meet their deadlines.

A Very Useful Special Case

In the special case of all harmonic periods, RMS allows you to use 100% of the processor's throughput and still meet all deadlines. In other words, the *utilization bound* is 1.0 for harmonic task sets. A task set is said to be harmonic if the periods of all its tasks are either integral multiples or sub-multiples of one another.

Assume a set of "n" periodic tasks, each with a period Ti and a worst case execution time Ci are given *rate-monotonic* priorities, that is, the task with a shorter period (higher rate) are assigned a fixed higher priority. If this simple rule is followed, then *all* tasks are *guaranteed* to meet their deadlines.

Examples of applications where the tasks tend to have harmonic periods include:

- Audio sampling in hardware.
- Audio sample processing.
- Video capture and processing.
- Feedback control (sensing and processing).
- Navigation.
- Temperature and speed monitoring.

These examples all have periodic arrivals with a fixed cost of processing.

A schedulable example

As a simple example of the utilization bound theorem, assume the following tasks:

- Task t1: C1=20; T1= 100; U1 = .2
- Task t2: C2=30; T2= 150; U2 = .2
- Task t3: C3=60; T3= 200; U3 = .3

The total utilization for this task set is .2 + .2 + .3 = .7. Since this is less than the 0.779 utilization bound for this task set, all deadlines will be met.

A two-task example that can't be scheduled

Consider the simple example of one task executing in a system as shown in Figure 8.16. This example shows a task with an execution interval of 10 (the scheduler invokes this task with a period of ten) and an execution time of 4 (ignore the time units for the time being).

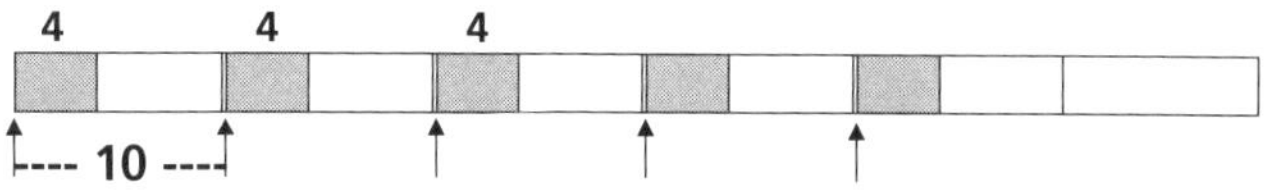

Figure 8.16 A simple task with an period of ten and execution time of 4

If a second task is added to the system, will the system still be schedulable? As shown in Figure 8.17, a second task of period 6 with an execution time of 3 is added to the system. If task 1 is given a higher priority than task 2, then task 2 will not execute until tasks 1 completes (assuming worst case task phasings which in these examples implies that all of the tasks become ready to run at the same time). As shown in the figure, task 2 starting after task 1 completes will not finish executing before its period expires, and the system is not schedulable.

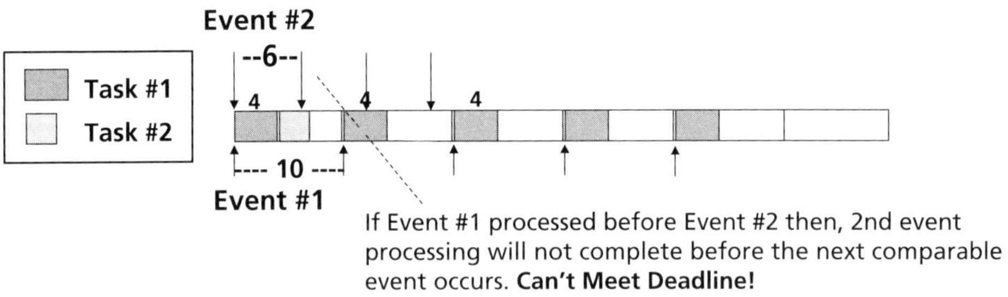

Figure 8.17 A two-task system which is not schedulable

Now assume that we switch the priorities of the two tasks shown in Figure 8.17–8.21. Now task 2 runs at a higher priority than task 1. Task 1 will run after task 2 completes. This scenario is shown in Figure 8.18. In this scenario, both tasks are allowed to run to completion without interruption. In this case, task 1, which is running at a lower priority than task 2, will not complete before task 2 must run again. The system is still not schedulable.

If task 1, which is running at a lower priority than task 2, is allowed to be preempted by task 1 when task 1 is ready to run, the system will become schedulable. This is shown in Figure 8.19. Task 2 runs to completion. Task 1 then begins to run. When task 2 becomes ready to run again, it interrupts task 1 and runs to completion. Task 1 then resumes execution and finishes processing.

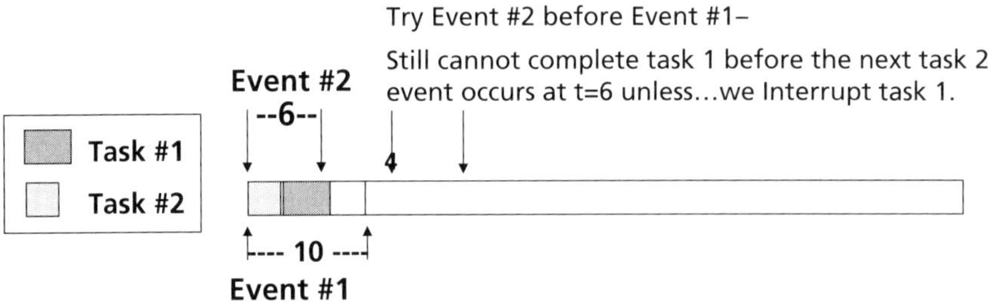

Figure 8.18 A two-task system that is still not schedulable

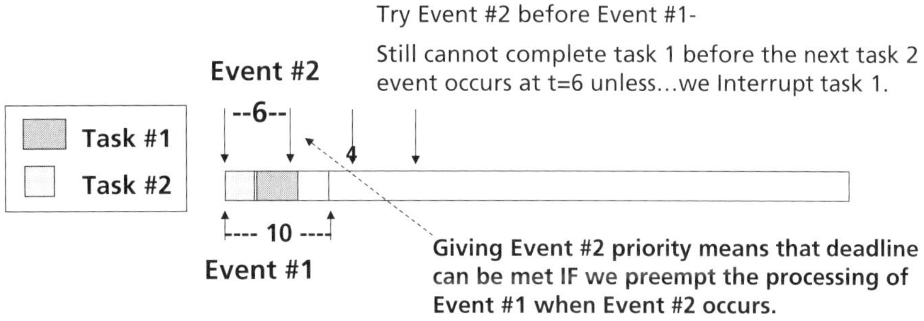

Figure 8.19 A two-task system that is schedulable using the rate monotonic scheduling approach

Task sets that exceed the utilization bound but whose utilization is less than 100% require further investigation. Consider the task set below:

- Task t1: C1=20; T1= 100; U1 = 20/100 = 0.2
- Task t2: C2=30; T2= 150; U2 = 30/150 = 0.2
- Task t3: C3=90; T3= 200; U3 = 90/200 = 0.45

The total utilization is .2 + .2 + .45 = .85. This is less than the Utilization bound for a three task set, which is 0.779. This implies that the deadlines for these tasks may or may not be meet. Further analysis in this situation is required.

Completion Time Theorem

The completion time theorem says that for a set of independent periodic tasks, if each task meets its first deadline when all tasks are started at the same time, then the deadlines will be met for any combination of start times. It is necessary to check the end of the first period of a given task, as well as the end of all periods of higher priority tasks.

For the given task set, a time line analysis is shown in Figure 8.20.

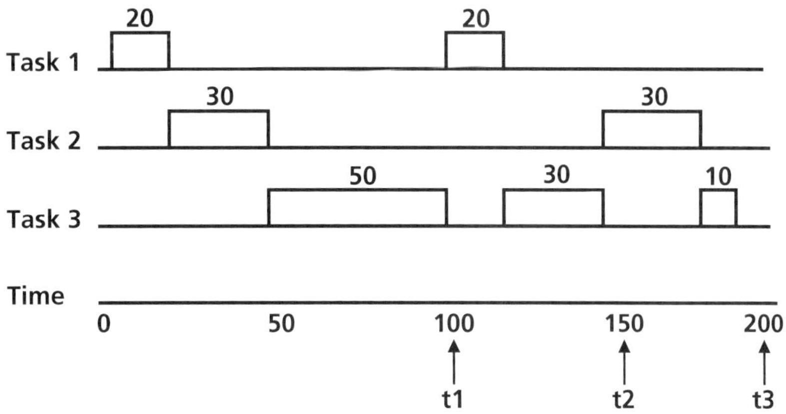

Figure 8.20 Time line analysis for a three-task set

The condition we will use for this analysis is a worst case task phasing, which implies that all three tasks become ready to run at the same time. The priorities of these tasks are assigned based on the rate monotonic scheduling approach. Task 1 has the highest priority because it has the shortest period. Task 2 has the next highest priority, followed by task 3. As shown in Figure 8.20, task 1 will meet its deadline easily, based on the fact that it is the highest priority and can pre-empt task 2 and task 3 when needed. Task 2, which executes after task 1 is complete, can also meet its deadline of 150. Task 3, the lowest priority task, must yield to task 1 and task 2 when required. Task 3 will begin execution after task 1 and task 2 have completed. At t1, task 3 is interrupted to allow task 1 to execute during its second period. When task 1 completes, task 3 resumes execution. Task 3 is interrupted again to allow task 2 to execute its second period. When task 2 completed, task 3 resumes again and finishes its processing before its period of 200. All three tasks have completed successfully (on time) based on worst case task phasings. The completion time theorem analysis has shown that the system is schedulable.

Example

Consider the task set shown below;

Task t1: C1=10; T1= 30; U1 = 10/30 = 0.33

Task t2: C2=10; T2= 40; U2 = 10/40 = 0.25

Task t3: C3=12; T3= 50; U3 = 12/50 = 0.24

The total utilization is .82. This is greater than the utilization bound for a three task set of 0.78, so the system must be analyzed further. The time line analysis for this task set is shown in Figure 8.21. In this analysis, tasks 1 and 2 make their deadline, but task 3, with a period of 50, misses its second deadline. This is an example of a timeline analysis that leads to a system that is not schedulable.

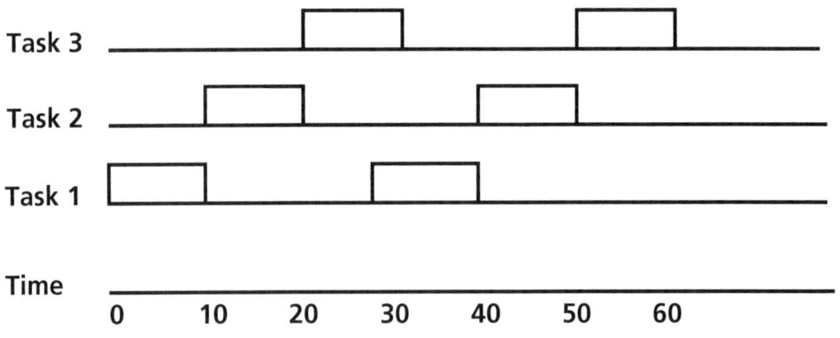

Figure 8.21 Timeline analysis of a three task set that misses a deadline

Response Time Analysis

The utilization based tests discussed earlier are a quick way to assess schedulability of a task set. The approach is not exact (if but not only if condition) and is not easily extensible to a general model.

An alternate approach is called *response time analysis*. In this analysis, there are two stages. The first stage is to predict the worst case response time of each task. The response time is then compared with the required deadline. This approach requires each task to be analyzed individually.

The response time analysis requires a worst case response time R, which is the time it takes a task to complete its operation after being posted. For this analysis, the same preemptive independent task model will be assumed;

Ri = Ci + Ii

where Ii is the maximum interference that task i can experience from higher priority tasks in any discrete time interval t, t + Ri. Like in the graphical analysis, the maximum interference will occur when all high priority tasks are released at the same time as task i. Therefore in the analysis we will assume all tasks are released a + t=0;

The interference from a task is computed as follows. Given a periodic task j, of higher priority than i, the number of times it is released between 0 and Ri is given by:

$$\text{Number\_of\_releases} = \left\lceil \frac{R_i}{T_j} \right\rceil$$

The ceiling function gives the smallest integer greater than the fractional number on which it acts. (this rounds up the fraction to an integer such that 7/4 is 2).

Each time it is released task i is delayed by its execution time Cj. This leads to the following maximum interference for a specific task.

$$\text{Maximum\_Interference} = \left\lceil \frac{R_i}{T_j} \right\rceil Cj$$

If we let hp(i) be the set of all higher priority tasks than task i, then the interference with task i can be represented as

$$I_i = \sum_{j \in hp(i)} \left\lceil \frac{R_i}{T_j} \right\rceil Cj$$

The response time is given by the fixed-point equation (note that Ri appears on both sides)

$$R_i = C_i + \sum_{j \in hp(i)} \left\lceil \frac{R_i}{T_j} \right\rceil Cj$$

To solve the response time equation, a recurrence relationship is formed;

$$w_i^{n+1} = C_i + \sum_{j \in hp(i)} \left\lceil \frac{w_i^n}{T_j} \right\rceil C_j$$

$$\text{where } w_i^0 = C_i$$

and the values of w are monotonically nondecreasing, because all numbers are positive and because of the nature of the ceiling function.

$$w_i^{n+1} = w_i^n$$

If Ri is less than the deadline than the task ti meets it deadline. If w becomes greater than the deadline, there is no solution.

Example

As an example of the response time calculation consider the following task set:

Task t1: C1=20; T1= 100; U1 = .2

Task t2: C2=30; T2= 150; U2 = .2

Task t3: C3=60; T3= 200; U3 = .3

Using the response time equation yields the following;

Example

A second example of the response time calculation is demonstrated using the following task set:

Task t1: C1=20; T1= 100; U1 = .2

Task t2: C2=30; T2= 150; U2 = .2

Task t3: C3=90; T3= 200; U3 = .45

Again, using the response time equation yields the following;

Calculations for task 1 and 2 are the same.

$$w_3^0 = 90$$

$$w_3^1 = 90 + \left\lceil \frac{90}{100} \right\rceil 20 + \left\lceil \frac{90}{150} \right\rceil 30 = 140$$

$$w_3^2 = 90 + \left\lceil \frac{140}{100} \right\rceil 20 + \left\lceil \frac{140}{150} \right\rceil 30 = 160$$

$$w_3^3 = 90 + \left\lceil \frac{160}{100} \right\rceil 20 + \left\lceil \frac{160}{150} \right\rceil 30 = 190$$

$$R_3 = w_3^4 = 90 + \left\lceil \frac{160}{100} \right\rceil 20 + \left\lceil \frac{160}{150} \right\rceil 30 = 190 < 200 \text{ meets deadline}$$

Example

One final example is now given. Consider the following task set:

Task t1: C1=10; T1= 30; U1 = .24

Task t2: C2=10; T2= 40; U2 = .25

Task t3: C3=12; T3= 50; U3 = .33

As shown below, this task set fails the response time test and the system is not schedulable under all worst case conditions.

$$w_3^0 = 12$$

$$w_3^1 = 12 + \left\lceil \frac{12}{30} \right\rceil 10 + \left\lceil \frac{12}{40} \right\rceil 10 = 32$$

$$w_3^2 = 12 + \left\lceil \frac{32}{30} \right\rceil 10 + \left\lceil \frac{32}{40} \right\rceil 10 = 42$$

$$w_3^3 = 12 + \left\lceil \frac{42}{30} \right\rceil 10 + \left\lceil \frac{42}{40} \right\rceil 10 = 52 > 50 \text{ never meets deadline}$$

Interrupt Latency

Hardware interrupts provide an effective means of notifying the application of the occurrence of external events. Interrupts are also used for sporadic I/O activities. The amount of time to service an interrupt varies based on the source of the interrupt. Interrupt handling in most processors, including DSPs, is divided into two steps; the immediate interrupt service and the scheduled interrupt service. The immediate interrupt service is executed at an interrupt priority level. The number of interrupt priority levels depends on the hardware. The TMS320C55 DSP family, for example, supports 32 interrupts with predefined priorities. All interrupt priority levels are higher than all task priorities. These interrupts are also higher in priority than the priority at which the scheduler executes as shown in Figure 8.22.

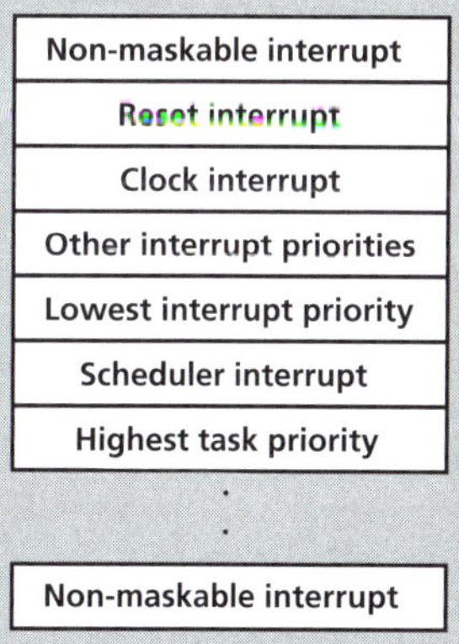

Figure 8.22 Priority levels for DSP interrupts

The total delay to service a DSP interrupt is the time the processor takes to complete the current instruction, do the necessary chores, and jump to the trap handler and interrupt dispatcher part of the kernel. The kernel must then disable external interrupts.

There may also be time required to complete the immediate service routines of higher priority interrupts, if any. The kernel must also save the context of the interrupted thread, identify the interrupting device, and get the starting address of the ISR. The sum of this time is called *interrupt latency*, it and it measures the responsiveness to external events via the interrupt mechanism. Many RTOSs provide the application with the ability to control when an interrupt is enabled again. The DSP can then control the rate at which external interrupts are serviced. The flow diagram to service a maskable and nonmaskable interrupt in the TMS320C55 DSP is shown in Figure 8.23.

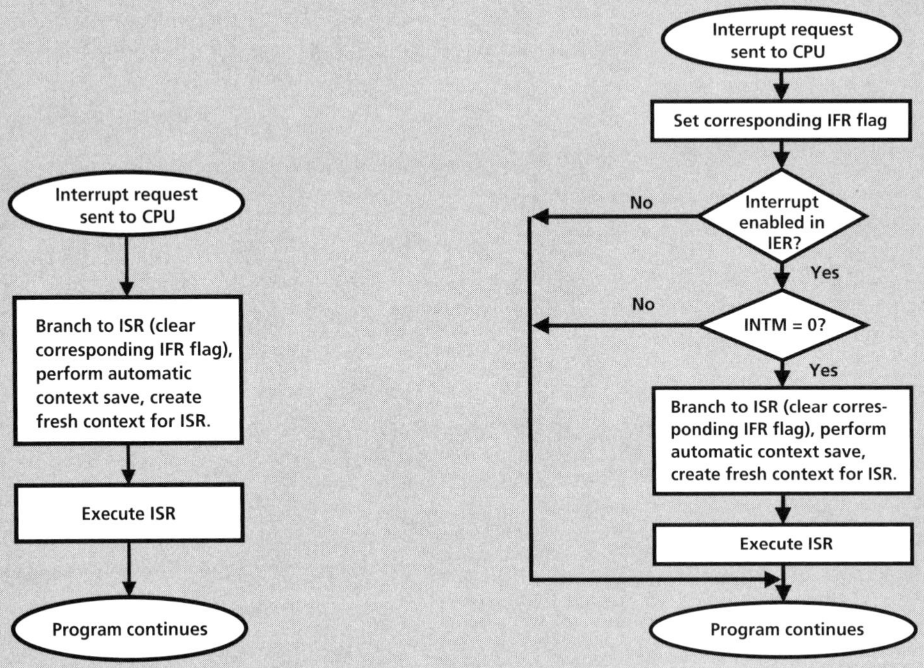

Figure 8.23 Flow for nonmaskable interrupts flow for maskable interrupts

Interrupt Flow in a TMS320C55 DSP

The second part of interrupt handling in a DSP is called the *scheduled interrupt service*. This is another service routine invoked to complete interrupt handling. This part of interrupt processing is typically pre-emptible (unless interrupts are specifically turned off by the DSP during this time).

Context Switch Overhead

In a task based system, it takes a finite amount of time for the RTOS scheduler to switch from one running thread to a different running thread. This delay is referred to as context switch overhead.

The worst case impact of context switching a task is two scheduling actions, one when the task begins to execute and one when the task completes. In calculating the utilization bound for a task, the context switch overhead must be taken into account. This will be referred to as CS in the equation below and represent the round-trip context switch time from one task to another.

$Ui = Ci/Ti + 2\ CSi$

The objective of a real-time system system builder is to keep 2*CS a small fraction of the task execution time T1 (the smallest period of all tasks in the system).

Analyzing More Complex Systems

Until now, the model used to determine the schedulability of tasks has been fairly simple:

- All tasks are periodic, with known periods.
- All tasks have a deadline equal to their period.
- Rate monotonic priority assignment.
- All tasks are completely independent of each other.
- Preemptive scheduling strategy.
- Application has a fixed set of priorities (static).
- All system overhead is included in the execution time.
- All tasks have a fixed worst-case execution time.

In reality, there are many examples of aperiodic as well as periodic events in real-time systems. An aperiodic task is assumed to execute once within some period Ta, which represents the minimum interarrival time of the event which activates the task. The CPU time Ca is the execution time caused by a single event. The utilization time is then a worst case utilization time and can be much less than the actual. From a schedulability analysis viewpoint, an aperiodic task is equivalent to a periodic task whose period is equal to Ta and whose execution time is Ca.

An aperiodic task often needs to respond much quicker than the minimum arrival time. An example would be an error signal which interrupts very infrequently, but when it does it needs to be handled with urgency. This breaks the assumption we have made that the deadline is equal to the period. Now the deadline is less than the period. This is referred to as rate monotonic priority inversion. A rate monotonic task i can only be pre-empted once by a task with rate monotonic priority inversion as the monotonic task has a shorter period. At most the task can be blocked for the execution time of the task with the rate monotonic priority inversion. The worst case assumption is made that every lower priority task tl is blocked once by every aperiodic task with monotonic priority inversion.

Deadline Monotonic Scheduling

Deadline monotonic priority assigns the highest priorities to the jobs with the shortest deadlines:

Di < Dj => Pi > Pj

The priorities are fixed and assigned before the application begins. It can be proven this is an optimal scheduling algorithm for deadlines less than or equal to the period T using the static scheduling model (Rate monotonic is a special case when all Di = T).

To test whether a task set is schedulable under the Deadline Monotonic policy, apply the following Deadline Monotonic scheduling to the task set;

```
For all jobs j(i), in the job set J {
   I = 0;
   do {
       R = I + C;
       if (R > D(i)) return unschedulable
       I = Σ R / T(j) x C(j) for j = i, i-1

   } while ( ( I + C(j) ) > R )
   return schedulable
}
```

The Deadline Monotonic scheduling algorithm tests if all jobs will meet their deadlines under worst case conditions (critical task phasings). The worst case response time includes the delays encountered by each task by preemption from other higher priority tasks in the system. This algorithm can easily be extended to include other delay factors such as context switch times, blocking due to waiting on system resources, and other scheduler overhead and latencies:

R(i) = ((Context Switch Time(i) × 2) + Scheduler latency(i) + Blocking(i)) + C(i) + I(i)

Example: dynamic vs. static scheduling

Rate Monotonic Scheduling has shown to be optimal among static priority policies. However, some task sets that aren't schedulable using RMS can be scheduled using dynamic strategies. An example is a task set where the deadline for completing processing is not the task period (the deadline is some time shorter than the task period). In this example, we'll show a task set that can be scheduled under the deadline monotonic priority policy, but not under RMS.

Consider the following task set;

Task	Deadline	Completion time	Period	Utilization	Priority
T1	5 ms	3 ms	20 ms	0.15	1
T2	7 ms	3 ms	15 ms	0.20	2
T3	10ms	4 ms	10 ms	0.40	3
T4	20 ms	3 ms	20 ms	0.15	4

Using the Deadline Monotonic approach to scheduling, the task execution profile is shown in Figure 8.24. All tasks meet their respective deadlines using this approach.

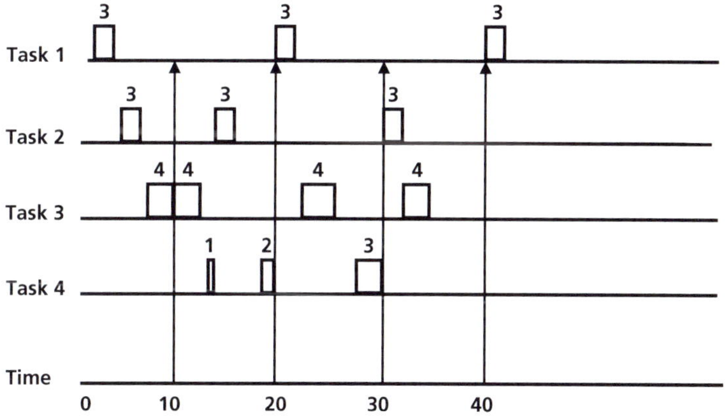

Figure 8.24 Example of deadline monotonic scheduling

Now consider the same task set, this time prioritized using the Rate Monotonic approach to scheduling. The task priorities change using the RMA approach, as shown below:

Task	Deadline	Completion time	Period	Utilization	Priority
T1	5 ms	3 ms	20 ms	0.15	3
T2	7 ms	3 ms	15 ms	0.20	2
T3	10ms	4 ms	10 ms	0.40	1
T4	20 ms	3 ms	20 ms	0.15	4

Same task set for rate monotonic scheduling

The timeline analysis using the RMA scheduling technique is shown in Figure 8.25. Notice that, using the RMA approach and the deadline constraints defined in the task set, that task 1 is now not schedulable. Although task 1 meets its period, it misses its defined deadline.

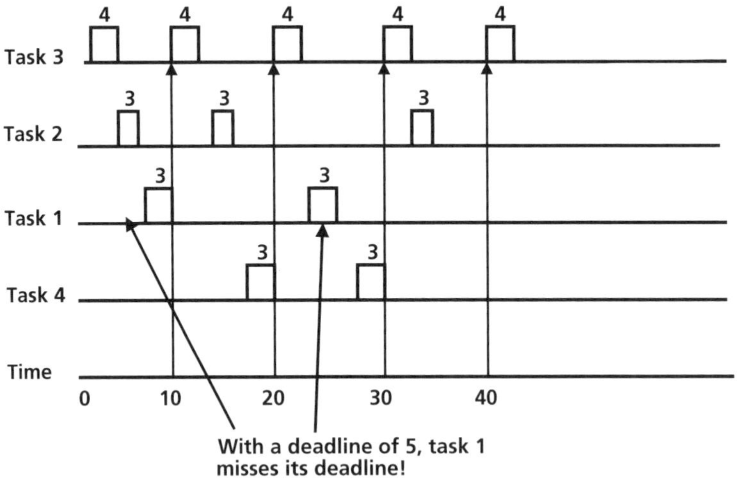

Figure 8.25 Same example with rate monotonic scheduling—task misses deadline

Other Dynamic Scheduling Algorithms

Other examples of dynamic scheduling algorithms include Earliest Deadline First and Least Slack Time. To successfully implement the Earliest Deadline First dynamic scheduling algorithm, the scheduler must know the deadline of the tasks. The scheduler runs the task with the closest deadline. One of he advantages of this algorithm is that the schedulable bound is 100%. Recall that, using RM scheduling, this bound approached 69% for large task sets. One of the main problems with this approach, however, is that there is no way to guarantee which tasks will fail in the case of a transient overload in the system. A transient overload is a situation where the system is temporarily overloaded beyond 100% utilization. In situations like this, it may be very important for the system to control which tasks fail and which tasks continue to exeucte. With the RM scheduling approach, the lowest priority tasks (which we know based in the a-priori assignments of task priorities) will fail while the high priority tasks will continue to exeucte. With a dynamic approach such as EDF, we do not know which tasks will fail and which ones will continue to execute.

Using the Least Slack Time algorithm, the scheduler must know the task deadline and remaining execution time. This algorithm assigns a "laxity" to each task in the system. The task with the minimum laxity is selected to execute next. The laxity term in this approach is determined by the following algorithm;

Laxity = deadline time – current time – CPU time still needed to complete processing

Laxity is basically a flexibility measure. It implies that a task can be delayed by the computed laxity time and still complete processing on time. A task that has a laxity time of zero means that the task must execute immediately or the task will fail. The least slack time approach also suffers from the limitation that there is no way to guarantee which task fails under system transient overload.

Figure 8.26 is an example of a dynamic scheduling algorithm. The three tasks t1, t2, and t3, have execution times c1, c2, and c3, and deadlines d1, d2, and d3, respectively. The EDF scheduling algorithm is used to schedule this task set. As you can see from the diagram, the tasks with the earliest deadline take priority of execution as opposed to just a static rate monotonic approach. This is illustrated in the diagram where t2 interrupts t1 to execute and t3 interrupts t2 to execute based on their respective deadlines being nearest in the future. After t3 executes, t2 finishes execution instead of t1 because if its deadline being nearer in the future than the deadline for t1.

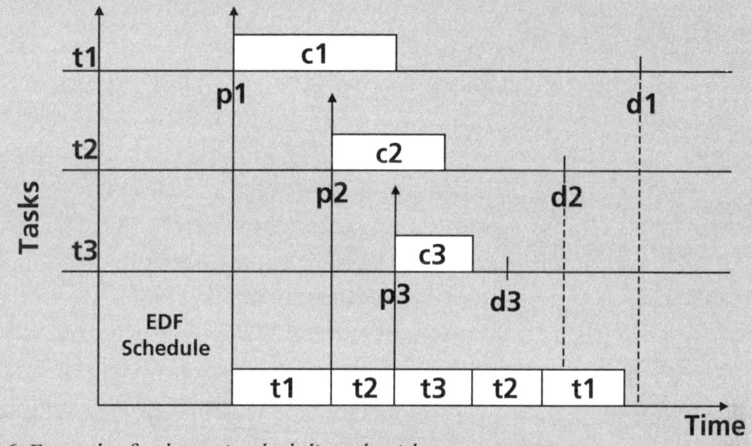

Figure 8.26 Example of a dynamic scheduling algorithm

Dynamic scheduling algorithms have several advantages. If the scheduler is allowed to change the priorities of the task set during execution, you can actually get a better overall utilization (in other words a lower idle time). Dynamic algorithms also do well with aperiodic tasks. Calculations of periodicity are not required to determine the priority.

There are some disadvantages to dynamic scheduling algorithms. Dynamic scheduling algorithms such as EDF also require the scheduler to know the period of each task, which may or may not be known. Dynamic scheduling algorithms also have a higher cost of evaluation during run time. Given the extra work that must be done in the scheduler in real-time, the task switch time could be longer adding to the latency of the system. Finally, dynamic scheduling algorithms cannot accurately control which tasks will be missed during system overload conditions. In static scheduling approaches, the lower priority tasks will be the ones that may not execute during overload conditions. But in dynamically scheduled systems, this is more unpredictable.

Scheduling with Task Synchronization

Independent tasks have been assumed so far, but this is very limiting. Task interaction is common in almost all applications. Task synchronization requirements introduce a new set of potential problems. Consider the following scenario; A task enters a critical section (it needs exclusive use of a resource such as IO devices or data structures). A higher priority task preempts and wants to use the same resource. The high priority task is then blocked until the lower priority task completes. Because the low priority task could be blocked by other higher priority tasks, this is unbounded. This example of a higher-priority task having to wait for a lower-priority task is call *priority inversion.*

Example: priority inversion

An example of priority inversion is shown in Figure 8.27. Tasklow begins executing and requires the use of a critical section. While in the critical section, a higher priority task, Taskh preempts the lower priority task and begins its execution. During execution, this task requires the use of the same critical resource. Since the resource is already owned by the lower priority task, Taskh must block waiting on the lower priority task to release the resource. Tasklow resumes execution only to be pre-empted by a medium priority task Taskmed. Taskmed does not require the use of the same critical resource and executed to completion. Tasklow resumes execution, finishes the use of the critical resource and is immediately (actually on the next scheduling interval) pre-empted by the higher priority task which executed its critical resource, completes execution and relinquishes control back to the lower priority task to complete. Priority inversion occurs while the higher priority task is waiting for the lower priority task to release the critical resource.

This type of priority inversion can be unbounded. The example showed a medium priority task pre-empting a lower priority task executing in a critical section and running to completion because the task did not require use of the critical resource. If there are many medium priority tasks that do not require the critical resource, they can all pre-empt the lower priority task (while the high priority task is still blocked) and execute to completion. The amount of time the high priority task may have to wait in scenarios likes this can become unbounded (Figure 8.28).

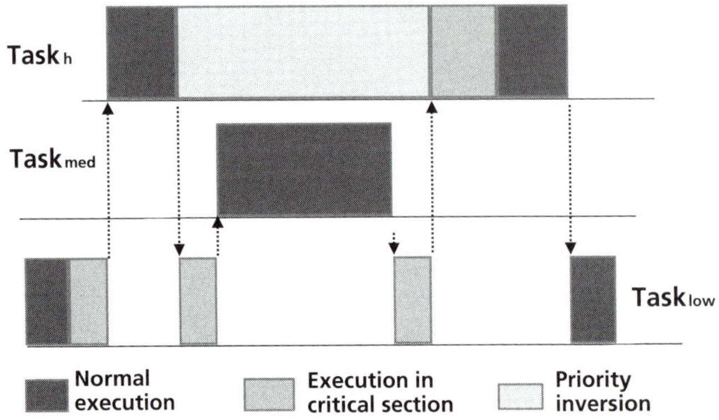

Figure 8.27 Example of priority inversion

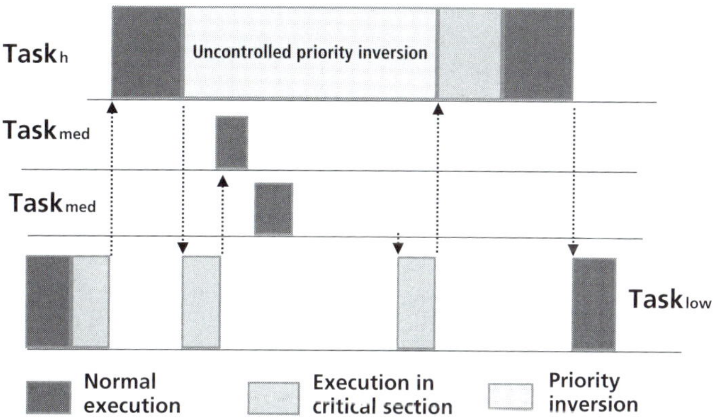

Figure 8.28 Unbounded priority inversion

Priority inheritance protocol

There are several approaches to bound the amount of priority inversion that can exist in a task based system. One approach is a protocol called the *priority inheritance protocol.* If a task p is suspended waiting for a task q to undertake some computation then the priority of q becomes equal to the priority of p. Using the priority inheritance protocol, the priority assigned to tasks is dynamic as opposed to static. This protocol should not be restricted to a single step. If task L4 is waiting for L3, but L3 is waiting for L2, then L2 as well as L3 would be given L4's priority. This implies that the run-time dispatcher has to handle tasks that are often changing priorities. This results in ineffective priority queues. The net result is that the dispatcher may be better off making the scheduling decisions at the time of action.

Example: priority inheritance

An example of PIP is shown in Figure 8.29. The same basic scenario applies. Tasklow begins executing and requires the use of a critical section. While in the critical section, a higher priority task, Taskh preempts the lower priority task and begins its execution. During execution this task requires the use of the same critical resource. Since the resource is already owned by the lower priority task, Taskh must block waiting on the lower priority task to release the resource. Tasklow resumes execution *but with a new priority, one equal to the priority of Taskh*! . This higher priority prevents Taskmed from pre-empting Tasklow. Tasklow runs to the end of the critical section and is then pre-empted by Taskh which runs to completion. Tasklow's priority is reset to the old priority level. This allows Taskmed to pre-empt Tasklow when Taskh completes execution and, in turn will run to completion. Tasklow is then allowed to finish processing.

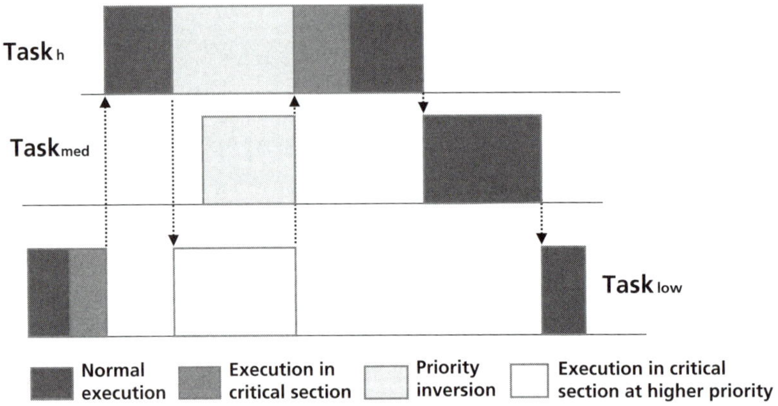

Figure 8.29 Example of priority inheritance protocol

Computing a priority inheritance bound

If a task has m critical sections that can lead to it being blocked by lower-priority tasks, then the maximum number of times it can be blocked is m. If there are only n (n<m) lower priority tasks then the maximum number of times it can be blocked is n. The worst case time a task can be blocked can therefore be calculated by summing the execution time of the critical sections of all the lower priority tasks that could be used by it or tasks with the same or higher priority (that sum is referred to as Bi in the equation below).

The response time calculation with blocking is now defined as;

$$R_i = C_i + B_i + \sum_{j \in hp(i)} \left\lceil \frac{R_i}{T_j} \right\rceil Cj$$

This formulation is now "if" but not "only if," since the blockage factor can be pessimistic. For example if all tasks are periodic with the same period, no preemption will take place although the blockage calculation will result in a positive number.

Priority ceiling protocols

Priority ceiling protocols (PCP) are more complex than the simple priority inheritance protocol but have a number of advantages. Using a PCP,a high-priority task can be blocked at most once during its execution by a lower-priority task. Chains of blocks are prevented. (Task 1 being blocked by Task 2 being blocked by Task 3). Using PCP, deadlocks are prevented. Mutual exclusive access to resources is ensured.

With PCP, each shared resource has a priority ceiling which is defined as the priority of the highest priority task that can ever access that shared resource. Using PCP a task runs at its original priority when it is outside a critical section. A task can lock a shared resource only if its priority is strictly higher than the priority ceilings of all shared resources currently locked by other tasks. Otherwise, the task must block. An interesting side effect to using PCP is that a task may not be able to lock a resource, even if it is available.

The PCP has the very useful property that a task can be blocked for at most the duration of one critical section of any lower priority task. This bounds the amount of priority inversion possible.

Using the priority ceiling protocol, each task in the system has a static default priority. Each resource also has a static ceiling value defined. This is the maximum priority of the tasks that use it. Each task in the system has a dynamic priority that is the maximum of its own static priority and the ceiling values of any resources it has locked. Using the PCP protocol, a task can suffer a block only at the beginning of its execution. Once a task starts running all resources it needs must be free. If this were not the case, then some other task would have an equal or higher priority and the beginning task would not begin.

Using the priority ceiling protocol, a task can be at most interrupted by one blockage. The worst case time a task can be blocked Bi can therefore be calculated by determining the execution time of the critical sections of the longest of the lower priority tasks that could be used by it or tasks with the same or higher priority. Otherwise the response time calculation is the same.

Example: priority ceiling

An example of the priority ceiling protocol is shown in Figure 8.30. Tasklow begins execution and enters a critical section. A high priority task, Taskh, preempts Tasklow and executes until it gets to the critical section which is locked by Tasklow. At this point, control is passed to Tasklow to execute its critical section, but at a higher priority. Any attempts by medium priority tasks to pre-empt Tasklow are not allowed because of the higher priority on Tasklow and the critical resource. The amount of priority inversion experienced by Taskh is limited to the amount of time Tasklow takes to complete execution of its critical section. No other task is allowed to run before Taskh completes.

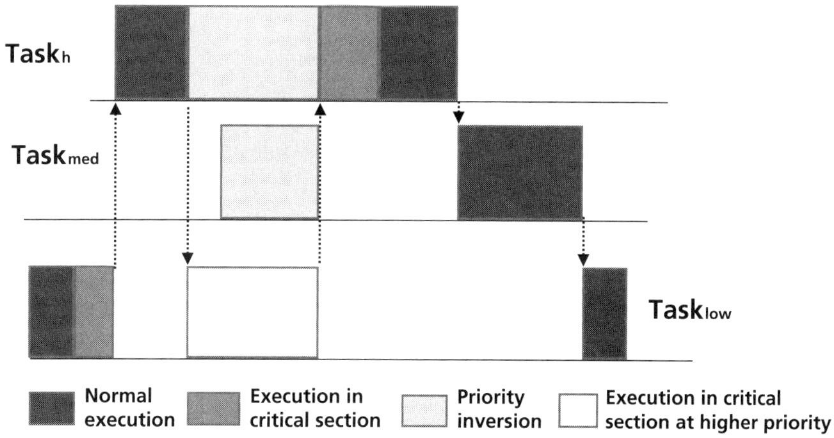

Figure 8.30 Example of the priority ceiling protocol

Summary

Many DSP RTOS's (and most commercial RTOS's in general) use static scheduling algorithms because they are simple to implement, easy to understand and support, and have a rich set of analysis techniques to draw from. DSP software developers can use Rate Monotonic Analysis to determine if the system task set is schedulable or not. If it is not schedulable, the analysis can be used to determine if there are any shared resource blocking issues which can be used to re-design the system task set and resource requirements and is especially useful for a first cut feasibility review. Most existing commercial RTOSs such as VxWorks, VRTX, and DSP RTOS's such as DSP/BIOS use Rate Monotonic Scheduling, with some form of priority inheritance or priority ceiling protocol to limit resource blocking and priority inversion. Despite the effectiveness of this approach, DSP software developers must be aware of the limitations of static scheduling, which include inefficient handling of nonperiodic processing and nonharmonic periods, as well as the requirement of allocation of resources based on worst case estimates.

References

Dijkstra, E. W. Solution of a problem in concurrent programming control. Communications of the ACM 8, 9 (September 1965), 569.

Knuth, Donald E. Addition comments on a problem in concurrent programming control. Communications of the ACM 9, 5 (May 1966), 321--322.

http://dspvillage.ti.com/docs/toolssoftwarehome.jhtml

How to Get Started With DSP/BIOS II, by Andy The and David W. Dart, Texas Instruments applications report SPRA697, October 2000

Real-time Systems, by Jane Liu, Prentice Hall, 2000

Hard Real-time Computing Systems; Predictable scheduling Algorithms and Applications, by Giorgio Buttazzo, Kluwer Academic Publishers, 1997

Designing with DSP/BIOS 2 by Karl Wechsler, DSP/BIOS developers conference, 2001

9

Testing and Debugging DSP Systems

In software development, perhaps the most critical, yet least predictable stage in the process is debugging. Many factors come into play when debugging software applications. Among these factors, time is of the utmost importance. The time required to set up and debug a software application can have significant impacts on time-to-market, meeting customer expectations, and the financial impact of a well developed product that succeeds in the market. The integration of an application follows a model of multiple spirals through the stages of build, load, debug/tune, and change, as shown in Figure 9.1.

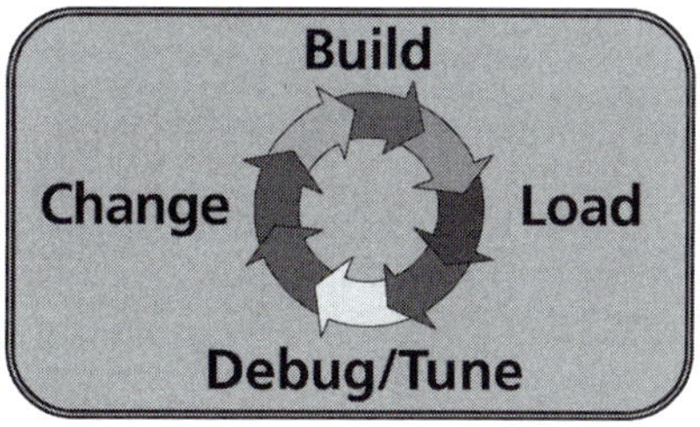

Figure 9.1 The integration and debug cycle. The goal is to minimize the number of times around this loop as well as minimizing the time spent in each segment.

Debugging embedded real-time systems is part art and part science. The tools and techniques used in debugging and integrating these systems have a significant impact on the amount of time spent in the debug, integration, and test phase. The more visibility we gain into the running system, the faster we are able to detect and fix bugs.

One of the more traditional and simplest ways of gaining visibility into the system is to add messages at certain points in the software to output information about the state of the system. These messages can be in the form of "printf" statements output to a monitor or the blinking of a LED or set of LEDs to indicate system status and health. Each function or task can begin by outputting a status message to indicate that the system has made it to a certain point in the program. If the system fails at some point, diagnosis of the output messages can help the engineer isolate the problem by knowing where the system was last "good." Of course, instrumenting the system in this way introduces overhead, which changes the behavior of the system. The engineer must either remove the instrumentation after the system has been tested and re-validate the system before shipping, or ship the system with the instrumented code in the system. The engineer must usually ship what is being tested (including this instrumentation) and test what is going to be shipped.

Engineers can use more sophisticated debug approaches to reduce the time spent in the integration and test phase. One approach is the use of a device called a "debug monitor." A debug monitor is a relatively small piece of code embedded in the target application or integrated into the micro controller or DSP core that communicates over a serial interface to a host computer[1]. These monitors provide the ability to download code, read and write DSP memory and registers, set simple and complex breakpoints, single step the program, and profile source code at some level.

For systems with ROM based software programs, another form of debug monitor, called a ROM emulator, is used. A ROM emulator is a plug-in replacement for the target system ROM devices. The plug-in device connects to the host computer over a link (serial, parallel, Ethernet, and so on). A ROM emulator provides the engineer with faster turnaround time during the debugging process. Instead of re-programming the ROM device using a ROM programmer for each software iteration, the code can instead be downloaded into fast RAM on the ROM emulator. The system is then run as if code was running out of the ROM device.

Debug monitors and ROM monitors certainly provide a large benefit to the embedded system debug phase. But, as embedded processors become faster and faster, and as systems migrate to system on a chip solutions, visibility into the internal processor presents challenges that require even more sophisticated debug solutions.

Integrating and debugging complex digital systems also requires the use of sophisticated and complex debug tools such as logic analyzers. A logic analyzer is a tool that allows the system integrator to capture and display digital signals in various formats such as bit, byte, and word formats. Using a logic analyzer, the system integrator can analyze digital behavior such as:

- Digital counters
- Complex state machines
- Buffers and FIFOs
- System buses
- Other system on a chip (SOC) functions such as FPGA, ASIC and standard cell implementations

Logic analyzers are powerful tools that are also portable and very versatile. These tools require a small learning curve as well as high initial investment (depending on

[1] Arnold Berger provides a very intuitive overview of debuggers in *Embedded Systems Design*, Chapter 6, CMP Books, copyright 2002.

how much capability is needed by the tool and what clock rates need to be supported). By using triggering mechanisms in the tool, the system integrator can capture data into large buffers within the logic analyzer. This data can be pre-trigger data, post-trigger data, or a combination. Traces can be saved and printed and the data can be filtered in a number of different ways.

A fundamental disadvantage to using a logic analyzer for embedded DSP software debug is that they are complex hardware debug tools being used for software debug. The degree of success using a logic analyzer is related to how hardware savvy the system integrator is since the tool is hardware-debug based and may require complex setup and configuration to get the right information to analyze.

Another disadvantage to using a logic analyzer for system debug is visibility of the signals. A logic analyzer needs to be connected to the pins of the DSP device in order to get visibility into the system. The visibility is limited by the types of pins on the DSP. As DSP devices become more integrated into system on a chip capability, the visibility to see what is going on inside the device diminishes.

Vanishing visibility

In 1988, the embedded system industry went through a change from conventional In Circuit Emulation[2] to scan based emulation. This was motivated by design cycle time pressures and newly available space on the embedded device for on-chip emulation. Scan-based, or JTAG, emulation is now widely preferred over the older and more expensive "in-circuit emulation," or "ICE" technology.

Debug Challenges for DSP

The have been a number of industry forces that have been changing the DSP system development landscape:

System level integration – As application complexity has increased and system on a chip complexity has led to smaller footprints, the visibility into the system components has diminished (Figure 9.2). Embedded system buses lead to an instrumentation challenge. Wider system buses also lead to system bandwidth issues. Program control in these environments is difficult.

In order to restore visibility, DSP vendors have addressed the issue on several fronts:

On-chip instrumentation – As systems become more integrated, on-chip visibility into the device operation is becoming blocked (Figure 9.3). Bus snooping logic analyzer functions have been implemented in on-chip logic. Examples of this include triggering logic to find the events of interest, trace collection and export logic to allowing the

[2] In-circuit emulation technology replaces the target processor with a device that acts like, or "emulates," the original device, but has additional pins to make internal structures on the device, like buses, visible. ICE modules allow full access to the programmer model of the processor. These devices also allow for hardware breakpoints, execution control, trace, and other debug functions.

viewing of events, and maximizing export bandwidth per available pin on the DSP core. Debug control is through an emulator which extracts the information of interest.

Off chip collection foundation – Once the data is exported from the DSP core, the data must be stored, processed, filtered, and formatted in such a way as to be useful to those test engineers to interpret the data meaningfully.

Data visualization capability – DSP integration capabilities include the ability to easily view the data in different configurations. The entire chain is shown in Figure 9.4. The logic analyzer functions are now on-chip, the control and instrumentation collection is primarily through the emulation controller (Figure 9.5), and the data is displayed on the host in a visualization container. The key challenge, then, is to properly configure the system to collct the right data at the right time to catch the right problem.

Application space diversity – DSP applications are becoming more diverse and this presents challenges to DSP test and integration engineers. This diverse application space spectrum requires different cost models for debug support:

DSP basestation applications require high bandwidth, high frequency debug capabilities

Voice over IP applications require MIPS density and many homogeneous processors per board.

Cell phone and other wireless applications require multiheterogeneous processors and very high system level integration.

Automotive DSP applications require low cost debug solutions where DSP chip pins are at a premium.

Figure 9.12 shows examples of DSP applications and the types of DSP debug capability that is used to address the integration challenges for these various application spaces.

User development environment; the development environment for DSP developers is changing and DSP debug technologies are changing to accommodate these new environments. DSP engineers are transitioning debug platforms from desktop PC systems to laptops that are portable to the field for debug in the customers environment. Portable remote applications require portable DSP debug environments.

Continued clock rate increases; as DSP core clock speeds increase more data is required to perform debugging. In fact, the amount of data required to perform debug and tuning is directly proportional to the DSP core clock speed. More DSP pins and more data per pin are required to maintain the required visibility into the behavior of the device.

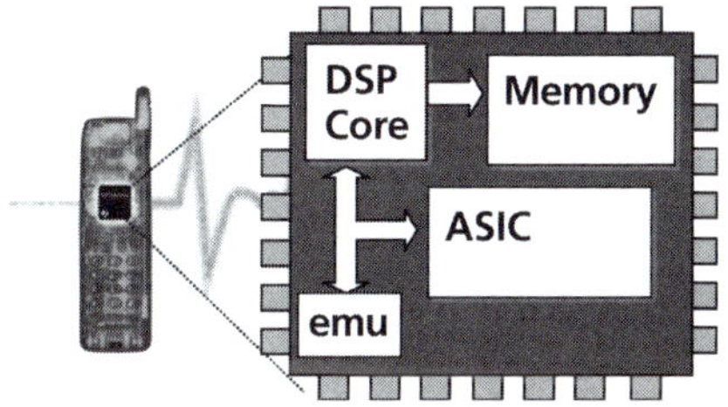

Figure 9.2 System level integration leads to diminishing visibility (courtesy of Texas Instruments)

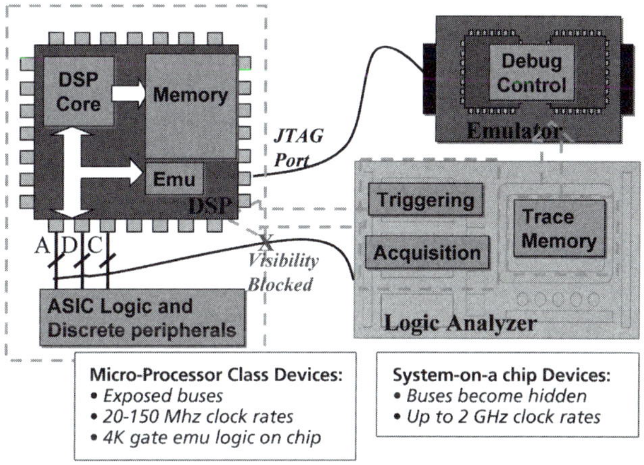

Figure 9.3 Vanishing visibility requires advanced debug logic on-chip (courtesy of Texas Instruments)

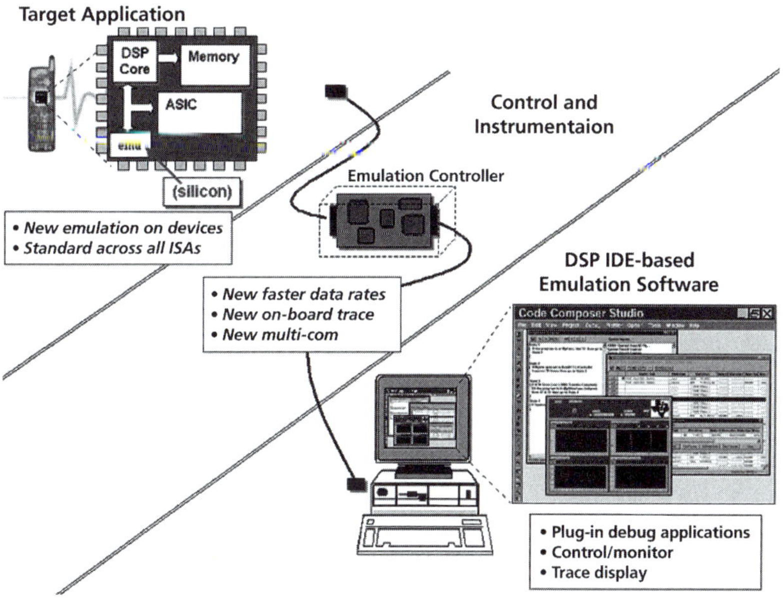

Figure 9.4 DSP tools are used to visualize debug data extracted from the DSP (courtesy of Texas Instruments)

The different levels of DSP debug capability provide a range of benefits in the integration process. The out of box experience allows the user to become productive as quickly as possible. Basic debug allows the DSP developer to get the application up and running. Advanced debug capabilities such as the ability to capture high bandwidth data in real-time allow the developer to get the application running in real time. Basic tuning capabilities provide the ability to perform code size and performance tuning.

The combined on and off chip emulation capabilities provide a variety of benefits. Execution control in real-time provides standard capabilities such as step, run, breakpoints (program counter) and data watchpoints. Advanced event triggering (AET) capabilities provide visibility and control of the programmer's model. Real-time data collection provides real-time visibility into algorithm behavior by tuning a stable program. Trace capabilities provide real-time visibility into the program flow through the process of debugging an unstable program.

Introduction to JTAG

One of the disadvantages of shrinking technology is that testing small devices effectively gets exponentially more complex. When circuit boards were large, boards were tested using techniques such as a bed-of-nails. This was a technique that used small spring-loaded test probes to make connections with solder pads on the bottom of the board. Such test fixtures were custom made, expensive, and inefficient; and still much of the testing could not be performed until the design was complete.

The problems with bed-of-nails testing were exacerbated as board dimensions got smaller and surface mount packaging technology improved. Further, if devices were mounted on both sides of a circuit board, there were no attachment points left for the test equipment.

Boundary scan

A group of European electronics companies joined forces in 1985 to find a solution to these problems. The consortium called themselves the Joint Test Action Group (JTAG). The result of their work was a specification for performing boundary scan hardware testing at the integrated circuit level. In 1990, that specification resulted in a standard (IEEE 1149.1) that specifies the details of access to any chip with a so-called JTAG port.

Boundary scan technology allows extensive debugging and diagnostics to be performed on an embedded system through a small number of dedicated test pins. Signals are scanned into and out of the I/O cells of a device(s) serially to control its inputs and test the outputs under various conditions. Today, boundary scan technology is probably the most popular and widely used design-for-test technique in the industry.

Test pins

Devices communicate to the world via a set of I/O pins. By themselves, these pins provide very limited visibility into what is going on inside the device. However, devices that support boundary scan contain a shift register cell for each signal pin of the device. These registers are connected in a dedicated path around the boundary (hence the name) of the device, as shown in Figure 9.5. This creates a virtual access capability that circumvents the normal inputs and provides direct control of the device and detailed visibility at its outputs.

During testing, I/O signals for the device enter and leave the chip through the boundary scan cells. The boundary scan cells can be configured to support external testing for testing the interconnection between chips or internal testing for testing the internal logic within the chip.

In order to provide the boundary scan capability, IC vendors have to add additional logic to each of their devices, including scan registers for each of the signal pins, a dedicated scan path connecting these registers, four (and an optional fifth) additional pins (more below), plus additional control circuitry.

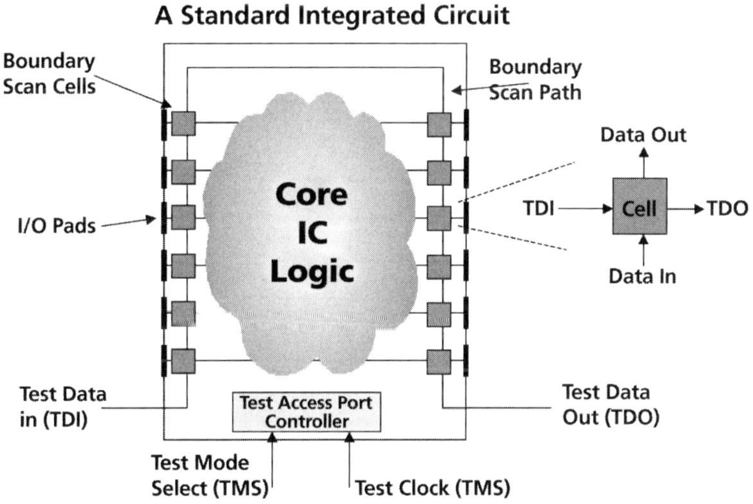

Figure 9.5 A standard integrated circuit with a JTAG bounday scan

The overhead for this additional logic is minimal and generally well worth the price in order to provide efficient test capability at the board level. The boundary scan control signals, collectively referred to as the test access port (TAP), define a serial protocol for scan based devices:

TCK/clock – To synchronize the internal state machine operations.

TMS/mode select – Sampled at the rising edge of TCK to determine the next state.

TDI/data in – When the internal state machine is in the correct state, this signal is sampled at the rising edge of TCK and shifted into the device's test or programming logic.

TDO/data out – When the internal state machine is in the correct state, this signal represents the data shifted out of the device's test or programming logic and is valid on the falling edge of TCK.

TRST/reset (optional) – When driven low, the internal state machine advances to the reset state asynchronously.

The TCK, TMS, and TRST input pins drive a 16-state TAP controller state machine. The TAP controller manages the exchange of data and instructions. The controller advances to the next state based on the value of the TMS signal at each rising edge of TCK. With proper wiring, multiple ICs or boards can be tested simultaneously. An external file known as the boundary-scan description language (BSDL) file defines the capabilities of any single device's boundary-scan logic.

Test process

The standard test process for verifying a device (or circuit board) using boundary scan technology is as follows:

The tester applies test or diagnostic data on the inputs pins of the device.

The boundary scan cells capture the data in the boundary scan registers monitoring the input pins.

Data is scanned out of the device via the TDO pin, for verification purposes.

Data can then be scanned into the device via the TDI pin.

The tester can then verify data on the output pins of the device.

Simple tests can find manufacturing defects such as open pins on the device, a missing device, incorrect, or rotated devices on a circuit board, and even a failed or dead device.

The primary advantage of boundary scan technology is the ability to observe data at the device inputs as well as the ability to control the data at the outputs independently of the application logic. Another key benefit is the ability to reduce the number of overall test points required for device access. With boundary scan there are no physical test points (boundary scan makes them redundant), which can help lower board fabrication costs and increase package density.

Boundary scan provides a better set of diagnostics than other test techniques. Conventional techniques apply test vectors (test patterns) to the inputs of the device and monitor the outputs. If there is a problem with the test, it can be time consuming to isolate the problem. Additional tests have to be run to isolate the failure. With boundary scan, the boundary scan cells observe device responses by monitoring the input pins of the device. This allows easy isolation of various classes of test failures, such as a pin not making contact with the circuit board.

Boundary scan can be used for functional testing and debugging at various levels, from internal IC tests to board-level tests. The technology is even useful for hardware/software integration testing.

Some test equipment and ASIC cell companies have put in extensions that use the JTAG capability to implement software debug functions. With the proper support built into a target CPU, this interface can be used to download code, execute it, and examine register and memory values. These functions cover the vast majority of the low-level functionality of a typical debugger. An inexpensive remote debugger can be run on a workstation or PC, to assist with software debug.

Boundary scan technology is also used for emulation. The emulator front-end acts as the scan manager, by controlling the delivery of the scan information to and from the target and the debugger window. Of course, when a host controls the JTAG scan information, it needs to be made knowledgeable about any other devices connected in the scan chain.

JTAG also allows the internal components of the device to be scanned (such as the CPU). This capability allows JTAG to be used to debug embedded devices by allowing access to all of the device that is accessible via the CPU and still perform testing at full speed. This has since become a standard emulation debug method used by silicon vendors. JTAG can also provide system level debug capability. The addition of extra pins to a device provides additional system integration capabilities such as benchmarking, profiling, and system level breakpoints.

Design cycle time pressure is created by these factors:

Higher integration levels – Many more functions are being integrated into one device as opposed to separate discrete devices.

Increasing clock rates – Electrical intrusiveness caused by external support logic.

More sophisticated packaging – This has caused external debug connectivity issues.

Today these same factors, with new twists, are challenging a scan based emulator's ability to deliver the system debug facilities needed by today's complex, higher clock rate, highly integrated designs. The resulting systems are smaller, faster, and cheaper. They are higher performance with footprints that are increasingly dense. Each of these positive system trends adversely affects the observation of system activity, the key enabler for rapid system development. The effect is called "*vanishing visibility.*"

The system on a chip (SOC) model implies that all the functionality found in a complete system is put together on a singl e piece of silicon chip. This includes processors, memory devices, logic elements, communications peripherals, and analog devices. The advantage of this approach is that by incorporating all of these components on a single chip, the physical space between components is reduced and the devices become smaller in size. This allows them to run faster and makes them simpler to manufacture. These advantages lead to improvements in reliability and lower cost.

Application developers prefer the optimum visibility level shown in Figure 9.6 as it provides visibility and control of all relevant system activity. The steady progression of integration levels and increases in clock rates steadily decrease the visibility and control available over time. These forces create a visibility and control gap, the difference between the desired visibility and control level and the actual level available. Over time this gap is sure to widen. Application development tool vendors are striving to minimize the gap's growth rate. Development tools, software and associated hardware components must do more with less and in different ways, tackling the ease of use challenge amplified by these forces. Advanced emulation technology is providing the required visibility into program behavior and execution. DSP-based systems require this visibility for many of the reasons mentioned above.

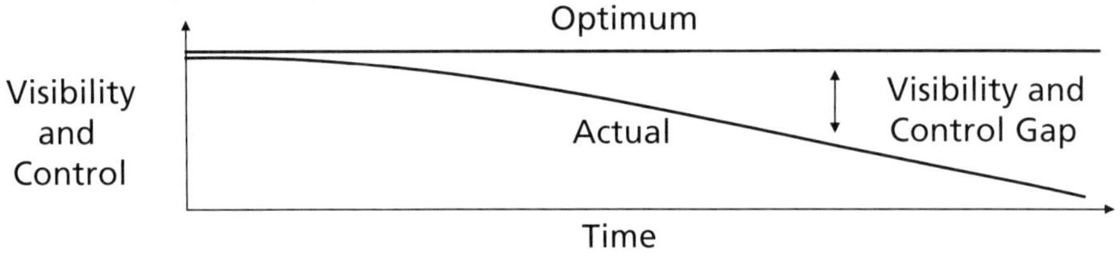

Figure 9.6 Visibility and control gap

Emulation Basics

Emulation is a technology used in the development of embedded systems. Emulation gives the developer the control and visibility needed to integrate hardware and software. This technology effectively imitates the DSP processor in its electrical characteristics and performance, while at the same time giving the engineer more visibility and control into the processor behavior than would otherwise be available from the processor itself.

An emulator contains both hardware and software technology. Emulation hardware consists of functionality on the DSP chip, which enables the collection of data. This data provides state behavior and other system visibility. Hardware is also required to extract this information from the DSP device at high rates and format the data. Emulator software provides additional higher level control and an interface with the host computer, usually in terms of an interface called a debugger. The debugger provides the development engineer an easy migration from the compilation process (compiling, assembling, and linking an application) to the execution environment. The debugger takes the output from the compilation process (for example, a .out file) and loads the image into the target system. The engineer then uses the debugger to interact with the emulator to control and execute the application and find and fix problems. These problems can be hardware as well as software problems. The emulator is designed to be a complete integration and test environment.

Emulator System Components

Emulator systems components

All emulator capability is created by the interaction of three main emulator components:

- On-chip debug facilities
- Emulation controller
- Debugger application program running on a host computer

These components are connected as shown in Figure 9.7. The host computer is connected to an emulation controller (external to the host) with the emulation controller also connected to the target system. The user controls the target application through the debugger in the IDE.

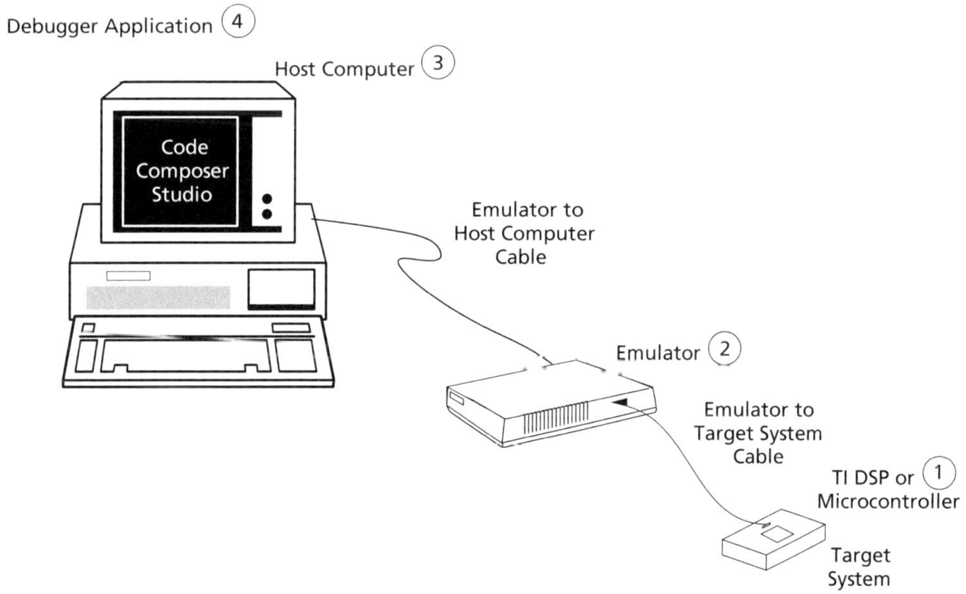

Figure 9.7 A basic emulation system (courtesy of Texas Instruments)

Emulation technology for DSP can be found throughout the development environment; from the processor, to the emulator, to the host platform.

On the DSP device itself, higher clock rates require the emulation logic to be on chip, so that it can run at top speed, and keep up with the processor rates. Higher levels of integration hide the buses within the chip, rather than allowing them to be visible through pins. This also pushes the emulation logic onto the chip, so that it can gain access to the system buses. DSP vendors have been aggressive in integrating advanced emulation capabilities on the DSP device.

The next element is the emulator, which is designed to connect the target board to the host platform, and provides the data transfer mechanism between the host and the target processor.

The third and final element is the debugger and emulation software. This software is self-configuring to the DSP device, and implements user interfaces to make debugging system-on-a-chip (SoC) devices as simple as possible. These IDEs also support the development of "plug-in" applications that can both control the processor, and visualize emulation data coming up from the processor via a high-speed data interface.

The host computer, where the debugger runs, can be a PC or workstation. The host computer can connect to the emulator using a variety of interfaces including Ethernet, universal serial bus (USB), Firewire (IEEE 1394), or parallel port.

The host computer can play an important role in determining the data bandwidth from the device to the host. The host to emulator communication plays a major role in defining the maximum sustained data bandwidth for certain data transfer protocols. In some emulation implementations, for example, the emulator must empty its receive data buffers as fast as they are filled. Secondly, the host client originating or receiving the data must have sufficient MIPs and/or disc bandwidth to sustain the preparation and transmission or processing and/or storing of the received data from the DSP. The point here is that a powerful PC or workstation will have a performance impact on the emulation system as a whole.

Emulator Physical Attributes

Physical attributes

Most emulation controllers are physically located outside the host computer. The emulator is partitioned into communication and emulation sections. The communication section supports host communication links communication while the emulation section interfaces to the target, managing target debug functions and the device debug port.

Emulator/Target Communication

Communicating with the target

The emulation controller communicates with the target through a target cable or cables. Debug, Trace, Triggers, and real-time transfer capabilities share the target cable, and in some cases the same device pins. More than one target cable may be required when the target system deploys a trace width that cannot be accommodated in a single cable. All trace, real-time data transfer, and debug communication occurs over this link. The emulator allows target/emulator separation of at least two feet to make it convenient for DSP developers in various environments.

On-Chip Emulation Capability

Given the vanishing visibility in today's sophisticated DSP processors, debug capability is moving onto the chip itself. This is referred to generally as on-chip debug. On-chip debugging is actually a combination of both hardware and software. The functionality that resides on the DSP itself is hardware implemented resources. These resources include capability available to the end user code such as breakpoint registers and other dedicated hardware. Communication between the chip and the debugger host requires additional pins on the DSP device. These may be the JTAG port pins and perhaps some additional pins dedicated to control or data.

On-chip debug also requires a host system to communicate with and control the debug session and extraction of data. The host software runs the debugger software and interfaces to the on-chip debug registers through the dedicated interface header. The host debugger provides the graphical display of the source code, the processor resources, memory locations, peripheral status, etc.

Some of the major functions available with on-chip debug include:

- Interrupt or break into debug mode on program and/or data memory address
- Interrupt or break into debug mode on a peripheral access
- Enter debug mode using a DSP microprocessor instruction
- Read or write any DSP core register
- Read or write peripheral memory mapped registers
- Read or write program or data memory
- Step one or more instructions
- Trace one or more instructions
- Read real-time instruction trace buffer

An example of the emulation hardware that is available on some higher performance DSP devices is shown in Figure 9.9. This logic resides on the DSP device and performs a number of enabling functions:

- *Watching* – Bus event detectors are used to watch for certain events occurring in the system. The specific events or conditions to watch for are programmable by the user from the debugger interface.
- *Remember* – Counters and state machines are used to remember events that have occurred in the system.

- *Direct and control* – Trigger builders are used to route the useful data captured with the counters and state machines.
- *Export* – Export functionality is used to export data from the system. For example, trace logic is used to export raw program counter and data trace information from the system.
- *Acceleration* – Local oscillators are used to increase the data transmission rates for devices running at high clock rates.
- *Import* – Logic is present to import data from a host computer. This capability allows the developer to input files of data used to debug and integrate a system.

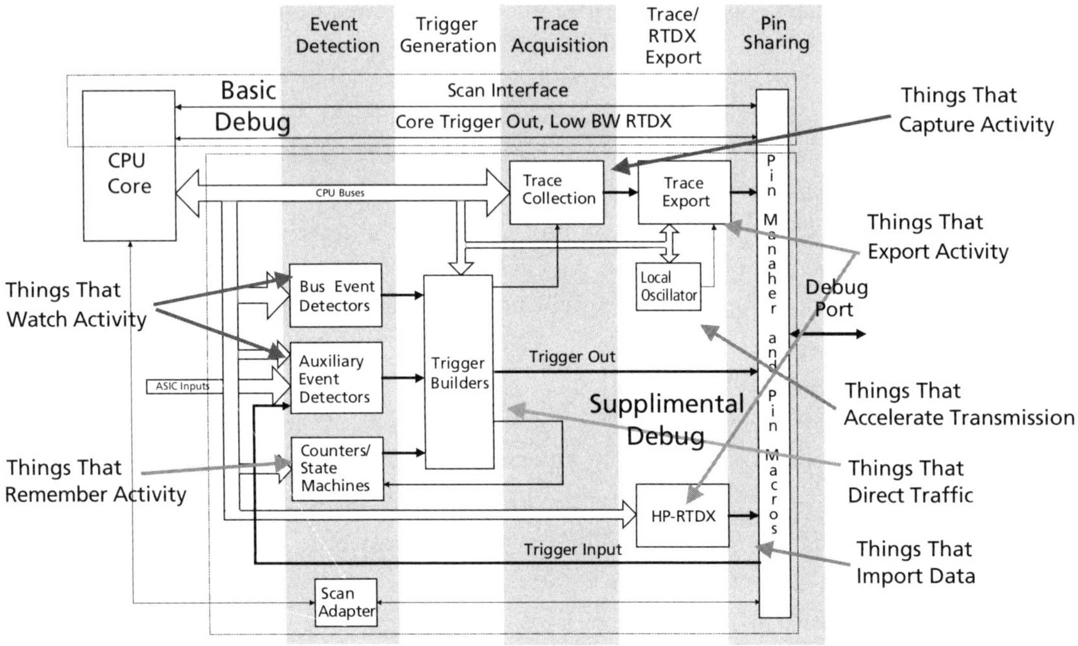

Figure 9.8 Emulation logic present on a DSP device enables the visibility needed for effective system integration (courtesy of Texas Instruments)

The emulation model also includes a host communications controller. This provides the connection to the host debugger which controls the overall process. The debugger can reside on either a PC or a workstation and may be integrated with a more integrated development environment or operate as standalone. The code built on the PC or workstation is loaded into the target via a communication link.

The emulator controller (along with the debugger) is a debug tool with two different functional groups of features. One group of features provides simple run control. This capability allows the DSP developer to control the operation of the DSP processor. Examples of run control actions include GO, HALT, STEP, and Hit Breakpoints at given memory locations.

Another group of features is used to capture and record DSP processor activity as shown on the processor bus. A triggering system allows the developer to specify conditions to control capturing of this trace information. The trace system records the DSP processor bus activity in high-speed RAM either in the system itself or possibly external to the system.

The debugger is a software component that executes on the host system. The debugger provides the capability to monitor and control the entire emulation session. Some of the common functions provided by the debugger are:

- Go/Run – This command will start the target CPU executing. Execution begins from the current program counter location and the current register values.
- Stop/Halt – This command is used to inform the emulator to stop the target CPU and halt the execution. When this command is executed, the current context of the target CPU and registers is saved. This is done so that when the processor starts running again, the execution can continue as if the stop had never occurred.
- Single-Step – This command is a special case of a Go or Run command but with a breakpoint set at the next instruction. This capability provides the user the ability to sequentially step through the code. At each step, the user can observe the registers, execution stack, and other important information about the state of the system. This is probably the most often used command to find problems in software or firmware modules.
- Step-Over/Step Through – This command is similar to a single step command with one significant difference. When stepping past a subroutine call, the step over/through commands will execute the subroutine but not step into the subroutine. If a single step command is used, the debugger will step into the subroutine and continue to execute the subroutine one instruction at a time. When the user does not desire to see every instruction in a subroutine or library function, this command allows the user to bypass this detail.
- Run To – This command provides the capability to set a breakpoint at a code location somewhere in the program deemed useful and then run until that point is reached. This prevents the user from having to single step many times to reach the same location.

DSP emulators provide visibility into the DSP processor, registers, and application software. This visibility allows the software engineer to understand what changes are taking place inside of the processor as the application executes. The software engineer can set breakpoints in the application based on hardware signal values or software locations within the application. At these breakpoints, the user can understand the state of the processor and data and determine if their application is still operating correctly. They can also perform benchmarking (timing analysis) and profiling (CPU loading) of their application software within the emulator. Multiprocessor debug can allow the user to debug software on several processors at the same time, and it provides a method of stopping one or multiple processors based on a condition set in another processor: allowing the user to capture the entire system state at the time in question. The capabilities mentioned, and many more within DSP debuggers, can greatly reduce debugging time in the software development cycle.

An emulator is connected directly to the DSP processor. Electrical signals are sent to the emulator which provides access to portions of the processor that a standard software debugger cannot. The engineer can view and modify registers that are unavailable to a standard software debugger. Hardware signaling also allows better run-time control. Emulators also provide the ability to record processor activity in real time. Therefore, if a problem occurs the developer have a history of system activity available to analyze.

Another advantage of an emulator over a standard software debugger is in debugging system startup code. A standard software debugger will usually require the target operating system in order to provide access to the system and the communication port. During the system initialization process, this is not available. Emulator provides its own communication port (usually JTAG). Emulators can access any part of the system, usually the same visibility as the CPU.

Another advantage of an emulator is in debugging systems that have crashed. If the target system crashes for any reason, the operating system usually suspends operation. This renders the software debugger inoperative. Emulators are not affected by these types of system crashes. An emulator can preserve valuable trace information as well as DSP processor state information such as the register values. This data can then be analyzed to help determine the situation that caused the crash.

Basic emulation components

When using a DSP debugger for software debugging on a hardware platform, it is necessary to perform a few setup procedures to ensure proper working of the target processor with the debugger. The emulation setup is comprised of two tools: the emulator itself (such as a TI XDS510 or XDS560), which controls the information flow to and from the target, and the debugger, which is the user interface to this information. Beyond the emulation setup is the target processor. The emulation logic within most DSP processors uses the Joint Test Action Group (JTAG) standard connection in procuring the debug information from within the processor.

Debug of the hardware is performed by stopping the DSP core to enable information to be scanned into and out of the device via the JTAG header. This information is transferred serially through the JTAG port following IEEE 1149.1 JTAG specifications. It is important to understand that this debug method is near real time, but is intrusive, as it may require that the core be halted to scan the information. While the connection to the JTAG header may be the same, the scan chains used for emulation purposes are different from those used for boundary scan. Internally to the processor, there are various serial scan chains which the information can be scanned into and out of. The control of which scan chain is used and what information is contained in each scan chain, is performed by a microprocessor. This "scan manager" has the task of controlling this information as it is scanned to and from the various processors in the scan chain, and directing it to and from the various debugger windows.

The host of the emulator acts as the scan manager as it controls the delivery of the scan information to and from the target and the debugger window. For example, the operating system may be a PC and the JTAG connection is made through an ISA card (Figure 9.1). Other configurations are possible as well. When the host CPU or a separate processor controls the JTAG scan information, the host needs to be supplied with information regarding the devices included in the scan chain.

Emulation Capabilities

Emulation provides a standard set of operations for use during the integration and debug phase. A few of these capabilities are described below.

Breakpoints

One of the common capabilities supported by emulation technology is the breakpoint. A breakpoint will interrupt the DSP and allow the developer to examine data or registers on the target system. The breakpoint capability is controlled by the emulator. The emulator executes the protocols to stop the CPU at the earliest possible point in the execution stream, as well as enabling the developer to continue execution from the current location when necessary. Most breakpoints are synchronous in that the transition from a running state to a paused state happens immediately.

A software breakpoint is one form of synchronous breakpoint. A software breakpoint function will save the instruction at the specified breakpoint location and replace it with a different instruction which creates an exception condition. This transfers control to the controller which saves the context of the important DSP status registers. Control is transferred to the host debugger and the developer can then "peek and poke" at registers and variables while the CPU is paused. The reverse is done to allow the CPU to continue execution form the current location. These type of breakpoint is useful for target systems which contain RAM to allow the writing and replacing of instructions.

An additional form of breakpoint is called a *hardware breakpoint*. This form of breakpoint is implemented using custom hardware on the target device. It is useful for DSP devices that have complicated instruction fetch sequences and for setting breakpoints in systems with ROM where replacing instructions using the "software" breakpoint technique is not possible. The hardware logic is designed to monitor a set of address and status signals on the device and stop execution of the device when a code fetch is performed from a specified location.

Event Detectors

The event detectors in Figure 9.9 provide an advanced form of target visibility and execution interruption. The bus event and auxiliary event detection logic in Figure 9.4 provides emulation break capability on a complicated set of events occurring in the system. In addition to the code execution breakpoints provided by hardware and software breakpoints, the event detectors provide breakpoint capability due to data accesses and other combinations of address, data, and other system status. The event detectors in Figure 9.9 consist of a set of comparators and other logic. The user, from the debugger interface, can program these comparators to look for a specific pattern of events in the system. The comparators trigger other event logic to perform certain actions such as stop execution, increment a counter tracking a specified occurrence of an event, or generating a signal on a pin that can be used by some other piece of equipment to perform some other operation.

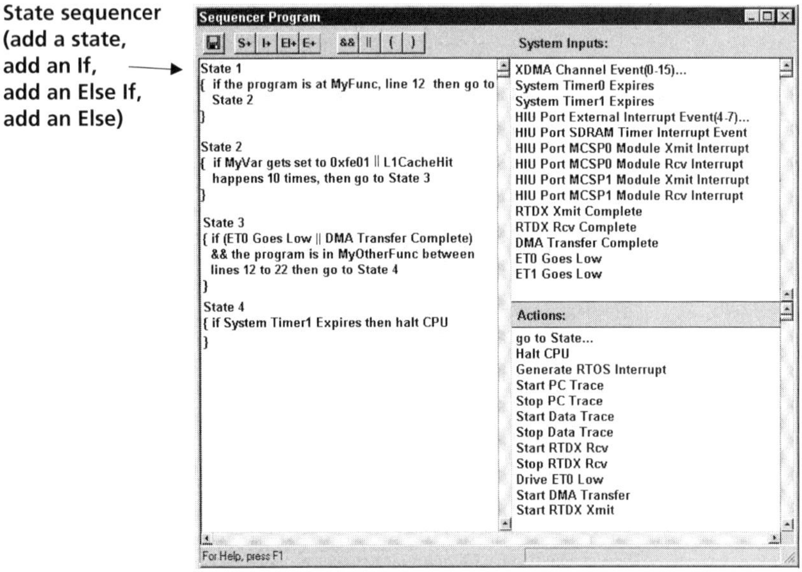

Figure 9.9 The developer can program the event counters and comparators from the user interface. The developer can program complicated scenarios and is limited only by the amount of logic on the device used for this capability. (Courtesty of Texas Instruments.)

Once programmed by the developer, the bus and auxiliary event system logic will monitor the running system for the conditions programmed into the logic. Once the condition is detected, the programmed response is generated. This may be a command to the emulator to halt execution or the setting of an output pin to inform another device or piece of test equipment. One of the advantages of on-chip logic like this is the ability to "see" what is going on inside the device. Using the external pins for visibility is limited to just those signals or conditions which the pins represent. DSP vendors have improved this capability by providing on-chip logic to provide visibility inside the DSP. This is key as DSPs and other embedded processors migrate to system on a chip architectures which, by definition, limit the amount of visibility into the device.

Because these event triggering capabilities are built directly into the DSP processor, they have no overhead in terms of either CPU cycles or memory. Thus when activated via user control, the event triggering device logic detects all events nonintrusively, without stopping the CPU. This preserves the real-time behavior of the system and can reduce debug time because the developer does not have to set breakpoints on each individual event, and then step repeatedly until the next complex event occurs.

Trace

The trace collection block of Figure 9.9 provides a powerful extension to the DSP emulation capability. This capability provides the means to extract the program counter, timing information, and raw data access information from the device at full clock rate speeds. The data is recorded and formatted externally in a large memory block. Using this data, the developer can gain visibility into what the processor is doing at an extremely detailed level. This is helpful for debugging many intermittent hardware and software problems on the target system. The trace capability can be controlled by the event logic to start only when a specific set of conditions has occurred in the system, such as a counter reaching a certain value, a specific module being executed, or when a specific variable is accessed. The program counter and/or data access information is usually tagged with a time stamp. This is useful for determining access times, interrupt latencies, module execution times and other valuable profiling data.

Continuous Execution Visibility

For some DSP applications, stopping the processor while continuing to service interrupts is a requirement. This is of primary interest for control applications such as hard-disk drive applications where the DSP is controlling the disk head positioning. In continuous execution visibility mode, when a physical interrupt occurs, the debugger gives up control and allows the application to service the hardware interrupt. Then upon returning from the hardware interrupt service routine (ISR), the processor is again halted. In these systems the DSP is used to control the servo to avoid a head crash in communications. During the debugging of the system, however, the developer may need to maintain synchronization with a communication system. This requires the DSP to continue to service interrupts while debugging the system. Special DSP emulation logic is required to enable this capability. This is just one example of how DSP emulation is customized for certain DSP families where the application domain drives the type of emulation capability required.

Source Level Debugging

Source level debugging is a useful capability that allows the developer to integrate systems at a higher level of abstraction. It allows the developer to tie the data extracted from the system to the high level source code in which the program was written. Access to system variables and other program locations can be done using the symbolic name from the source code or the raw memory address where the data is physically located. Often, the source code can be displayed along with the assembly code. This is useful for seeing what assembly code has been produced by the compiler. This is important when optimizing compilers are used to produce the code. DSP compilers provide high performance optimizing capability at various levels. Visibility into the assembly code for each high level language statement is very important when using code optimization switches. The developer can access the various program variables, structures, and segments using a set of symbols that are produced by the compiler and linker, depending on the debug options chosen when building the system. The symbol information is loaded into the emulator prior to each debug session.

Another useful capability of the emulation function is the visibility of the assembly language instructions being executed. Since the object file built through the compilation process is a binary file, there must be a conversion process from the binary object code (the machine code) back to the assembly instructions. This conversion process is referred to as "disassembly." The disassembly function takes the binary numbers loaded into memory and disassembles it to allow the user to view the actual assembly language stream that generated the machine language codes.

The trace capability also requires a disassembly operation. Trace disassembly must take trace frames and disassemble the code that was actually executed from this raw trace data. This data is very useful to the developer. It can provide an accurate indication of what actually occurred in the system during the period of time in which the data was collected. Using the capability of data trace, the user can view not only the instructions that were actually executed (instead of the instructions that were suppose to execute!) but also the data accesses performed during by those instructions.

High-Speed Data Collection and Visualization

The process of stopping an application with a breakpoint to exchange data "snapshots" with the host computer in a technique that's called "stop-mode debugging." This is an intrusive approach to debugging systems and can even be misleading, because the isolated snapshot of a halted high-speed application cannot show the system's real-world operation.

DSP vendors have solved this problem using technology that allows real-time data exchange with the processor. This technology goes by different names (TI calls it real-time data exchange [RTDX]). This capability gives designers continuous, real-time visibility into their applications by providing asynchronous exchange of data between the target and the host, without stopping the target application. This is important for debugging many real-time applications.

The technology is a data link that provides a "data pipe" between the DSP application and the host. The bidirectional capability provides the capability to access data from the application for real-time visibility, or to simulate data input to the DSP, perhaps before real-time sensor hardware is available. This can give the developer a more realistic view of the way their systems operate.

This technology requires support from the standard emulation technology discussed earlier. The data rate performance is dependent on the emulator used and the host computer. The data rate requirement is also application dependent (Figure 9.10).

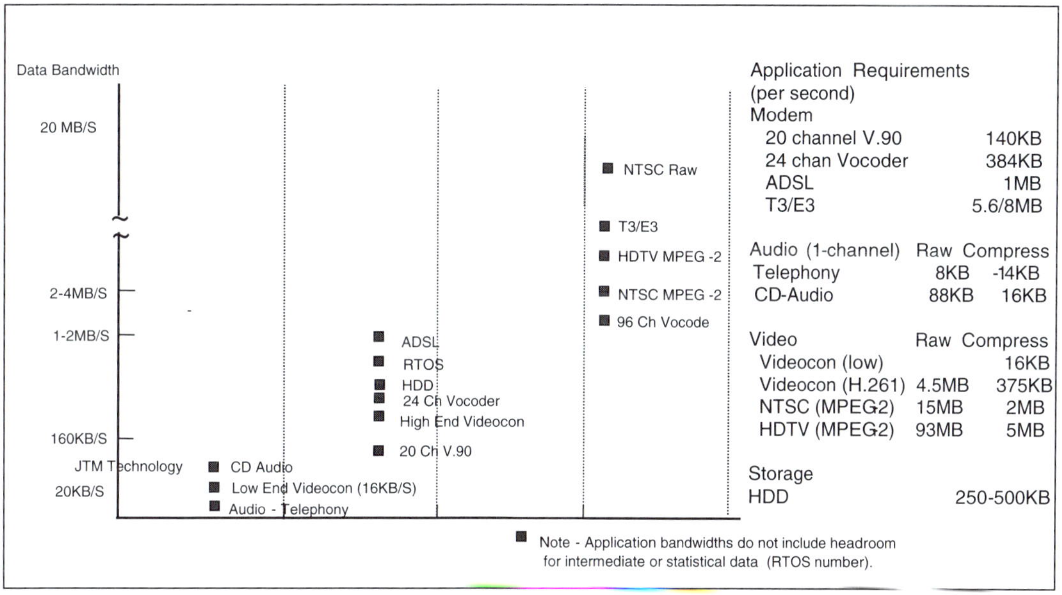

Figure 9.10 Data bandwidth required for debug and validation is dependent on the application.

Applications such as CD audio and low-end video conferencing may require up to 10K bytes per second. Higher end applications such as streaming video and ADSL will require much more, perhaps over 2 Mbytes per second.

This data transfer capability requires some level of support and will vary depending on the DSP vendor. When planning to use this technology, verify the following:

- A special high-speed peripheral on the target DSP device is present, if needed.
- A supporting emulator is available.
- The application is linked with the appropriate library to enable the capability. The TI technology, for example, is implemented with a small monitor (2 to 4 Kbytes of memory) that is included as a part of the RTOS. There is a minimal amount of CPU overhead required to implement the capability.

Scalability is Essential

Since these on-chip capabilities affect the chip's recurring cost, the scalability of debug solutions is of primary importance. "Pay only for what you need" is the guiding principle for on-chip tools deployment. In this new paradigm, the system architect specifies the on-chip debug facilities along with the remainder of functionality, balancing chip cost constraints and the debug needs of the product development team. Figure 9.11 shows examples of DSP applications and the types of DSP debug capability that are used to address the integration challenges for these various application spaces.

	/-----Control ---------	\ /----- Visibility -------		\ / -Usage.. \
	Exe. Cntl.	Event Triggers	Trace	Realtime Data	Configuration
Basestation (High Perf, High Frequency)	Realtime	Precise Triggering (Constrain High BW Trace)	PC & Data Trace High BW/Pin	High Bandwidth Point-to-point	Long Cables (Rack Mount Equip) Ganging Emulator Tools (High BW)
VOIP (Many Homogeneous Processors)	Realtime Homogeneous Multi-processing Global Commands	Shared Triggering Components Cross-triggering	Multi-channel Concurrent PC Trace Selective Data Trace	Multi-drop Realtime Data Exchange (JTAG)	Very wide trace Ganging Emulator Tools (Many Chips per Board
Wireless (Multiple Heterogeneous Processors)	Realtime Hterogeneous Multi-processing Global Commands	Shared Triggering Components Cross-triggering	Multi-channel Concurrent PC Trace Selective Data Trace	Multi-drop Realtime Data Exchange (JTAG)	Small Connection Footprint (in-System Debug) Short Cables (Desk Top, Small System Debug)
Automotive (Extreme Cost Sensitivity)	Realtime	Effective Triggering with MInimal Gates	Effective PC Trace with Low Pin Counts	Low Cost Realtime Data (JTAG)	Small Connection Footprint (In-system Debug)

Figure 9.11 The DSP debug technology needed depends on the application space (courtesy of Texas Instruments)

This data transfer technology runs asynchronously, in real-time, between the target and the host. Once the target data is in the host, it can be viewed in various ways, using custom or commercial viewers such as MathWorks or Labview.

These viewers can be used to check the correctness of embedded DSP algorithms. This can be done by comparing the results of the embedded algorithm with those obtained from a known correct algorithm from a standard data analysis or signal processing algorithm. Direct comparisons of the results can be done visually or a script can be used to check the accuracy of the data at a finer level of accuracy.

Compiler and Linker Dependencies

Most, if not all, DSP compilers provide options to the user that allow them to pass information needed for debugging to the linker as well as the debugger. This information includes, but is not limited to, symbol definitions and other information to allow the debugger to map addresses to line numbers in the source code. This is what ultimately provides the capability to perform source level debug and code display.

DSP compilers provide code optimization at various levels. When the compiler performs optimization, the resulting code may not look like the code as originally written! This is because the compiler makes decisions as to how to arrange the code statements in order to take advantage of the DSP hardware in a more optimal way, yet provide the correctness of the original high level source code. Some code may actually be eliminated altogether if the compiler determines that it is not needed. Other code may be relocated for more efficient execution. All this makes source level debugging more difficult for the developer.

As an example, consider the following code snippet:

```
for (i=0; i<10; i++)

  {

     array_A[i] = 4 * constant_X * array_B[i];

  }
```

A optimizing DSP compiler will recognize this loop as inefficient. In particular, the compiler will recognize that the calculation "4 * constant_X" is a redundant calculation that will be performed unnecessarily each loop iteration. This calculation will be pulled outside the loop and performed just once, with the result being applied to each iteration of the loop. The compiler, depending on the DSP architecture, will also attempt to unroll the loop to perform multiple calculations each iteration. The amount of loop unrolling in dependent on the DSP architecture. Finally, the DSP compiler will probably recognize that the array calculations can be done using pointers instead of array indices which requires multiple accesses to memory which is extremely expensive. Using pointers requires a index increment instead of a memory access to fetch the next array element. These optimizations will be made to reduce the total amount of multiplications and additions in the loop. This is, obviously advantageous for real-time requirements for DSP applications. In the end, the performance will be greatly improved. However the resulting code will look very different than what the programmer developed. If the developer attempts to debug this code, the source code in the debugger (the disassembly and source code) will look totally different from the source code file in the project directory.

Work is being done to improve this overall scenario but DSP developers should take caution when attempting to debug optimized code. This can be a very frustrating experience. One approach used by DSP developers is to debug the code without turning optimizations on first and get the code working functionally correct. Then go back and turn on optimizations and re-verify the code. This is more time consuming but less frustrating.

The debugging process is also affected by the linker technology used to build the DSP program. The link process is the final step in preparing the DSP code for the emulation and debugging phase. The linker is primarily responsible for merging all of the individual compiled code segments as well as required library routines and other reusable components into a single downloadable executable image file. The image is a binary representation which the linker formats in such a way as to make it readable by the debugger. The debugger, for example, needs to locate the code in the proper memory sections of the target system. The linker also produces the symbol file (sometimes called a map file) which contains the information needed by the debugger along with the program counter, to accurately communicate where in the source code the system is actually executing, as well as provide the necessary information to do source level debugging talked about earlier. DSP linker technology, like other embedded system devices, allows for various linker modes and switches, which produces various kinds of information. Each device is different so it may take some experimentation to get the right information for your debug session in the output file.

When selecting tools for DSP development activities, it is important to match the output of the build process to the expected debugger input. For example, if the DSP build toolset produces an output file in .coff format, then it is important to either select a debugger technology that can read .coff format files or find the appropriate conversion utility to convert to the format needed by the debugger. Otherwise you will not be able to load the code into the emulator!

Real-Time Embedded Software Testing Techniques

Testing for real-time systems

The specific characteristics of real-time systems make them a major challenge to test. The time-dependent nature of real-time applications adds a new and difficult element to testing. Not only does the developer have to apply the standard black and white box testing techniques, but also the timing of system data and the parallelism of the system tasks must be considered as well. In many situations, test data for real-time embedded systems may result in errors when the system is in one state but to in others. Comprehensive test cases design methods for real-time systems must be employed to achieve the proper coverage and system quality. DSP systems fall into this real-time system category. The basic testing approaches for DSP real-time systems involve (Figure 9.12):

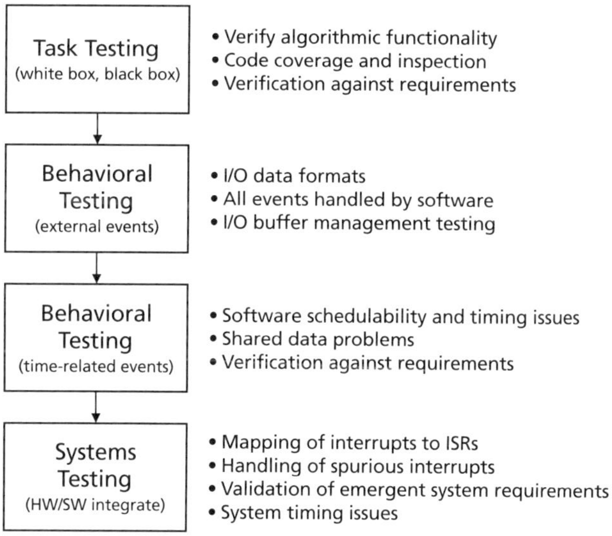

Figure 9.12 A Real-time DSP testing process

- *Task testing* – This involves testing each of the system tasks independently
- *Behavioral testing* – Test the behavior and actions of the real-time DSP system as a result of external events.
- *Inter-task testing* – This testing phase involves *time-related* external events. The system behavior is examined as a result of real-world frequency of external events into the system.
- *Systems testing* – In this final phase, software and hardware are integrated and a full set of systems tests are executed to detect errors at the software and hardware interfaces of the system.

As we all know, testing is expensive. Testing progress can also be hard to predict and the fact that embedded real-time systems have different needs as described above, we need to plan the testing process to achieve the most effective and efficient plan. The key goal is to know what you are looking for and how to effectively locate problems. You must plan to succeed but also manage risk accordingly and optimize and customize the testing process as necessary. The key question to answer is "What are we looking for?"

As you know, the consequences of a bug vary depending on the severity; production stop, critical, nuisance, "nice to have" are all categories of software bugs that demand different forms of attention. How your organization categorizes and manages software bugs will vary but the severity is usually always on this sort of graduated scale.

Many bugs are found in the source code and are referred to as implementation bugs. These result from improper implementation of algorithms and other processing, data bugs, real-time bugs (which is a category unique to real-time embedded systems) and other system related bugs.

Bugs are found in the source code but there are also plenty of sources of nonimplementation bugs that cause damage as well. Sources of nonimplementation bugs include errors in the specification, design, the hardware and even the compiler (remember the compiler is a software program with bugs just like other software programs!).

Algorithm bugs

DSP systems are comprised of complicated signal processing algorithms. We looked at some of these algorithms earlier. It is easy to make implementation errors in these algorithms because of the inherent complexity. Common mistakes in signal processing algorithms include:

- Improper return codes from complex functions
- Parameter passing errors
- Reentrancy mistakes
- Incorrect control flow through the algorithm stream
- Logical errors
- Mathematical computation errors
- Fence post or "off by one" errors

For example, we discussed the structure of many DSP algorithms in Figure 3.12.

The summations indicate looping operations in the software. Many errors introduced by improper looping limits—these are off-by-one errors that introduce subtle errors into the data stream. The simple expression:

```
for ( x = 0;, x <= 1000; x++)
```

will execute 1001 times and not 1000. This is a common mistake in these looping constructs for DSP algorithms.

When initializing arrays of filter coefficients or input data samples, the same mistake could occur:

```
for ( x = array_min; x <= array_max; x++)
```

this code snippet does not initialize the last element of the array as was intended because of a simple logic error in the loop bounds.

Many DSP algorithms are "wrapped" with control structures that manage the signal processing algorithm execution. These control structures can be complicated Boolean expressions like:

```
If (( a AND b ) OR ( c AND d )
AND NOT ( x AND y ))
```

can be hard to understand and implement correctly the first time. Remember George Miller's seven plus/minus two. His research has shown that, when someone is trying to solve a new problem or do an unfamiliar task, the number of individual variables that they can handle is relatively small; four variables are difficult, while five are nearly impossible. If you consider each logic element in the example above to be one of these variables, then you can see how easy it can be to implement logic like this incorrectly.

DSP algorithm errors also include those of underflow and overflow. These are common in DSP systems because many DSP systems are prototypes using floating-point processors to assess accuracy and performance and then ported over to fixed-point processors for cost reasons. This conversion or porting exercise leads to mistakes in mathematical underflow and overflow in the DSP algorithms unless a good regression test suite is used.

DSP systems are also potential victim to a class of real-time bugs. Given the real-time nature of DSP systems and the high use of operating systems to control task complexity, these bugs are becoming more and more common in DSP systems.

Task Synchronization and Interrupt Bugs

We discussed the issues associated with scheduling problems in the earlier chapter on real-time operating systems. These included problems such as priority inversion, race conditions and task scheduling policies which, if not implemented correctly, could lead to scheduling issues. Recall that, for a real-time DSP system, timely results are just as important as correct results. Therefore, schedulability issues can be a source of real-time bugs.

Interrupts are also a fundamental component for real-time systems. When used correctly they provide efficient mechanisms for dealing with high priority tasks in a real-time system. But when used incorrectly, interrupts can cause various classes of bugs in the system. Examples include:

Interrupt service routines that are too long; this can cause long latency delays in the system since interrupts are usually disabled during ISRs. The guideline is to write short ISRs that do the absolute minimum while deferring the rest of the processing to the main routines.

Shared data corruption; in most real time systems, there are several classes of critical resources that must be shared amongst various tasks in the system. These include the CPU, memory, and peripheral. When sharing data (memory) in a real-time system, data can be corrupted if the access to the shared data is not synchronized and managed. To illustrate this, I will use the classic example from Simon[1] that shows the shared data problem in action. The code below shows an ISR called vReadTemperatures that reads two values from hardware into variable iTemperature. The main( ) routine also modifies the same data structure. If main( ) is modifying the iTemperatures structure and gets interrupted in the middle of this update which vectors to the vReadTemperatures ISR, this routine will also update the data structure, potentially leading to a corrupted

1 David Simon's book, Embedded Systems Primer, is a highly recommended book on embedded systems.

array by the time main( ) regains execution control and then proceeds to complete its own update of the data structure.

```
static int iTemperatures[2];

void interrupt vReadTemperatures (void)
{
        iTemperatures[0] = !! Read in value from hardware register
        iTemperatures[1] = !! Read in value from hardware register
}

void main (void)
{
        int iTemp0, iTemp1;
        while (TRUE)
        {
                iTemp0 = iTemperatures[0];
                iTemp1 = iTemperatures[1];
                if (iTemp0 != iTemp1)
                        !! Set off alarm
        }
}
```

This is a simple example of how data structures can become corrupted due to improper data protection. There are mechanisms to alleviate this problem, including disabling interrupts and using semaphores. But these protection mechanisms are the responsibility of the programmer so there is the possibility of these bugs propagating in the system if this is not done right.

Conclusion

As the complexity of DSPs, and embedded applications in general, grows, the hardware/software integration phase is becoming the biggest bottleneck in the entire development activity. Integration tools support has become one of the most important drivers for time to market for complex embedded DSP projects.

DSP emulation and debug capability can provide a significant advantage during the integration and test phase of complex applications. Using the right combination of hardware debug capability as well as software debug capability can aid the DSP developer in many ways including:

- The reconstruction of events that lead to a system crash.
- Visibility of specific instructions sequences during execution.
- Quick download time from the host to the target reducing the overall time spent in the debug and integration "loop."
- Debug of code that resides in ROM or Flash memory using hardware assist debug.

- Tracking application behavior under realistic environmental conditions in a non-intrusive way.
- Acquisition of performance data to aid in improving overall system execution performance and efficiency.

References

Sensible Software Testing, by Sean Beatty

http://www.embedded.com/2000/0008/0008feat3.htm

David Simon, (1999), *An Embedded Software Primer*, Addison Wesley

10

Managing the DSP Software Development Effort

Overview

This chapter will cover several topics related to managing the DSP software development effort. The first section of this chapter will discuss DSP system engineering and problem framing issues that relate to DSP and real-time software development. High level design tools are then discussed in the context of DSP application development. Integrated development environments and prototyping environments are also discussed. Specific challenges to the DSP application developer such as code tuning, profiling and optimization are reviewed. At the end of the chapter, there are several white papers that discuss topics related to DSP and real-time system development, integration, and analysis.

Software development using DSPs is subject to many of the same constraints and development challenges that other types of software development face. These include a shrinking time to market, tedious and repetitive algorithm integration cycles, time intensive debug cycles for real-time applications, integration of multiple differentiated tasks running on a single DSP, as well as other real-time processing demands. Up to 80% of the development effort is involved in analysis, design, implementation, and integration of the software components of a DSP system.

Early DSP development relied on low level assembly language to implement the most efficient algorithms. This worked reasonably well for small systems. However, as DSP systems grow in size and complexity, assembly language implementation of these systems has become impractical. Too much effort and complexity is involved to develop large DSP systems within reasonable cost and schedule constraints. The migration has been towards higher level languages like C to provide the maintainability, portability, and productivity needed to meet cost and schedule constraints. Other real-time development tools are emerging to allow even faster development of complex DSP systems (Figure 10.1).

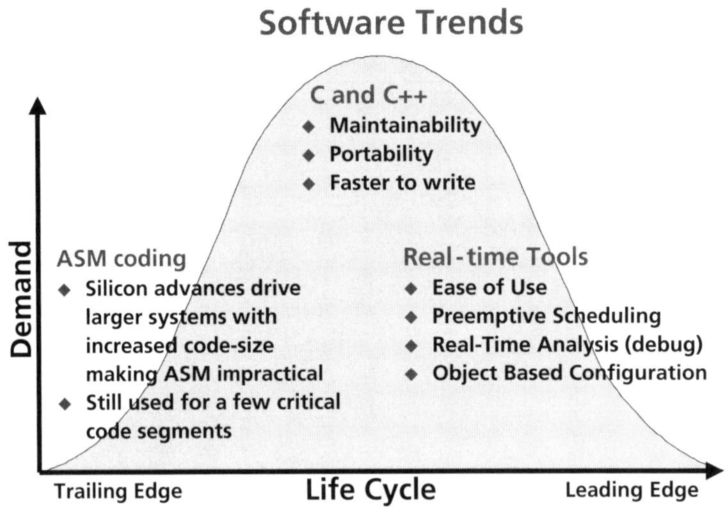

Figure 10.1 Programming in the large has forced DSP developers to rely on higher level languages and real-time tools to accomplish their tasks (courtesy of Texas Instruments)

DSP development environments can be partitioned into host tooling and target content (Figure 10.2). Host tooling consists of tools to allow the developer to perform application development tasks such as program build, program debug, data visualization and other analysis capabilities. Target content refers to the integration of software running on the DSP itself, including the real-time operating system (if needed), and the DSP algorithms that perform the various tasks and functions. There is a communication mechanism between the host and the target for data communication and testing.

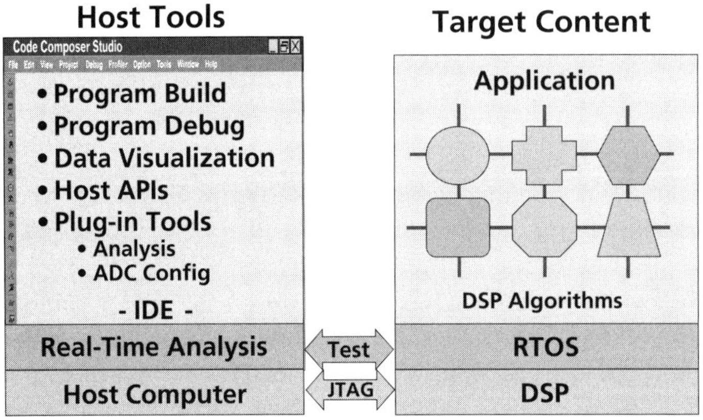

Figure 10.2 DSP development can be partitioned into a host development environment and a target environment (courtesy of Texas Instruments)

DSP development consists of several phases, as shown in Figure 10.3. During each phase, there exists tooling to help the DSP developer quickly proceed to the next stage.

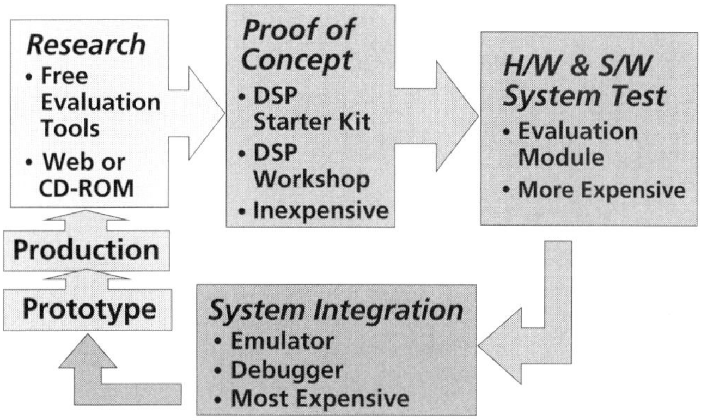

Figure 10.3 Phases of DSP development (courtesy of Texas Instruments)

Challenges in DSP Application Development

The implementation of software for embedded digital signal processing applications is a very complex process. This complexity is a result of increasing functionality in embedded applications; intense time-to-market pressures; and very stringent cost, power and speed constraints.

The primary goal for most DSP application algorithms focuses on the minimization of code size as well as the minimization of the memory required for the buffers that implement the main communication channels in the input dataflow. These are important problems, because programmable DSPs have very limited amounts of on-chip memory, and the speed, power, and cost penalties for using off-chip memory are prohibitively high for many cost-sensitive embedded applications. Complicating the problem further, memory demands of applications are increasing at a significantly higher rate than the rate of increase in on-chip memory capacity offered by improved integrated circuit technology.

To help cope with such complexity, DSP system designers have increasingly been employing high-level, graphical design environments in which system specification is based on hierarchical dataflow graphs. Integrated development environments (IDEs) are also being used in the program management and code build and debug phases of a project to manage increased complexity.

The main goal of this chapter is to explain the DSP application development flow and review the tools and techniques available to help the embedded DSP developer analyze, build, integrate, and test complex DSP applications.

Historically, digital signal processors have been programmed manually using assembly language. This is a tedious and error-prone process. A more efficient approach is to generate code automatically using available tooling. However, the auto-generated code must be efficient. DSPs have scarce amounts of on-chip memory and its use must be managed carefully. Using off-chip memory is inefficient due to increased cost, increased power requirements, and decreased speed penalty. All these drawbacks have a significant impact on real-time applications. Therefore, effective DSP-based code generation tools must specify the program in an imperative language such as C or C++ and use a good optimizing compiler.

The DSP Design Process

The DSP design process is similar in many ways to the standard software and system development process. However, the DSP development process also has some unique challenges to the development process that must be understood in order to develop efficient, high performance applications.

A high level model of the DSP system design process is shown in Figure 10.4. As shown in this figure, some of the steps in the DSP design process are similar to the conventional system development process.

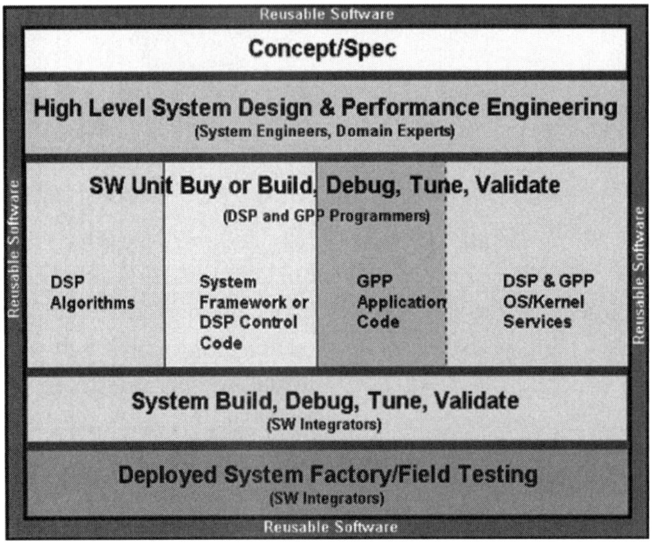

Figure 10.4 A block diagram of the general system design flow

Concept and Specification Phase

The development of any signal processing system begins with the establishment of requirements for the system. In this step, the designer attempts to state the attributes of the system in terms of a set of requirements. These requirements should define the characteristics of the system expected by external observers such as users or other systems. This definition becomes the guide for all other decisions made during the design cycle.

DSP systems can have many types of requirements that are common in other electronic systems. These requirements may include power, size, weight, bandwidth, and signal quality. However, DSP systems often have requirements that are unique to digital signal processing. Such requirements can include sample rate (which is related to bandwidth), data precision (which is related to signal quality), and real-time constraints (which are related to general system performance and functionality).

The designer devises a set of specifications for the system that describes the sort of system that will satisfy the requirements. These specifications are a very abstract design for the system. The process of creating specifications that satisfy the requirements is called *requirements allocation*.

A specification process for DSP systems

During the specification of the DSP system, using a real-time specification process like Sommerville's six step real-time design process is recommended[1]. The six steps in this process include:

- Identify the stimuli to be processed by the system and the required responses to these stimuli.
- For each stimulus and response, identify the timing constraints. These timing constraints must be quantifiable.
- Aggregate the stimulus and response processing into concurrent software processes. A process may be associated with each class of stimulus and response.
- Design algorithms to process each class of stimulus and response. These must meet the given timing requirements. For DSP systems, these algorithms consist mainly of signal processing algorithms.
- Design a scheduling system that will ensure that processes are started in time to meet their deadlines. The scheduling system is usually based on a preemptive multitasking model, using a rate monotonic or deadline monotonic algorithm.
- Integrate using a real-time operating system (especially if the application is complex enough).

Once the stimuli and responses have been identified, a software function can be created by performing a mapping of stimuli to responses for all possible input sequences. This is important for real time systems. Any unmapped stimuli sequences can cause behavioral problems in the system. Although the details of this are beyond the scope of this chapter, an example stimuli sequence mapping for a cell phone is shown in Appendix D.

Algorithm development and validation

During the concept and specification phase, a majority of the developer's time is spent doing algorithm development. During this phase, the designer focuses on exploring approaches to solve the problems defined by the specifications at an abstract level. During this phase, the algorithm developer is usually not concerned with the details of

[1] *Software Engineering* Version 7 by Ian Sommerville, Chapter 16.

how the algorithm will be implemented. Moreover, the developer focuses on defining a computational process which can satisfy the system specifications. The partitioning decisions of where the algorithms will be hosted (DSP, GPP, hardware acceleration such as ASIC or FPGA) are not the number one concern at this time.

A majority of DSP applications require sophisticated control functions as well as complex signal processing functions. These control functions manage the decision-making and control flow for the entire application. (for example, managing the various functional operations of a cell phone as well as adjusting certain algorithm parameters based on user input). During this algorithm development phase, the designer must be able to specify as well as experiment with both the control behavior and the signal processing behavior of the application.

In many DSP systems, algorithm development first begins using floating-point arithmetic. At this point, there is no analysis or consideration of fixed-point effects resulting from running the application on a fixed-point processor. Not that this analysis is not important. It is very critical to the overall success of the application and is considered shortly. But the main goal is to get an algorithm stream working, providing the assurance that the system can indeed work! When it comes to actually developing a productizable system, a less expensive fixed-point processor may be the choice. During this time transition, the fixed-point effects must be considered. In most cases it will be advantageous to implement the productizable system using simpler, smaller numeric formats with lower dynamic range to reduce system complexity and cost. This can only be found on fixed-point DSPs.

For many kinds of applications, it is essential that system designers have the ability to evaluate candidate algorithms running in real-time, before committing to a specific design and hardware/software implementation. This is often necessary where subjective tests of algorithm quality are to be performed. For example, in evaluating speech compression algorithms for use in digital cellular telephones, real-time, two-way communications may be necessary.

DSP Algorithm Standards and Guidelines

Digital Signal Processors are often programmed like "traditional" embedded microprocessors. They are programmed in a mix of C and assembly language, they directly access hardware peripherals, and, for performance reasons, almost always have little or no standard operating system support. Thus, like traditional microprocessors, there is very little use of commercial off-the-shelf (COTS) software components for DSPs. However, unlike general-purpose embedded microprocessors, DSPs are designed to run sophisticated signal processing algorithms and heuristics. For example, they may be used to detect important data in the presence of noise, to or for speech recognition in a noisy automobile traveling at 65 miles per hour. Such algorithms are often the result of many years of research and development. However, because of the lack of consistent standards, it is not possible to use an algorithm in more than one system without significant reengineering. This can cause significant time to market issues for DSP developers. Appendix F provides more detail on DSP algorithm development standards and guidelines.

High-Level System Design and Performance Engineering

High level system design refers to the overall partitioning, selection and organization of the hardware and software components in a DSP system. The algorithms developed during the specification phase are used as the primary inputs to this partitioning and selection phase. Other factors are considered are:

- performance requirements,
- size, weight, and power constraints,
- production costs,
- nonrecurring engineering (engineering resources required),
- time-to-market constraints,
- reliability.

This phase is critical, as the designer must make trade-offs among these often conflicting demands. The goal is to select a set of hardware and software elements to create an overall system architecture that is well-matched to the demands of the application.

Modern DSP system development provides the engineer with a variety of choices to implement a given system. These include:

- custom software,
- optimized DSP software libraries,
- custom hardware,
- standard hardware components.

The designer must make the necessary system trade-offs in order to optimize the design to meet the system performance requirements that are most important. (performance, power, memory, cost, manufacturability, and so on).

Performance engineering

Software performance engineering (SPE) aims to build predictable performance into systems by specifying and analyzing quantitative behavior from the very beginning of a system, through to its deployment and evolution. DSP designers must consider performance requirements, the design, and the environment in which the system will run. Analysis may be based on various kinds of modelling and design tools. SPE is a set of techniques for:

- gathering data,
- constructing a system performance model,
- evaluating the performance model,
- managing risk of uncertainty,

- evaluating alternatives,
- verifying the models and results.

SPE requires the DSP developer to analyze the complete DSP system using the following information[2]:

- *Workload* – Worst case scenarios
- *Performance objectives* – Quantitative criteria for evaluating performance (CPU, memory, I/O)
- *Software characteristics* – Processing steps for various performance scenarios (ADD, simulation)
- *Execution environment* – Platform on which the proposed system will execute, partitioning decisions
- *Resource requirements* – Estimate of the amount of service required for key components of the system
- *Processing overhead* – Benchmarking, simulation, prototyping for key scenarios

Appendix A is a case study of developing DSP-based systems using software performance engineering.

Software Development

Most DSP systems are developed using combinations of hardware and software components. The proportion varies depending on the application (a system that requires fast upgradability or changability will use more software; a system that deploys mature algorithms and requires high performance may use hardware). Most systems based on programmable DSP processors are software-intensive.

Software for DSP systems comes in many flavors. Aside from the signal processing software, there are many other software components required for programmable DSP solutions:

- Control software
- Operating system software
- Peripheral driver software
- Device driver software
- Other board support and chip support software
- Interrupt routine software

There is an growing trend towards the use of reusable or "off the shelf" software components. This includes the use of reusable signal processing software components, application frameworks, operating systems and kernels, device drivers, and chip support software. DSP developers should take advantage of these reusable components whenever possible. The topic of reusable DSP software components is a topic of a later chapter.

---

[2] For more information on software performance engineering, see the excellent reference textbook, *Performance Solutions: A Practical Guide to Creating Responsive, Scalable Software*, (1st Edition) by Connie U. Smith and lloyd G. Williams, Addison-Wesley Professional; 2001.

System Build, Integration and Test

As DSP system complexity continues to grow, system integration becomes paramount. System integration can be defined as the progressive linking and testing of system components to merge their functional and technical characteristics into a comprehensive interoperable system. System integration is becoming more common as DSP systems can contain numerous complex hardware and software subsystems. It is common for these subsystems to be highly interdependent. System integration usually takes place throughout the design process. System integration may first be done using simulations prior to the actual fabrication of hardware.

System integration proceeds in stages as subsystem designs are developed and refined. Initially, much of the system integration may be performed on DSP simulators interfacing with simulations of other hardware and software components. The next level of system integration may be performed using a DSP evaluation board (this allows the software integration with device drivers, board support packages, kernel, and so on). A final level of system integration can begin once the remaining hardware and software components are available.

Factory and Field Test

Factory and field test includes the remote analysis and debugging of DSP systems in the field or in final factory test. This phase of the life cycle requires sophisticated tools which allow the field test engineer to quickly and accurately diagnose problems in the field and report those problems back to the product engineers to debug in a local lab.

Design Challenges for DSP Systems

What defines a DSP system are the signal processing algorithms used in the application. These algorithms represent the numeric recipe for the arithmetic to be performed. However, the implementation decisions for these algorithms are the responsibility of the DSP engineer. The challenge for the DSP engineer is to understand the algorithms well enough to make intelligent implementation decisions what endure the computational accuracy of the algorithm while achieving "full technology entitlement" for the programmable DSP in order to achieve the highest performance possible.

Many computationally intensive DSP systems must achieve very rigorous performance goals. These systems operate on lengthy segments of real-world signals that must be processed in real-time. These are hard real time systems that must always meet these performance goals, even under worst case system conditions. This is orders of magnitude more difficult that a soft real time system where the deadlines can be missed occasionally.

DSPs are designed to perform certain classes of arithmetic operations such as addition and multiplication very quickly. The DSP engineer must be aware of these advantages and be able to use the numeric formats and type of arithmetic wisely to have a significant influence on the overall behavior and performance of the DSP system. One important choice is the selection of fixed-point or floating-point arithmetic. Floating-point arithmetic provides much greater dynamic range than fixed-point arithmetic. Floating-point also reduces the probability of overflow and the need for the programmer to worry about scaling. This alone can significantly simplify algorithm and software design, implementation and test. The drawback to floating-point processors (or floating-point libraries) is that they are slower and more expensive than fixed-point. DSP engineers must perform the required analysis to understand the dynamic ranges needed throughput the application. It is highly probable that a complex DSP system will require different levels of dynamic range and precision at different points in the algorithm stream. The challenge is to perform the right amount of analysis in order to use the required numeric representations to give the performance required from the application.

In order to test a DSP system properly, a set of realistic test data is required. This test data may represent calls coming into a basestation or data from another type of sensor that represents realistic scenarios. These realistic test signals are needed to verify the numeric performance of the system as well as the real-time constraints of the system. Some DSP applications must be tested for long periods of time in order to verify that there are no accumulator overflow conditions or other "corner cases" that may degrade or break the system.

High Level Design Tools for DSP

System-level design for DSP systems requires both high-level modeling to define the system concept and low level modeling to specify behavior details. The DSP system designer must develop complete end-to-end simulations and integrate various components such as analog and mixed signal, DSP, and control logic. Once the designer has modeled the system, the model must be executed and tested to verify performance to specifications. These models are also used to perform design trade-offs, what-if analysis, and system parameter tuning to optimize performance of the system (Figure 10.5).

DSP modeling tools exist to aid the designer in the rapid development of DSP-based systems and models. Hierarchical block-diagram design and simulation tools are available to model systems and simulate a wide range of DSP components. Application libraries are available for use by the designer that provide many of the common blocks found in DSP and digital communications systems. These libraries allow the designer to build complete end-to-end systems. Tools are also available to model control logic for event driven systems.

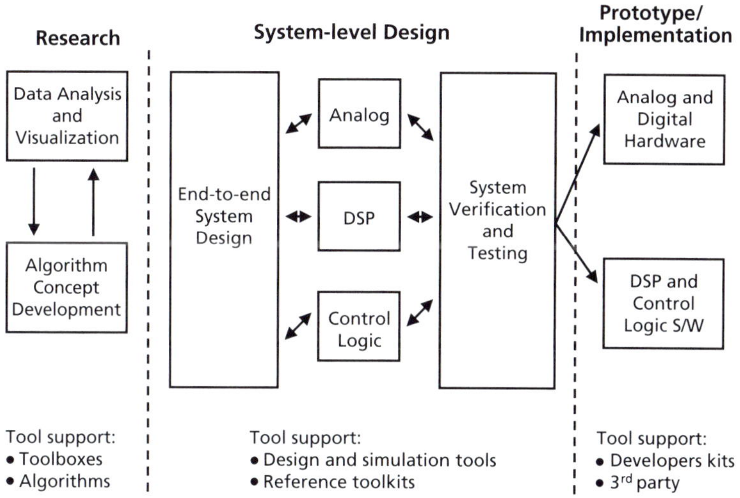

Figure 10.5 DSP system development requires phases of research, system development and prototyping (Reprinted with permission from MathWorks, Inc.)

DSP Toolboxes

DSP toolboxes are collections of signal processing functions that provide a customizable framework for both analog and digital signal processing. DSP toolboxes have graphical user interfaces that allow interactive analysis of the system. The advantage of these toolbox algorithms is that they are robust, reliable and efficient. Many of these toolboxes allow the DSP developer to modify the supplied algorithm to more closely map to the existing architecture as well as the ability for the developer to add custom algorithms to the toolbox.

DSP system design tools also provide the capability for rapid design, graphical simulation, and prototyping of DSP systems. Blocks are selected from available libraries and interconnected in various configurations using point and click operations. Signal source blocks are used to test models. Simulations can be visualized interactively and passed on for further processing. ANSI standard C code is generated directly from the and model. Signal processing blocks, including FFT, DFT; window functions; decimation/interpolation; linear prediction; multirate signal processing are available for rapid system design and prototyping. A Kalman Adaptive Filter, for example, is represented using the block diagram in the form shown in Figure 10.6.

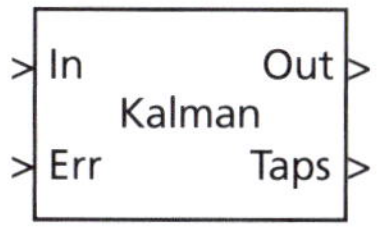

Figure 10.6 Block diagram form of a signal processing function (Reprinted with permission from MathWorks, Inc.)

Block diagram systems also have a physical realization. The Kalman filter block diagram function is based on the physical realization of a dynamical system shown in Figure 10.7. In this example, the Kalman Adaptive Filter block computes the optimal linear minimum mean-square estimate of the FIR filter coefficients using a one-step predictor algorithm.

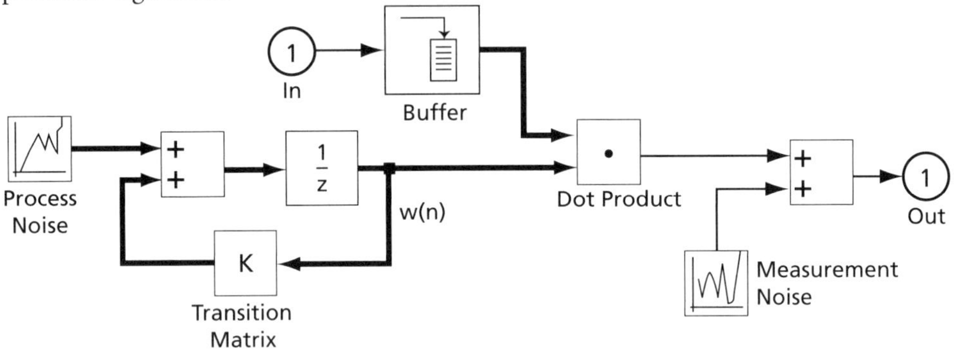

Figure 10.7 Physical realization of a block diagram signal processing function (Reprinted with permission from MathWorks, Inc.)

Host Development Tools for DSP Development

As mentioned earlier, there are many software challenges facing the real-time DSP developer:

- Life cycle costs are rising
- Simple system debug is gone
- Software reuse is minimal
- Systems are increasingly complex

A robust set of tools to aid the DSP developer can help speed development time, reduce errors, and more effectively manage large projects, among other advantages. Integrating a number of different tools into one integrated environment is called an *integrated development environment* (IDE). An IDE is a programming environment that has been packaged as an application program. A typical IDE consists of a code editor, a compiler, a debugger, and a graphical user interface (GUI) builder. IDEs provide a user-friendly framework for building complex applications. PC developers first had access to IDEs. Now applications are large and complex enough to warrant the same development environment for DSP applications. Now, DSP vendors have IDEs to support their development environments. DSP development environments support some but not all of the overall DSP application development life cycle. As shown in Figure 10.8, the IDE is mainly used after the initial concept exploration, systems engineering, and partitioning phases of the development project. Development of the DSP application from inside the IDE mainly addressed software architecture, algorithm design and coding, the entire project build phase and the debug and optimize phases.

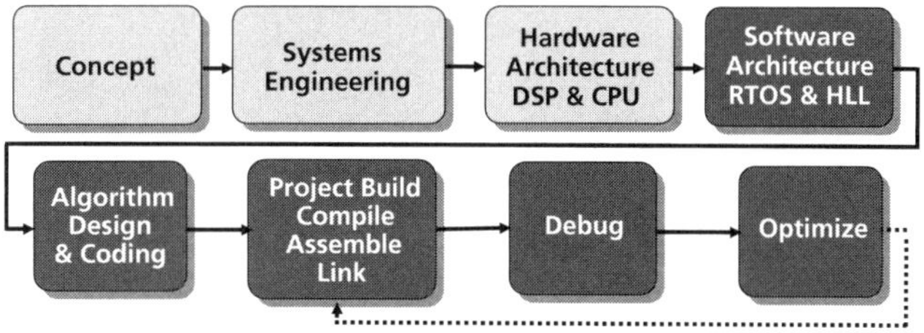

Figure 10.8 The DSP IDE is useful for some, but not all, of the DSP development life cycle

A typical DSP IDE consists of several major components (Figure 10.9):

- Code generation (compile, assemble, link)
- Edit
- Simulation
- Real-time analysis
- Debug and emulation
- Graphical user interface
- Other "plug in" tools
- Efficient connection to a target system

A DSP IDE is similar to a Microsoft Visual C++ IDE but optimized for DSP development. DSP development is different enough from other development to warrant a set of DSP-centric options within the IDE:

- Advanced real-time debugging which includes advanced breakpoints, C-expression-based conditional breakpoints, and simultaneous view of source and dis-assembly
- Advanced watch window
- Multiprocessor debug
- Global breakpoints
- Synchronized control over groups
- Probe points (advanced break points) provide oscilloscope-like functions
- File I/O with advanced triggering injects or extracts data signals

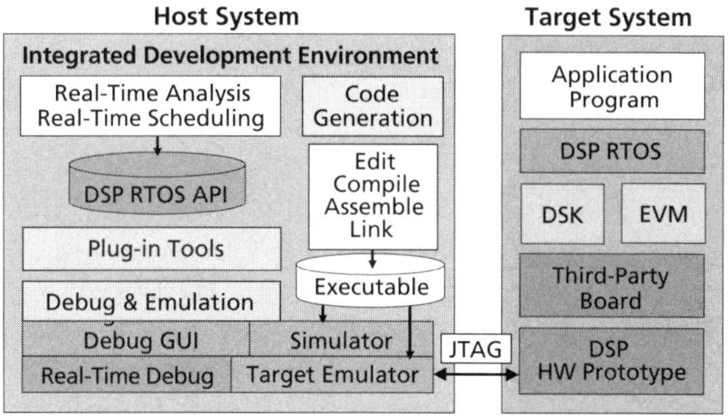

Figure 10.9 A DSP integrated development environment (courtesy of Texas Instruments)

Data Visualization, for example, allows the DSP developer to perform graphical signal analysis. This gives the developer the ability to view signals in native format and change variables on the fly to see their effects There are numerous application-specific graphical plots available including FFT waterfall, eye diagram, constellation plot, and image displays (Figure 10.10).

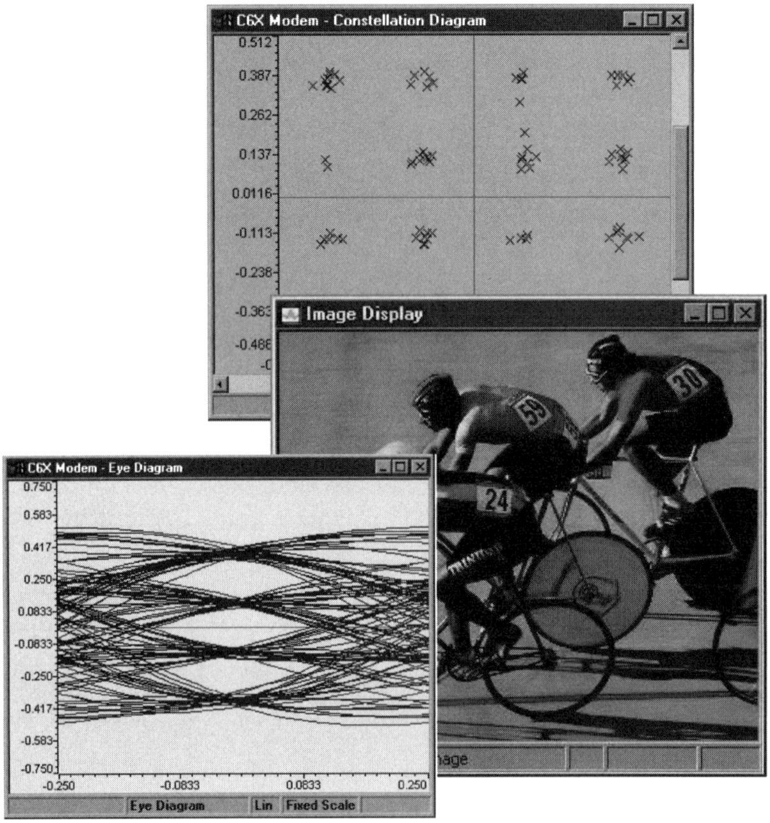

Figure 10.10 Visual displays are useful for DSP developers (courtesy of Texas Instruments)

As DSP complexity grows and systems move from being cyclic executive to task based execution, more advanced tools are required to facilitate the integration and debug phase of development. DSP IDEs provide a robust set of "dashboards" to help analyze and debug complex real-time applications. Figure 10.13 shows an example of such a dashboard. This display shows the task execution history for the various tasks executing in a system. This "real-time analysis" dashboard shows the state of each task (running, ready, etc) over time. If more advanced task execution analysis is desired, a third party plug in capability can be used. For example, if the DSP developer needs to know *why* and *where* a task is blocked, not just *if* a task is blocked, a third party rate monotonic analysis tool can be used to perform more detailed analysis.

DSP applications require real-time analysis of the system as it runs. DSP IDE's provide the ability to monitor the system in real-time with low overhead and interference of the application. Because of the variety of real-time applications, these analysis capabilities are user controlled and optimized. Analysis data is accumulated and sent to the host in the background (a low priority nonintrusive thread performs this function, sending data over the JTAG interface to the host). These real-time analysis capabilities act as a software logic analyzer, performing tasks that, in the past, were performed by hardware logic analyzers. The analysis capability can show CPU load percentage (useful for finding hot spots in the application), the task execution history (to show the sequencing of events in the real-time system), a rough estimate of DSP MIPS (by using an idle counter) and the ability to log best and worst case execution times (Figure 10.11). This data can help the DSP developer determine whether the system is operating within its design specification, meeting performance targets, and whether there are any subtle timing problems in the run time model of the system[3].

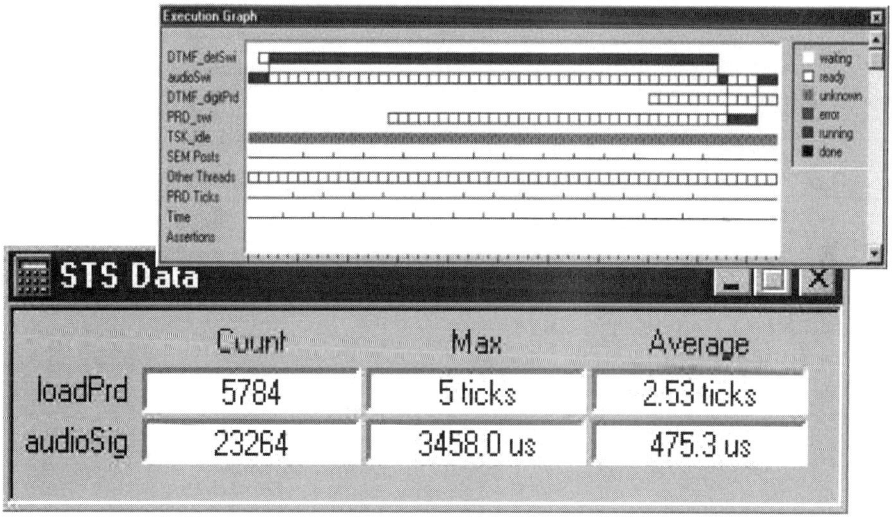

Figure 10.11 Real-time analysis capabilities give the DSP developer valuable data for real-time system integration (courtesy of Texas Instruments)

System configuration tools (Figure 10.12) allow the DSP developer to prioritize system functions quickly and perform what if analysis on different run time models.

[3] If you can't see the problem, you can't fix the problem.

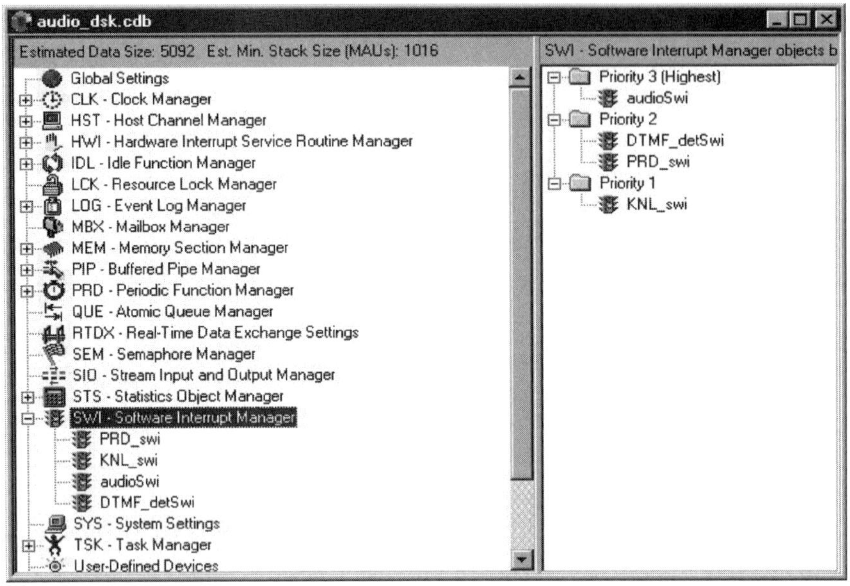

Figure 10.12 System configuration tools allow the DSP developer to quickly configure and prioritize the application resources and tasks (courtesy of Texas Instruments)

The diagram in Figure 10.13 illustrates development tool flow starting from the most basic of requirements. The editor, assembler, and linker are, of course, are the most fundamental blocks. Once the linker has built an executable, there must be some way to load it into the target system. To do this, there must be run control of the target. The target can be a simulator (SIM), which is ideal for algorithm checkout when the hardware prototype is not available. The target can also be a starter kit (DSK) or an evaluation board (EVM) of some type. An evaluation board lets the developer run on real hardware, often with a degree of configurable I/O. Ultimately, the DSP developer will run the code on a prototype, and this requires an emulator.

Another important component of the model in Figure 10.13 is the debugger. The debugger controls the simulator or emulator and gives the developer low level analysis and control of the program, memory, and registers in the target system. This means the DSP developer must debug systems without complete interfaces, or ones with I/O, but that don't have real data available[4]. This is where file I/O helps the debug process. DSP debuggers generally have the ability to perform data capture as well as graphing capability in order to analyze the specific bits in the output file. Since DSP is all about code and application performance, it is also important to provide a way to measure code speed with the debugger (this is generally referred

[4] A case study of the hardware/software co-design as well as the analysis techniques for embedded DSP and micro devices is at the end of the chapter.

to as profiling). Finally, as the generic data flow in Figure 10.13 shows, a real-time operating system capability allows the developer to build large complex applications and the Plug-In interface allows third parties (or the developer) to develop additional capabilities that integrate into the IDE.

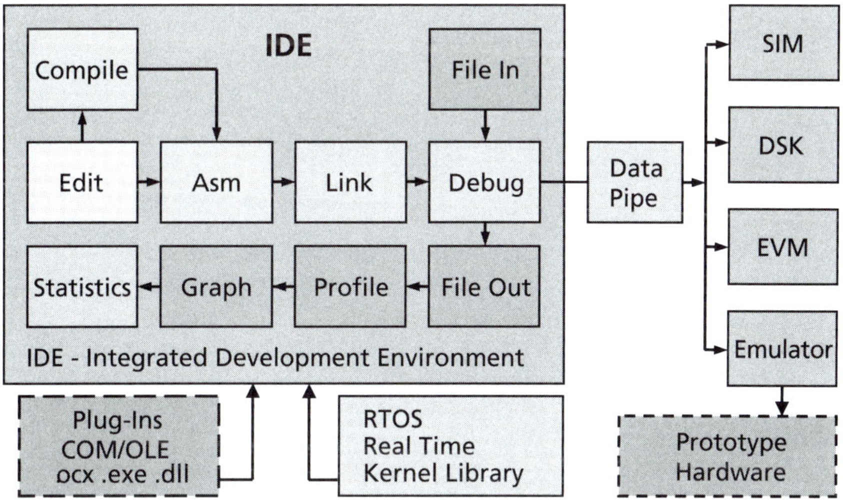

Figure 10.13 Basic components of a DSP development environment (courtesy of Texas Instruments)

A Generic Data Flow Example

This section will describe a simple example that brings together the various DSP development phases and tooling to produce a fully integrated DSP application. Figure 10.14 shows the simple model of a software system. Input is processed or transformed into output. From a real-time system perspective, the input will come from some sensor in the analog environment and the output will control some actuator in the analog environment. The next step will be to add data buffering to this model. Many real-time systems have some amount of input (and output) buffering to hold data while the CPU is busy processing a previous buffer of data. Figure 10.15 shows the example system with input and output data buffering added to the application.

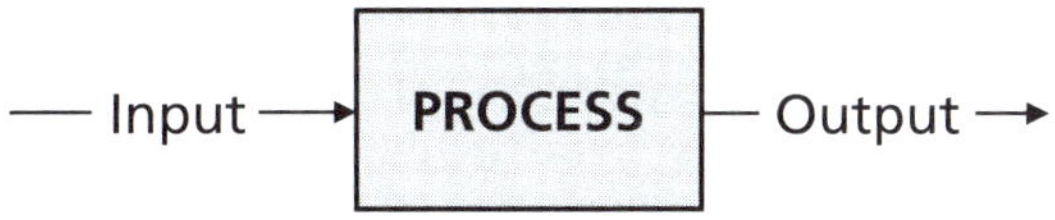

Figure 10.14 A generic data flow example

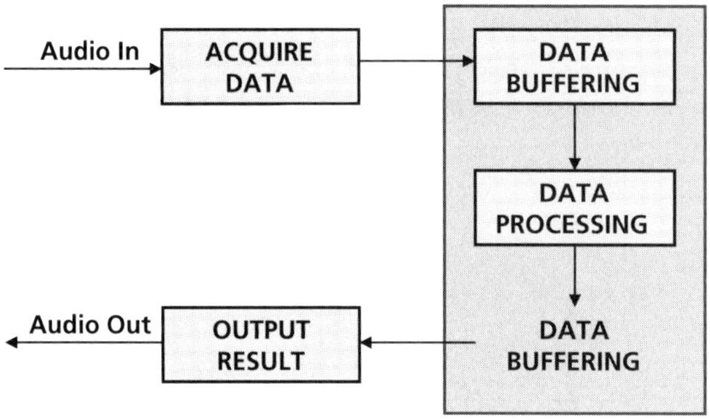

Figure 10.15 A data buffering model for a DSP application (courtesy of Texas Instruments)

A more detailed model for data buffering is a double buffer model as shown in Figure 10.16.

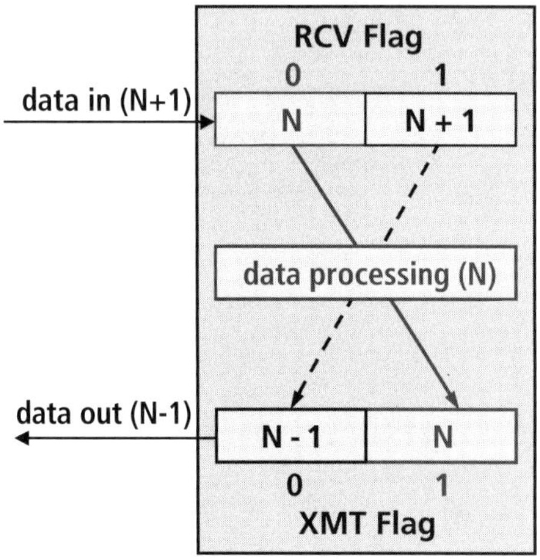

Figure 10.16 – A data buffering model for processing real-time samples from the environment (courtesy of Texas Instruments)

A double buffer is necessary because, as the CPU is busy processing one buffer, data from the external sensors must have a place to go and wait until the CPU is done processing the current buffer of data. When the CPU finishes one buffer, it will begin processing the next buffer. The "old" buffer (the buffer that was previously processed by the CPU) is now the current input buffer to hold data from the external sensors. When data is ready to be received (RCV "ping" is empty, RCV flag = 1), RCV "pong" buffer is emptied into the XMT "ping" buffer (XMT flag = 0) for output and vice versa. This repeats continuously as data is input to and output from the system.

DSPs are customized to perform the basic operations shown in the block diagram in Figure 10.17. The analog data from the environment is input using the multichannel buffered serial port interface (McBSPs). The external direct memory access controller (EDMA) manages the input of data into the DSP core and frees the CPU to perform other processing. The dual buffer implementation is realized in the on-chip data memory, which acts as the storage area. The CPU performs the processing on each of the buffers.

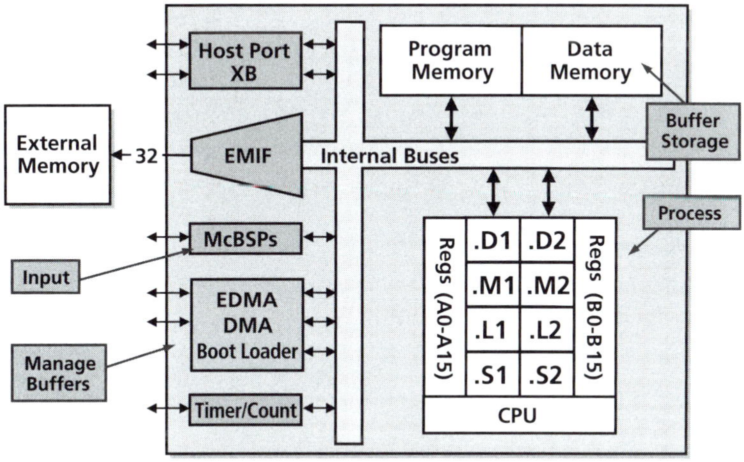

Figure 10.17 The DSP engine showing core components that manage the input, storage, and processing of data samples from the environment (courtesy of Texas Instruments)

A combination of these two views is shown in Figure 10.18. This is a system block diagram view shown mapped to a model of a DSP starter kit or evaluation board. The DSP starter kit in Figure 10.19 has a CODEC which performs a transformation on the data before being sent to the McBSP and on into the DSP core using the DMA function. The DMA can be programmed to input data to either the Ping or Pong buffer and switch automatically so that the developer does not have to manage the dual buffer mechanism on the DSP explicitly.

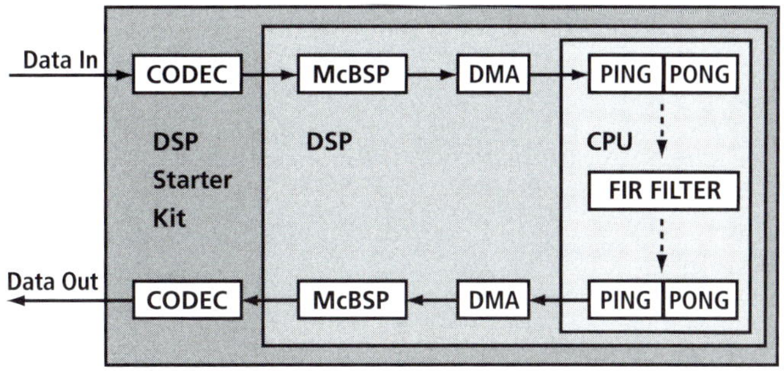

Figure 10.18 System block diagram of a DSP application mapped to a DSP starter kit (courtesy of Texas Instruments)

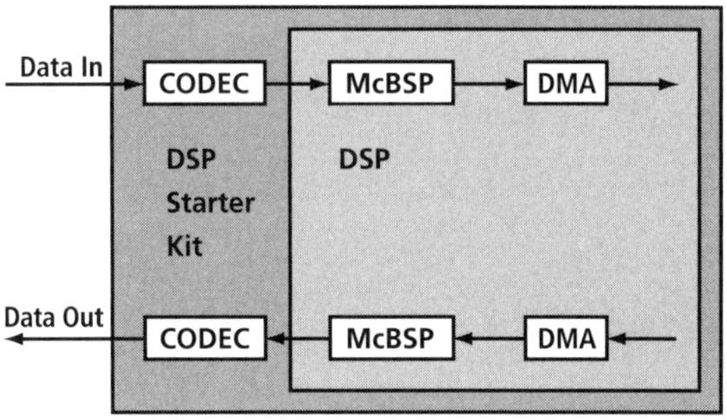

Figure 10.19 The peripheral processes required to get data in and out of the DSP on-chip memory (courtesy of Texas Instruments)

In order to perform the coding for these peripheral processes, the DSP developer must perform the following steps (Figure 10.20):

- direct DMA to continuously fill ping-pong buffers & interrupt CPU when full;
- select McBSP mode to match CODEC;
- select CODEC mode & start data flow via McBSP;
- route McBSP send & receive interrupts to DMA syncs.

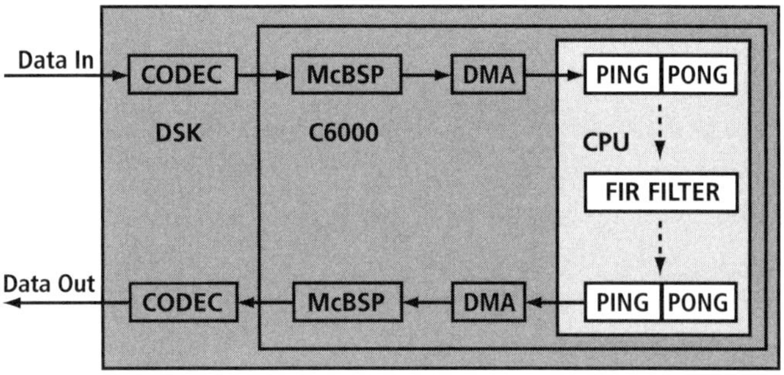

Figure 10.20 The main() function, buffer logic and filter code are the primary DSP functions (courtesy of Texas Instruments)

The main software initialization processes and flow includes:

1. Hardware Reset
   - Reset Vector
   - System Init
   - Launch “Main”
2. Main
   - set up Application
   - set up Interrupts
   - enter infinite Loop
   - If New Data In, then
     - o Clear Flag
   - “Process Data”

3. Process Data
   - Update Buffer Pointers
   - Filter Data

The reset vectors operation performs the following tasks;

- Link hardware events to software response
- Establish relative priorities of hardware events

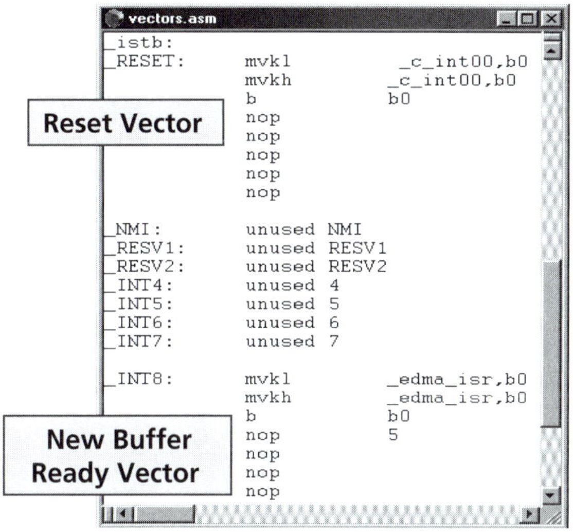

Figure 10.21 Sample code for the reset vectors operations (courtesy of Texas Instruments)

This is a device specific task that is manually managed. Figure 10.21 shows some sample code within the DSP IDE editor for the reset vectors function. Figure 10.22 shows sample code for the "main()" loop for the application. This sample code sets up the application, initializes the interrupts, and enters an infinite loop checking for new data and processes the new data when it is available. Figure 10.23 shows additional sample code for the initApplication() function. This function performs the peripheral function declarations, declares the ping pong buffers to manage the input data, and performs peripheral and buffer initialization.

Runtime support libraries are available to aid the DSP developer in developing DSP applications. This runtime software provides support for functions that are not part of the C language itself through inclusion of the following components:

- ANSI C standard library
- C I/O library—including printf()
- low-level support functions for I/O to the host operating system
- intrinsic arithmetic routines
- system startup routine, _c_int00
- functions and macros that allow C to access specific instructions

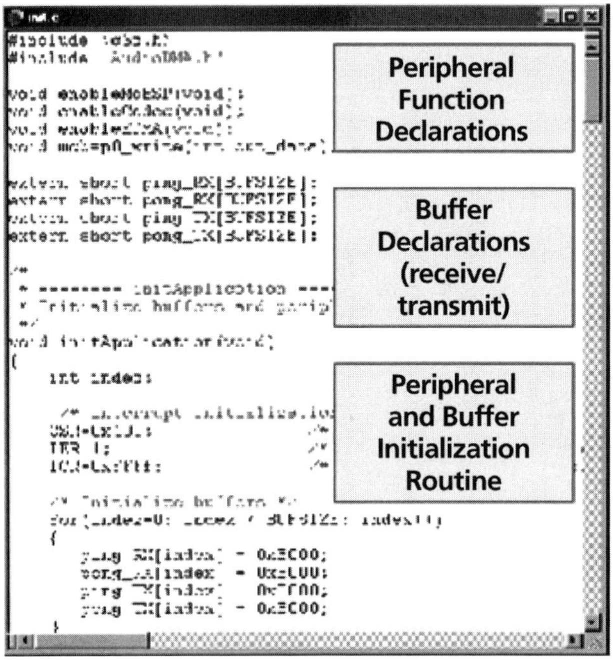

Figure 10.22 Sample code for the "main()" loop for the sample application (courtesy of Texas Instruments)

```
void processBuffer(void);

void main()
{
    initApplication();
    initInterrupts();

    while(1)
    {
        if(dataReadyFlag)
        {
            dataReadyFlag = 0;
            processBuffer();
            printf("Loop count = %d\n",i++);
        }

    }
}
```

Figure 10.23 Sample code from the main loop that performs initialization and checks for new data arriving (courtesy of Texas Instruments)

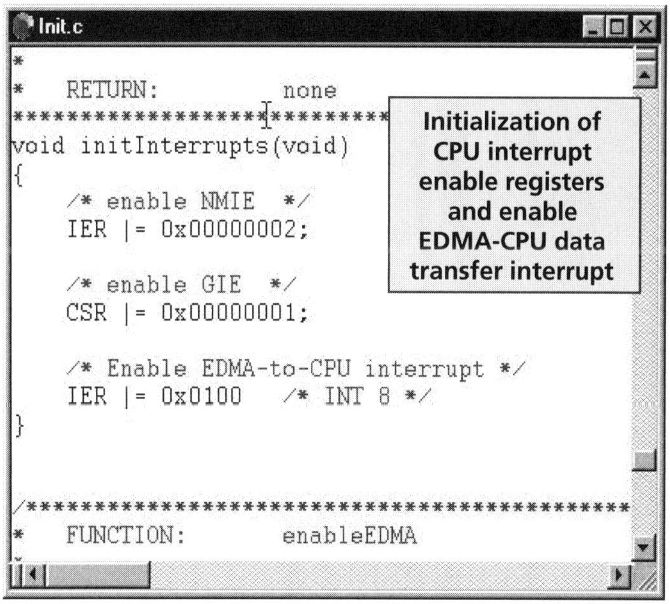

Figure 10.24 Sample code for the initInterrupts() function that performs initialization of CPU interrupt registers (courtesy of Texas Instruments)

The main loop in this DSP application checks the dataReadyFlag and calls the processBuffer() function when data is ready to be processed. The sample code for this is shown in Figure 10.25.

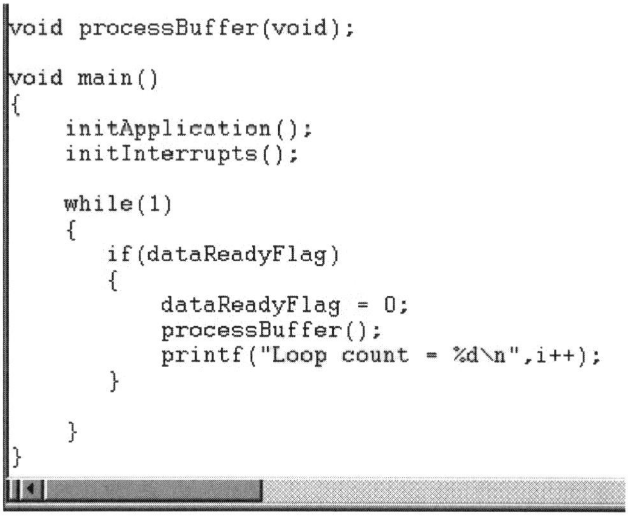

Figure 10.25 The main loop which checks for data ready in the ping pong input buffers (courtesy of Texas Instruments)

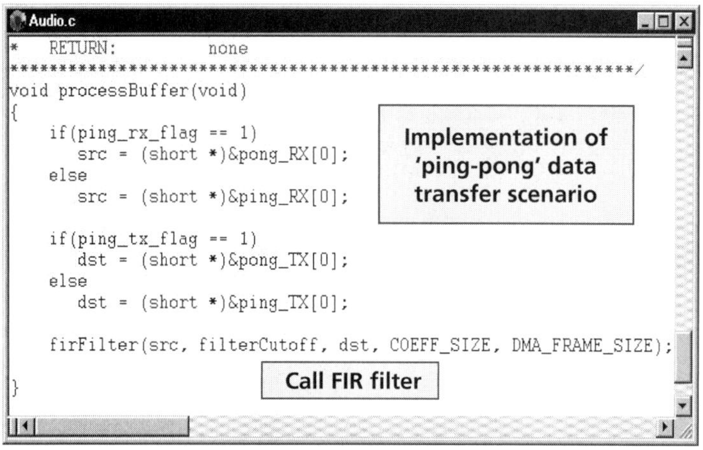

Figure 10.26 Sample code for the processBuffer function which manages the input buffers (courtesy of Texas Instruments)

The sample code for the processBuffer function is shown in Figure 10.26. This function implements the ping pong buffer data transfer functionality as well as the call to the FIR filter routine to perform the filtering function on each buffer of data. Finally, Figure 10.27 shows the sample code for the FIR filter function. This function performs the filtering function on the input data. The FIR filter coefficients are also declared.

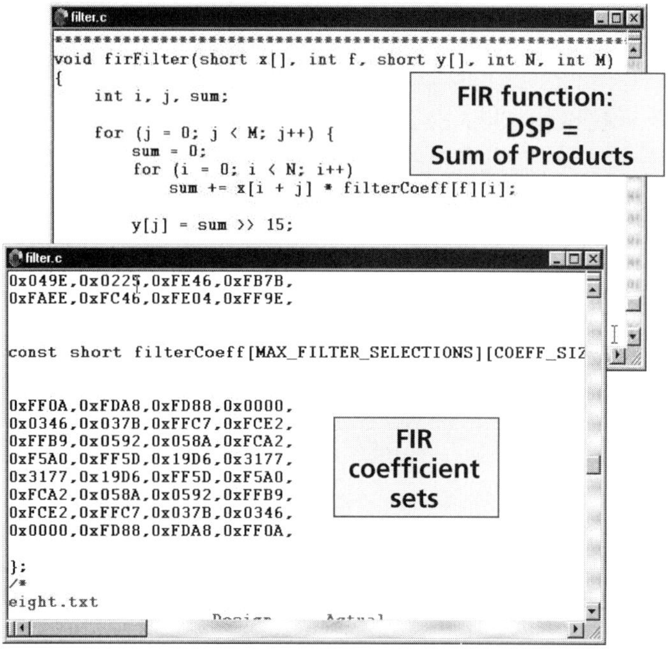

Figure 10.27 Sample code for the FIR filter function (courtesy of Texas Instruments)

The next step is to link the various software components into the hardware configuration. Figure 10.28 shows how the link process maps the input files for the application to the memory map available on the hardware platform. Memory consists of various types including vectors for interrupts, code, stack space, global declarations, and constants.

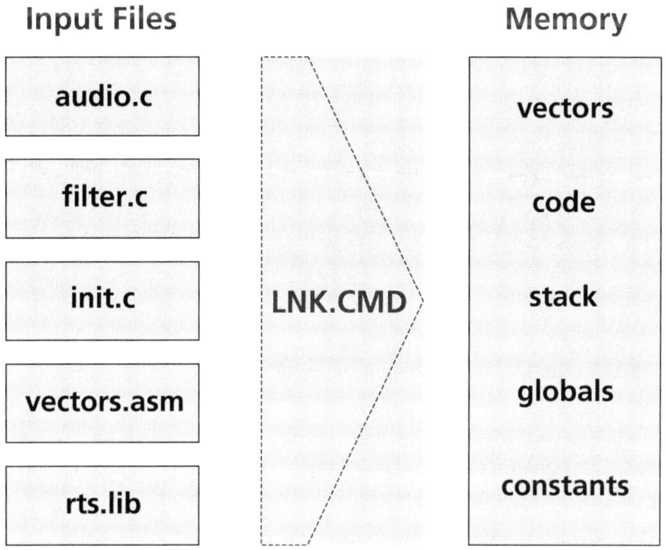

Figure 10.28 Mapping the input files to the memory map for the hardware requires a link process (courtesy of Texas Instruments)

The link command file, LNK.CMD, is used to define the hardware and usage of the hardware by the input files. This allocation is shown in Figure 10.29. The MEMORY section lists the memory that the target system contains and the address ranges for those sections. The example in Figure 10.29 is specific to a particular device, in this case the TMS320C6211 DSP memory layout. The SECTIONS directive defines where the software components are to be placed in memory. Some section types include:

.text;	program code
.bss;	global variables
.stack;	local variables
.sysmem;	heap

The next step in our process is to set up the debugger target. Modern DSP IDEs support multiple hardware boards (usually within the same vendor family). Most of the setup can be done easily with drag and drop within the IDE (Figure 10.30). DSP developers can configure both DSP and nonDSP devices easily using the IDE.

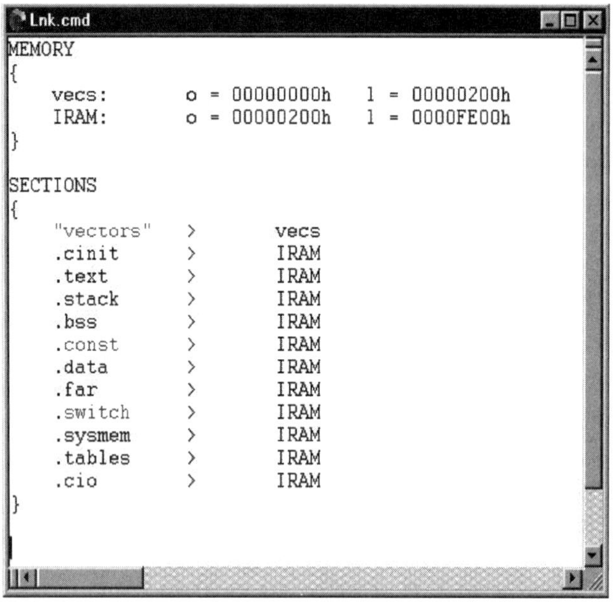

Figure 10.29 Sample code for the link command file mapping the input files to the target configuration (courtesy of Texas Instruments)

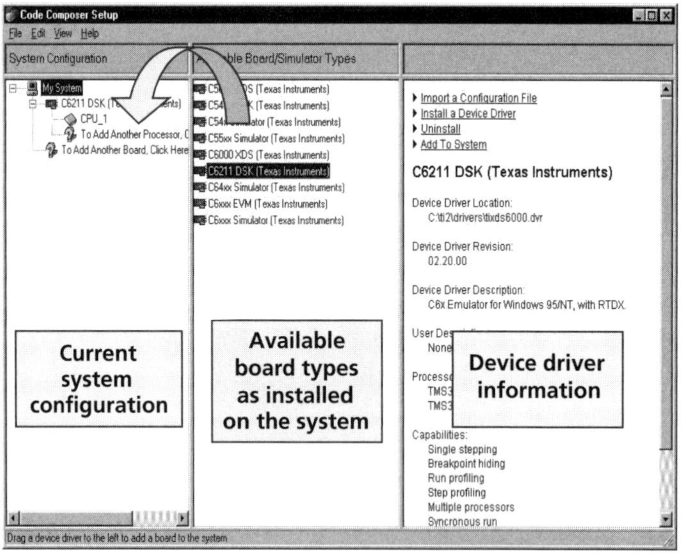

Figure 10.30 Setting up the debugger for the target application (courtesy of Texas Instruments)

DSP IDEs allow projects to be managed visually (Figure 10.31). Component files are placed into the project easily using drag and drop. Dependency listings are also maintained automatically. Project management is supported in DSP IDEs and allows easy configuration and management of a large number of files in a DSP application (Figure 10.32).

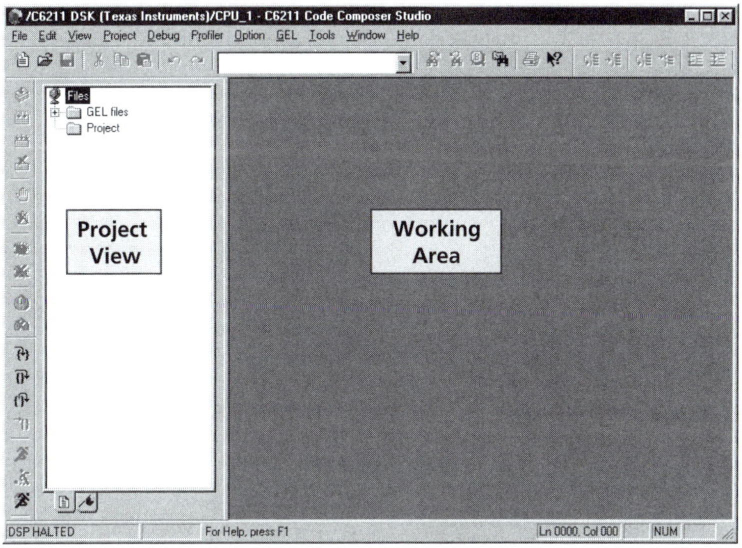

Figure 10.31 Project view within a DSP IDE (courtesy of Texas Instruments)

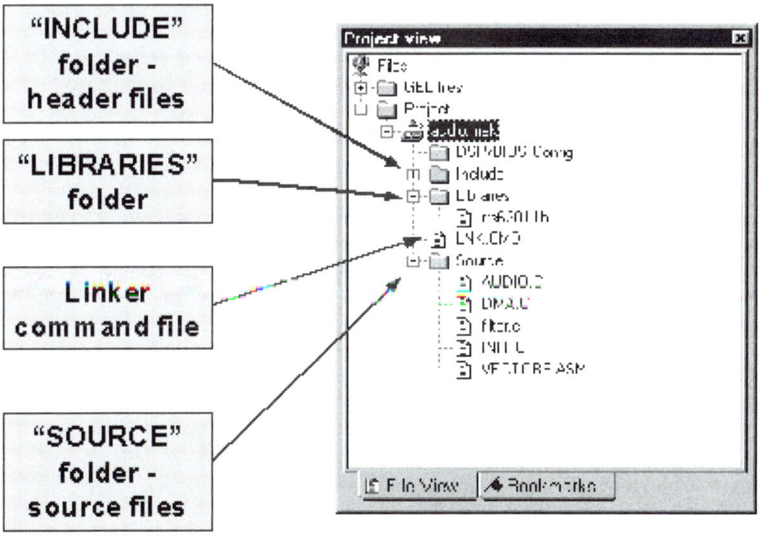

Figure 10.32 Project management of a large number of files is provided in a DSP IDE (courtesy of Texas Instruments)

The plug in capability provided by DSP IDEs allow the DSP developer to customize the development environment to the specific needs of the application and developer. Within the development environment, the developer can customize the input and output devices to be used to input and analyze data, respectively. Block diagram tools and custom editors and build tools can be used with the DSP IDE to provide valuable extensions to the development environment (Figure 10.33).

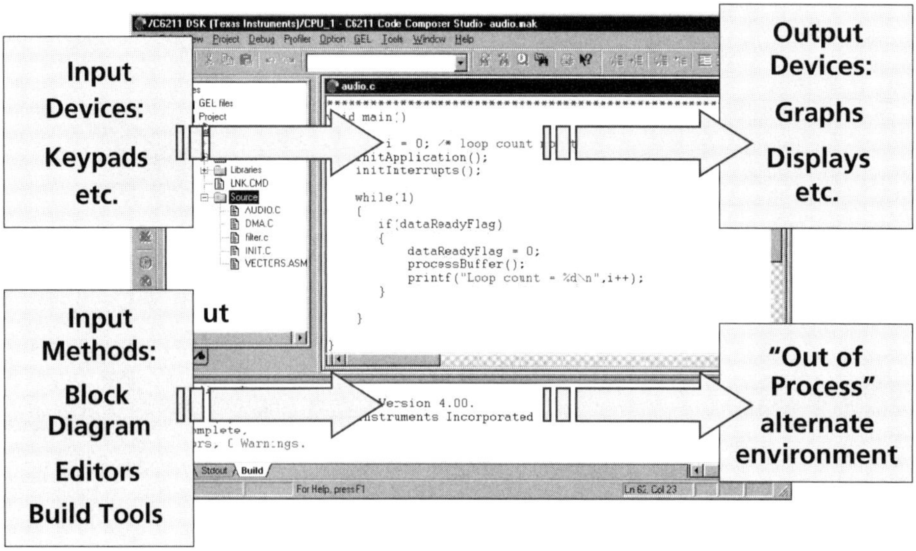

Figure 10.33 Plug in capabilities can extend the development environment (courtesy of Texas Instruments)

Debug—verifying code performance

DSP IDEs also support the debug stage of the software development life cycle. During the phase, the first goal is to verify the system is logically correct. In the example being discusses, this phase is used to ensure that the filter operates properly on the audio input signal. During this phase, the following steps can be performed in the IDE to set up and run the system to verify the logical correctness of the system:

- **Load** program to the target DSP
- **Run** to 'main()' or other functions
- Open **Watches** on key data
- Set and run to **Breakpoints** to halt execution at any point
- Display data in **Graphs** for visual analysis of signal data
- Specify **Probe Points** to identify when to observe data
- Continuously run to breakpoint via **Animate**

The debug phase is also used to verify the temporal or real-time goals of the system are being met. During this phase, the developer determines of the code is running as efficiently as possible and if the execution time overhead is low enough without sacrificing the algorithm fidelity. Probe points inserted in the tool can be used in conjunction with visual graphing tools are helpful during this phase to verify both the functional as well as temporal aspects of the system (Figure 10.34). The visual graphs are updated each time the probe point is encountered in the code.

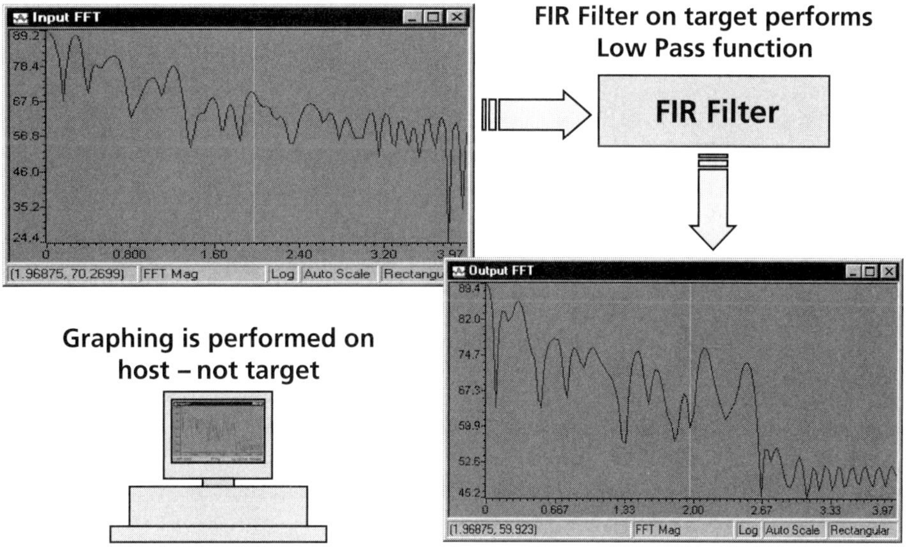

Figure 10.34 Analyzing the FIR function using visual graphing capabilities (courtesy of Texas Instruments)

Code Tuning and Optimization

One of the main differentiators between developers of nonreal-time systems and real-time systems is the phase of code tuning and optimization. It is during this phase that the DSP developer looks for "hot spots" or inefficient code segments and attempts to optimize those segments. Code in real-time DSP systems is often optimized for speed, memory size, or power. DSP code build tools (compilers, assemblers, and linkers) are improving to the point where developers can write a majority, if not all, of their application in a high level language like C or C++. Nevertheless, the developer must provide help and guidance to the compiler in order to get the technology entitlement from the DSP architecture. DSP compilers perform architecture specific optimizations and provide the developer with feedback on the decisions and assumptions that were made during the compile process. The developer must iterate in this phase to address the decisions and assumptions made during the build process until the performance goals are met. DSP developers can give the DSP compiler specific instructions using a number of compiler options. These options direct the compiler as to the level of aggressiveness to use when compiling the code, whether to focus on code speed or size, whether to compile with advanced debug information, and many other options. Figure 10.35 shows how the DSP IDE allows the developer to manage these options easily using point and click functionality. Figure 10.36 shows an example of the feedback provided by the compiler in the output file.

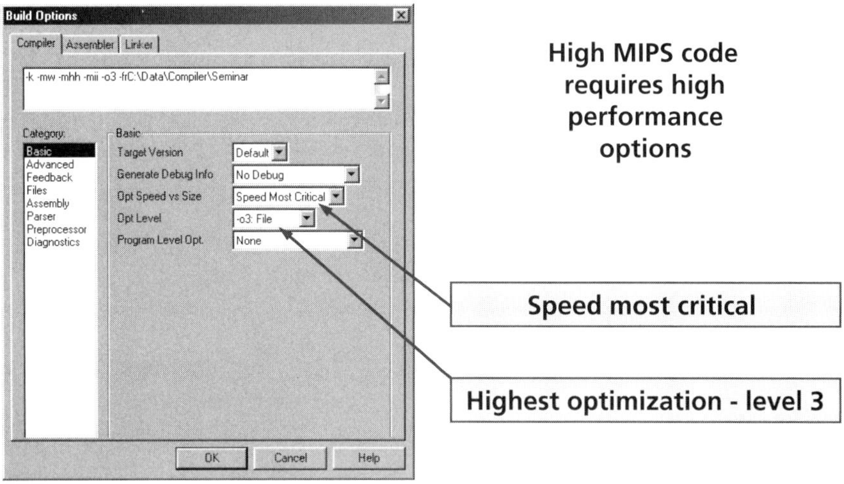

Figure 10.35 Compile options can be easily set by the developer through the IDE (courtesy of Texas Instruments)

Given the potentially large number of degrees of freedom in compile options and optimization axes (speed, size, power), the number of trade-offs available during the optimization phase can be enormous (especially considering that every function or file in an application can be compiled with different options. Profile based optimization (Figure 10.37) can be used to graph a summary of code size vs. speed options. The developer can choose the option that meets the goals for speed and power and have the application automatically built with the options that yield the selected size/speed trade-off selected.

```
;*----------------------------------------------------------------------------*
;*   SOFTWARE PIPELINE INFORMATION
;*
;*      Known Minimum Trip Count         : 1
;*      Known Max Trip Count Factor      : 1
;*      Loop Carried Dependency Bound(^) : 0
;*      Unpartitioned Resource Bound     : 1
;*      Partitioned Resource Bound(*)    : 1
;*      Resource Partition:
;*                                A-side   B-side
;*      .L units                     0        0
;*      .S units                     0        1*
;*      .D units                     1*       1*
;*      .M units                     1*       0
;*      .X cross paths               1*       0
;*      .T address paths             1*       1*
;*      Long read paths              0        0
;*      Long write paths             0        0
;*      Logical  ops (.LS)           0        0     (.L or .S unit)
;*      Addition ops (.LSD)          1        1
;*      Bound(.L .S .LS)             0        1*
;*      Bound(.L .S .D .LS .LSD)     1*       1*
;*
;*      Searching for software pipeline schedule at ...
;*         ii = 1  Schedule found with 8 iterations in parallel
;*      done
;*
;*      Collapsed epilog stages     : 7
;*      Prolog not entirely removed
;*      Collapsed prolog stages     : 2
;*
;*      Minimum required memory pad : 14 bytes
;*
;*      Minimum safe trip count     : 1
;*----------------------------------------------------------------------------*
```

Key Information for Loop

`ii = 1` (iteration interval = 1 cycle)
Means: Single Cycle Inner Loop

`B side .M unit not used`
Means: Only one MPY per cycle

Figure 10.36 Example of compiler feedback provided to the user (courtesy of Texas Instruments)

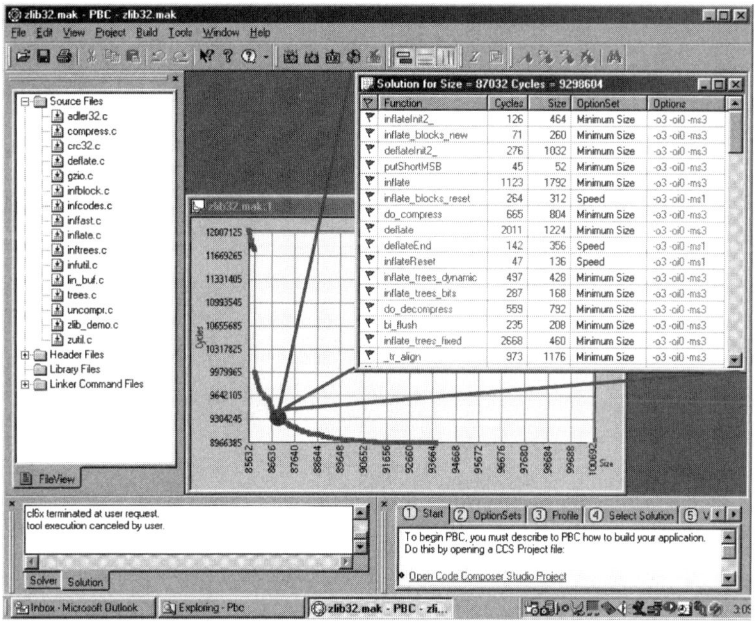

Figure 10.37 Profile based compilation allows the DSP developer to make code size/speed trade-offs quickly and easily (courtesy of Texas Instruments)

Typical DSP development flow

DSP developers follow a development flow that takes them through several phases:

- *Application definition* – It's during this phase that the developer begins to focus on the end goals for performance, power, and cost.
- *Architecture design* – During this phase, the application is designed at a systems level using block diagram and signal flow tools if the application is large enough to justify using these tools.
- *Hardware/software mapping* – In this phase a target decision is made for each block and signal in the architecture design.
- *Code creation* – This phase is where the initial development is done, prototypes are developed and mockups of the system are performed.
- *Validate/debug* – Functional correctness is verified during this phase.
- *Tuning/optimization* – This is the phase where the developer's goal is to meet the performance goals of the system.
- *Production and deployment* – Release to market.
- *Field testing.*

Developing a well tuned and optimized application involves several iterations between the validate phase and the optimize phase. Each time through the validate phase the developer will edit and build the modified application, run it on a target or simulator, and analyze the results for functional correctness. Once the application is functionally correct, the developer will begin a phase of optimization on the functionally correct code. This involves tuning the application towards the performance goals of the system (speed, memory, power, for example), running the tuned code on the target or simulator to measure the performance, and evaluation where the developer will analyze the remaining "hot spots" or areas of concern that have not yet been addressed, or are still outside the goals of performance for that particular area (Figure 10.38). Once the evaluation is complete, the developer will go back to the validate phase where the new, more optimized code is run to verify functional correctness has not been compromised. If not, and the performance of the application is within acceptable goals for the developer, the process stops. If a particular optimization has broken the functional correctness of the code, the developer will debug the system to determine what has broke, fix the problem, and continue with another round of optimization. Optimizing DSP applications inherently leads to more complex code, and the likelihood of breaking something that used to work increased the more aggressively the developer optimizes the application. There can be many cycles in this process, continuing until the performance goals have been met for the system.

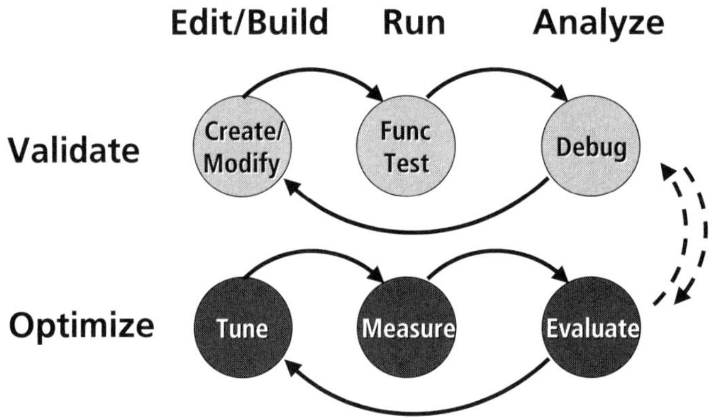

Figure 10.38 DSP developers iterate through a series of optimize and validate steps until the goals for performance are achieved

Generally, a DSP application will initially be developed without much optimization. During this early period, the DSP developer is primarily concerned with functional correctness of the application. Therefore, the "out of box" experience from a performance perspective is not that impressive, even when using the more aggressive levels of optimization in the DSP compiler. This initial view can be termed the "pessimistic" view in the sense that there are no aggressive assumptions made in the compiled output, there is no aggressive mapping of the application to the specific DSP architecture,, and there is no aggressive algorithmic transformation to allow the application to run more efficiently on the target DSP.

Significant performance improvements can come quickly by focusing on a few critical areas of the application:

- Key tight loops in the code with many iterations
- Ensuring critical resources are in on-chip memory
- Unrolling key loops where applicable

The techniques to perform these optimizations where discussed in the chapter on optimizing DSP software. If these few key optimizations are performed, the overall system performance goes up significantly. As Figure 10.39 shows, a few key optimizations on a small percentage of the code early leads to significant performance improvements. Additional phases of optimization get more and more difficult as the optimization opportunities are reduced as well as the cost/benefit of each additional optimization. The goal of the DSP developer must be to continue to optimize the application until the performance goals of the system are met, not until the application is running at its theoretical peak performance. The cost/benefit does not justify this approach.

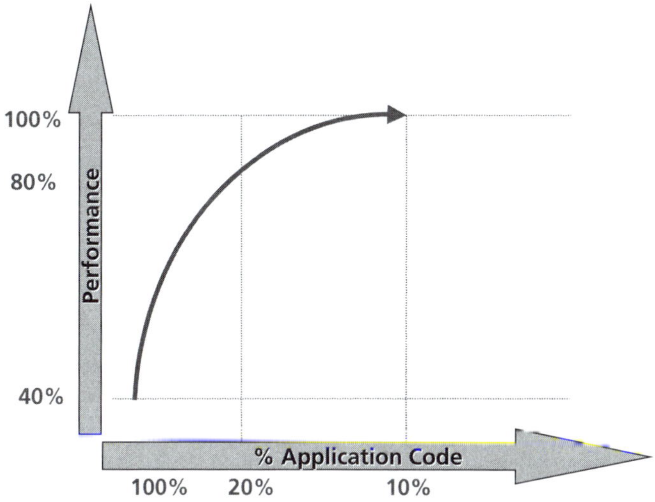

Figure 10.39 Optimizing DSP code takes time and effort to reach desired performance goals

DSP IDEs provide advanced profiling capabilities that allow the DSP developer to profile the application and display useful information about the application such as code size, total cycle count, number of times the algorithm looped through a particular function, etc. This information can then be analyzed to determine which functions are optimization candidates (Figure 10.40).

An optimal approach to profiling and tuning a DSP application is to attack the right areas first. These represent those areas where the most performance improvement can be gained with the smallest effort. A Pareto ranking of the biggest performance areas (Figure 10.41) will guide the DSP developer towards those areas where the most performance can be gained.

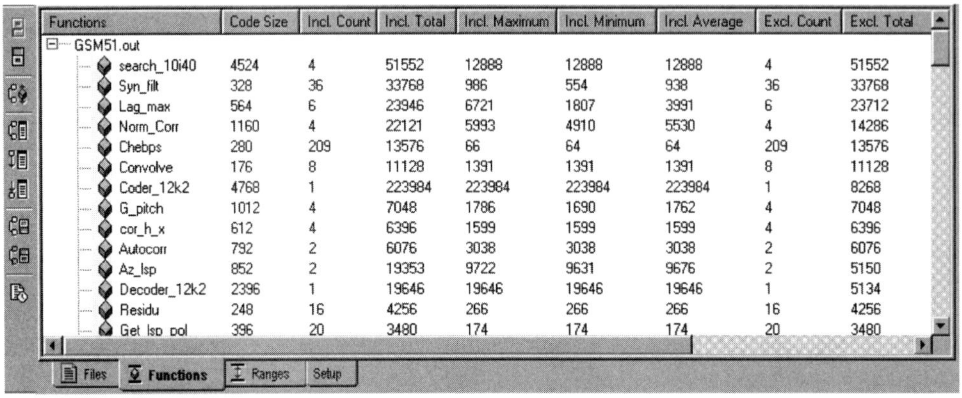

Functions	Code Size	Incl. Count	Incl. Total	Incl. Maximum	Incl. Minimum	Incl. Average	Excl. Count	Excl. Total
GSM51.out								
search_10i40	4524	4	51552	12888	12888	12888	4	51552
Syn_filt	328	36	33768	986	554	938	36	33768
Lag_max	564	6	23946	6721	1807	3991	6	23712
Norm_Corr	1160	4	22121	5993	4910	5530	4	14286
Chebps	280	209	13576	66	64	64	209	13576
Convolve	176	8	11128	1391	1391	1391	8	11128
Coder_12k2	4768	1	223984	223984	223984	223984	1	8268
G_pitch	1012	4	7048	1786	1690	1762	4	7048
cor_h_x	612	4	6396	1599	1599	1599	4	6396
Autocorr	792	2	6076	3038	3038	3038	2	6076
Az_lsp	852	2	19353	9722	9631	9676	2	5150
Decoder_12k2	2396	1	19646	19646	19646	19646	1	5134
Residu	248	16	4256	266	266	266	16	4256
Get_lsp_pol	396	20	3480	174	174	174	20	3480

Figure 10.40 Profiling data for a DSP application. This data helps the DSP developer focus on those functions that provide the best cost/benefit ratio during the optimization phase.

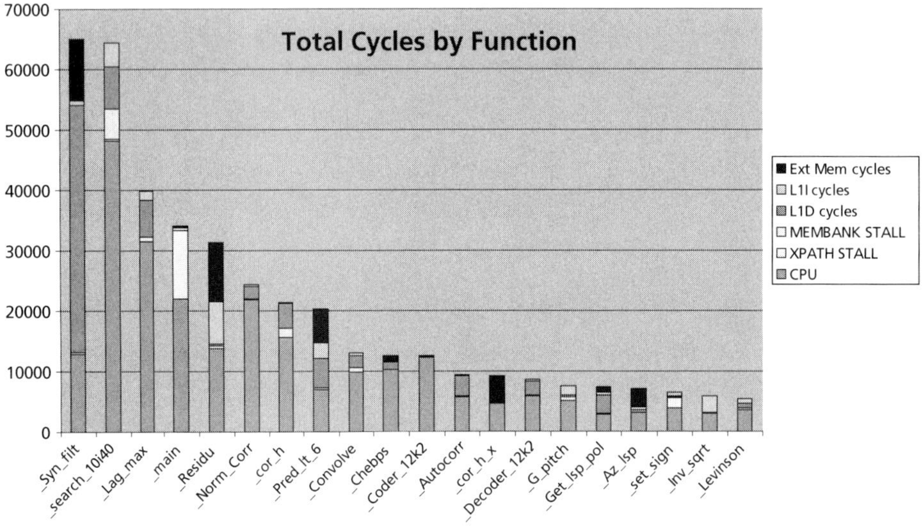

Figure 10.41 A pareto analysis of a DSP function allows the DSP developer to focus on the most important areas first

Getting Started

A DSP starter kit (Figure 10.42) is easy to install and allows the developer to get started writing code very quickly. The starter kits usually come with a daughter card for expansion slots, the target hardware, software development tools, a parallel port interface for debug, a power supply, and the appropriate cables.

Figure 10.42 A DSP starter kit allows the DSP developer to get started quickly (courtesy of Texas Instruments)

Putting it all Together

Figure 10.43 shows the entire DSP development flow. There are five major stages to the DSP development process:

- System Concept and Requirements; this phase includes the elicitation of the system level functional and nonfunctional (sometimes called "quality") requirements. Power requirements, quality of service (QoS), performance and other system level requirements are elicited. Modeling techniques like signal flow graphs are constructed to examine the major building blocks of the system.
- System algorithm research and experimentation; during this phase, the detailed algorithms are developed based on the given performance and accuracy requirements. Analysis is first done on floating-point development systems to determine if these performance and accuracy requirements can be met. These system are then ported, if necessary, to fixed-point processors for cost reasons. Inexpensive evaluation boards are used for this analysis.
- System Design; during the design phase, the hardware and software blocks for the system are selected and/or developed. These systems are analyzed using prototyping and simulation to determine if the right partitioning has been performed and whether the performance goals can be realized used the given hardware and software components. Software components can be custom developed of reused, depending on the application.

- System implementation; during the system implementation phase, inputs from system prototyping, trade off studies, and hardware synthesis options are used to develop a full system co-simulation model. Software algorithms and components are used to develop the software system. Combinations of signal processing algorithms and control frameworks are used to develop the system.
- System Integration; during the system integration phase, the system is built, validated, tuned if necessary and executed in a simulated environment or in a hardware in the loop simulation environment. The scheduled system is analyzed and potentially re-partitioned if performance goals are not being met.

In many ways, the DSP system development process is similar to other development processes. Given the increased amount of signal processing algorithms, early simulation-based analysis is required more for these systems. The increased focus on performance requires the DSP development process to focus more on real-time deadlines and iterations of performance tuning. We have discussed some of the details of these phases throughout this book.

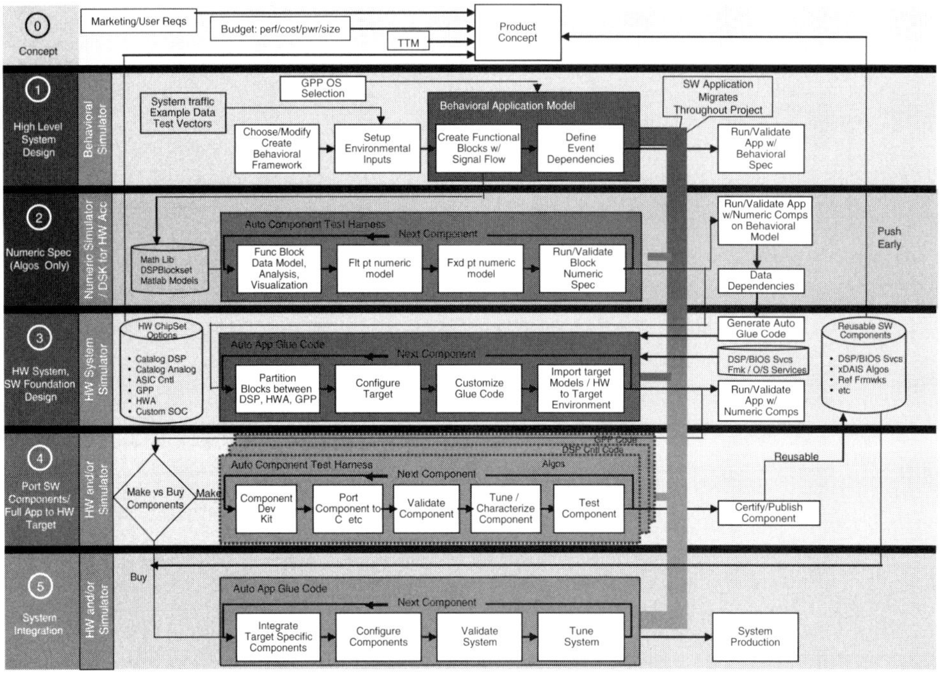

Figure 10.43 The DSP development flow (courtesy of Texas Instruments

References

1. Hakkarainen, Harri, "Evaluating the TMS320C62xx for Comm Applications," *Communication Systems Design*, October 1997.
2. Blalock, Garrick, "General-Purpose uPs for DSP applications: consider the trade-offs," *EDN*, October 23, 1997.
3. Levy, Markus, "Virtual Processors and the Reality of Software Simulation," *EDN*, January 15, 1998.
4. Mayer, John H., "Sophisticated tools bring real-time DSP applications to market," *Military and Aerospace Electronics*, January 1998.
5. Stearns, Samuel D. and Ruth David, *Signal Processing Algorithms in MATLAB*, Prentice Hall 1996.

11

Embedded DSP Software Design Using Multicore System-on-a-Chip (SoC) Architectures

Multicore System-on-a-Chip

Designing and building embedded systems is a difficult task, given the inherent scarcity of resources in embedded systems (processing power, memory, throughput, battery life, and cost). Various trade-offs are made between these resources when designing an embedded system. Modern embedded systems are using devices with multiple processing units manufactured on a single chip, creating a sort of multicore system-on-a-chip (SoC) can increase the processing power and throughput of the system while at the same time increasing the battery life and reducing the overall cost. One example of a DSP based SoC is shown in Figure 11.1. Multicore approaches keep hardware design in the low frequency range (each individual processor can run at a lower speed, which reduces overall power consumption as well as heat generation), offering significant price, performance, and flexibility (in software design and partitioning) over higher speed single-core designs.

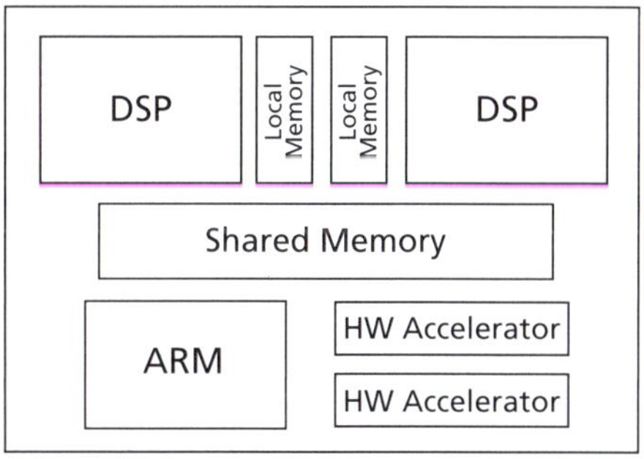

Figure 11.1 Block diagram of a DSP SoC

There are several characteristics of SoC that we will discuss (reference 1). I will use an example processor to demonstrate these characteristics and how they are deployed in an existing SoC.

1. *Customized to the application* – Like embedded systems in general, SoC are customized to an application space. As an example, I will reference the video application space. A suitable block diagram showing the flow of an embedded video application space is shown in Figure 11.2. This system consists of input capture, real-time

signal processing, and output display components. As a system there are multiple technologies associated with building a flexible system including analog formats, video converters, digital formats, and digital processing. An SoC processor will incorporate a system of components; processing elements, peripherals, memories, I/O, and so forth to implement a system such as that shown in Figure 11.2. An example of an SoC processor that implements a digital video system is shown in Figure 11.3. This processor consists of various components to input, process, and output digital video information. More about the details of this in a moment.

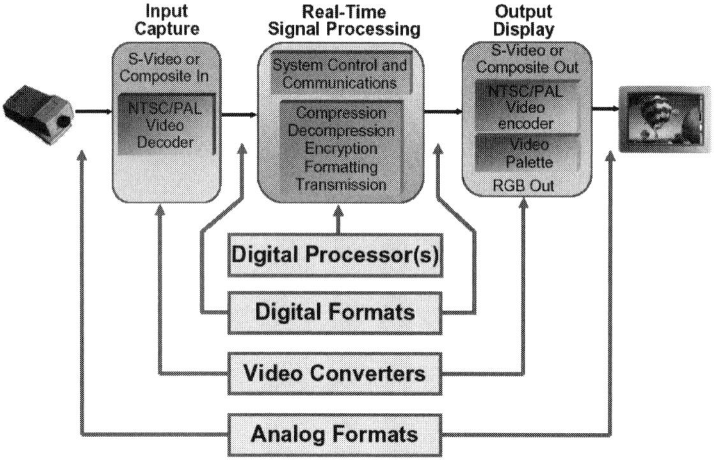

Figure 11.2 Digital video system application model (courtesy of Texas Instruments)

2. *SoCs improve power/performance ratio* – Large processors running at high frequencies consume more power, and are more expensive to cool. Several smaller processors running at a lower frequency can perform the same amount of work without consuming as much energy and power. In Figure 11.1, the ARM processor, the two DSPs, and the hardware accelerators can run a large signal processing application efficiently by properly partitioning the application across these four different processing elements.
3. *Many apps require programmability* – SoC contains multiple programmable processing elements. These are required for a number of reasons:
   - *New technology* – Programmability supports upgradeability and changeability easier than nonprogrammable devices. For example, as new video codec technology is developed, the algorithms to support these new standards can be implemented on a programmable processing element easily. New features are also easier to add.
   - *Support for multiple standards and algorithms* – Some digital video applications require support for multiple video standards, resolutions, and quality. Its easier to implement these on a programmable system.
   - *Full algorithm control* – A programmable system provides the designer the ability to customize and/or optimize a specific algorithm as necessary which provides the application developer more control over differentiation of the application.

- *Software reuse in future systems* – By developing digital video software as components, these can be reuse/repackaged as building blocks for future systems as necessary.

4. *Constraints such as real-time, power, cost* – There are many constraints in real-time embedded systems. Many of these constraints are met by customizing to the application.
6. *Special instructions* – SoCs have special CPU instructions to speed up the application. As an example, the SoC in Figure 11.3 contains special instructions on the DSP to accelerate operations such as:
   - 32-bit multiply instructions for extended precision computation
   - Expanded arithmetic functions to support FFT and DCT algorithms
   - Improve complex multiplications
   - Double dot product instructions for improving throughput of FIR loops
   - Parallel packing Instructions
   - Enhanced Galois Field Multiply

   Each of these instructions accelerate the processing of certain digital video algorithms. Of course, compiler support is necessary to schedule these instructions, so the tools become an important part of the entire system as well.
7. *Extensible* – Many SoCs are extensible in ways such as word size and cache size. Special tooling is also made available to analyze systems as these system parameters are changes.
8. *Hardware acceleration* – There are several benefits to using hardware acceleration in an SoC. The primary reason is better cost/performance ratio. Fast processors are costly. By partitioning into several smaller processing elements, cost can be reduced in the overall system. Smaller processing elements also consume less power and can actually be better at implementing real-time systems as the dedicated units can respond more efficiently to external events.

   Hardware accelerators are useful in applications that have algorithmic functions that do not map to a CPU architecture well. For example, algorithms that require a lot of bit manipulation require a lot of registers. A traditional CPU register model may not be suited to efficiently execute these algorithms. A specialized hardware accelerator can b built that performs bit manipulation efficiently which sits beside the CPU and used by the CPU for bit manipulation operations. Highly responsive I/O operations are another area where a dedicated accelerator with an attached I/O peripheral will perform better. Finally, applications that are required to process streams of data, such as many wireless and multimedia applications, do not map well to the traditional CPU architecture, especially those that implement caching systems. Since each streaming data element may have a limited lifetime, processing will require the constant thrashing of cache for new data elements. A specialized hardware accelerator with special fetch logic can be implemented to provide dedicated support to these data streams.

Hardware acceleration is used on SoCs as a way to efficiently execute classes of algorithms. We mentioned in the chapter on power optimization, how the use of accelerators if possible can lower overall system power since these accelerators are customized to the class of processing and, therefore, perform these calculations very efficiently. The SoC in Figure 11.3 has hardware acceleration support. In particular, the video processing sub-system (VPSS) as well as the Video Acceleration block within the DSP subsystem are examples of hardware acceleration blocks used to efficiently process video algorithms. Figure 11.4 shows a block diagram of one of the VPSS. This hardware accelerator contains:

A front end module containing:

- CCDC (charge coupled device)
- Previewer
- Resizer (accepts data from the previewer or from external memory and resizes from ¼x to 4x)

And a back end module containing:

- Color space conversion
- DACS
- Digital output
- On-screen display

This VPSS processing element eases the overall DSP/ARM loading through hardware acceleration. An example application using the VPSS is shown in Figure 11.5.

9. *Heterogeneous memory systems* – Many SoC devices contain separate memories for the different processing elements. This provides a performance boost because of lower latencies on memory accesses, as well as lower power from reduced bus arbitration and switching.

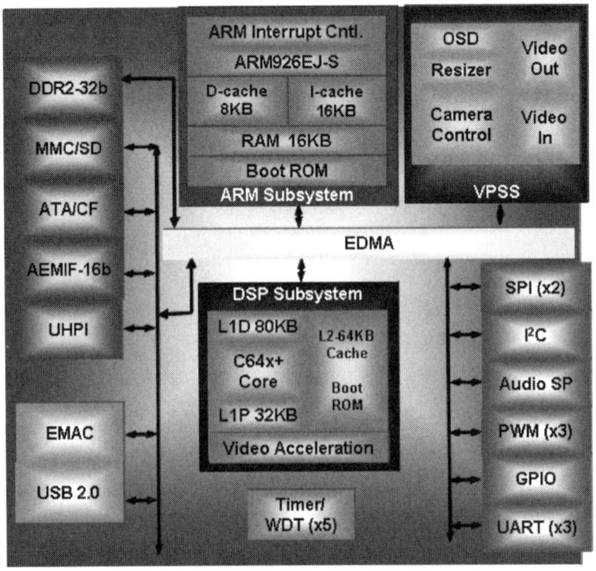

Figure 11.3 A SoC processor customized for Digital Video Systems (courtesy of Texas Instruments)

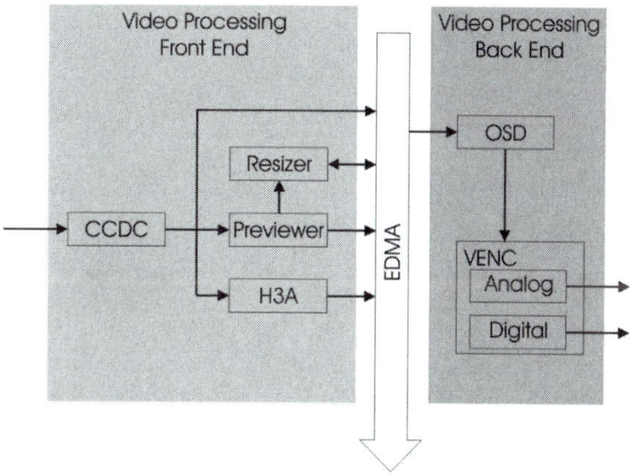

Figure 11.4 Block diagram of the video processing subsystem acceleration module of the SoC in Figure 11.3 (courtesy of Texas Instruments)

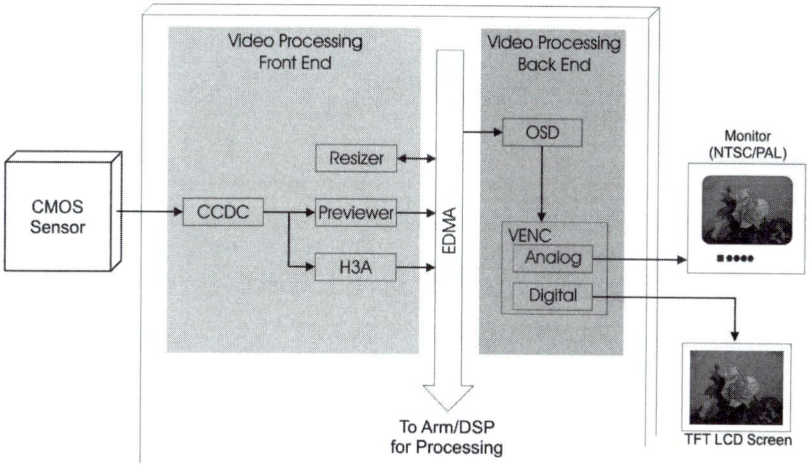

Figure 11.5 A Video phone example using the VPSS acceleration module (courtesy of Texas Instruments)

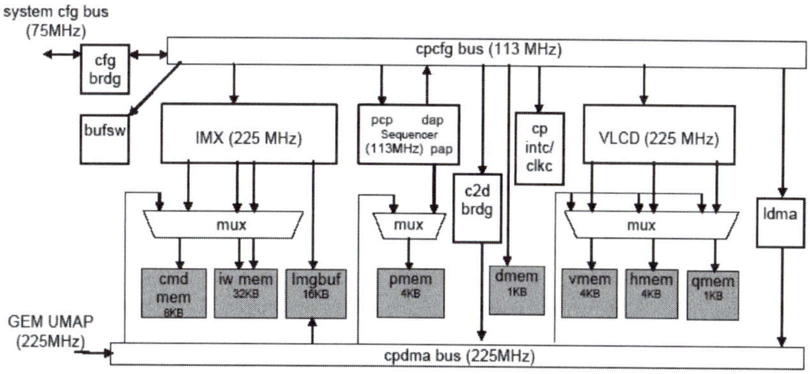

Figure 11.6 A hardware accelerator example; video and imaging coprocessor (courtesy of Texas Instruments)

This programmable coprocessor is optimized for imaging and video applications. Specifically, this accelerator is optimized to perform operations such as filtering, scaling, matrix multiplication, addition, subtraction, summing absolute differences, and other related computations.

Much of the computation is specified in the form of commands which operate on arrays of streaming data. A simple set of APIs can be used to make processing calls into this accelerator. In that sense, a single command can drive hundreds or thousands of cycles.

As discussed previously, accelerators are used to perform computations that do not map efficiently to a CPU. The accelerator in Figure 11.6 is an example of an accelerator that performs efficient operations using parallel computation. This accelerator has an 8-parallel multiply accumulate (MAC) engine which significantly accelerates classes of signal processing algorithms that requires this type of parallel computation. Examples include:

- JPEG encode and decode
- MPEG-1/2/4 encode and decode
- H.263 encode and decode
- WMV9 decode
- H.264 baseline profile decode

The variable length code/decode (VLCD) module in this accelerator supports the following fundamental operations very efficiently;

- Quantization and inverse quantization (Q/IQ)
- Variable length coding and decoding (VLC/VLD)
- Huffman tables
- Zigzag scan flexibility

The design of this block is such that it operates on a macroblock of data at a time (max 6 8x8 blocks, 4:2:0 format). Before starting to encode or decode a bitstream, the proper registers and memory in the VLCD module must first be initialized by the application software.

This hardware accelerator also contains a block called a *sequencer* which is really just a 16-bit microprocessor targeted for simple control, address calculation, and loop control functions. This simple processing element offloads the sequential operations from the DSP. The application developer can program this sequencer to coordinate the operations among the other accelerator elements including the iMX, VLCD, System DMA, and the DSP. The sequencer code is compiled using a simple macro using support tools, and is linked with the DSP code to be later loaded by the CPU at run time.

One of the other driving factors for the development of SoC technology is the fact that there is an increasing demand for programmable performance. For many applications, performance requirements are increasing faster than the ability of a single CPU to keep pace. The allocation of performance, and thus response time, for complex real-time systems is often easier with multiple CPUs. And dedicated CPUs in peripherals or special accelerators can offload low-level functionality from a main CPU, allowing it to focus on higher-level functions.

Software Architecture for SoC

Software development for SoC involves partitioning the application among the various processing elements based on the most efficient computational model. This can require a lot of trial and error to establish the proper partitioning. At a high level the SoC partitioning algorithm is as follows:

- Place the state machine software (those algorithms that provide application control, sequencing, user interface control, event driven software, and so on) on a RISC processor such as an ARM.
- Place the signal processing software on the DSP, taking advantage of the application specific architecture that a DSP offers for signal processing functions.
- Place high rate, computationally intensive algorithms in hardware accelerators, if they exist and if they are customized to the specific algorithm of consideration.

As an example, consider the software partitioning shown in Figure 11.7. This SoC model contains a general-purpose processor (GPP), a DSP, and hardware acceleration. The GPP contains a chip support library which is a set of low level peripheral APIs that provide efficient access to the device peripherals, a general-purpose operating system, an algorithmic abstraction layer and a set of API for and application and user interface layer. The DSP contains a similar chip support library, a DSP centric kernel, a set of DSP specific algorithms and interfaces to higher level application software. The hardware accelerator contains a set of APIs for the programmer to access and some very specific algorithms mapped to the acceleration. The application programmer is responsible for the overall partitioning of the system and the mapping of the algorithms to the respective processing elements. Some vendors may provide a "black box" solution to one or more of these processing elements, including the DSP and the hardware accelerators. This provides another level of abstraction to the application developer who does not need to know the details of some of the underlying algorithms. Other system developers may want access to these low level algorithms, so there is normally flexibility in the programming model for these systems, depending on the amount of customization and tailoring required.

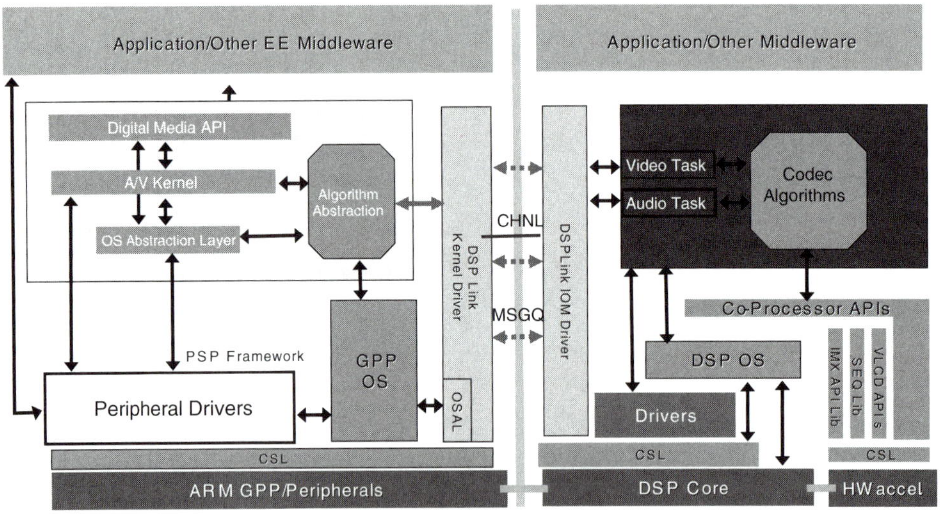

Figure 11.7 Software Architecture for SoC (courtesy of Texas Instruments)

Communication in an SoC is primarily established by means of software. The communication interface between the DSP and the ARM in Figure 11.7, for example, is realized by defining memory locations in the DSP data space as registers. The ARM gains read/write access to these registers through a host interface. Both processors can asynchronously issue commands to each other, no one masters the other. The command sequence is purely sequential; the ARM cannot issue a new command unless the DSP has sent a "command complete" acknowledgement.

There exist two register pairs to establish the two-way asynchronous communication between ARM and DSP, one register pair is for the sending commands to ARM, and the other register pair is for the sending commands to DSP. Each register pair has:

- a command register, which is used pass commands to ARM or DSP;
- a command complete register, which is used to return the status of execution of the command;
- each command can pass up to 30 words of command parameters;
- also, each command execution can return up to 30 words of command return parameters.

An ARM to DSP command sequence is as follows:

- ARM writes a command to the command register
- ARM writes number of parameters to number register
- ARM writes command parameters into the command parameter space
- ARM issues a Nonmaskable interrupt to the DSP
- DSP reads the command
- DSP reads the command parameters
- DSP executes the command
- DSP clears the command register

- DSP writes result parameters into the result parameter space
- DSP writes "command complete" register
- DSP issues HINT interrupt to ARM

The DSP to ARM command sequence is as follows:

- DSP writes command to command register
- DSP writes number of parameters to number register
- DSP writes command parameters into the command parameter space
- DSP issues an HINT interrupt to the DSP
- ARM reads the command
- ARM reads the command parameters
- ARM executes DSP command
- ARM clears the command register
- ARM writes result parameters into the result parameter space
- ARM writes "command complete" register
- ARM sends an INT0 interrupt to the DSP

Communication between the ARM and the DSP is usually accomplished using a set of communication APIs. Below is an example of a set of communication APIs between a general-purpose processor (in this case an ARM) and a DSP. The detailed software implementation for these APIs is given at the end of the chapter.

```
#define  ARM_DSP_COMM_AREA_START_ADDR  0x80
         Start DSP address for ARM-DSP.
#define  ARM_DSP_COMM_AREA_END_ADDR  0xFF
         End DSP address for ARM-DSP.
#define  ARM_DSP_DSPCR  (ARM_DSP_COMM_AREA_START_ADDR)
         ARM to DSP, parameters and command from ARM.
#define  ARM_DSP_DSPCCR  (ARM DSP_COMM_AREA_START_ADDR+32)
         ARM to DSP, return values and completion code from DSP.
#define  ARM_DSP_ARMCR  (ARM_DSP_COMM_AREA_START_ADDR+64)
         DSP to ARM, parameters and command from DSP.
#define  ARM_DSP_ARMCCR  (ARM_DSP_COMM_AREA_START_ADDR+96)
         DSP to ARM, return values and completion code from ARM.
#define  DSP_CMD_MASK  (Uint16)0x0FFF
         Command mask for DSP.
#define  DSP_CMD_COMPLETE  (Uint16)0x4000
         ARM-DSP command complete, from DSP.
#define  DSP_CMD_OK  (Uint16)0x0000
         ARM-DSP valid command.
#define  DSP_CMD_INVALID_CMD  (Uint16)0x1000
         ARM-DSP invalid command.
#define  DSP_CMD_INVALID_PARAM  (Uint16)0x2000
         ARM-DSP invalid parameters.
```

Functions

STATUS	ARMDSP_sendDspCmd (Uint16 cmd, Uint16 *cmdParams, Uint16 nParams) *Send command, parameters from ARM to DSP.*
STATUS	ARMDSP_getDspReply (Uint16 *status, Uint16 *retParams, Uint16 nParams) *Get command execution status, return parameters sent by DSP to ARM.*
STATUS	ARMDSP_getArmCmd (Uint16 *cmd, Uint16 *cmdParams, Uint16 nParams) *Get command, parameters sent by DSP to ARM.*
STATUS	ARMDSP_sendArmReply (Uint16 status, Uint16 *retParams, Uint16 nParams) *Send command execution status, return parameters from ARM to DSP.*
STATUS	ARMDSP_clearReg () *Clear ARM-DSP communication area.*

SoC System Boot Sequence

Normally, the boot image for DSP is part of the ARM boot image. There could be many different boot images for the DSP for the different tasks DSP needs to execute. The sequence starts with the ARM downloading the image related to the specific task to be executed by the DSP. ARM resets then the DSP (via a control register) and then brings the DSP out of reset. At this stage the DSP begins execution at a pre-defined location, usually in ROM. The ROM code at this address initializes the DSP internal registers and places the DSP into an idle mode. At this point ARM downloads the DSP code by using a host port interface. After it completes downloading the DSP image, the ARM can send an interrupt to the DSP, which wakes it up from the idle mode, vectors to a start location and begins running the application code loaded by the ARM. The DSP boot sequence is given below:

- ARM resets DSP and then brings it out of reset.
- DSP gets out of reset and load its program counter (PC) register with a start address.
- The ROM code in this location branches the DSP to an initialization routine address.
- A DSP status register is initialized to move the vector table to a dedicated location, all the interrupts are disabled except for a dedicated unmaskable interrupt and the DSP is set to an mode.
- While DSP is in its mode, the ARM loads the DSP Program/Data memory with the DSP code/data.
- When the ARM finishes downloading the DSP code, it wakes up DSP from the mode by asserting an interrupt signal.
- The DSP then branches to a start address where the new interrupt vector table is located. The ARM should have loaded this location with at least a branch to the start code.

Tools Support for SoC

SoC, and heterogeneous processors in general, require more sophisticated tools support. A SoC may contain several programmable debuggable processing elements that require tools support for code generation, debug access and visibility, and real-time data analysis. A general model for this is shown in Figure 11.8. A SoC processor will have several processing elements such as an ARM and DSP. Each of these processing elements will require a development environment that includes mechanisms to extract, process, and display debug and instrumentation streams of data, mechanisms to peak and poke at memory and control execution of the programmable element, and tools to generate, link, and build executable images for the programmable elements.

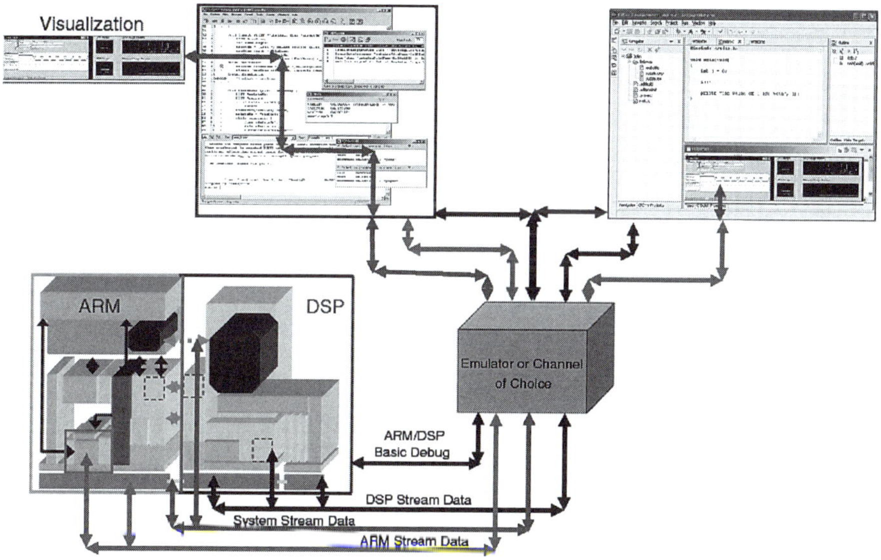

Figure 11.8 An SoC tools environment (courtesy of Texas Instruments)

SoC tools environments also contain support for monitoring the detailed status of each of the processing elements. As shown in Figure 11.9, detailed status reporting and control of the processing elements in an Soc allows the developer to gain visibility into the execution profile of the system. Also, since power sensitive SoC devices may power down some or all of the device as the application executes, it is useful to also understand the power profile of the application. This can also be obtained using the proper analysis tools.

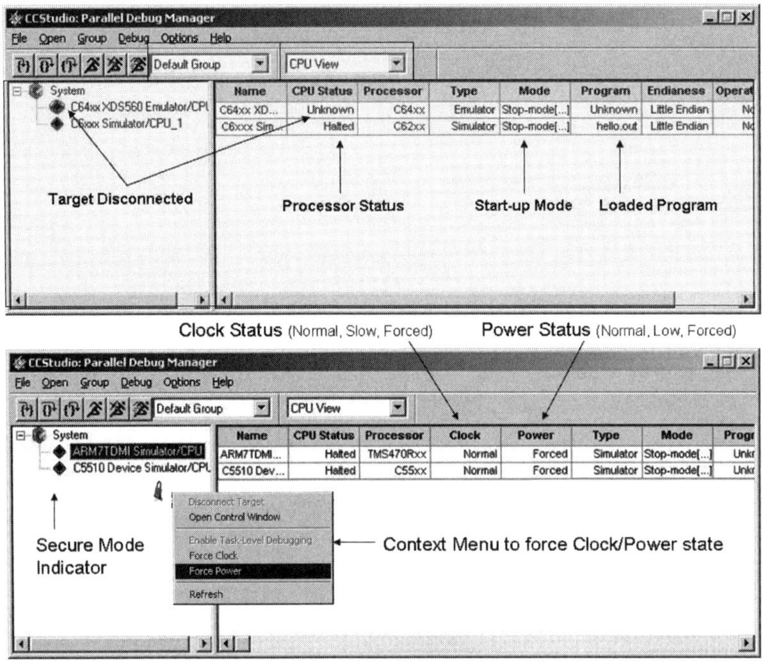

Figure 11.9 Tools support provide visibility into the status of each of the SoC processing elements (courtesy of Texas Instruments)

A Video Processing Example of SoC

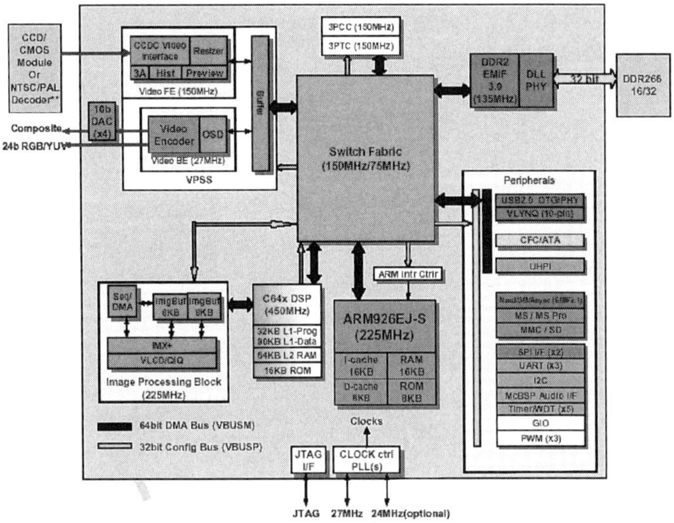

Figure 11.10 A SoC designed for video and image processing using a RISC device (ARM926) and a DSP (courtesy of Texas Instruments)

Video processing is a good example of a commercial application requiring a system on a chip solution. Video processing applications are computationally intensive and demand a lot of MIPS to maintain the data throughput required for these applications. Some of the very compute-intensive algorithms in these applications include:

– Image pipe processing and video stabilization
– Compression and Decompression
– Color conversion
– Watermarking and various forms of encryption

To perform a 30 frame per second MPEG-4 algorithm can take as much as 2500 MIPS depending on the resolution of the video.

The Audio channel processing is not as demanding but still requires enough overall MIPS to perform audio compression and decompression, equalization and sample rate conversion.

As these applications become even more complex and demanding (for example new compression technologies are still being invented), these SoC will need to support not just one but several different compression standards. SoCs for video applications include dedicated instruction set accelerators to improve performance. The SoC programming model and peripheral mix allows for the flexibility to support several formats of these standards efficiently.

For example the DM320 SoC processor in Figure 11.10 has an on chip SIMD engine (called iMX) dedicated to video processing. This hardware accelerator can perform the common video processing algorithms (Discrete Cosine Transform (DCT), IDCT, Motion Estimation, Motion Correlation to name a few)

The VLCD (variable length coding/decoding) processor is built to support variable length encoding & decoding as well as quantization of standards such as JPEG, H.263, MPEG-1/2/4 video compression standards.

As you can see from the figure, an SoC solution contains appropriate acceleration mechanisms, specialized instruction sets, hardware co-processors, etc that provide efficient execution of the important algorithms in DSP applications. We discussed an example of video processing but you will find the same mechanisms supporting other applications such as wireless basestation and cellular handset.

The code listings below implement the ARM-side APIs that talk to the DSP Controller module that manages the ARM/DSP interface across the DSP's Host Port Interface. These APIs are used to boot and reset the DSP and load the DSP code from the ARM, since the DSP can only execute code from internal memory that ARM loads.

```
/**
        \ DSP Control Related APIs
*/
static STATUS DSPC_hpiAddrValidate(Uint32 dspAddr, Uint8 read);
```

```
/**
      \ Reset the DSP, Resets the DSP by toggling the DRST bit of HPIB Con-
trol Register. \n
*/
STATUS DSPC_reset() {
      DSPC_FSET( HPIBCTL, DRST, 0);
      DSPC_FSET( HPIBCTL, DRST, 1);
      return E_PASS;
}

/**
      \ Generate an Interrupt to the DSP. Generates either INT0 or NMI in-
terrupt to the DSP depending on which one is specified.

      \param  intID     DSP interrupt ID : INT0 - interrupt 0 NMI - NMI interrupt
      \return if success, \c E_PASS, else error code
*/
STATUS DSPC_strobeINT(DSP_INT_ID intID) {

 STATUS status = E_PASS;
  switch(intID){
    case INT0:
             DSPC_FSET( HPIBCTL, DINT0, 0);
             DSPC_FSET( HPIBCTL, DINT0, 1);
          status = E_PASS;
          break;
      case NMI:
          DSPC_FSET( HPIBCTL, HPNMI, 0);
          DSPC_FSET( HPIBCTL, HPNMI, 1);
          status = E_PASS;
          break;
      default:
          status = E_INVALID_INPUT;
          break;
  }
  return (status);
}

/**
      \ Assert the hold signal to the DSP
*/
STATUS DSPC_assertHOLD() {
      DSPC_FSET( HPIBCTL, DHOLD, 0);
      return E_PASS;
}

/**
      \ Release the hold signal that was asserted to the DSP
*/
STATUS DSPC_releaseHOLD() {
      DSPC_FSET( HPIBCTL, DHOLD, 1);
      return E_PASS;
```

```
}

/**
	\ Check if HOLD acknowledge signal received from DSP
*/
DM_BOOL   DSPC_checkHOLDACK() {
	return((DM_BOOL)( DSPC_FGET( HPIBSTAT, HOLDA ) == 0 ? DM_TRUE : DM_
FALSE));
}

/**
	\ Enable/Disable byte swapping when transferring data over HPI inter-
face

	\param enable Byte swap, DM_TRUE: enable, DM_FALSE: disable
*/
STATUS DSPC_byteSwap(DM_BOOL enable) {
	DSPC_FSET( HPIBCTL, EXCHG, ((enable == DM_TRUE) ? 1 : 0));
	return E_PASS;
}

/**
	\ Enable/Disable HPI interface

	\param enable HPI interface, DM_TRUE: enable, DM_FALSE: disable
*/
STATUS DSPC_hpiEnable(DM_BOOL enable) {
	DSPC_FSET( HPIBCTL, EXCHG, ((enable == DM_TRUE) ? 1 : 0));
	return E_PASS;
}

/**
	\ Get HPI interface status register HPIBSTAT

	\return register HPIBSTAT (0x30602)
*/
Uint16 DSPC_hpiStatus() {
	return DSPC_RGET( HPIBSTAT );
}

/**
	\ Write data from ARM address space to DSP address space

	Memory map in DSP address space is as follows:
	\code
	Address              Address Access           Description
	Start          End
	0x60           0x7F    R/W                DSP specific memory area (32W)
	0x80           0x7FFF R/W                 DSP on-chip RAM, mapped on
both program and data space        (~32KW)
	0x8000         0xBFFF R/W                 DSP on-chip RAM, mapped on
```

```
data space only     (16KW)
      0x1C000              0x1FFFF      R/W                 DSP on-chip RAM,
mapped on program space only (16KW)
\endcode

      \param address             Absolute address in ARM address space, must
                                 be 16-bit aligned
      \param size                Size of data to be written, in units of 16-
                                 bit words
      \param dspAddr             Absolute address in DSP address space, 0x0
                                 .. 0x1FFFF

      \return if success, \c E_PASS, else error code
*/
STATUS DSPC_writeData(Uint16 *address, Uint32 size, Uint32 dspAddr) {

      if(size==0)
            return E_PASS;

      if((Uint32)address & 0x1 )
            return E_INVALID_INPUT;

      if( DSPC_hpiAddrValidate(dspAddr, 0) != E_PASS )
            return E_INVALID_INPUT;

      {
            Uint16 *hpiAddr;
            Uint16 *armAddr;

            hpiAddr=(Uint16*)HPI_DSP_START_ADDR;
            armAddr=(Uint16*)address;

            if(((dspAddr >= 0x10000) && (dspAddr < 0x18000)) || (dspAddr >=
 0x1C000 ))
            {
                  hpiAddr += (dspAddr - 0x10000);
            }else if((dspAddr >= 0x0060) && (dspAddr < 0xC000)){
                  hpiAddr += dspAddr;
            }else {
                  hpiAddr = (Uint16*)COP_SHARED_MEM_START_ADDR;
                  hpiAddr += (dspAddr - 0xC000);
            }

            while(size--)
                  *hpiAddr++ = *armAddr++;
      }
      return E_PASS;
 }

 /**
      \ Read data from DSP address space to ARM address space
```

```
        Memory map in DSP address space is as follows:
        \code
        Address                 Address Access              Description
        Start           End
        0x60            0x7F   R/W                  DSP specific memory area (32W)
        0x80            0x7FFF R/W                  DSP on-chip RAM, mapped on
                                                    both program and data space
                                                    (~32KW)
        0x8000          0xBFFF R/W                  DSP on-chip RAM, mapped on
                                                    data space only     (16KW)
        0x1C000         0x1FFFF       R/W           DSP on-chip RAM, mapped on
                                                    program space o
\endcode

        \param address              Absolute address in ARM address space, must
                                    be 16-bit aligned
        \param size                 Size of data to be read, in units of 16-bit
                                    words
        \param dspAddr              Absolute address in DSP address space, 0x0
                                    .. 0x1FFFF

        \return if success, \c E_PASS, else error code
*/
STATUS DSPC_readData(Uint16 *address, Uint32 size, Uint32 dspAddr) {

        if(size==0)
                return E_PASS;

        if((Uint32)address & 0x1 )
                return E_INVALID_INPUT;

        if( DSPC_hpiAddrValidate(dspAddr, 1) != E_PASS )
                return E_INVALID_INPUT;

        {
                Uint16 *hpiAddr;
                Uint16 *armAddr;

                hpiAddr=(Uint16*)HPI_DSP_START_ADDR;
                armAddr=(Uint16*)address;

                if(((dspAddr >= 0x10000) && (dspAddr < 0x18000)) || (dspAddr >=
0x1C000 ))
                {
                        hpiAddr += (dspAddr - 0x10000);
                }else if((dspAddr >= 0x0060) && (dspAddr < 0xC000)){
                        hpiAddr += dspAddr;
                }else {
                        hpiAddr = (Uint16*)COP_SHARED_MEM_START_ADDR;
                        hpiAddr += (dspAddr - 0xC000);
                }
```

```
            while(size--)
                  *armAddr++ = *hpiAddr++;
      }
      return E_PASS;
}

/**
      \     Similar to DSPC_writeData(), except that after writting it
verifies the contents written to the DSP memory

      Memory map in DSP address space is as follows:
      \code
      Address            Address Access           Description
      Start        End
      0x60         0x7F   R/W                 DSP specific memory area (32W)
      0x80         0x7FFF R/W                 DSP on-chip RAM, mapped on
                                              both program and data space
                                              (~32KW)
      0x8000       0xBFFF R/W                 DSP on-chip RAM, mapped on
                                              data space only     (16KW)
      0x1C000      0x1FFFF      R/W           DSP on-chip RAM, mapped on
                                              program space o
\endcode

      \param address              Absolute address in ARM address space, must
                                  be 16-bit aligned
      \param size                 Size of data to be written, in units of 16-
                                  bit words
      \param dspAddr              Absolute address in DSP address space, 0x0
                                  .. 0x1FFFF
      \param retryCount           Number of times to retry in case of failure
                                  in writting data to DSP memory

      \return if success, \c E_PASS, else error code
*/
STATUS DSPC_writeDataVerify(Uint16 *address, Uint32 size, Uint32 dspAddr,
Uint16 retryCount) {

      if(size==0)
            return E_PASS;

      if((Uint32)address & 0x1 )
            return E_INVALID_INPUT;

      if( DSPC_hpiAddrValidate(dspAddr, 0) != E_PASS )
            return E_INVALID_INPUT;

      {
            volatile Uint16 *hpiAddr;
            volatile Uint16 *armAddr;

            hpiAddr=(Uint16*)HPI_DSP_START_ADDR;
```

```
            armAddr=(Uint16*)address;

            if(((dspAddr >= 0x10000) && (dspAddr < 0x18000)) || (dspAddr >=
0x1C000 ))
            {
                  hpiAddr += (dspAddr - 0x10000);
            }else if((dspAddr >= 0x0060) && (dspAddr < 0xC000)){
                  hpiAddr += dspAddr;
            }else {
                  hpiAddr = (Uint16*)COP_SHARED_MEM_START_ADDR;
                  hpiAddr += (dspAddr - 0xC000);
            }

            {
                  Uint16 i;
                  volatile DM_BOOL error;

                  while(size--) {
                        error = (DM_BOOL)DM_TRUE;
                        for(i=0;i<retryCount;i++) {
                              *hpiAddr = *armAddr;
                              if(*hpiAddr==*armAddr) {
                                    error=(DM_BOOL)DM_FALSE;
                                    break;
                              }
                        }
                        if(error==DM_TRUE)
                              return E_DEVICE;
                        hpiAddr++;
                        armAddr++;
                  }
            }
      }
      return E_PASS;
}



/**
      \ Download code to DSP memory

      \param pCode  code to be dowloaded

      \see DSPCODESOURCE
*/
STATUS DSPC_loadCode(const DSPCODESOURCE* pCode) {

   if ( pCode == NULL || pCode->size == 0 )
      return E_INVALID_INPUT;

      // reset DSP
   DSPC_reset();
```

```
    // download the code to DSP memory
    while ( pCode->size != 0 ) {
        Uint16 nRetry=5;

        if( DSPC_writeDataVerify((Uint16 *)pCode->code, pCode->size, pCode-
>address, nRetry) != E_PASS  )
            return E_DEVICE;
        pCode++;
    }

    // let DSP go
    DSPC_strobeINT(INT0);

    return E_PASS;
}

static STATUS DSPC_hpiAddrValidate(Uint32 dspAddr, Uint8 read) {
// even if dspAddr <= 0x80 allow write
       if(dspAddr >= 0x60 && dspAddr <= 0xFFFF )
              return E_PASS;

       if(dspAddr >= 0x10000 && dspAddr <= 0x17FFF )
              return E_PASS;

       if(dspAddr >= 0x1c000 && dspAddr <= 0x1FFFF )
              return E_PASS;

       return E_INVALID_INPUT;
}




/**
       \ ARM-DSP Communication  APIs
*/
/*
/**
       \      Send command, parameters from ARM to DSP

       This routine also triggers the NMI interrupt to DSP

       \param  cmd                 command to be sent to DSP
       \param  cmdParams           pointer to paramters
       \param  nParams             number of parameters to be sent 0..30, \n
 if \c nParams < 30, then remaining ARM-DSP register set is filled with 0's

       \return if success, \c E_PASS, else error code
*/
STATUS ARMDSP_sendDspCmd(Uint16 cmd, Uint16* cmdParams, Uint16 nParams) {
```

```
        DSPC_writeData( &cmd, 1, ARM_DSP_COMM_AREA_START_ADDR);
        DSPC_writeData( &nParams, 1, ARM_DSP_COMM_AREA_START_ADDR+1);
        DSPC_writeData( cmdParams, nParams, ARM_DSP_COMM_AREA_START_ADDR+2);
        DSPC_strobeINT(NMI);
        return E_PASS;
}

/**
        \       Get command execution status, return parameters sent by DSP to
ARM

        \param  status              command status received from DSP
        \param  retParams           pointer to return paramters
        \param  nParams             number of parameters to be fetched from
                                    ARM-DSP communication area, 0..30

        \return if success, \c E_PASS, else error code
*/
STATUS ARMDSP_getDspReply( Uint16* status, Uint16* retParams, Uint16 nParams
) {
        DSPC_readData( status, 1, ARM_DSP_COMM_AREA_START_ADDR+32);
        DSPC_readData( retParams, nParams, ARM_DSP_COMM_AREA_START_ADDR+34);
        return E_PASS;
}

/**
        \       Get command, parameters sent by DSP to ARM

        \param  cmd                 command received from DSP
        \param  cmdParams           pointer to paramters
        \param  nParams             number of parameters to be fetched from
                                    ARM-DSP communication area, 0..30

        \return if success, \c E_PASS, else error code
*/
STATUS ARMDSP_getArmCmd( Uint16* cmd, Uint16* cmdParams, Uint16 nParams) {
        DSPC_readData( cmd, 1, ARM_DSP_COMM_AREA_START_ADDR+64);
        DSPC_readData( cmdParams, nParams, ARM_DSP_COMM_AREA_START_ADDR+66);
        return E_PASS;
}

/**
        \       Send command execution status, return parameters from ARM to
DSP

        This routine also triggers the NMI interrupt to DSP

        \param  status              command execution status to be sent to DSP
        \param  retPrm              pointer to return paramters
        \param  nParams             number of parameters to be sent 0..30, \n
if \c nParams < 30, then remaining ARM-DSP register set is filled with 0's
```

```
        \return if success, \c E_PASS, else error code
*/
STATUS ARMDSP_sendArmReply( Uint16 status, Uint16* retParams, Uint16 nParams
) {
        DSPC_writeData( &status, 1, ARM_DSP_COMM_AREA_START_ADDR+96);
        DSPC_writeData( retParams, nParams, ARM_DSP_COMM_AREA_START_ADDR+98);
        DSPC_strobeINT(INT0);
        return E_PASS;
}

/**
        \       Clear ARM-DSP communication area

        \return if success, \c E_PASS, else error code
*/
STATUS ARMDSP_clearReg() {

        Uint16 nullArray[128];

        memset((char*)nullArray, 0, 256);

        if(DSPC_writeData(nullArray, 128, ARM_DSP_COMM_AREA_START_ADDR) !=
E_PASS )
                return E_DEVICE;

        return E_PASS;
}
```

References

1. Multiprocessor systems-on-chips, by Ahmed jerraya, hannu tenhunen and Wayne Wolf, page 36, *IEEE Computer*, July 2005.
2. Embedded Software in Real-Time Signal Processing Systems: Design Technologies, *Proceedings of the IEEE*, vol. 85, no. 3, March 1997.
3. A Software/Hardware Co-design Methodology for Embedded Microprocessor Core Design, *IEEE* 1999.
4. *Component-Based Design Approach for Multicore SoCs*, Copyright 2002, ACM.
5. A Customizable Embedded SoC Platform Architecture, *IEEE IWSOC'04* <- International Workshop on System-on-Chip for Real-Time Applications.
6. How virtual prototypes aid SoC hardware design, Hellestrand, Graham. *EEdesign.com* May 2004.
7. Panel Weighs Hardware, Software Design Options, Edwards, Chris. *EETUK.com* Jun 2000.
8. Back to the Basics: Programmable SoCs, Zeidman, Bob. *Embedded.com* July 2005.
9. *Computer as Components*, Wayne Wolf, Morgan Kauffman.

12

The Future of DSP Software Technology

Contributed by Bob Frankel, TI Fellow, Texas Instruments

Changes in DSP Technology—Hardware and Software

The last decade has witnessed significant advances in DSP hardware technology, yielding the unprecedented levels of price/performance, power dissipation, and silicon integration which characterize today's DSP devices; and all indicators remain positive that DSP hardware capabilities will continue to rise at comparable rates for many years to come. By contrast, DSP software technology has progressed at a somewhat slower pace during the past decade, evidenced in part by the fact that software now represents as much as 80% of the total engineering costs incurred when developing applications for today's DSPs; software (not hardware) has consequently emerged as *the* critical factor affecting timely delivery of robust DSP-based end-equipments to the marketplace.

The movement from low-level assembly code to high-level languages such as C or C++ has tracked a corresponding growth in DSP programmer productivity, illustrated in part by the sheer size and complexity of application software targeted for today's DSP devices. The simple, single-function DSP programs of the past comprising a few hundred instructions have ballooned into full-scale software sub-systems comprising upwards of 100,000 lines of high-level language source code that execute a multitude of system control as well as signal processing functions. To cope with programs of this magnitude, developers have simultaneously embraced integrated visual environments whose constituent tools address the unique requirements of generating and debugging DSP-centric programs predominantly written in high-level language but supplemented with hand-optimized assembly code where necessary.

Notwithstanding significant improvements in tools for generating and debugging ever larger DSP programs written in high-level languages such as C or even C++, developers still often (re-)create much of this target software with each new application; extensive re-use of software elements—especially third-party content—from one application to the next remains the exception rather than the rule across the industry. Projecting current trends and tendencies forward in time, we risk introducing an insurmountable gap between hardware and software capability that will effectively impede overall growth of the DSP market: simply stated, we will never enjoy full entitlement to the underlying silicon technology if our customers cannot develop new applications for new devices in a timely manner.

To avoid this growing problem, developers will turn more and more to component-based designs. But, before component-based designs can deliver their full promise, the industry must move from a "closed" systems model to a more open one.

Foundations for Software Components

DSP vendors are beginning to break new ground by promoting re-usable target content in the form of basic kernel services common to all DSP applications, as well as through programming standards that facilitate interoperability among third-party algorithms. By offering customers the choice of deploying modular DSP software elements from a variety of vendors, the engineering costs and risks associated with developing new applications for new devices will diminish dramatically; a large catalog of "ready-to-run" target content will also make DSP devices more attractive to a broader population of application developers and system integrators, who lack deep expertise in DSP fundamentals and have historically avoided the technology. This trend should accelerate the rate of innovation across the industry by enabling more developers to create more applications in less time and with less effort.

The next stage in the ascent of programming technology for DSP—broad support for pre-fabricated, interoperable software components—may well drive a paradigm shift across the industry even more fundamental than the move to high-level language programming triggered by the advancement of robust optimizing compilers. As the industry evolves from the environment of today in which programmers continually "re-invent the wheel," the significant metric of adoption will become the percentage of software component content present in deployed DSP applications. Pre-fabricated, interoperable software components will have a similar impact upon DSP customers, enabling them to leverage their own limited resources more efficiently and more effectively. Components alone, however, are enough to spawn this transfromation. For developers to capture the full benefit of component-based development, their development tools will need to become component aware and design focused. Ultimately, we will also need better tools for efficient product deployment and maintenance.

Component-Aware IDEs

Notwithstanding its fundamental simplicity, realizing the full potential of this basic model for DSP software components will demand comprehensive support from all elements contained within the next-generation IDE environment—much as today's program generation and debug tools fully comprehend the use of high-level language within the target application. Starting with component-awareness on the part of the compiler, linker, and debugger, supplementary infrastructure capabilities integrated within the environment would include a database of program components that maintains configuration parameter settings, as well as a server for arbitrating host interaction with individual components during execution of the target application program; and the next-generation of DSP RTOS—itself an amalgamation of independently configurable components—would make available a rudimentary set of run-time kernel services to any other DSP software components embedded within the target program.

At a higher plane, opportunities abound to field sophisticated visual tools that not only would automate much of the repetitive effort in producing, packaging, and publishing individual components, but would especially simplify programming for a burgeoning base of component consumers through a new breed of application-centric development tools featuring intuitive "drag-and-drop" user interfaces that mask a plethora of lower-level details. By elevating the traditional *compile-link-debug* programming cycle into one of *configure-build-analyze* using a set of pre-fabricated interoperable software components, more application developers and system integrators can effectively deploy larger DSP programs by fully leveraging the efforts and expertise of a relatively small population of component producers.

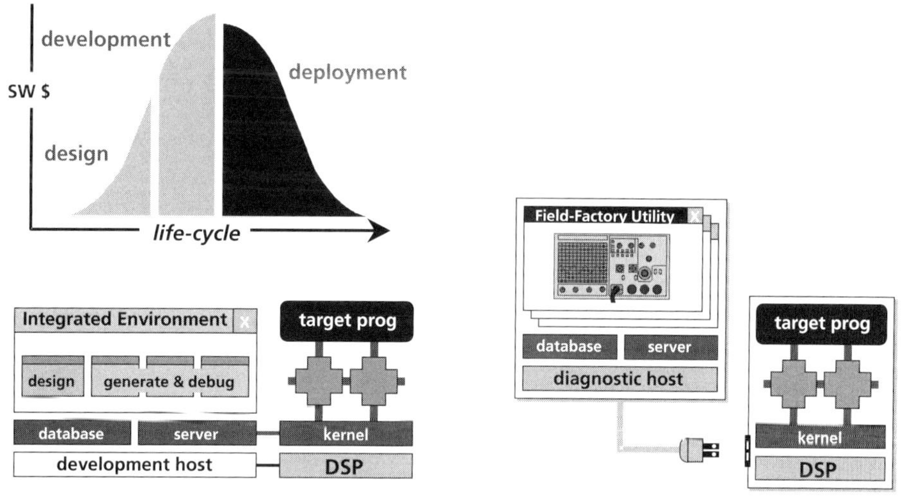

Figure 12.1 Most of the cost of developing complex DSP systems occurs after the software has been developed. Factory DSP components lead to higher productivity and higher quality systems (courtesy of Texas Instruments)

Design-focused IDEs

DSP IDE environments currently facilitate many of the activities conventionally associated with the software *development* cycle: editing source code, managing project files, building executable programs, configuring DSP platforms, debugging target applications, analyzing real-time behavior, and so on. While absolutely critical to rapid production of robust DSP application software, these development activities may represent no more than a third of the total software costs incurred over the lifetime of the customer's end product. As these products become more sophisticated and complex, customers will also engage in explicit system *design* activities prior to initiating the software development cycle: formally capturing product requirements, analyzing hardware/software trade-offs, modeling or simulating system performance, and so forth. With comprehensive visual tools for hardware/software co-design already available from leading vendors such as Cadence and Synopsys, the challenge becomes one of integrating these capabilities within the DSP IDE—itself an open and extensible environment—and creating a more seamless flow from high-level system design to conventional software development. The model of re-usable interoperable software components in fact resonates with this vision, as many of the hardware/software co-design tools available today already embrace a metaphor of configurable and connectable "black-boxes" that represent standard system functions subject to a variety of synthesized implementations.

Better Deployment and Maintenance

Still, even the most advanced environments for system design and software development cannot begin to address the challenges that come about during the product *deployment* cycle. What we conventionally lump under the general heading of "maintenance"—basically, any related software activities that occur *after* first product shipment—represents a resource liability that typically equals (and sometimes exceeds) the total engineering costs incurred during the preceding design and development cycles combined; and with DSP application programs growing ever larger, developers become especially vulnerable to these hidden costs on the tail-end of their own product life-cycles. To illustrate the point, modern DSP applications that perform a variety of independent *real-time* signal processing and system control functions will often exhibit critical software "glitches" due to timing anomalies that surface when these tasks coexist within a single target program; and to make matters worse, these seemingly random program bugs may only rear their head during product deployment within the end-customers' environment. As software faults such as these begin to dominate the overall failure rate of DSP-based systems, developers ability to identify, analyze, diagnose, and service these sorts of malfunctions in the factory and the field may well emerge as *the* controlling variable in moving tomorrow's DSP applications from first shipment to full-scale production.

As pre-fabricated interoperable software components populate a larger percentage of tomorrow's DSP applications, new paradigms may emerge for servicing operational DSP-based products during deployment. In particular, as target application programs embed more and more of these "black-box" software elements—produced by independent third-parties—the industry will simultaneously demand that each of these program components expose what amounts to a *software test connector* via a standardized abstract interface; leveraging the underlying infrastructure, hosted utilities could extract diagnostic data on a per-component basis and render this information in a format suited to the job at hand. Further support from the infrastructure, developers could also find themselves fielding product utilities that carry out on-the-fly updates of faulty or obsolete software components within an operative system (Figure 12.2).

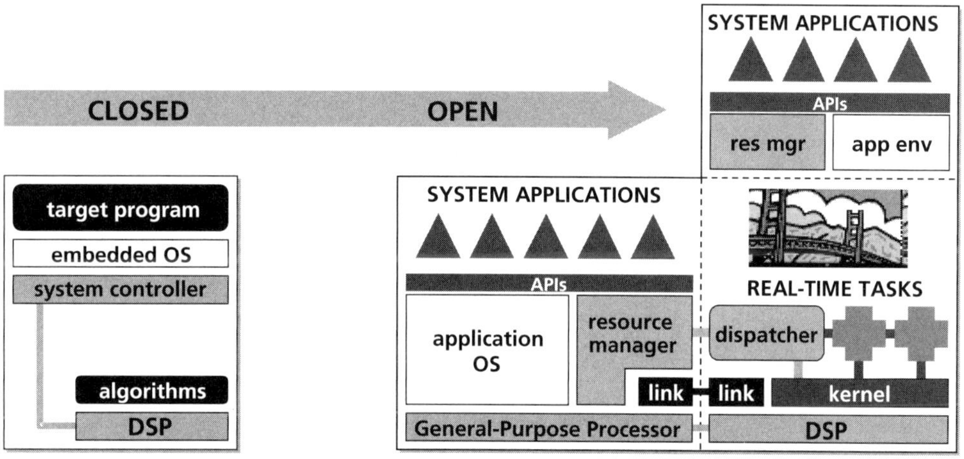

Figure 12.2 Migrating to an open architecture for DSP application development (courtesy of Texas Instruments)

From Closed to Open Embedded Systems

The future of DSP software and systems development may be a fundamental paradigm shift in the design of end-equipment employing programmable DSPs in conjunction with general-purpose processors (GPPs), moving from *closed* embedded systems in which the DSP's software functions largely remain static to *open* application platforms in which the pool of DSP resources available to GPP programs can perpetually expand and mutate. On the DSP side, these open platforms would incorporate target infrastructure elements such as basic real-time kernel services and run-time support for interoperable software components. On the GPP side, however, these platforms would feature a dramatically different set of capabilities built around one of the prevalent operating systems for general-purpose embedded applications (be it WinCE, VxWorks, or Java)—an environment familiar to a very broad base of developers. Candidate end-equipments for potentially deploying open GPP•DSP platforms range from the new wave of "smart" consumer appliances to the latest trend towards "intelligent" network gateways, which together represent some of the largest market segments for programmable DSPs.

Enabling innovative applications that capitalize upon the strengths of both processors in an open, heterogeneous GPP•DSP platform effectively requires a *software bridge* that spans these otherwise dissimilar communities of developers and their respective operating environments. Rather the forcing homogeneity, however, the DSP industry must respect the unique demands of each domain and look instead to use the bridge as a software liaison between them. In the limit, this sort of connectivity solution will only further amplify the impact of re-usable DSP software components on producers and consumers alike: the same small clique of expert component producers can now leverage their intellectual property across a large set of GPP application developers; and the latter, who know virtually nothing about signal processing technology, unwittingly become component consumers without ever writing a line of DSP code.

Away From Undifferentiated Hacking

Why re-invent the wheel when you don't have to? Why make the same mistakes that have already been made over and over again by others? Why not learn from the best practices of other engineers in the field? These are all good questions and should be asked before taking on a software development project using DSPs or any other device.

The software community recognized some time ago that software development solutions follow specific patterns, similar to building architectures. Providing a solution abstraction that follows a general pattern can enable the software community to customize a solution to their specific need while following the patterns that have been proved to be efficient and correct by others who have solved the problem in the past[1]. There are various levels of abstraction to software patterns including architectural patterns, design patterns, and idioms, which are low-level implementation solutions. Software patterns contain information about the problem including a name that describes the pattern, the problem to be solved by the pattern, the context, or settings, in which the problem occurs, the solution proposed to the problem, the context for the solution, and sometimes sample code related to the solution.

Embedded system vendors are beginning to recognize the need to help accelerate the software development process by providing software components and frameworks, the later being similar to a design pattern. In the DSP space, vendors provide hundreds of DSP-centric software components through their third party program for DSP designers to use in product and system development. They have also introduced a series of DSP software reference frameworks. These design-ready software frameworks are getting-started solutions for designers in the early stages of application development. The RFs contain easy-to-use source code that is common to many applications. Much of the initial low-level design decisions have been eliminated allowing developers more time to focus on the code that truly differentiates their products. Designers can choose the specific framework that best meets their system needs and then populate the framework with algorithms (either DSP COTS algorithms sold by other vendors or their own algorithms) to create specific applications for a range of end-equipments such as broadband, voice, video imaging, biometrics and wireless infrastructure. These frameworks are usually 100% C source code. This further eliminates the need for the developer to spend time implementing capability that is undifferentiated – things that provide no value-add to the overall application. As shown in Figure 12.3, a DSP application consists of various layers of software:

- Drivers
- A target support library
- Operating system
- Standard DSP algorithm components
- A generic application framework
- A custom application

[1] "Design Patterns: Elements of Reusable Object-Oriented Software," by E. Gamma, R. Helm, R. Johnson, and J. Vlissides.

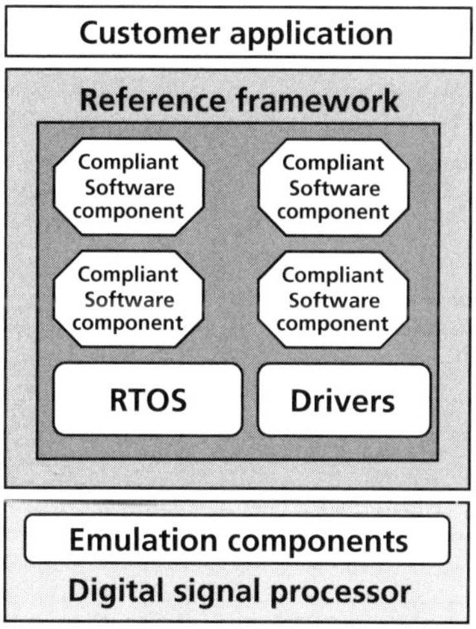

Figure 12.3 A target side DSP software stack. Many of these software components are reusable components that avoid "undifferentiated hacking," (courtesy of Texas Instruments)

Many of these layers; the drivers, target support library, operating system, a subset of the DSP algorithm components, and a portion of the application framework do not add any differentiating value to the application. Every application needs this capability but the implementation should not change from one application to the next.

A reference design that includes these software components pre-packaged and easy to extend and tailor provides the DSP developer more time to focus on those innovative capabilities that truly add value to the application and gets the application to market more quickly. Figure 12.4 shows a "bare bones" reference design which provides a standard ping-pong buffer framework, a control module and an DSP algorithm "plug" with a default software component (the FIR block in the diagram below) inserted into the plug. Application notes provide simple instruction for replacing the default algorithm with another software component. This reference design is a compact static design for relatively simple systems. The reference design comes RTOS enabled if required, and contains all the chip support libraries and drivers needed for the specific DSP device.

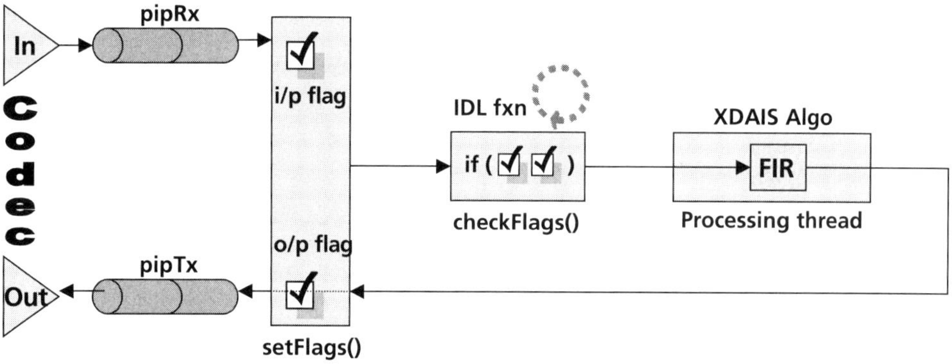

Figure 12.4 A simple reference design (courtesy of Texas Instruments)

Figure 12.5 shows a reference design for a more complicated system. This reference design addresses the key issue of ease of use in multichannel, multialgorithm systems. This reference design targets medium range systems of up to ten processing channels. This reference design also comes packaged with a set of RTOS modules, DSP software components, source code for the framework, chip support libraries and drivers, and application notes which describe how to put it all together and customize the application to the specific needs of the developer. A reference design user would modify the source code to create their own application.

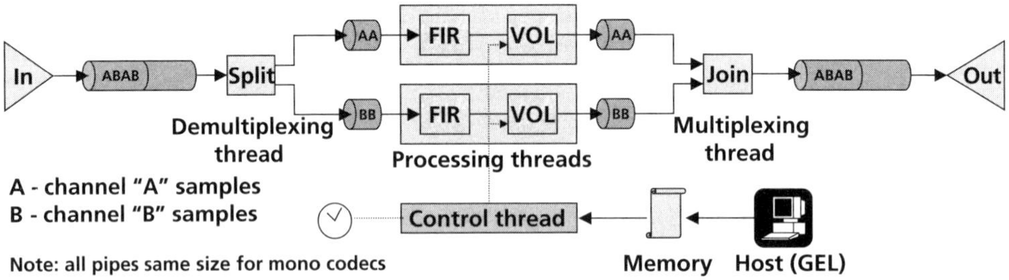

Figure 12.5 A more advanced reference design (courtesy of Texas Instruments)

Conclusion

Harnessing the power of today's DSPs for tomorrow's applications demands dramatic discontinuities in the underlying software technology, both to reduce engineering costs and risks when implementing these systems as well as to make DSP accessible to a broader population of application developers.

Building upon the migration from low-level assembly code to high-level language over the past decade, a comprehensive model for developing and deploying independent *software components and reference designs* represents the next major plateau in the evolution of programming technology for DSPs.

With half of a product's software costs incurred *after* the program development cycle winds down (Figure 12.1) and with DSP application programs growing ever larger DSP software technology must incorporate the necessary tools and infrastructure to address the product deployment cycle as well.

As open GPP•DSP platforms become widespread, there must be a bridge connecting a pair of distinct software domains such that a large base of GPP application developers can transparently control and communicate with real-time signal processing tasks executing on adjacent DSPs and hardware accelerators. This bridge must be able to initiate and control tasks on the DSP, exchange messages with the DSP, stream data to and from the DSP, and perform status queries. By providing a seamless and abstracted view of the programming model, the application developer can develop on the OMAP platform in the same manner as if they were developing on a single RISC processor.

A

Software Performance Engineering of an Embedded DSP System Application

Based on "Winning Teams; Performance Engineering Through Development" by Robert Oshana, which appeared in *IEEE Computer*, June 2000, Vol 33, No 6. ©2000 IEEE.

Introduction and Project Description

Expensive disasters can be avoided when system performance evaluation takes place relatively early in the software development life cycle. Applications will generally have better performance when alternative designs are evaluated prior to implementation. Software performance engineering (SPE) is a set of techniques for gathering data, constructing a system performance model, evaluating the performance model, managing risk of uncertainty, evaluating alternatives, and verifying the models and results. SPE also includes strategies for the effective use of these techniques. Software performance engineering concepts have been incorporated into a Raytheon Systems Company program developing a digital signal processing application concurrently with a next generation DSP-based array processor. Algorithmic performance and an efficient implementation were driving criteria for the program. As the processor was being developed concurrently with the software application a significant amount of the system and software development would be completed prior to the availability of physical hardware. This led to incorporation of SPE techniques into the development life cycle. The techniques were incorporated cross-functionally into both the systems engineering organization responsible for developing the signal processing algorithms and the software and hardware engineering organizations responsible for implementing the algorithms in an embedded real-time system.

Consider the DSP-based system shown in Figure A.1. The application is a large, distributed, multiprocessing embedded system. One of the sub-systems consists of two large arrays of digital signal processors (DSP). These DSPs execute a host of signal processing algorithms (various size FFTs and digital filters, and other noise removing and signal enhancing algorithms). The algorithm stream being implemented includes both temporal decomposition of the processing steps as well as spatial decomposition of the data set. The array of mesh-connected DSPs is used because the spatial decomposition required maps well to the architecture. The required throughput of the system drives the size of the array. The system is a data driven application, using interrupts to signal the arrival of the next sample of data. This system is a "hard" real-time system in the sense that missing one of the data processing deadlines results in a catastrophic loss of system performance.

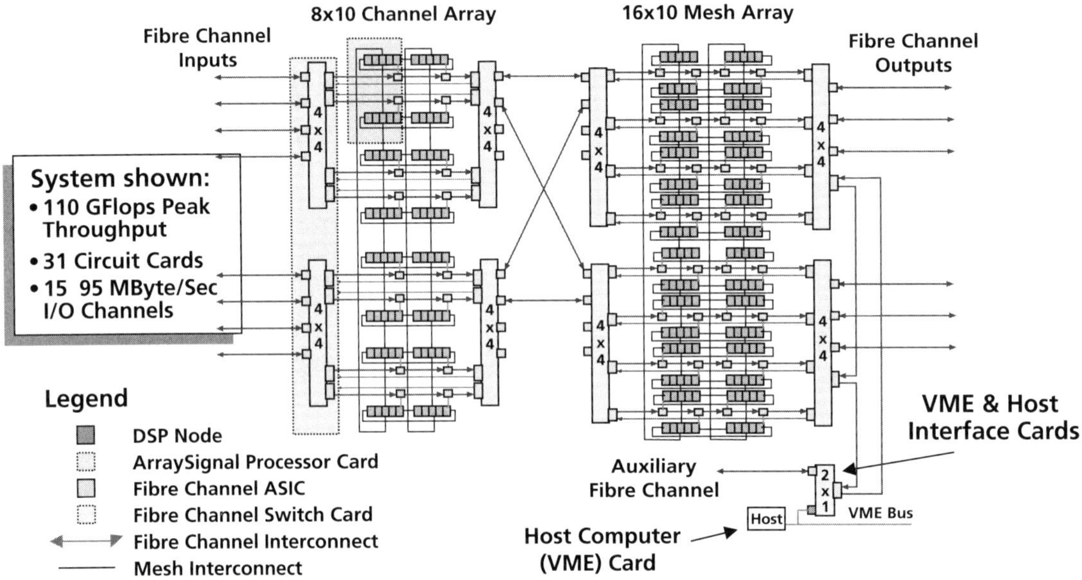

Figure A.1. DSP array architecture in mesh configuration

This system was a hardware-software co-design effort. This involved the concurrent development of a new DSP-based array processor using high performance DSP devices. In this project, the risk of the delivered system not meeting performance requirements was a serious concern. To further complicate matters the algorithm stream was being enhanced and revised as part of the development effort. The incorporation of SPE techniques into the development processes of the various functional organizations was deemed critical to mitigating these risks.

The issue of performance was addressed from the inception of the program throughout its development phases. The main measures of performance are captured in three metrics:

- Processor throughput
- Memory utilization
- I/O bandwidth utilization

These are the metrics of choice because monthly reporting of these metrics was a customer requirement for the program. Initial estimates of these metrics were made prior to the start of the program and updated monthly during the development effort. Uncertainties associated with key factors driving these estimates were identified. Plans for resolving these uncertainties during the development effort were developed and key dates identified. Updating the metrics and maintaining the associated risk mitigation plans was a cross-functional collaborative effort involving systems engineering, hardware engineering and software engineering.

Initial performance estimates and information requirements

The information generally required for a SPE assessment is [1]:

Workload – The expected use of the system and applicable performance scenarios. We chose performance scenarios that provided the array processors with the worst case data rates. These worst case scenarios were developed by interfacing with the users and our system engineers.

Performance objectives – This represents the quantitative criteria for evaluating performance. We used CPU utilization, memory utilization, and I/O bandwidth because of the customer requirement that we report on these monthly.

Software characteristics – This describes the processing steps for each of the performance scenarios and the order of the processing steps. We had accurate software characteristics due to an earlier prototype system using a similar algorithm stream. We also had an algorithms description document detailing the algorithmic requirements for each of the functions in the system. From this a discrete event simulation was developed to model the execution of the algorithms.

Execution environment – This describes the platform on which the proposed system will execute. We had an accurate representation of the hardware platform due to involvement in the design of the I/O peripherals of the DSP as well as some of the DSP core features as well. The other hardware components were simulated by the hardware group.

Resource requirements – This provides an estimate of the amount of service required for the key components of the system. Our key components were CPU, memory and I/O bandwidth for each the DSP software functions.

Processing overhead – This allows us to map software resources onto hardware or other device resources. The processing overhead is usually obtained by benchmarking typical functions (FFTs, filters) for each of the main performance scenarios.

CPU throughput utilization was the most difficult metric to estimate and achieve. Therefore, the rest of this paper will focus primarily on the methods we used to develop an accurate estimate for the CPU throughput utilization metric.

Developing the initial estimate

The process used to generate the initial performance metric estimates is shown in Figure A.2. This flow was used throughout the development effort to update the metrics. The algorithm stream is documented in an algorithm document. From this document, the systems engineering organization developed a static spreadsheet model of the algorithm stream which provided estimates of throughput and memory utilization for each of the algorithms in the algorithm requirements document. The spreadsheet includes allowances for operating system calls and inter-processor communication. The systems engineering organization used a current generation DSP processor to perform algorithm prototyping and investigation activities. The results of this work influenced algorithm implementation decisions and were used to develop the discrete event simulations used to estimate the performance metrics. A discrete event simulation was used to model the dynamic performance of the algorithm stream. The simulation model included allowances for operating system task switches and associated calls. The initial algorithm spreadsheet of resource allocations for each algorithm and discrete event simulation processes provide the system engineering 'algorithm' performance metrics. At this point the metrics reflect the throughput, memory, and I/O bandwidth required to perform the algorithms defined in the algorithm document and implemented using the prototype implementations. The software engineering organization then updates the performance metrics to reflect the costs of embedding the algorithm stream in a robust, real-time system. These metric adjustments include the effects of system-level real-time control, built-in-test, formatting of input and output data, and other 'overhead' functions (processing overhead) required for the system to work. The results of this process are the reported processor throughput, memory utilization and I/O utilization performance metrics.

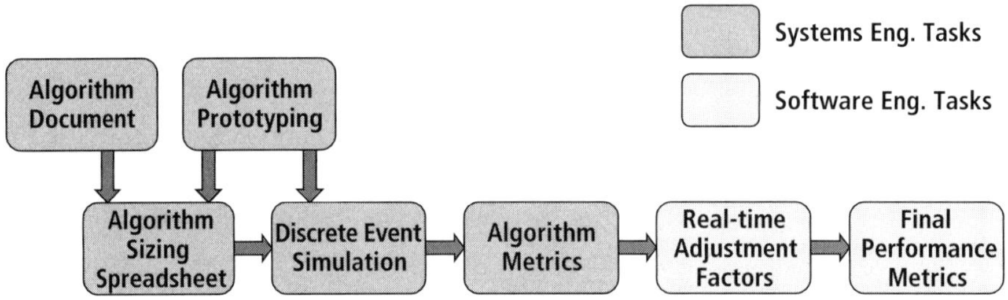

Figure A.2 Performance metric calculation flow

Key factors in the spreadsheet that influence the processor throughput metric are:

- The quantity of algorithms to implement
- Elemental operation costs (measured in processor cycles)
- Sustained throughput to peak throughput efficiency
- Processor family speed-up

The quantity of algorithms to perform is derived from a straightforward measurement of the number of mathematical operations required by the functions in the algorithm stream. The number of data points to be processed is also included in this measurement. The elemental operation costs measures the number of processor cycles required to perform multiply accumulate operations, complex multiplies, transcendental functions, FFTs, and so on. The sustained throughput to peak throughput efficiency factor de-rates the "marketing" processor throughput number to something achievable over the sustained period of time a real world code stream requires. This factor allows for processor stalls and resource conflicts encountered in operation. The processor family speed-up factor was used to adjust data gained from benchmarking on a current generation processor. This factor accounted for the increase in clock rate and number of processing pipelines in the next generation device compared to its current generation predecessor.

Key factors in the spreadsheet that influence the memory utilization metric are:

- Size and quantity of intermediate data products to be stored
- Dynamic nature of memory usage
- Bytes/data product
- Bytes/instruction
- Size and quantity of input and output buffers based on worst case system scenarios (workloads)

The size and quantity of intermediate data products is derived from a straightforward analysis of the algorithm stream. A discrete event simulation was used to analyze memory usage patterns and establish high water marks. The bytes/data product and bytes/instruction were measures used to account for the number of data points being processed and storage requirement for the program load image.

All of these areas of uncertainty are the result of the target processor hardware being developed concurrently with the software and algorithm stream. While prototyping results were available from the current generation DSP array computer, translating these results to a new DSP architecture (superscalar architecture of the C40 vs. the very long instruction word (VLIW) of the C67 DSP), different clock rate, and new memory device technology (synchronous DRAM vs. DRAM) required the use of engineering judgement.

Tracking the reporting the metrics

The software development team is responsible for estimating and reporting metrics related to processor throughput and memory. These metrics are reported periodically to the customer, and are used for risk mitigation. Reserve requirements are also required to allow for future growth of functionality (our reserve requirement was 75% for CPU and memory). Throughout the development life cycle, these estimates varied widely based on the different modeling techniques used in the estimation and hardware design decisions which influenced the amount of hardware available to execute the suite of algorithms as well as measurement error. Figure A.3 shows the metric history for throughput and memory for the first array processor application. There is a wide variability in the throughput throughout the life cycle, reflecting a series of attempts to lower the throughput estimate followed by large increases in the estimate due to newer information. In Figure A.3, the annotations describe the increases and decreases in the estimate for the CPU throughput measurement. Table A.1 describes the chronology of this estimate over the course of the project (not completed as of this writing).

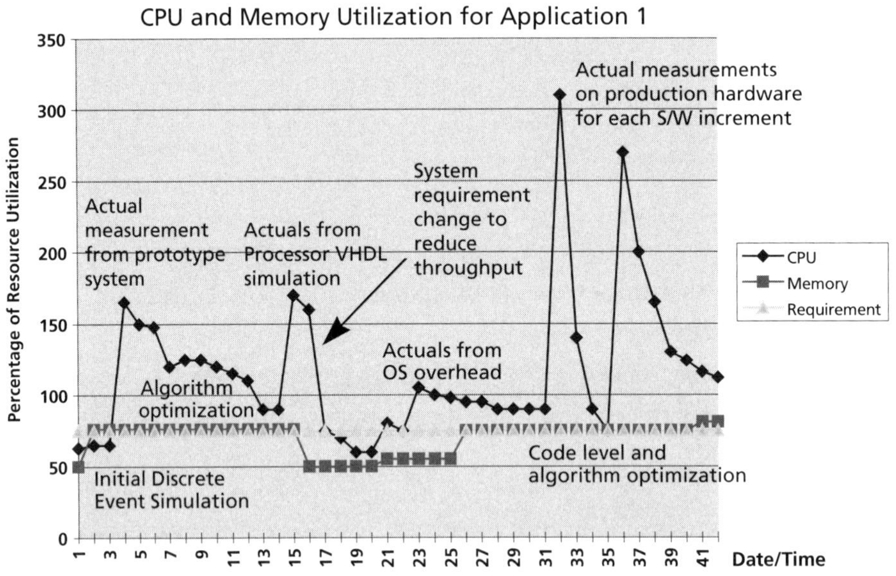

Figure A.3 Resource utilization metric for application 1

Increase or Decrease in Metric	Explanation
Initial discrete event simulation was used as the starting point for the metric estimation	Discrete event simulation was built using algorithm cycle estimations and first order modeling for task iterations due to context switching, and so on.
Measurement on prototype C40 based array	Prototype code was ported to a C40 based DSP small scale array and measured. The measurement was then scaled based on the speedup of the C67 based DSP full scale array.
Algorithm level optimization	Algorithms were made more efficient using algorithm restructuring methods and reducing complexity in other areas of the algorithm stream.
Processor VHDL measurement	Big increase in throughput measurement was due to unexpected high cost of accessing data from external (off-chip) memory. Several benchmarks were performed and scaled to the entire application
System level requirement change	Project decision was made to change a system level parameter. This caused significant algorithm restructuring and was an unpopular decision with the customer.
OS level overhead measured	Because the processor was new, the COTS OS was not immediately available. This point indicated the first time to run the application in a multitasking environment with the OS.
Actuals on production hardware array of DSPs for each software increment	The production code was initially developed without code optimization techniques in place (Make it work right and then make it work fast). Initial measurement for the full algorithm stream was not entirely optimized when we first took the measurement.
Continued code and algorithm level optimization	Dedicated team in place to work code optimization and other algorithm transformation techniques to reduce CPU throughput (i.e., taking advantage of symmetry in the algorithms and innovative techniques to reduce communications between DSPs which were expensive).

Table A.1 Chronology of CPU Throughput Reduction for Application 1

The first large increase in the estimate came as a result of implementing the algorithm stream on a prototype current generation processor. These measurements were then scaled based on the anticipated performance of the next generation processor. An effort was then undertaken to optimize the implementation of the algorithm stream to lower the throughput estimate.

The next unexpected increase came from running representative benchmarks on the next generation cycle accurate simulator. This simulation allowed us to estimate the true cost of external memory accesses, pipeline stalls, and other processor characteristics that increased the cost of executing the algorithms. These results led the development teams to undertake another significant effort to optimize the algorithm stream for real-time operation. The main techniques undertaken during this phase included instrumentation of the direct memory access (DMA) to stage data on and off chip, re-structuring of code to allow critical loops to pipeline, assembly language implementation of critical algorithm sections, and efficient use and management of on-chip memory where memory access time is much shorter.

The representative benchmarks showed us that we could reduce the throughput using code-level optimization techniques (use of on-chip memory, pipelining of important loops, etc) but we were still in danger of not meeting our overall throughput requirement. It was at this time that a system requirement was modified to reduce throughput. Although a very unpopular decision with the customer (the change reduced data rate and performance of the algorithms), it allowed us to save money by not having to add additional hardware to the system (which is more cost per unit delivered). Algorithm studies also showed that we could still meet system performance by improvements in other areas of the system.

The third major increase came when we measured the full application on the target array of DSPs. The main reason for the increase was due to the fact that many of the algorithms were not optimized. Only a small percentage of algorithms were benchmarked on the processor VHDL simulator (representative samples of the most commonly used algorithms such as the FFTs and other algorithms called inside major loops in the code). The software group still needed to employ the same optimization techniques for the remaining code for each of the software increments being developed. By this time the optimization techniques were familiar to the group and the process went fairly fast.

The memory estimate, although not as severe as the throughput estimate, continued to grow throughout the development cycle. The main reasons for the increase in memory were:

- additional input and output buffers required for a real time system to operate;
- additional memory was required for each section of code that is instrumented to use the DMA (although this does save on throughput cycles);
- additional memory is needed for code optimization techniques such as loop unrolling and software pipelining which cause the number of instructions to increase.

The life cycle throughput estimates for the second array processor application is shown in Figure A.4. A similar pattern in the reported numbers is seen here due to the same basic issues. Table A.2 shows the chronology of this CPU utilization estimation.

Once again the initial discrete event simulation proved to be inaccurate and the prototype system measurements were much higher than anticipated due to overly aggressive estimates of the CPU throughput, failure to account for realistic overhead constraints, etc. A long process of code and algorithm optimization was able to bring the estimate back down close to the goal before the VHDL simulation measurements uncovered some other areas that made us increase the estimate. The increase in the estimate in this application resulted in several risk management activities to be triggered.

The estimate in month 5 was high enough and was made early enough in the program schedule that the program was able to add more hardware resources to reduce the algorithm distribution and lower the throughput estimate. This was made at the expense of more power and cooling requirements as well as more money for the hardware (no new designs were required, just more boards). These increases in power and cooling had to be offset by sacrifices elsewhere to maintain overall system requirements on these parameters.

The measurement in month 19 was caused consternation among the managers as well as the technical staff. Although we felt continued optimization at the code level would drop the number significantly, meeting the application requirement of 75% CPU throughput (25% reserved for growth) would be hard to accomplish.

One contributor to the CPU throughput estimate increase was a result of an underestimation of a worst case system scenario which led to an increase in data rate for the processing stream,. This resulted in several algorithm loops being executed more frequently which increased the overall CPU utilization.

The decision was made to move some of the software functionality being done in the DSPs into a hardware ASIC to reduce the throughput significantly (there were a sufficient number of unused gates in the ASIC to handle the increased functionality). With this decision coming so late on the development cycle, however, significant re-design and re-work of the ASIC and the interfaces was required which was extremely expensive for the hardware effort as well as delays in the system integration and test phase.

The last increase in CPU utilization was a result of scaling the algorithms from the small (single node) DSP benchmark to the full array of DSPs. The increase was mainly due to a mis-estimation in the overhead associated with inter-processor communication. Once again, the development teams were faced with the difficult challenge of demonstrating real-time operation given these new parameters. At this late date in the development cycle, there are not many options left for the system designers. The main techniques used at this point to reduce the throughput estimate were additional code optimization, assembly language implementation of additional core algorithms, additional limited hardware support, and a significant restructuring of the algorithm control flow to circumvent the use of slow operating system functions. For example, we eliminated some of the robustness in the node to node communication API in order to save valuable CPU cycles.

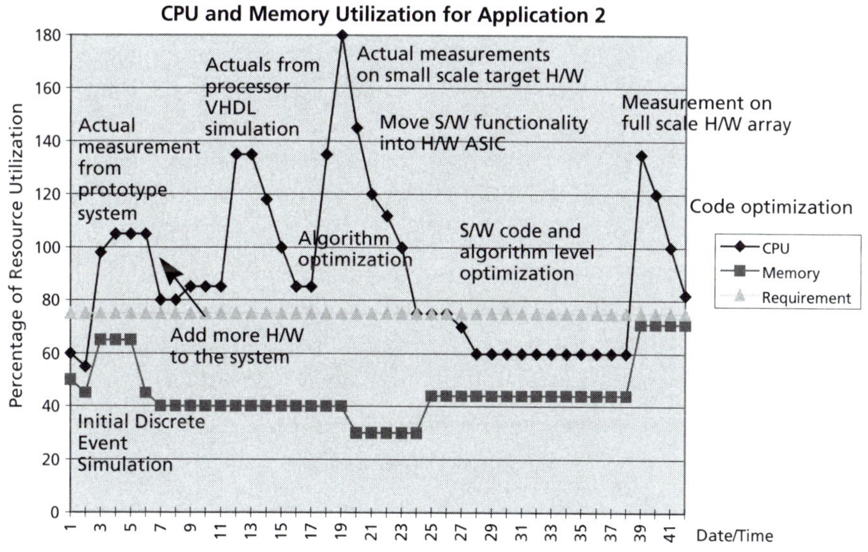

Figure A.4. Resource utilization metric for application 2

Increase or Decrease in Metric	Explanation
Initial discrete event simulation was used as the starting point for the metric estimation	Discrete event simulation was built using algorithm cycle estimations and first order modeling for task iterations due to context switching, and so on.
Measurement on prototype C40 based array	Prototype code was ported to a C40 based DSP small scale array and measured. The measurement was then scaled based on the speedup of the C67 based DSP full scale array.
Add more hardware to the system	Number of DSP nodes was increased by adding more DSP boards. Good hardware design made scalability relatively easy.
Processor VHDL measurement	Big increase in throughput measurement was due to unexpected high cost of accessing data from external (off-chip) memory. Several benchmarks were performed and scaled to the entire application.
Algorithm optimization	Because of the nature of the algorithms, we were able to significantly cut CPU throughput utilization by restructuring the algorithms to pipeline the major loops of the algorithm stream.
Actual measurement on small scale target Hardware	In our hardware/software co-design effort, we did not have full scale hardware until late in the cycle. Initial benchmarking for this application was performed on a single node prototype DSP card.
Move software functionality into hardware ASIC	Decision was made for risk mitigation purposes to move part of the algorithm stream into a hardware ASIC in another sub-system, saving significant CPU cycles in the application software.
Software code and algorithm level optimization	Dedicated team in place to work code optimization and other algorithm transformation techniques to reduce CPU throughput.
Measurement on full-scale hardware	Measuring the application CPU throughput on the full scale hardware showed that we had underestimated the overhead for communication among all the array nodes. We developed a tailored comm API to perform intra-node communications more quickly.

Table A.2 Chronology of CPU throughput reduction for application 2

It did not take long for management to realize that these "spikes" in the CPU throughput utilization would continue until all of the application had been measured on the target system under worst case system loads. Rather than periodically being surprised by a new number (we were optimizing the code in sections so every few months or so we would have actuals for a new part of the algorithm stream) we were asked to develop a plan of action and milestones (POA&M) chart which predicted when we would have new numbers and the plan for reducing the throughput after each new measurement that would support completion by the program milestone. In the plan we predicted the remaining spikes in the estimate and the plan for working these numbers down (Figure A.5). This new way of reporting showed management that we knew increases were coming and had a plan for completing the project.

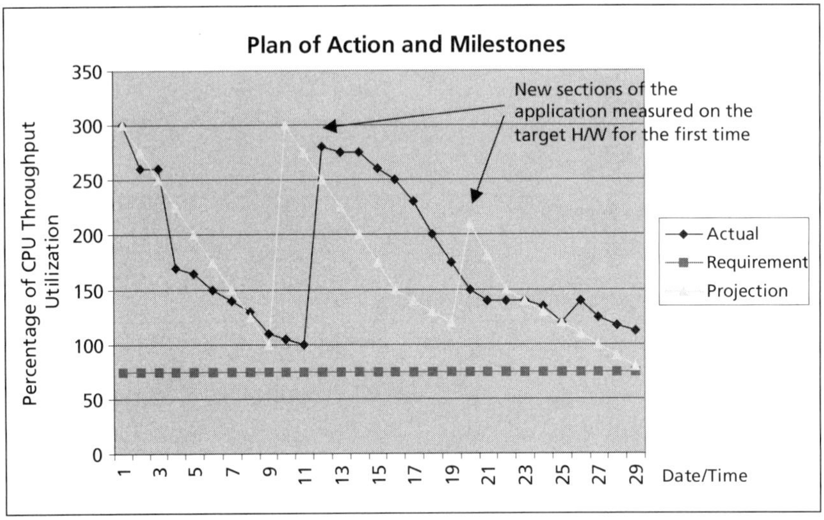

Figure A.5 Plan of action and milestones for application 1

Reducing the measurement error

The performance engineering plan detailed when hardware and software tools would become available which could be used to reduce the errors in the performance metrics. These availability dates when combined with the system development schedule provided decision points at which design trade-offs would be performed balancing algorithm, hardware, and software design points to yield a system that met cost and performance goals. Table A.3 lists the tools identified and the error factors reduced by them.

Tool	Error Factors Resolved
Code Generation Tools (Compiler, Assembler, Linker)	Compiler efficiency Quality of generated assembly code Size of load image
Instruction Level Processor Simulator	Utilization of dual processor pipelines Cycle counts for elemental operations

Tool	Error Factors Resolved
Cycle-accurate Device Level VHDL Model	Effect of external memory access times Instruction Caching effects Device resource contention between processor and DMA channels
Single DSP Test Card	Validate VHDL results Runtime interrupt effects
Multi-DSP Test Card	Inter-processor communication resource contention effects

Table A.3 Tools identified in performance plan and errors resolved by them

As these tools became available, benchmark code was executed using them and the performance metrics updated accordingly. This data was used to support program level decision points to review the proposed computer design. This review included hardware resources in the computer, algorithmic functionality assigned to the computer, and the proposed software architecture. At various decision points all of these areas were modified. The computer hardware resources were increased through the addition of more DSP processor nodes. The clock rate of the DSP was increased by 10%. Some algorithms were moved to other portions of the system. The software architecture was reworked to reduce overhead by eliminating extraneous interrupts and task switches. All aspects of the design were considered and adjusted as appropriate to meet the performance and cost objectives.

The performance plan also included the use of analytical tools to address the overall schedulability and large-scale performance of the array processor. We attempted to use rate monotonic analysis (RMA) to validate the schedulability of the software architecture [3,4,5]. RMA is a mathematical approach to determining schedulability under worst case task phasings and allows the designer to determine ahead of time whether the system will meet its timing requirements. RMA has the advantage over discrete event simulation in that the model is easier to develop and change and the model provides a conservative answer that guarantees schedulability (using a simulation, it becomes hard to predict how long to execute the model before a certain set of task phasings causes the system to break). One powerful feature of RMA tools is the ability to identify blocking conditions. Blocking and preemption are the most common reasons for missing deadlines and are one of the main focuses of most RMA tools. We were interested in using RMA because the model could identify potential timing problems even before the system is built. Alternative designs could be analyzed quickly before actually having to implement design. Our attempt at using RMA provided only a high level look at schedulability but not the details. The tools used did not scale well to large systems with thousands of task switch possibilities and nonpreemptible sections (one of the compiler optimization techniques produced software pipelined loops which, because of the nature of the processor pipeline turns off interrupts during the pipelined loop, thereby creating a small nonpreemptible section. Attempting to input and model thousands of these conditions proved to be too cumbersome for our application without becoming too abstract for our purposes).

As the array computer hardware was being developed concurrently with the software, the software team did not have target hardware available to them until late in the development life cycle. To enable the team to functionally verify their software prior to the availability of hardware an environment was developed using networked Sun workstations running Solaris. Using features of the Solaris operating system, the environment enabled small portions of the array computer to be created with interprocessor communication channels logically modeled. The application code was linked to a special library that implemented the DSP operating system's API using Solaris features. This enabled the team to functionally verify the algorithms, including intertask and inter-processor communications, prior to execution on the target hardware.

The fundamental approach was to make the application work correctly and then attempt to add efficiency to the code, "Make it work right—then make it work fast!" We felt this was required for this application for the following reasons:

- Given the hardware/software co-design effort, the processor (and user documentation) was not available so the development team did not thoroughly understand the techniques required to optimize the algorithm stream.
- The algorithms themselves were complicated and hard to understand and this was seen as a risk by the development team. Making the algorithm stream run functionally correct was a big first step for a development team tackling a new area.
- Optimization of an algorithm stream should be performed based on the results of profiling the application. Only after the development team knows where the cycles are being spent can they effectively optimize the code. It does not make sense to optimize code that is executed infrequently. Removing a few cycles from a loop that executes thousands of times, however, can result in a bigger savings in the bottom line.

Conclusions and lessons learned

Estimating throughput may not be exact science but active attention to it during life cycle phases can mitigate performance risks and enable time to work alternatives while meeting overall program schedules and performance objectives. This needs to be a collaborative effort across multiple disciplines. System performance is the responsibility of all parties involved. There are no winners on a losing team.

Processor CPU, memory, and I/O utilization are important metrics for a development effort. They give early indications as to problems and provide ample opportunity for the development teams to take mitigation actions early enough in the life cycle. These metrics also give management the information necessary to manage system risks and allocate reserve resources (that is, money and schedule) where needed. Often, one or more of these metrics will become an issue at some point during the development cycle. To obtain a system solution to the problem, sensitivity analysis is usually performed, examining various alternatives that trade off throughput, memory, I/O bandwidth, as well as cost, schedule, and risk. When performing this analysis, it is essential to understand the current accuracy in the metric estimates. Early in the life cycle the accuracy will be less than in later stages, where the measurements are much more aligned to the actual system due to the simple fact that more information is available (Figure A.6).

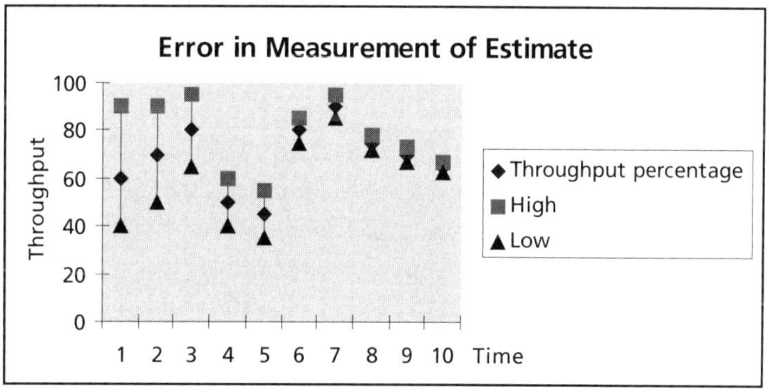

Figure A.6. Improvement in Accuracy in Estimate over time

There were several lessons learned in this experience:

- *Prototype early in the development life cycle* – Several of the surprises we encountered could have been uncovered earlier if the proper level and type of prototyping was performed. Although prototyping was scheduled early in the life cycle, schedule pressures forced the development team to dedicate a limited amount of resources to this early in the development cycle.
- *Benchmark* – Don't rely totally on the processor marketing information. Most processors will never achieve the throughput proposed in the literature. The numbers are often theoretical. In many cases, actual is much lower and very dependent on how the application maps to the processor architecture (DSPs run DSP-like algorithms very well but they are not very good at finite state machines and other "control" software). Marketing information for processors will show how well they perform on the best mapped algorithms there processors support.
- *Analyze the functions executing most often* – These areas are where the hidden cycles can come back to haunt a development team. Eliminating just a few cycles from a function that executes many times will have significant impact on the overall throughput.
- *Don't ignore the interfaces* – Real-time systems carry an inherent "overhead" that never seems to be accounted for in throughput estimates. Although the signal processing algorithms may be where the main focus is from a system requirements and functionality point of view, real-time systems also need throughput for interrupt handling, data packing and unpacking, data extraction, error handling, and other management functions that are easily overlooked in throughput estimates. Many battles were fought over how much of the timeline should be devoted to overhead tasks.
- *Benchmarks of discrete algorithms don't scale well to real-time systems* – Benchmarking an individual algorithm inherently implies that algorithm has complete control and

use of all the processor resources including internal and external memory, the DMA controller, and other system resources. In reality, there may be other tasks competing for these same resources. Assumptions made when benchmarking individual algorithms may not apply when the system is put together and running under full system loads. Resource conflicts result in additional overhead that can easily be overlooked when forming throughput estimates.

- *Keep management informed* – As we approach the completion of the code level optimization effort, it appears the model we established early in the project was a relatively accurate estimate. However, it took a substantial amount of resources (schedule and budget) to accomplish this goal. Along the way, the estimate periodically rose and fell as we optimized and measured our algorithm stream. The reporting period for these metrics was short enough to catch these spikes which caused premature concern from management. A longer reporting interval may have "smoothed" some of these spikes.
- *Budget accordingly* – The two pass approach of functional correctness followed by code optimization will take more time and more resources to accomplish. This needs to be planned. A one pass approach to code level optimization at the same time the functionality is being developed should be attempted only by staffs experienced in the processor architecture and the algorithms.

References

1. Smith, Connie U., "Performance Engineering for Software Architectures," *21st Annual Computer Software and Applications Conference*, 1997, pp. 166–167.
2. Baker, Michelle and Warren Smith, "Performance Prototyping: A Simulation Methodology for Software Performance Engineering," *Proceeding of the Computer Systems and Software Engineering*, 1992, pp. 624–629.
3. Oshana, Robert , "Rate Monotonic Analysis Keeps Real-Time Systems On Track," *EDN*, September 1, 1997.
4. Liu, C and J Layland, "Scheduling algorithms for multiprogramming in a hard real time environment," *Journal of the Association for Computing Machinery*, January 1973.
5. Obenza, Ray, "Rate Monotonic Analysis for Real-Time Systems," March 1993.

B

More Tips and Tricks for DSP Optimization

The first stage is getting the first 60–70% of optimization. This includes setting the right compiler switches to enable pipelining, keeping the loop from being disqualified, and other "simple" optimizations. This stage requires very basic knowledge of the architecture and compiler. By the end of this stage, the code should be running in real-time. This application note will provide the technical background and optimization tips needed to get to the end of the first stage of optimization.

The second stage is the last 30–40%. This stage requires an in-depth knowledge of how to interpret compiler feedback to reduce loop carried dependency bounds, reduce resource bounds, partition intelligently between inner and outer loop, and so on. Optimization at this stage may also require intelligent memory usage. In fact, the cache architecture of the DA6xx DSP may have a big impact on performance and should be considered when trying to optimize an algorithm fully. Optimal cache usage will not be addressed in this application note. However, there are many other papers on this topic. Please see the "References" section for pointers. This application note will not address techniques for second stage code optimizations either. However, references will be provided throughout this application note and in the "References" section on where to find more information on these second stage optimizations.

While any ANSI C code can run on a TI C6000 DSP, with the right compiler options and a little reorganization of the C code, big performance gains can be realized. For existing C code, one will need to retrofit the code with these quick optimizations to gain better performance. However, for new algorithm development, one should incorporate these optimizations into his/her "programming style," thus eliminating at least one optimization step in the development cycle.

1.0 Software Development Cycle

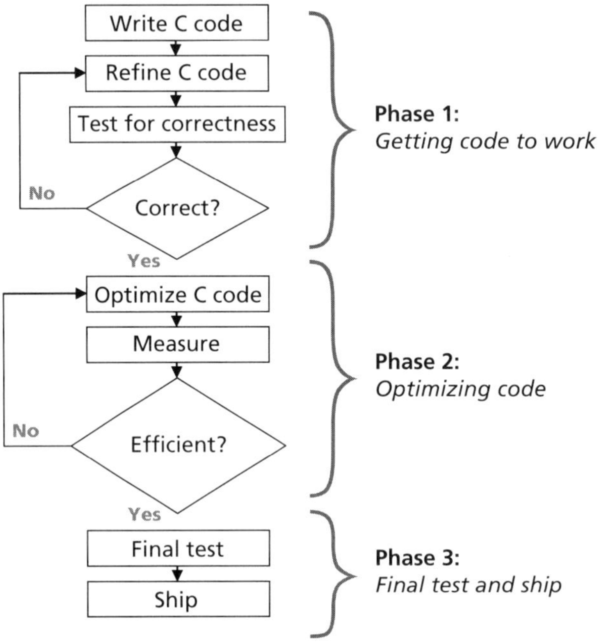

Figure B.1 Software Development Flow

The concept of a software development cycle should not be new. The reason that it is mentioned explicitly here is to keep in mind that "getting code to work" and "optimizing code" are two very distinct steps and should not overlap. Optimization work should only begin after the code is working correctly.

In general, the compiler can be most aggressive when it does not have to retain any debug information. The C6000 compiler can be thought of having three aggressiveness levels that correspond to the three phases of software development.

In Phase 1, using appropriate compiler switches, the compiler should be set to its least aggressive setting. At this level, the compiler will perform no optimizations. Each line of C code will have corresponding assembly code, and the assembly code will be ordered in the same manner as the C source. This level of aggressiveness is appropriate when "getting the code to work."

In Phase 2, using appropriate compiler switches, the compiler should be set to its mid aggressive setting. At this level, the compiler will perform many optimizations such as simplifying functions, eliminating unused assignments, converting arrays to pointers, and software pipelining loops (discussed later). The net result of these optimizations is assembly code that does not match with C source code line for line. This makes it more difficult to step through code or set breakpoints in C, for example, making debugging more difficult at this stage. Enough information is retained to profile on a function-by-function basis, however. Keep in mind that these optimizations do not alter the functionality of the original source code (assuming the original source is correct ANSI C and not relying on undefined behavior).

In Phase 3, using appropriate compiler switches, the compiler should be set to it most aggressive setting. At this level, the compiler has visibility into the entire program, across different functions and files. The compiler performs optimizations based on this program-wide knowledge, collapsing the output of multiple C source files into a single intermediate module. Only system-level debug and profiling can be performed at this stage, as the code will be reduced and reordered (depending on how much optimization was possible) such that it matches the C source in functionality only.

Note that after multiple iterations of C optimization, if the desired performance is still not achieved, the programmer has the option of using linear assembly. Linear assembly can be thought of as pseudo-assembly coding where the programmer uses assembly mnemonics but codes without immediate regard to some of the more complex aspects of the CPU architecture, that is, the programmer does not attend to scheduling of instructions. The compiler and assembly optimizer will manage the scheduling and optimization of the linear assembly code. In general, writing linear assembly is more tedious than writing C code but may produce slightly better performing code.

If the desired performance is still not achieved after linear assembly coding, the programmer has the option of hand coding and hand optimizing assembly code with no help from the compiler or assembly optimizer.

DSP Technical Overview

The CPU of the C6000 family of DSPs was developed concurrently with the C compiler. The CPU was developed in such a way that it was conducive to C programming, and the compiler was developed to maximally utilize the CPU resources. In order to understand how to optimize C code on a C6000 DSP, one needs to understand the building blocks of the CPU, the functional units, and the primary technique that the compiler uses to take advantage of the CPU architecture, software pipelining.

Functional Units

While thorough knowledge of the CPU and compiler is beneficial in writing fully optimized code, this requires at least a four-day workshop and a few weeks of hands-on experience. This section is no replacement for the myriad application notes and workshops on the subject (see Chapter 5.0 for references), but rather it is meant to provide enough information for the first round of optimization.

The CPU of the DA610 processor, in its most simplified form, looks like this:

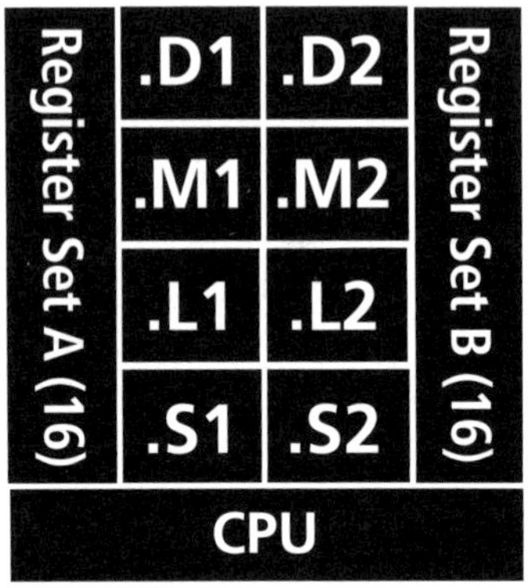

Figure B.2 C6000 CPU Architecture

Of course this diagram does not show all the memory buses and interconnects, but provides enough information for this discussion. The CPU has two sets of 4 distinct functional units and two sets of 16 (32-bit) registers. For the most part, each functional unit can execute a single instruction in a single cycle (there are a few exceptions). All functional units are independent, in other words, all functional units can operate in parallel. Therefore, the ideal utilization is all eight units executing on every cycle. (The memory architecture and CPU are organized such that eight instructions can be fetched and fed to the CPU on every cycle.)

The goal of the compiler is to reach ideal utilization of eight instructions per cycle. The goal of the programmer is to organize code and provide information to the compiler to help the compiler reach its goal. The programmer will achieve his/her goal by using appropriate compiler switches and structuring C code to exploit the powerful CPU. The Optimization Techniques section of this document will discuss such compiler switches and C coding techniques.

What "function" does each functional unit perform?

Below is a simplified breakdown of the function of each functional unit (each functional unit can execute many more instructions than listed here, see Fig. B.2).

Unit	Floating-point	Integer
M	Multiplies	Multiplies
L	Add/Conversion	Add/Compare
S	Compares/Reciprocal	Add/Shift
D	Load/Store	Add/Load/Store

Example:

Look at the simple example of a dot product.

```
Y = Σ ai * xi
```

In simple C code, we could write this as:

```
int dotp(short *a, short *m, int count)
{
                    for (i=0; i<count; i++)
                        sum += a[i] * m[i];
}
```

The DA610 has the ability to perform a sustained dual MAC operation on such a loop. For example, the pseudo-assembly code for such a loop could look like (In reality, the above C code would not generate the following assembly code as-is. This example is for illustrative purposes only.):

```
loop:  PIPED LOOP KERNEL
    LDDW    .D1   *a, A7:A6    ; double-word (64-bit) load of a[i] and a[i+1]
||  LDDW    .D2   *m, B7:B6    ; double-word (64-bit) load of m[i] and m[i+1]
||  MPYSP   .M1   A6, B6, A5   ; single-precision (32-bit) multiply of a[i]*m[i]
||  MPYSP   .M2   A7, B7, B5   ; single-precision (32-bit) multiply of
                                 a[i+1]*m[i+1]
||  ADDSP   .L1   A5, A8, A8   ; accumulate the even multiplies in register A8
||  ADDSP   .L2   B5, B8, B8   ; accumulate the odd multiplies in register B8
||  [A1] SUB.S1   A1, 1, A1    ; loop counter stored in A1, if (A1!=0), A1--
||  [A1] B  .S2   loop         ; if (A1!=0), branch back to beginning of loop
```

The horizontal bars (||) in the first column indicate that the instruction is executed in parallel with the instruction before. (In this case, all eight instructions are executed in parallel.) The second column contains the assembly mnemonic. The third column contains the functional unit used. The fourth column contains the operands for the instruction. The last column contains the comments preceded by a semi-colon (;). The square brackets ([]) indicate that the instruction is conditional on a register. For example, [A1] indicates that the instruction will only be executed if A1!=0. All instructions can be executed conditionally.

From this example, it can be seen that the CPU has the capability of executing two single-precision (32-bit) floating-point multiplies, two single-precision (32-bit) floating-point additions, and some additional overhead instructions on each cycle. The .D functional units each have the ability to load 64 bits, or the equivalent of two single-precision (32-bit) floating-point elements, each cycle. The .M units each can execute a single-precision (32-bit) multiply each cycle. The .L units can each execute a single-precision (32-bit) add each cycle. The .S units can perform other operations such as flow control and loop decrement to help sustain maximum throughput.

Figure B.3 shows a comprehensive list of the instructions performed by each functional unit.

.S Unit		
ADD	NEG	ABSSP
ADDK	NOT	ABSDP
ADD2	OR	CMPGTSP
AND	SET	CMPEQSP
B	SHL	CMPLTSP
CLR	SHR	CMPGTDP
EXT	SSHL	CMPEQDP
MV	SUB	CMPLTDP
MVC	SUB2	RCPSP
MVK	XOR	RCPDP
MVKH	ZERO	RSQRSP
		RSQRDP
		SPDP

.L Unit		
ABS	NOT	ADDSP
ADD	OR	ADDDP
ABD	SADD	SUBSP
CMPEQ	SAT	SUBDP
CMPGT	SSUB	INTSP
CMPLT	SUB	INTDP
LMBD	SUBC	SPINT
MV	XOR	DPINT
NEG	ZERO	SPRTUNC
NORM		DPTRUNC
		DPSP

.D Unit	
ADD	NEG
ADDAB (B/H/W)	STB (B/H/W)
LDB (B/H/W)	SUB
LDDW	SUBAB (B/H/W)
MV	ZERO

.M Unit		
MPY	SMPY	MPYSP
MPYH	SMPYH	MPYDP
MPYLH		MPYI
MPYHL		MPYID

NO UNIT USED		
	NOP	IDLE

Figure B.3 Assembly instructions associated with each functional unit.

From Figure B.3, it can be seen that each functional unit can perform many operations and in some cases the same instruction can be executed in more than one type of unit. This provides the compiler with flexibility to schedule instructions. For example, the .S, .L, and .D units can each perform an ADD operation. Practically, this means that the CPU can execute six ADD operations, as well as two MPY operations, each cycle.

Software Pipelining

In most traditional DSP code, the majority of the cycles are in looped code, that is, contained in for() loops. This is especially true in audio DSP code where the algorithms are inherently block-oriented. Software pipelining is a technique that the compiler uses to optimize looped code to try to achieve maximum utilization of the CPU's functional units. Software pipelining is enabled at the mid-level of compiler aggressiveness and higher.

Software pipelining is best illustrated by an example. (This example is not complete as it eliminates some concepts such as delay slots. It is meant to illustrate the concept of software pipelining only.)

How many cycles does it take to execute the following loop five times?

```
   LDW    ; load a 32 bit value
|| LDW    ; load a 32 bit value
   MPYSP  ; multiply two Single Precision (32-bit) floating point values
   ADDSP  ; add two Single Precision (32-bit) floating point values
```

The parallel vertical bars (||), at the beginning of line 2, indicate that the assembly instruction is executed in parallel with the instruction before. In this case, lines 1 and 2 are executed in parallel; in other words, two 32-bit values are loaded into the CPU on a single cycle.

The following table illustrates the execution of this code **without** software pipelining.

cycle	.D1	.D2	.M1	.M2	.L1	.L2	.S1	.S2
1	LDW	LDW						1 iteration = 3 cycles
2			MPYSP					
3					ADDSP			
4	LDW	LDW						
5			MPYSP					
6					ADDSP			
7	LDW	LDW						
8			MPYSP					

From the table, one iteration of the loop takes three cycles. Therefore five iterations of the loop will take 5 × 3 = **15 cycles**.

By using software pipelining, the compiler tries to maximize utilization of the functional units on each cycle. For example, in some cases, the result of an instruction will not appear on the cycle following the instruction. It may not appear for two or three cycles. These cycles spent waiting for the result of an instruction are known as *delay slots*. Since the functional unit is not occupied during these "delay slots," to maximize efficiency, the compiler will use these "delay slots" to execute other independent instructions.

Software pipelining can also be used if one part of an algorithm is not dependent on another part. For example, in looped code, if each iteration is independent of the previous, then the execution of the two iterations can overlap.

In this example, the delay slots are not shown, but we can still pipeline the code by overlapping execution of successive iterations. Since iteration 2 is not dependent on the completion of iteration 1, the CPU can start executing iteration 2 before iteration one is complete. Also, the CPU can start executing iteration 3 before iteration 2 and iteration 1 are complete, etc.

The following table illustrates the execution of this code **with** software pipelining.

cycle	.D1	.D2	.M1	.M2	.L1	.L2	.S1	prolog .S2
1	LDW	LDW						
2	LDW	LDW	MPYSP					kernel
3	LDW	LDW	MPYSP		ADDSP			
4	LDW	LDW	MPYSP		ADDSP			
5	LDW	LDW	MPYSP		ADDSP			epilog
6			MPYSP		ADDSP			
7					ADDSP			
8								

From the table, the total execution time for five iterations is 7 **cycles**. By using software pipelining, the execution time was reduced by more than 50%. Also note that the execution time for six iterations is 8 cycles, i.e., each additional iteration costs only 1 cycle. For a large number of iterations, the average cost per iteration approaches one cycle.

A few terms need to be defined to discuss software pipelined loops:

Prolog: The overhead needed before the kernel can fill the pipeline.

Kernel: The core of the loop. In the kernel, the maximum number of functional units is utilized in parallel and the maximum number of iterations is being executed in parallel. In this example, in the kernel, four instructions are executed in parallel (LDW, LDW, MPYSP, ADDSP), four functional units are utilized at once (.D1, .D2, .M1, .L1), and three iterations of the loop are being executed in parallel. The kernel is executed over and over for most of the duration of the loop.

The number of cycles needed to execute one iteration of the kernel is called the *iteration interval (ii)*. This term will appear again in the optimization section. In this example, each iteration of the kernel (LDW, LDW, MPYSP, ADDSP) executes in one cycle, so ii=1.

Epilog: The overhead at the end of a pipelined loop to empty out the pipeline.

To take advantage of the performance gains provided by software pipelining, the programmer needs to perform two important actions. First, software pipelining must be enabled through appropriate selection of compiler switches (explained in Optimization Techniques). Second, the C code within the for() loop needs to be structured such that the compiler is able to pipeline it. There are a few conditions that prevent certain code from being software pipelined. The programmer needs to make sure that the code is not "disqualified" from software pipelining (see section "Algorithm Specific Optimizations" for more information on software pipeline disqualification).

After the for() loop is able to be software pipelined, further optimization techniques will improve the performance of the software pipeline. This is also addressed in the section "Optimization Techniques."

3.0 Optimization Techniques

At this point, the first phase of the development cycle should be complete. The algorithm has been designed, implemented in C, and tested for correctness. The code should work as expected, although maybe not as fast as expected. We are now ready to optimize. Until this point, during the "getting the code to work" phase, the lowest level of aggressiveness should have been used, that is, the only compiler switches selected should be -g and –k (and –ml0 to indicate far data model, to be safe).

Low level of aggressiveness:

-g full debug information

-k keep generated assembly files

-ml0 far aggregate data memory model (to be safe)

("Far aggregate data memory" model indicates that the entire .bss section is bigger than 32 kbytes. That is, the total space for all static and global data is more than 32 kbytes. If it is known that the .bss section is smaller than 32 kbytes, then the "near" data memory model should be indicated. The "near" data model is indicated by not using –ml at all. The "near" model uses a slightly more efficient manner to load data than the "far" model. If the "near" model is selected and the .bss section is actually bigger than 32k, an error will be flagged at build time.

Now, the functions that need to be optimized should be identified. These are the functions that consume the most cycles, or functions that need to meet critical deadlines. Within these critical functions, the main for() loop, that is, the most cycle intensive for() loop, should be identified. (There are many sophisticated profiling methods enabled by Code Composer Studio but are beyond the scope of this document. Please see Code Composer Studio User's Guide, spru328 for more information on profiling.) Optimization effort should focus on this for() loop. Two things need to be done before optimization, a baseline cycle count needs to be obtained, and a simple test for correctness needs to be developed.

Get a baseline cycle count

Cycles will be the metric we use to measure how well optimized the code is. One can measure the cycle count of a particular loop without ever running code. This is a very crude way of measurement, but allows us a quick way of gauging optimization effort without the need for hardware or simulator.

Assume that the loop to be optimized is in function func() which is in a file foo.c. The compiler will produce a file (assuming –k option is selected) named foo.asm. In foo.asm, one can scroll down to _func to find the assembly version of func().

Under _func, in foo.asm, there will be a message such as:

```
;*----------------------------------------------------------------------------*
;*   SOFTWARE PIPELINE INFORMATION
;*      Disqualified loop: software pipelining disabled
;*----------------------------------------------------------------------------*
```

This is the beginning of the for() loop. Count the number of cycles between here and the comment:

```
; BRANCH OCCURS                                    ; |60|
```

which marks the end of the loop. Remember that parallel instructions count as one cycle, and NOP *x* counts as *x* cycles.

Many other sophisticated techniques exist in Code Composer Studio for profiling code. They are beyond the scope of this document.

Develop a test for "correctness"

When the mid and high levels of aggressiveness are used, the compiler will aggressively optimize the code using techniques such as removing unused functions, removing unused variables, and simplifying instructions. In some cases, improperly written code may be strictly interpreted by the compiler producing incorrect results. For example, a global variable may be tested in a loop in one file, but may be set elsewhere in the program (in an interrupt handler for example). When the compiler optimizes the loop, it will assume that the variable stays constant for the life of the loop and eliminate the read and test of that variable, unless the variable is marked with the "volatile" keyword.

To catch such problems, a sanity check—whatever test makes sense for the system under development—should be run after each round of optimization.

Compiler Optimizations

These optimizations are not algorithm specific. They can be applied equally to all code.

Tip: Set mid-level aggressiveness

To set mid-level aggressiveness, change the –g option to –gp, add the –o3 option, and retain the –k option. At this level of aggressiveness, the compiler attempts to **software pipeline** the loop. The –k option has nothing to do with compiler aggressiveness, it just tells the compiler to save the intermediate assembly files. We will need this for the "Software Pipeline Information" feedback.

Mid level aggressiveness

–gp Function Profile Debug

–o3 Opt. Level: File

–k keep generated .asm files

–ml0 far aggregate data memory model (to be safe)

(See the beginning of Optimizing Techniques for an explanation of –ml.)

Use a profiling method for measuring the optimization gain. The simple technique mentioned above, opening the .asm file and counting cycles, is acceptable. Compare this cycle count to the baseline count.

C Code Optimizations

At this point, looking in the assembly output of the compiler (e.g., foo.asm), we should see one of two possible comments at the beginning of the for() loop:

A

```
;*----------------------------------------------------------------------------*
;*   SOFTWARE PIPELINE INFORMATION
;*      Disqualified loop: ...
;*----------------------------------------------------------------------------*
```

OR

B

```
;*----------------------------------------------------------------------------*
;*   SOFTWARE PIPELINE INFORMATION
;*
;*      Loop source line                 : 41
;*      Loop opening brace source line   : 42
;*      Loop closing brace source line   : 58
;*      Known Minimum Trip Count         : 1
;*      Known Maximum Trip Count         : 65536
;*      Known Max Trip Count Factor      : 1
;*      Loop Carried Dependency Bound(^) : 14
;*      Unpartitioned Resource Bound     : 4
;*      Partitioned Resource Bound(*)    : 4
;*      Resource Partition:
;*                                A-side   B-side
;*      .L units                     3        3
;*      .S units                     0        1
;*      .D units                     2        3
;*      .M units                     1        1
;*      .X cross paths               2        2
;*      .T address paths             2        3
```

If A is observed, the for() was disqualified from software pipelining, i.e., the compiler did not even attempt to pipeline the loop. If B is observed, the for() loop was not disqualified from software pipelining, i.e., the compiler did attempt to pipeline the loop. It is not necessary to understand the contents of B at this point (this is explained below). It is enough to know that if B is observed, the loop was not disqualified. Without some prior knowledge of the TI C compiler, it is not uncommon for A to occur.

If A is observed, it means the compiler tried to software pipeline the loop in question, but some code structure prevented it from being pipelined. There are a few reasons why a loop would be disqualified from software pipelining (for more information, see spru187). Below are a few of the more common reasons.

Loop contains a call: If there is a function call in the loop, and the compiler cannot inline that function, the loop will be disqualified.

Bad loop structure: Examples of bad loop structure are:

- Assembly statement within the C loop using "asm()" (intrinsics are OK)
- GOTO statements within the loop
- `break` statements within the loop
- Nested `if()` statements
- Complex condition code requiring more than five condition registers.

Software pipelining disabled: Pipelining is disabled by a compiler switch. Pipelining is disabled if the **–mu** option is selected, if **–o2** or **–o3** is not selected, or if **–ms2** or **–ms3** is selected.

Too many instructions: Too many instructions in the loop to software pipeline. This is a problem because this code usually requires simultaneous use of more registers than are available. A workaround may be to divide the loop into consecutive smaller loops or use intrinsics (see explanation of intrinsics in the section Tip: Use intrinsics).

Uninitialized trip counter: The compiler could not conclusively identify the initialization instructions for the trip counter (loop counter).

Suppressed to prevent code expansion: If using the –ms1 option or higher, the compiler will be less aggressive about trading off code size for speed. In the cases where software pipelining increases code size with not much gain in speed, software pipelining is disabled to prevent code expansion. To always enable pipelining, use –ms0 option or do not use –ms option at all.

Loop carried dependence bound too large: A loop carried dependence is any dependence a later loop iteration has on an earlier iteration. The loop carried dependence bound is the minimum number of cycles between the start of iteration "n" of a loop and iteration "n+1" of a loop due to dependences between the two iterations.

If the compiler detects that a loop carried dependence is large enough that it cannot overlap multiple loop iterations in a software pipeline, then it will disqualify the loop from pipelining. Software pipelining cannot improve the performance of the loop due to the dependence.

The most likely (but not the only) cause of this problem is a "memory alias" dependence between a load instruction near the top of the loop and a store instruction near the bottom. Using the "restrict" keyword judiciously can help.

Cannot identify trip counter: This message indicates that the trip counter could not be identified or was used incorrectly in the loop body. The loop counter should not be modified in the body of the loop.

The labels above appear in the Software Pipeline Information comments of the assembly file. For example, if a loop was disqualified because the "Loop contains a call," then the comments in the assembly file would be:

```
;*----------------------------------------------------------------------------*
;*   SOFTWARE PIPELINE INFORMATION
;*      Disqualified loop: Loop Contains a Call
;*----------------------------------------------------------------------------*
```

Tip: Remove function calls from within a loop

Function calls from within for() loops are the most common reason for pipeline disqualification in audio DSP code, therefore this should be the first thing to look for when a loop is disqualified from software pipelining. Removing function calls may sound simple, but there may be a few tricky situations.

At first pass, remove all the obvious calls like printf()'s or calls to other user functions. One way to eliminate debug calls like printf()'s is to save the information into temporary variables and then print them out at a future time.

Another way of eliminating the prinf() calls is to guard them with an #ifdef or wrap them in a macro. For example, the following macro can be used:

```
#ifdef DEBUG
#  define dprintf(x) printf x
#else
#  define dprintf(x)
#endif
```

To use this macro, write your printf's as follows (note the double parentheses):

```
dprintf(("Iterations %2d: x=%.8X y=%.8X\n", i, x, y));
```

If the code is compiled with a –dDEBUG compiler switch, then the printf is enabled. Without it, the printf disappears.

To eliminate calls to other (possibly user-defined) functions, try and inline the called function into the calling function.

One nonobvious function call that may be in the loop is a call to the modulo operator, %. When writing audio DSP code, use of circular buffers is common. Programmers unfamiliar with TI DSPs may be tempted to use the modulo operator to perform circular buffer addressing. For example, to update the index of a circular buffer, one may write:

```
sample = circ_buffer[index];      /* access the circular buffer via index */

index = (index + step) % circ_buffer_size; /* update the pointer */
```

In most cases, this syntax will be perfectly acceptable. (Please see section, *TIP: Address circular buffers intelligently,* for more details on circular buffer addressing techniques.) However, in a few cases, this syntax will produce less efficient code. In particular, when the argument of the modulo operator, in this case 'circ_buffer_size', is variable, then the modulo operator (%) will trigger a call to the run-time support (RTS) library to perform the modulo operation. This function call will disqualify the loop from software pipelining. For suggestions on what to put in place of the modulo operator, please see the section, *TIP: Address circular buffers intelligently.*

Other nonobvious function calls are: division by a variable, structure assignment (such as doing "x=y" where x is a "struct foo"), and long->float or float->long type conversion.

After removing all the function calls, the loop should now be qualified for software pipelining. This can be verified by looking in the assembly output file for the comments such as B above. If the loop is still disqualified, check the other causes for pipeline disqualification listed above. If the code is successfully qualified, measure the cycle count again to see the performance improvement.

Tip: Address circular buffers intelligently

Use of circular buffers in audio DSP algorithms is quite common. Unfortunately, when programming in C on a DA6xx DSP, the circular buffer addressing hardware is not accessible to the programmer. (When programming in assembly, this circular buffer addressing hardware is available.) This forces the programmer to manually check the buffer index for "end of buffer" conditions and to wrap that pointer back to the beginning of the buffer. Although this results in a little extra work on the part of the programmer (an extra one or two lines of code vs. using circular addressing hardware), the good news is that when done carefully, this manual manipulation of the buffer index will achieve the same performance as circular addressing hardware in most cases.

A typical line of audio DSP code using circular addressing hardware would look something like this:

```
sum += circ_buffer[index += step];
```

As "index" is advanced by an amount of "step," the circular buffer addressing hardware would take care of wrapping "index" around back to the beginning of the buffer. Of course, prior to this line of code, the circular addressing hardware would need to be programmed with the buffer size.

On a TI DSP, to achieve the same functionality, the programmer would need an extra line or two of code. There are a number of different ways of writing the code to take care of wrapping the pointer back to the beginning of the circular buffer. Some are more efficient than others depending on the circumstances.

The most efficient way to write this code on a TI DSP would be:

```
sum += circ_buffer[index];
index = (index + step) % size;
```

"size" is the length of the circular buffer.

As can be seen, only one extra line of source code was added from the first case. There are a few restrictions to writing the code in this manner in order to gain maximum efficiency from the compiler.

– "size" must be a constant (cannot be variable).
– "size" must be a power of 2.
– "step" and "size" must both be positive integers with 'step' < "size"
– The compiler knows that the initial value for "index" is 0 <= "index" < "size."

The first two points are fairly straightforward, but the last point may need more explanation. It is very probable that the starting value for index will be in the range of 0 <= "index" < "size," but how to inform the compiler of this fact?

This is done through an intrinsic called _nassert(). More information on _nassert() can be found in the section "Tip: Use _nassert()." For now, it is enough to know that _nassert() is only used to provide information to the compiler and does not generate any code. In this case, we need to inform the compiler about the initial value of "index." At some point in the code, before "index" is used (in a loop) as in the example above, it must be initialized. At this point, _nassert() is used to inform the compiler of that initial value. The syntax would look like this:

```
int    index;
...
_nassert(startingIndex >= 0);     // inform compiler that initial value of
index is >= 0
indcx = startingIndex % size;     // set 'index' to its initial value
...
for (i=0; i<loop_count; i++) {
...
       sum += circ_buffer[index];
       index = (index + step) % size;
...
}
```

Of course, this code is most efficient if the conditions are met. In some cases, the conditions may not be met, especially the first two. There may be a case where 'size' is not a constant or "size" is not a power of 2. In this case, use of the modulo operator (%) is not recommended. In fact, if these two conditions are not met, the modulo operator will actually trigger a call to the run-time support (RTS) library, thereby disqualifying the loop from software pipelining (see section "Tip: Remove function calls from within a "loop" for more information). In this case, an alternative is needed for updating the circular buffer index.

If the modulo operator (%) cannot be used because the conditions are not met, then the following code should be used to update the circular buffer index:

```
sum += circ_buffer[index];
index+=step;
if (index >= size)
index -= size;
```

Although this code may seem inefficient at first glance, it is only three extra lines of C source code compared to the case where circular buffer addressing hardware is used. In those extra three lines of code, the operations performed are 1 ADD, 1 COMPARE, and 1 SUB. Recall from the section "Functional Units" that the ADD and SUB instructions can each be executed on any one of six different functional units (L, S, or D units on either side of CPU). Also, the COMPARE can be done on any one of two units (L units on either side of CPU). Thus, these three extra instructions can easily be executed in parallel with other instructions in the for() loop including any MPY instructions.

The net result is that updating the index of the circular buffer manually adds none to very little overhead vs. using circular addressing hardware. This is true despite the fact that a few more lines of source code are required.

Tip: More C code does not always produce less efficient assembly code

Since the CPU functional units are independent, when code is broken down into atomic operations, the compiler has maximum flexibility as to when to schedule an operation and on which functional unit. In the example above, a complex operation such as modulo with a variable argument could be reduced to three atomic operations, add, compare, and zero. This allowed the compiler to produce much more efficient code.

At this point, we need to learn a little bit about how to read the **Software Pipeline Information**. *Here is an example:*

```
    ;*----------------------------------------------------------------------------*
    ;*   SOFTWARE PIPELINE INFORMATION
    ;*
1   ;*      Loop source line                     : 51
2   ;*      Loop opening brace source line       : 52
3   ;*      Loop closing brace source line       : 70
4   ;*      Known Minimum Trip Count             : 1
5   ;*      Known Maximum Trip Count             : 65536
6   ;*      Known Max Trip Count Factor          : 1
7   ;*      Loop Carried Dependency Bound(^)     : 17
8   ;*      Unpartitioned Resource Bound         : 4
9   ;*      Partitioned Resource Bound(*)        : 4
10  ;*      Resource Partition:
    ;*                                A-side   B-side
11  ;*      .L units                     3        3
12  ;*      .S units                     0        1
```

```
13  ;*      .D units                     3        3
14  ;*      .M units                     2        0
15  ;*      .X cross paths               1        1
16  ;*      .T address paths             3        3
17  ;*      Long read paths              1        1
18  ;*      Long write paths             0        0
19  ;*      Logical  ops (.LS)           0        0     (.L or .S unit)
20  ;*      Addition ops (.LSD)          6        1     (.L or .S or .D unit)
21  ;*      Bound(.L .S .LS)             2        2
22  ;*      Bound(.L .S .D .LS .LSD)     4*       3
23  ;*
24  ;*      Searching for software pipeline schedule at ...
25  ;*         ii = 17 Schedule found with 3 iterations in parallel
26  ;*      done
27  ;*
28  ;*      Collapsed epilog stages     : 2
29  ;*      Prolog not entirely removed
30  ;*      Collapsed prolog stages     : 1
31  ;*
32  ;*      Minimum required memory pad : 0 bytes
33  ;*
34  ;*      For further improvement on this loop, try option -mh2
35  ;*
36  ;*      Minimum safe trip count     : 1
37  ;*----------------------------------------------------------------------------*
```

Lines 1–3: *Information on where the C code is located.*

Lines 4–6: *Information compiler has on number of times the loop is executed. This is important because the more information the compiler has on the loop count (trip count), the more aggressively it can perform optimizations such as loop unrolling. In this example, the compiler has no information on the loop count other than it will run at least once and at most 2^16 times.*

If some information on the loop counter for any particular loop is known by the programmer, but not by the compiler, it is useful for the programmer to feed this information to the compiler. This can be done through the use of a `#pragma`*. A* `#pragma` *is only used to provide information to the compiler. It does not produce any code. (For more information, see Tip: Use #pragma.)*

Note that when the compiler is set to the highest level of optimization, it can usually figure out this loop count information on its own, without the use of a pragma. By examining lines 4-6, the programmer can determine if the compiler has enough information about the loop as it possibly could have. If not, then using a pragma is appropriate.

Line 7: *This line measures the longest data dependence between consecutive loop iterations. This is important because the kernel of the loop could never be less than this number. The loop carried dependence constrains how aggressively multiple loop iterations can be overlapped due to dependences between the consecutive iterations. Instructions that are part of the loop carried bound are marked with the ^ symbol in the assembly file.*

Lines 8–9: *These lines indicate a limit (bound) on the minimum size of the kernel due to resource (functional units, data paths, etc.) constraints. "Unpartitioned"'indicates the total resource constraint, before the resources are assigned to the A side or B side of the CPU. 'Partitioned' indicates the resource constraint after the resources have been allocated to either the A or B side. This is important because the cycle count of the kernel of the loop can never be less than the larger of lines 8 and 9.*

By comparing lines 8–9 (resource limitation) to line 7 (loop carried dependency limit), it can be quickly determined if the reduction in kernel size is limited by the resources used or the data dependencies in the loop. The bigger of the two is the limiting item. This helps guide further optimization efforts to either reduce resource usage or eliminate data dependencies.

Lines 10–22: *These lines are a further elaboration on lines 8–9. Lines 10–22 show the actual usage for each resource on each iteration of the kernel. These lines provide a quick summary of which resources are used the most (thus causing a resource bound/limit) and which are under-utilized. If the kernel is resource constrained, optimization can be done to try and move operations from over-used resources to less used resources. The '*' indicates the most-used resource, that is, the limiting resource.*

.LS refers to operations that can be executed on either the L or S unit. Similarly, .LSD refers to operations that can be executed on each of the L, S, or D units.

"Bound (.L .S .LS)" is the resource bound value as determined by the number of instructions that use the L and S units. Bound(.L .S .LS) = ceil((.L + .S + .LS)/2).

"Bound (.L .S .D .LS .LSD)" is the resource bound value as determined by the number of instructions that use the L, S, and D units. Bound(.L .S .D .LS .LSD) = ceil((.L + .S + .D + .LS + .LSD)/3).

This information is useful if the iteration interval (ii) is limited by resources (not dependences). If this is the case, one should try to re-write code to use the unused resources and relieve the heavily used resources.

Line 25: ii is iteration interval. *This is the length of the kernel in number of cycles. The kernel is the part of the loop that repeats over and over for the duration of the loop counter. It is where the bulk of the processing occurs.*

In some cases, depending on the information provided to the compiler about the loop, the compiler may unroll the loop. When the loop is unrolled, one iteration of the kernel will execute multiple iterations of the original loop. For example, if the compiler unrolls the loop by a factor of 4, then each iteration of the kernel will execute 4 iterations of the original loop.

If a loop was unrolled by the compiler, a comment will appear within the Software Pipeline Information that looks like:

```
;*       Loop Unroll Mutliple      : 4x
```

*For large loop counts, one can calculate the approximate cycle time for one iteration of the original source loop by dividing ii by the loop unroll multiple. (In the case where the loop was not unrolled, the loop unroll multiple is 1.) So, ii/(loop unroll multiple) will approximate the number of cycles for each iteration of the original source loop, that is, the total cycle count of the for() loop will be approximately loop count * (ii/loop unroll multiple).*

The goal in optimizing the loop is to make the kernel as small as possible, that is, **make ii/loop unroll multiple as small as possible**. *By measuring ii after each optimization step, we will be able to gauge how well optimized the loop is.*

Schedule found with 3 iterations in parallel means that on any iteration of the loop, before iteration n is completed, iteration n+1 and n+2 will have started.

Lines 28–30: *As much as possible, the compiler will try to optimize for both code size and execution speed. To this end, the compiler can sometimes remove the prolog and/or epilog of the loop by executing the kernel more times (for prolog/epilog definition, see section on Software Pipelining). These lines indicate how successful the compiler was at trying to eliminate the prolog and epilog sections. No actions by the programmer are necessary based on information from these lines.*

Line 32: *Sometimes the compiler can reduce code size by eliminating the prolog and/or epilog and running the kernel more times. For example, if the epilog were removed, the kernel may run one or more extra iterations, but some of the results from the extra iterations may not be used. An extra memory pad may be required on the data buffer, both before and after the buffer, to guard against these "out-of-bounds" accesses. The data in the pad region will not be modified. This line in the pipeline feedback indicates how much of a memory pad is required. This technique is known as "speculative load."*

Note that on the DA610, the overflow part of the buffer (the memory pad) may not reside in L2 Cache. Incorrect behavior may result if it does. Buffers for loops in which the speculative load technique is used may only reside in L2SRAM and must be at least "memory pad" bytes away from the end of L2SRAM (and the beginning of L2 Cache). (See spru609 for more details on memory architecture.)

The "speculative load threshold" value is a measure of how aggressively the compiler uses this technique to collapse code. This speculative load threshold value can be adjusted with the –mh<n> compiler option. If a memory pad were suggested for the loop, a second line of comment would have been added to inform the programmer of the minimum required threshold value to attain that level of collapse.

Line 34: *This line provides an optimization suggestion in the form of a higher threshold value that may be used to cause even more collapsing of the loop.*

Line 36: *The minimum number of times the loop needs to run, that is, the minimum loop count, to guarantee correct results.*

Tip: Mind your pointers

Inform the compiler as much as possible about the nature of pointers/arrays.

One easy C code optimization that can be done in accordance with this tip is to include the size of an array in a function declaration that uses that array as an argument, if the array size is known. For example, if it is known that the size of an array is always 256 elements, then a function declaration using this array would look like:

```
void function(int array[256]);
```

This helps the compiler improve its analysis of the function. In general, the compiler prefers to see array declarations, particularly when the size of the array is known.

Another C optimization in accordance with this tip occurs if there are two pointers passed into the same function. The compiler can better optimize if it knows if the two pointers could point to the same memory location or not.

Here is an example prototype for an audio algorithm:

```
void delay(short *inPtr, short *outPtr, short count)
{...}
```

The function is passed a pointer to the input buffer (inPtr) and a pointer to the output buffer (outPtr). This is common in audio algorithms. The algorithm will read data from the input buffer, process it, and write data to the output buffer.

With no additional information, the compiler must be conservative and assume that inPtr and outPtr access the same buffer, in other words, a write by one pointer can affect a read by the other pointer. The 'restrict' keyword can be used as a type qualifier on a pointer to guarantee to the compiler that, within the scope of the pointer declaration, the object pointed to can only be accessed by that pointer. This helps the compiler determine memory dependencies hence more aggressively optimize. The 'restrict' keyword can also be used with arrays.

In this case, using the 'restrict' keyword provides a guarantee to the compiler that inPtr and outPtr do not point to overlapping memory regions. Rewriting the above function with the 'restrict' keyword helps the compiler produce more optimized code:

```
void delay(short* restrict inPtr, short* restrict outPtr, short count)
{...}
```

Another problem that may occur with pointers is a little more subtle, but still common. What happens if the pointers are pointing to overlapped spaces, that is, two pointers do point to the same buffer? This is common in audio algorithms when using a delay buffer. One pointer may be reading from one part of the delay buffer while another pointer may be storing a value to another part of the buffer. For example,

```
for (i=0; i<loop_count; i++)
{
...
```

```
delay_buffer[current] = constant * delay_buffer[old];
out[i] = in[i] + (gain * delay_buffer[old]);
...
}
```

From looking at the code, `delay_buffer[old]` could be read once and used in both lines of code. From our knowledge of the CPU, we would assume that `delay_buffer[current]` and `out[i]` could be calculated in parallel.

However, in this case, `delay_buffer[current]` and `delay_buffer[old]` physically exist the same buffer. Without any other information, the compiler has to be conservative and assume that there is a possibility of `current==old`. In this case, the code could look like:

```
delay_buffer[old] = constant * delay_buffer[old];
out[i] = in[i] + (gain * delay_buffer[old]);
```

The result of the first line is used in the second line, which means that the second line can't start executing until the first line finishes. If we can guarantee that `current!=old` always, then we need to specify this to the compiler so that it can be more aggressive and execute the two lines in parallel.

This is where the subtlety comes into play. If delay_buffer[old] and delay_buffer[current] never point to the same location for the entire duration of the function (ideal situation), then they should be passed into the function as two distinct pointers with one of them specified as "restrict."

For example, if loop_count=100, and we know that (old > current+100), then delay_buffer[old] and delay_buffer[current] will never point to the same location fo the duration of loop_count. In this case, the declaration of the function should be:

```
void delay(int* restrict oldPtr, int* currentPtr, int count)
```

and the function should be called using the following:

```
delay(&(delay_buffer[old], &(delay_buffer[current], 100)
```

This will inform the compiler that the two pointers never point to the same memory location for the duration of the loop.

The second scenario (nonideal) is that we cannot guarantee that delay_buffer[old] and delay_buffer[current] will not point to the same memory location for the duration of the loop. However, if we can guarantee that they will not point to the same memory location for each iteration of the loop, then we can still optimize. In this situation, we do not modify function is defined, but we do modify the code slightly within the loop.

If we know that `current!=old` always for each iteration of the loop, then we could re-write the code this way:

```
float temp_delay;  //define a temp variable for unchanging delay value

for (i=0; …
{
…
temp_delay = delay_buffer[old];

delay_buffer[current] = feedback * temp_delay;
out[i] = in[i] + (gain * temp_delay);
…
}
```

Now the compiler will produce code to execute the two lines in parallel.

***NOTE: If a situation arises where* `current==old`*, then this code will produce incorrect results!* **This is not checked by the compiler.** *The programmer must guarantee that this scenario is not possible.*

Tip: Call a float a float

Floating-point constants are treated as double-precision (64-bit) unless specified as single-precision (32-bit). For example, 3.0 would be treated as a 64-bit constant while 3.0f or (float)3.0 would be treated as single-precision (32-bit).

One of the benefits of the DA610 is being able to write floating-point code. However, when using constants, one has to know how the compiler will interpret the number. In the C6000 family, the following names are used for floating-point data types:

float – (IEEE) 32-bit floating-point
double – (IEEE) 64-bit floating-point

Care needs to be taken when using floating-point constants. Look at the following lines of code.

```
#define M 4.0

float y,x;
x=3.0;
y= M * x;
```

In this example, the code has specified to the compiler that x and y are to be single precision (32-bit) floating-point values (by defining them as float), but what about M? All that is indicated to the compiler is that it is to be floating-point (by virtue of the decimal point), but we haven't specified single-precision (32-bit) or double-precision (64-bit). As usual, without any information, the compiler will be conservative and assume double-precision.

The resulting assembly code from the above code will be:

```
ZERO       .D1    A0              ;
||MVKH     .S1    0x40100000,A1 ; put constant (4.0) into register A1
                                ; A1:A0 form DP (64-bit) constant (4.0)
MVKH       .S1    0x40400000,B4 ; put x (3.0) into register B4
SPDP       .S2    B4,B5:B4      ; convert x to DP (64-bit)
MPYDP      .M1X   B5:B4,A1:A0,A1:A0 ; DP multiply of x (B5:B4) and M (A1:A0)
NOP             9               ; wait 9 delay slots for the multiply
DPSP       .L1    A1:A0,A0      ; convert the DP (64-bit) result (y) to SP
                                (32-bit)
```

This is more code than what would be expected for a 32 × 32 floating-point multiply. The problem is that even though a 32 × 32 multiply was intended, we did not inform the compiler of this. To correct the situation, indicate to the compiler that the constant is single-precision. There are two ways to do this:

i) Place an "f" after the constant, such as:

```
#define M 4.0f
```

ii) Typecast the constant as single-precision:

```
#define M (float)4.0
```

Changing the above code to:

```
#define M 4.0f // or (float)4.0

float y,x;
x=3.0f;
y= M * x;
```

produces the following assembly output:

```
MVKH    .S1     0x40400000,B4      ; move x (3.0) into register B4
MVKH    .S1     0x40800000,A0      ; move M (4.0) into register A0
MPYSP   .M2X    B4,A0,B4           ; (single-precision) multiply of M*x
```

Making sure to specify constants as 'float', when single precision floating-point is desired, will save both code space and cycles.

In general, the programmer always has to be mindful of data types. Mixing floating-point and fixed-point data types in a calculation when not required will cause unnecessary overhead.

Example:

```
int a;
int b;
#define C 1.0
b = C * a;
```

In this case, the compiler will treat C as a floating-point value but a and b are declared as fixed-point. This forces the compiler to add instructions to convert C from floating-point to fixed-point before doing the addition. Instead, the code should read:

```
int a;
int b;
#define C 1
b= C * a;
```

The resulting compiled code will be more efficient.

Tip: Be careful with data types

The following labels are interpreted as follows by the C6000 compiler. Be careful to assign the types appropriately. Each of the following data types can be signed or unsigned.

char	8-bit fixed-point
short	16-bit fixed-point
int	32-bit fixed-point
long	40-bit fixed-point
float	32-bit floating-point
double	64-bit floating-point

Do not confuse the *int* and *long* types. The C6000 compiler uses *long* values for 40-bit operations. This may cause the generation of extra instructions, limit functional unit selection, and use extra registers.

Use the *short* data type for fixed-point multiplication inputs whenever possible because this data type provides the most efficient use of the 16-bit multiplier in the C6000 (1 cycle for "*short* * *short*" vs. 5 cycles for "*int* * *int*").

Use *int* type for loop counters, rather than *short* data type, to avoid unnecessary sign-extension instructions.

Do not use unsigned integer types for loop counters or array subscript variables. i.e., use *int* instead of *unsigned int*.

When using floating–point instructions on a floating-point device, such as the C6700, use the -mv6700 compiler switch so the code generated will use the device's floating-point hardware instead of performing the task with fixed-point hardware. For example, if the -mv6700 option is not used the RTS floating–point multiply function will be called instead of the MPYSP instruction. Specifically, the flag "–mv6710" should be used for DA6xx and C671x devices and "-mv6210" should be used for C621x devices.

Tip: Limit type conversions between fixed and floating-point

Any conversions from fixed-point to floating-point (e.g., short->float or int->float) or from floating-point to fixed-point (float->short or float->int) use the valuable L unit in the CPU. This is the same hardware unit that performs floating-point ADDs, for instance. Many type conversions along with many floating-point ADDs will overstress the L unit.

If a variable needs to be converted from int to float, for example, and then will be used many times, convert the variable once and store in a temporary variable rather than converting on each usage.

Assume that part of the loop looks like the following:

```
void delay(short *in, short *out, int loop_count)

float feedback, gain, delay_value;
...
for (i=0; i<loop_count; i++)
{...
buffer[current] = (float)in[i] + feedback * delay_value ;
out[i] = (short)( (float)in[i] + ( gain * delay_value) );
...
}
```

In this case, the arrays in[] and out[] both contain fixed-point values. Thus, in[i] needs to be converted to floating-point before being added to (floating-point) delay_value. This needs to be done twice, once for each ADD. It would be more efficient to do the conversion from fixed to floating-point once, then use that converted in value in both calculations. In other words, the more efficient way to code this would be:

```
void delay(short *in, short *out, unsigned short loop_count)

float feedback, gain, delay_value;
float temp_in;     //temporary variable to store in[i] after conversion
to floating point
...
for (i=0; i<loop_count; i++)
{...
temp_in = (float)in[i];//do the conversion once only

buffer[current] = temp_in + feedback * delay_value ;
out[i] = (short)( temp_in + ( gain * delay_value) );
...
}
```

Tip: Do not access global variables from within a for() loop

This is a general tip on C coding style that should be incorporated into the programmer's programming style. If global variables need to be used in a for() loop, copy the variables into local variables before the loop and use the local variables in the loop. The values can be copied back to the global variables after the loop, if the values have changed. The assembler will always load global variables from the data pointer area, resulting in less efficient code.

Tip: Use #pragma

In some cases the programmer may have knowledge of certain variables that the compiler cannot determine on its own. In these cases, the programmer can feed information to the compiler through the use of #pragma. #pragma does not produce any code, it is simply a directive to the compiler to be used at compile time.

For example, for looped code, often the loop counter is passed in as a variable to the function that contains the looped code. In this case, the compiler may have difficulty determining any characteristics about the loop counter. The programmer can feed information about the loop counter to the compiler through the use of the MUST_ITERATE pragma.

The syntax for the pragma is:

```
#pragma MUST_ITERATE(min, max, factor)
```

`min` – minimum number of iterations possible for the loop to execute

`max` – maximum number of iterations possible for the loop to execute

`factor` – known factor for the loop counter, for example, if the loop will always run an even number of times, then factor could be set to 2.

It is not necessary to have all three pieces of information to use this pragma. Any of the arguments can be omitted. For example, if all that is known about a particular loop is that any time it runs, it will run for at least 4 iterations and it will always run an even number of iterations, then we can use the pragma in the following way:

```
#pragma MUST_ITERATE(4, , 2)
```

This line says that the loop will run at least 4 times, and it will run an even number of times (i.e., 4, 6, 8, …). This line should be inserted immediately preceding the `for()` loop in the C source file.

Another use of #pragma is to direct the compiler to compile the code in a specific way. One of the more useful pragmas for looped code is the UNROLL pragma. This pragma specifies to the compiler how many times a loop should be unrolled. It is useful since the resources of the CPU can be spread over multiple interations of the loop, thereby utilizing the resources more efficiently.

For example, say that the original source code requires three multiplies for every iteration of the loop. Since there are two multiply hardware units, two cycles per iteration will be required to execute the three multiplies. If the loop were unrolled such that two iterations of the original source code were executed at once, we would be required to execute six multiplies per iteration. Since there are two multiply hardware units, three cycles would be required to execute the six multiplies. In other words, three cycles are required to execute two iterations of the original source code for an average of 1.5 cycles per iteration. By unrolling the loop by a factor of two, we have reduced the execution time from two cycles per iteration to 1.5 cycles per iteration.

A few things need to be considered when using this pragma. It will only work if the optimizer (–o1, –o2, or –o3) is invoked. The compiler has the option of ignoring this pragma. No statements (but other pragmas are ok) should come between this pragma and the beginning of the for() loop.

The syntax for the pragma is:

```
#pragma UNROLL(n)
```

which indicates to the compiler that the loop should be unrolled by a factor of n. In other words, one iteration of the kernel should execute n iterations of the original loop.

To increase the chances that the loop is unrolled, the compiler needs to know the smallest possible number of iterations of the loop, the largest possible number of iterations of the loop, and that the loop will iterate a multiple of x times. The compiler may be able to figure out this information on its own, but the programmer can pass this information to the compiler through the use of the MUST_ITERATE pragma.

```
#pragma MUST_ITERATE(32, 256, 4)
#pragma UNROLL(4)

for (i=0; i < loop_count; i++) …
```

These lines indicate that the loop will run at least 32 times and at most 256 times and that loop_count will be a multiple of four (32, 36, 40, …). This also indicates to the compiler that we want four iterations of the loop to run for every one iteration of the compiled loop, that is, we want the loop to be unrolled by a factor of 4.

Tip: Use _nassert()

The _nassert() intrinsic generates no code and tells the optimizer that the statement contained in its argument is true. This gives a hint to the optimizer as to what optimizations are valid.

A common use for _nassert() is to guarantee to the optimizer that a pointer or array has a certain alignment. The syntax for giving this pointer alignment is:

```
_nassert((int) pointer % alignment == 0);
```

For example, to inform the compiler that a pointer is aligned along an 8-byte boundary ("double-word aligned"):

```
_nassert((int) buffer_ptr % 8 == 0);
```

This intrinsic along with the MUST_ITERATE pragma gives the compiler a lot of knowledge to use when optimizing the loop.

In the use of _nassert(), one question may arise: How do we know that a pointer is aligned? The answer is that we can force the compiler to align data on certain boundaries by using the DATA_ALIGN pragma. The syntax for this pragma is:

```
#pragma DATA_ALIGN(symbol, constant);
```

In this case, "symbol" is the name of the array to be aligned and "constant" is the byte boundary to which it will be aligned. For example, to be aligned on a 16-bit (short or two-byte) boundary, we would use a constant of 2. This would guarantee that the first address of the array will be a multiple of 2 (in other words, the compiler will place the array such that the starting address is 0x0, or 0x2, or 0x4, ...). To be aligned on a 64-bit (i.e., double-word or eight-byte) boundary, we would use a constant of 8. This would guarantee that the first address of the array will be a multiple of 8 (i.e., the compiler will place the array such that the starting address is 0x0 or 0x8 or 0x10 or 0x18, ...). This pragma should be inserted directly before the declaration of the array. For example:

```
#pragma DATA_ALIGN(buffer, 8);
float buffer[256];
```

This will declare a buffer of size 256 floats (1024 bytes), and align this buffer to an 8-byte boundary. Now the _nassert intrinsic can be used to inform the compiler that 'buffer' is on a double-word boundary.

Tip: Get more feedback

Using –on2, the compiler produces an .nfo file. This file provides more information on advanced optimizations for getting better C performance. The suggestions will usually be in the form of different compiler switches, or additional compiler directives. If compiler directives are suggested, the .nfo file will show what code to include and where to include it in the C source file.

Other System Level Optimizations

There are some optimizations that are not necessarily for a specific loop but can help increase overall DSP performance.

Tip: Use Fast run-time support (RTS) library

The fast run-time support (RTS) library is a set of math functions provided by TI and optimized to run on TI's floating-point DSPs. The FastRTS library implements functions such as arctangent, cos, sin, exponent (base e, 10, and 2), log (base e, 10, and 2), power, reciprocal, divide, and square root. The compiler will automatically make calls to the FastRTS library when the function is used in your code.

In order to use FastRTS, the library must be explicitly added to the project. With the project open in CCS, go to the Project menu, select Add Files to Project..., navigate to fastrts67x.lib and click Add. The library will generally be installed at: C:\ti\c6000\cgtools\lib.

(For information on downloading FastRTS library, please see http://dspvillage.ti.com/ and navigate to C6000 software libraries.)

Tip: Use floating-point DSP library (DSPLIB)

Another library that will provide better system performance is the 67x-optimized DSP libray (DSPLIB). At the writing of this app note, the library is still being developed. (Check the same website as for the FastRTS library.) This dsp library will implement floating-point optimized code for common DSP functions such as FIR, IIR, FFT and others.

Tip: Use intrinsics

The compiler provides special functions that can be called from C code but are directly mapped to assembly instructions. All functions that are not easily supported in C are supplied as intrinsics. Intrinsics provide a way to quickly and easily optimize C code.

Example:

This code implements a saturated addition function. The resultant compiled code will require multiple cycles to execute.

```
int sadd(int a, int b)
{
  int result;
  result = a + b;
  if (((a ^ b) & 0x80000000) == 0)
  {
    if ((result ^ a) & 0x80000000)
    {
      result = (a < 0) ? 0x80000000 : 0x7fffffff;
    }
  }
  return (result);
```

Using intrinsics, this code can be replaced by:

```
result = _sadd(a,b);
```

This will compile to a single SADD instruction. The hardware can support up to two SADD instructions per cycle.

Final Compiler Optimization

Tip: Use highest level of aggressiveness

Once the code is near the "ship" stage, there is one last optimization that can be applied. The compiler should be cranked up to its highest level of aggressiveness.

Unfortunately, the reason this level provides more optimization is the same reason why it is difficult to profile. When using the highest level of aggressiveness, all debug and profile information is removed from the code. A more system-level approach needs to be adopted for profiling. Instead of profiling on a loop-by-loop basis, or function-by-function basis, you may profile on an interrupt-to-interrupt basis or input to output. Similarly, to test the final code, a system-level approach is required. Instead of stepping through a loop or a function, one would have to capture an output bitstream from a reference input, and compare it to a reference output, for example.

Set the highest level of aggressiveness by removing –gp, keeping –o3, and adding –pm and –op0.

High level aggressiveness

–o3: File level optimization

–pm –op0: Program level optimization

–ml0 far calls/data memory model (to be safe)

The total compiler switches should be: –o3 –pm –op0 –ml0. A switch other than –op0 can be used for more optimization, but –op0 is a safe place to start.

Using –pm and –o3 enables program-wide optimization. All of the source files are compiled into a single intermediate file, a *module*, before passing to the optimization stage. This provides the optimizer more information than it usually has when compiling on a file-by-file basis. Some of the program level optimizations performed include:

- If a function has an argument that is always the same value, the compiler will hard code the argument within the function instead of generating the value and passing it in as an argument every place that it appears in the original source.
- If the return value of a function is never used, the compiler will delete the return portion of the code in that function.
- If a function is not called, either directly or indirectly, the compiler will remove that function.

NOTE: If a file has **file-specific compiler options** *different from the rest of the files in the project, it will be* **compiled separately**, *i.e., it will not participate in the program level optimization.*

The –op option is used to control the program level optimization. It is used to indicate to what extent functions in other modules can call a module's external functions or access a module's external variables. The –op0 switch is the safest. It indicates that another module will access both external functions and external variables declared in this module.

The amount of performance gained from this step depends on how well optimized your C code was before you turned on the highest level of aggressiveness. If the code was already very well optimized, then you will see little improvement from this step.

Certain building blocks appear often in audio processing algorithms. We will examine three common ones and the effect of optimization on each.

Delay

A delay block can be used in effects such as Delay, Chorus, or Phasing. Here is an example implementation of a Delay algorithm.

```
void delay(delay_context *delHandle, short *in, short *out,unsigned short Ns)
{
        short i;
        int delayed,current,length;
        float *buffer = delHandle->delBuf;
        float g = delHandle->gain;
        float fb = delHandle->feedback;

        current = delHandle->curSamp;
        length = delHandle->sampDelay;
        for(i=0; i<Ns; i++)
        {
                delayed = (current+1)%length;

                buffer[current] = (float)in[i] + fb * buffer[delayed] ;
                out[i] = (short)( (float)in[i] + ( g * buffer[delayed]) );

                current = delayed;
        }
        delHandle->curSamp = current;

}
```

Assuming that the code is "correct," the first tip, set mid-level aggressiveness, can be applied to get a baseline cycle count for this code.

Tip: Set mid-level aggressiveness

The above code was compiled (using codegen v4.20, the codegen tools the come with CCS 2.1) with the following options (mid-level aggressiveness):

–gp –k –o3

Here is the assembly output:

```
;*----------------------------------------------------------------------------*
;*   SOFTWARE PIPELINE INFORMATION
;*      Disqualified loop: loop contains a call
```

```
;*----------------------------------------------------------------------------*
L1:
       .line  12
           B        .S1      __remi              ; |66|
           MVKL     .S2      RL0,B3              ; |66|
           MVKH     .S2      RL0,B3              ; |66|
           ADD      .D1      1,A3,A4
           NOP               1
           MV       .D2      B7,B4               ; |66|
RL0:       ; CALL OCCURS                         ; |66|

           LDH      .D2T2    *B5,B4              ; |66|
||         LDW      .D1T1    *+A7[A4],A5         ; |66|

           SUB      .D1      A0,1,A1
           SUB      .S1      A0,1,A0
           NOP               2

           MPYSP    .M1      A8,A5,A5            ; |66|
||         INTSP    .L2      B4,B4               ; |66|

           NOP               3
           ADDSP    .L1X     A5,B4,A5            ; |66|
           LDH      .D2T2    *B5++,B4            ; |67|
           NOP               2
           STW      .D1T1    A5,*+A7[A3]         ; |66|
           LDW      .D1T1    *+A7[A4],A3         ; |67|
           NOP               4

           INTSP    .L2      B4,B4               ; |67|
||         MPYSP    .M1      A9,A3,A3            ; |67|

           NOP               3
           ADDSP    .L2X     A3,B4,B4            ; |67|
           MV       .D1      A4,A3               ; |69|
           NOP               1
   [ A1]   B        .S1      L1                  ; |70|
           SPTRUNC  .L2      B4,B4               ; |67|
           NOP               3
       .line  19
           STH      .D2T2    B4,*B6++            ; |67|
           ; BRANCH OCCURS                       ; |70|
;** --------------------------------------------------------------------------*
           LDW      .D2T2    *++SP(8),B3         ; |73|
           MVKL     .S1      _delay1+12,A0       ; |71|
           MVKH     .S1      _delay1+12,A0       ; |71|
           STW      .D1T1    A3,*A0              ; |71|
           NOP               1
           B        .S2      B3                  ; |73|
       .line  22
           NOP               5
           ; BRANCH OCCURS                       ; |73|
       .endfunc      73,000080000h,8
```

There are two things that are immediately obvious from looking at this code. First, the Software Pipeline Feedback indicates that the code was disqualified from pipelining because the loop contained a function call. From looking at the C code, we see that a modulo operator was used which triggers a function call (`remi()`) to the run-time support (RTS) library. Second, there are almost no parallel bars (||) in the first column of the assembly code. This indicates a very low utilization of the functional units.

By hand-counting the instructions in the kernel of the loop (in the .asm file), we determine that the kernel takes approximately **38 cycles,** not including calls to the run-time support library.

Based on the tips in this app note, the code was optimized. These are the tips that were used and why.

Tip: Remove function calls from with a loop

Tip: More C code does not always produce less efficient assembly code

The first optimization is to remove the function call from within the loop that is causing the pipeline disqualification. The modulo operator on the update to the buffer index 'delayed' is causing the function call. Based on the two tips above, the index 'delayed' was updated by incrementing the pointer and manually checking to see if the pointer needed to be wrapped back to the beginning of the buffer.

Tip: Mind your pointers

This tip was applied in two ways. First, it removed the dependency between `*in` and `*out` by adding the keyword `restrict` in the function header. Second, it removed the dependency between `buffer[current]` and `buffer[delayed]` by assigning `buffer[delayed]` to a temporary variable. Note that for this to work, the programmer must guarantee that `current!=delayed` ALWAYS!

Tip: Be careful with data types

The loop counter 'I' was changed from a *short* data type to an *int*.

Tip: Limit type conversions between fixed and floating-point types

The statement '(float)in[i]' was used twice in the original source code causing two type conversions from *short* to *float* (in[] is defined as *short*). A temporary variable was created, 'temp_in', and the converted value '(float)in[i]' was stored in this temporary variable. Then 'temp_in' was used twice in the subsequent code. In this way, it is only necessary to do the type conversion from *short* to *float* once.

Tip: Use #pragma

Two pragmas were added to provide more information to the compiler. The MUST_ITERATE pragma was used to inform the compiler about the number of times the loop would run. The UNROLL pragma was used to suggest to the compiler that the loop could be unrolled by a factor of 4 (as it turns out, the compiler will ignore this directive).

Tip: Use _nassert

The _nassert intrinsic was used three times. The first two times was to provide alignment information to the compiler about the in[] and out[] arrays that were passed in to the compiler. In this case, we informed the compiler that these arrays are on an int (32-bit) boundary. (Note that these _nassert statements are coupled with DATA_ALIGN pragmas to force the in[] and out[] arrays to be located on 4-byte boundaries. That is, a statement such as '#pragma DATA_ALIGN(in, 4);' was placed where the in[] array was declared. This is not shown below.)

The _nassert intrinsic was also used to inform the compiler about the array index 'delayed' which is the index for the circular buffer 'buffer[]'. The value 'delayed' is set to 'current+1' so the _nassert is used to tell the compiler that current is >=0, thus 'delayed' is >=0.

Below, the code is rewritten with these optimizations (changes shown in bold):

```
void delay(delay_context *delHandle, short * restrict in, short * restrict out, int Ns)
{
        int i;
        int delayed,current,length;
        float *buffer = delHandle->delBuf;
        float g = delHandle->gain;
        float fb = delHandle->feedback;

        float temp_delayed;             /* add temp variable to store buffer[delayed] */
        float temp_in;                          /* add temp variable to store in[i] */

        current = delHandle->curSamp;
        length = delHandle->sampDelay;

        _nassert((int) in % 4 == 0);   /* inform compiler of in[] pointer alignment */
        _nassert((int) out % 4 == 0);  /* inform compiler of out[] pointer alignment */

#pragma MUST_ITERATE(8, 256, 4);
#pragma UNROLL(4);

        for(i=0; i<Ns; i++)
        {
                _nassert(current>=0);           /* inform compiler that current is >=0 */

                delayed = (current+1);          /* manual update circular buffer pointer
*/
                if (delayed>=length) delayed=0;/* this will eliminate function call
caused by % */

                temp_in = (float) in[i]; /* do the type conversion once and store in
temp var */
                temp_delayed = buffer[delayed];

                buffer[current] = temp_in + fb * temp_delayed ;
                out[i] = (short)( temp_in + ( g * temp_delayed) );

                current = delayed;
        }
        delHandle->curSamp = current;
```

```
}
```

Again, the code was compiled (using codegen v4.20 – CCS2.1), with the following options (mid-level aggressiveness):

-gp -k –o3

Here is the assembly output:

```
;*----------------------------------------------------------------------------*
;*   SOFTWARE PIPELINE INFORMATION
;*
;*      Loop source line                 : 88
;*      Loop opening brace source line : 89
;*      Loop closing brace source line : 103
;*      Known Minimum Trip Count          : 64
;*      Known Maximum Trip Count          : 64
;*      Known Max Trip Count Factor       : 64
;*      Loop Carried Dependency Bound(^) : 14
;*      Unpartitioned Resource Bound      : 4
;*      Partitioned Resource Bound(*)     : 4
;*      Resource Partition:
;*                                A-side   B-side
;*      .L units                     3        3
;*      .S units                     0        1
;*      .D units                     2        2
;*      .M units                     1        1
;*      .X cross paths               2        2
;*      .T address paths             2        2
;*      Long read paths              1        1
;*      Long write paths             0        0
;*      Logical  ops (.LS)           0        0     (.L or .S unit)
;*      Addition ops (.LSD)          7        1     (.L or .S or .D unit)
;*      Bound(.L .S .LS)             2        2
;*      Bound(.L .S .D .LS .LSD)     4*       3
;*
;*      Searching for software pipeline schedule at ...
;*         ii = 14 Schedule found with 2 iterations in parallel
;*      done
;*
;*      Epilog not removed
;*      Collapsed epilog stages     : 0
;*      Collapsed prolog stages     : 1
;*      Minimum required memory pad : 0 bytes
;*
;*      Minimum safe trip count     : 1
;*----------------------------------------------------------------------------*
L1:    ; PIPED LOOP PROLOG
;** --------------------------------------------------------------------------*
L2:    ; PIPED LOOP KERNEL
      .line  22
          NOP             1

          INTSP   .L1X    B7,A0              ; |99|
||        INTSP   .L2     B7,B7              ; |96|
```

```
||          MPYSP   .M2X    B4,A0,B8          ;   ^ |99|
||          MPYSP   .M1     A0,A4,A6          ; |100|

            NOP             3

            ADDSP   .L1X    B8,A0,A6          ;   ^ |99|
||          ADDSP   .L2X    A6,B7,B7          ;

            MV      .D1     A5,A0             ; Inserted to split a long life
||          ADD     .S1     1,A5,A5           ; @|93|

   [ B0]    SUB     .D2     B0,1,B0           ; |103|
||          CMPLT   .L1     A5,A3,A1          ; @|94|

   [ B0]    B       .S2     L2                ; |103|

   [!A2]    MV      .S1     A0,A8             ; Inserted to split a long life
|| [!A2]    STW     .D1T1   A6,*+A7[A8]       ;   ^ |99|
||          SPTRUNC .L2     B7,B7             ; |100|
|| [!A1]    ZERO    .L1     A5                ; @|94|

            LDH     .D2T2   *B6++,B7          ; @|96|
||          LDW     .D1T1   *+A7[A5],A0       ; @ ^ |97|

            NOP             2
        .line   36

   [ A2]    SUB     .D1     A2,1,A2           ;
|| [!A2]    STH     .D2T2   B7,*B5++          ; |100|

;** -----------------------------------------------------------------------*
```

(The Epilog is not shown here as it is not interesting for this discussion.)

Now the loop was successfully software pipelined and the presence of more parallel bars (||) indicates better functional unit utilization. However, also note that the presence of 'NOP' indicates that more optimization work can be done.

Recall that ii (iteration interval) is the number of cycles in the kernel of the loop. For medium to large loop counts, ii represents the average cycles for one iteration of the loop, i.e., the total cycle count for all iterations of the for() loop can be approximated by (Ns * ii).

From the Software Pipeline Feedback, **ii=14 cycles** (was 38 cycles before optimization) for a performance **improvement of ~63%.** (The improvement is actually much greater due to the elimination of the call to remi() which was not taken into account.)

By looking at the original C source code (before optimization), it is determined that the following operations are performed on each iteration of the loop:

- 2 floating-point multiplies (M unit)
  - fb * buffer[delayed]
  - g * buffer[delayed]

- 2 floating-point adds (L unit)
  - in[i] + (fb *...)
  - in[i] + (g * ...)
- 1 integer add (L, S, or D unit)
  - delayed = (current + 1) % length
- 2 array loads (D unit)
  - in[i] and buffer[delayed]
- 2 array stores (D unit)
  - buffer[current] and out[i]

From this analysis, it can be seen that the most heavily used unit is the D unit, since it has to do 2 loads and 2 stores every cycle. In other words, the D unit is the resource that is limiting (bounding) the performance.

Since 4 operations need to run on a D unit every iteration, and there are 2 D units, then the best case scenario is approximately 2 cycles per iteration. This is theoretical minimum (best) performance for this loop due to resource constraints.

In actuality, the minimum may be slightly higher than this since we did not consider some operations such as floating-point -> integer conversion. However, this gives us an approximate goal against which we can measure our performance. Our code was compiled to ii = 14 cycles and the theoretical minimum is approximately ii = 2 cycles, so much optimization can still be done.

The Software Pipeline Feedback provides information on how to move forward. Recall that ii is bound either by a **resource** constraint or by a **dependency** constraint.

The line with the * needs to be examined to see if the code is **resource** constrained (recall '*' indicates the most limiting resource):

```
;*        Bound(.L .S .D .LS .LSD)       4*        3
```

This line indicates that the number of operations that need to be performed on an L or S or D unit is 4 per iteration. However, we only have three of these units per side of the CPU (1 L, 1 S and 1 D), so it will require at least two cycles to execute these four operations. Therefore, the minimum ii can be due to resources is 2.

One line needs to be examined to find the **dependency** constraint:

```
;*        Loop Carried Dependency Bound(^) : 14
```

From this line, it can be determined that the smallest ii can be, due to dependencies in the code, is 14. Since ii for the loop is 14, the loop is constrained by dependencies and not lack of resources. In the actual assembly code, the lines marked with ^ are part of the dependency constraint (part of the loop carried dependency path).

Future optimizations should focus on reducing or removing memory dependencies in the code (see spru187 for information on finding and removing dependencies).

If no further code optimizations are necessary, that is, the performance is good enough, the compiler can be set to the high level of aggressiveness for the final code. Remember to perform a sanity check on the code after the high level of aggressiveness is set.

Comb Filter

The comb filter is a commonly used block in the reverb algorithm, for example. Here is an implementation of a comb filter in C code:

```
void combFilt(int *in_buffer, int *out_buffer, float *delay_buffer, int sample_count)
{
    int samp;
    int sampleCount = sample_count;
    int *inPtr;
    int *outPtr;
    float *delayPtr = delay_buffer;
       int read_ndx, write_ndx;

    inPtr = (int *)in_buffer;
    outPtr = (int *)out_buffer;

    for (samp = 0; samp < sampleCount; samp++)
    {

        read_ndx = comb_state;              // init read index
        write_ndx = read_ndx + comb_delay;  // init write index
        write_ndx %= NMAXCOMBDEL;           // modulo the write index

        // Save current result in delay buffer
        delayPtr[write_ndx] =
        delayPtr[read_ndx] * (comb_gain15/32768.) + (float)*inPtr++;

        // Save delayed result to input buffer
       *outPtr++ = (int)delayPtr[read_ndx];

        comb_state += 1;                    // increment and modulo state index
        comb_state %= NMAXCOMBDEL;

    }
}
```

The code is assumed to be correct, so one tip is applied before taking a baseline cycle count:

Tip: Set mid-level aggressiveness

The above code was compiled (using codegen v4.20 – CCS2.1) with the following options (mid-level aggressiveness):

–gp –o3 –k

The following is what appears in the generated assembly file:

```
;*----------------------------------------------------------------------------*
```

```
;*   SOFTWARE PIPELINE INFORMATION
;*      Disqualified loop: loop contains a call
;*----------------------------------------------------------------------------*
L1:
       .line  14

           B        .S1      __remi              ; |60|
||         LDW      .D2T1    *+DP(_comb_delay),A4 ; |60|

           NOP               3
           MVKL     .S2      RL0,B3              ; |60|

           ADD      .S1X     B6,A4,A4
||         MVKH     .S2      RL0,B3              ; |60|
||         MV       .D2      B8,B4               ; |60|

RL0:       ; CALL OCCURS                         ; |60|
           LDW      .D2T2    *+DP(_comb_gain15),B4 ; |60|
           NOP               4
           INTDP    .L2      B4,B5:B4            ; |60|
           NOP               2
           ZERO     .D1      A9                  ; |60|
           MVKH     .S1      0x3f000000,A9       ; |60|
           MPYDP    .M1X     A9:A8,B5:B4,A7:A6 ; |60|
           NOP               2
           LDW      .D2T2    *+B7[B6],B4         ; |60|
           NOP               4
           SPDP     .S2      B4,B5:B4            ; |60|
           LDW      .D1T2    *A10++,B9           ; |60|
           MPYDP    .M2X     A7:A6,B5:B4,B5:B4 ; |60|
           NOP               3
           INTSP    .L2      B9,B9               ; |60|
           NOP               3
           SPDP     .S2      B9,B1:B0            ; |60|
           NOP               1
           ADDDP    .L2      B1:B0,B5:B4,B5:B4 ; |60|
           NOP               6
           DPSP     .L2      B5:B4,B4            ; |60|
           NOP               2
           MV       .S2X     A4,B5               ; |60|
           STW      .D2T2    B4,*+B7[B5]         ; |60|
           LDW      .D2T1    *+B7[B6],A4         ; |64|
           NOP               4

           B        .S1      __remi              ; |66|
||         SPTRUNC  .L1      A4,A4               ; |64|

           NOP               3

           STW      .D1T1    A4,*A3++            ; |64|
||         MVKL     .S2      RL2,B3              ; |66|

           MV       .D2      B8,B4               ; |66|
||         ADD      .S1X     1,B6,A4
```

```
||          MVKH    .S2     RL2,B3              ; |66|

RL2:        ; CALL OCCURS                       ; |66|
            SUB     .D1     A0,1,A1
   [ A1]    B       .S1     L1                  ; |69|
            NOP             4
        .line  30

            MV      .S2X    A4,B6
||          SUB     .D1     A0,1,A0

            ; BRANCH OCCURS                     ; |69|
;** --------------------------------------------------------------------------*
            STW     .D2T2   B6,*+DP(_comb_state)
            LDW     .D2T2   *+SP(4),B3          ; |70|
;** --------------------------------------------------------------------------*
L2:
            LDW     .D2T1   *++SP(8),A10        ; |70|
            NOP             3
            B       .S2     B3                  ; |70|
        .line  31
            NOP             5
            ; BRANCH OCCURS                     ; |70|
        .endfunc      70,000080400h,8
```

From the Software Pipeline Feedback, it is determined that the loop was disqualified from pipelining because it contained a function call. Heuristically, it is determined that the loop is not well optimized since there are very few parallel bars (||) in the first column indicating very low functional unit utilization.

By hand-counting the instructions in the kernel of the loop (in the .asm file), it is determined that the kernel takes approximately **68 cycles**, not including calls to the run-time support library.

Now, the tips are employed to improve this code.

Tip: Address circular buffers intelligently

Tip: Remove function calls from with a loop

Tip: More C code does not always produce less efficient assembly code

From looking at the C source code, it can be seen that the modulo operator was called twice in the loop. The modulo operator was used to increment an index used by "delayPtr," a pointer to the circular "delay_buffer." Recall that there are four conditions for using the modulo operator. This code fails on the second condition. The length of the buffer is "NMAXCOMBDEL." In this code, "NMAXCOMBDEL" is a constant but happens to not be a power of 2. This means that the modulo operator is triggering a function call to the run time support library hence disqualifying the loop from pipelining. The two modulo operators are removed and replaced with code to manually update the pointers and check for wrap-around conditions.

Tip: Mind your pointers

Second, by adding the keyword `restrict` to `in_buffer` and `out_buffer`, it is indicated to the compiler that they do not point to overlapping spaces.

Tip: Call a float a float

The floating-point constant 32768. is changed to be typecast as `float`. In the original code, without this typecast, the compiler assumed the constant was a `double` and used costly double precision operations in the loop.

Tip: Do not access global variables from within a for() loop

In the original C source code, three global variables were accessed directly from within the for() loop: 'comb_state', 'comb_delay', and 'comb_gain15'. Local copies were made of all these variables, and the local copies were used in the for() loop instead. Since 'comb_state' was updated in the loop, the local copy had to be copied back to the global variable after exiting the for() loop.

In a similar vein, a local copy is made of 'local_comb_gain15/(float)32768.' before entering the for() loop, since these values are constant over the multiple iterations of the loop. By doing this, the divide calculation does not have to be done within the loop itself.

Tip: Use #pragma

Assume that some information is known by the programmer about the number of times the loop will run. The MUST_ITERATE can then be used to feed this information to the compiler. In this case, the programmer knows that the loop will run at least 8 times and at most 256 times, and the loop count will always be a factor of 4.

```
#pragma MUST_ITERATE(8, 256, 4)
```

Also, another pragma can be used to try and force the loop to unroll by a factor of 4. (As it turns out, the compiler will ignore this advice.)

```
#pragma UNROLL(4)
```

Tip: Use _nassert()

The _nassert intrinsic was used to inform the compiler that the buffers 'in_buffer', 'out_buffer', and 'delay_buffer' are aligned on an 8-byte boundary. Actually, the argument of the _nassert intrinsic is the local copy of the pointers to these buffers 'inPtr', 'outPtr', and 'delayPtr' respectively.

```
_nassert((int)inPtr % 8 == 0);
_nassert((int)outPtr % 8 == 0);
_nassert((int)delayPtr % 8 == 0);
```

(Corresponding to this code, statements were added earlier in the program to actually do the alignment of these buffers. For example, these earlier statements looked like:

```
#pragma DATA_ALIGN(in_buffer, 8);
int in_buffer[NUM_SAMPLES];
```

This was done everywhere that these buffers were declared.)

The code was rewritten with all of these above optimizations (changes shown in bold):

```
void combFilt(int * restrict in_buffer, int * restrict out_buffer, float *delay_buffer,
int sample_count)
{
    int samp;
    int sampleCount = sample_count;
    int *inPtr;
    int *outPtr;
    float *delayPtr = delay_buffer;
    int read_ndx, write_ndx;

    //make local copies of global variables
    int local_comb_delay = comb_delay;
    int local_comb_state = comb_state;
    int local_comb_gain15 = comb_gain15;

    //calculate constant and store in local variable so not done inside the loop
    float temp_gain = (local_comb_gain15/(float)32768.);

    inPtr = (int *)in_buffer;
    outPtr = (int *)out_buffer;

        _nassert((int)inPtr % 8 == 0);  /* indicate that the pointer is 8-byte aligned */
        _nassert((int)outPtr % 8 == 0);
        _nassert((int)delayPtr % 8 == 0);

        #pragma MUST_ITERATE(8, 256, 4);     /* feed loop count information to compiler */
        #pragma UNROLL(4);                    /* suggest to compiler to unroll loop */

    for (samp = 0; samp < sampleCount; samp++)
    {

        read_ndx = local_comb_state;   // init read index
        write_ndx = (read_ndx + local_comb_delay);     // init write index

        //manually update circular buffer index and check for wrap-around
 if (write_ndx >= NMAXCOMBDEL)  write_ndx-=NMAXCOMBDEL;

        // Save current result in delay buffer
        delayPtr[write_ndx] =
        delayPtr[read_ndx] * (temp_gain) + (float)*inPtr++;

        // Save delayed result to input buffer
       *outPtr++ = (int)delayPtr[read_ndx];

        local_comb_state += 1; // increment and modulo state index

        //manually check for wrap-around condition
        if (local_comb_state >= NMAXCOMBDEL)   local_comb_state=0;
```

```
    }

        //copy local_variables back to globals if necessary
        comb_state = local_comb_state;
}
```

The code was again compiled (using codegen v4.20 – CCS2.1) with the following options (mid-level aggressiveness):

-gp -k –o3

Here is the relevant section of the generated assembly output:

```
;*----------------------------------------------------------------------------*
;*   SOFTWARE PIPELINE INFORMATION
;*
;*      Loop source line                 : 113
;*      Loop opening brace source line : 114
;*      Loop closing brace source line : 129
;*      Known Minimum Trip Count         : 8
;*      Known Maximum Trip Count         : 256
;*      Known Max Trip Count Factor      : 4
;*      Loop Carried Dependency Bound(^) : 14
;*      Unpartitioned Resource Bound     : 4
;*      Partitioned Resource Bound(*)    : 5
;*      Resource Partition:
;*                                A-side   B-side
;*      .L units                    2        3
;*      .S units                    0        1
;*      .D units                    3        2
;*      .M units                    1        0
;*      .X cross paths              1        2
;*      .T address paths            3        2
;*      Long read paths             1        1
;*      Long write paths            0        0
;*      Logical  ops (.LS)          1        0      (.L or .S unit)
;*      Addition ops (.LSD)         7        3      (.L or .S or .D unit)
;*      Bound(.L .S .LS)            2        2
;*      Bound(.L .S .D .LS .LSD)    5*       3
;*
;*      Searching for software pipeline schedule at ...
;*         ii = 14 Schedule found with 2 iterations in parallel
;*      done
;*
;*      Epilog not removed
;*      Collapsed epilog stages     : 0
;*      Collapsed prolog stages     : 1
;*      Minimum required memory pad : 0 bytes
;*
;*      Minimum safe trip count     : 1
;*---------------------------------------------------------------------------*
L1:    ; PIPED LOOP PROLOG
;** --------------------------------------------------------------------------*
L2:    ; PIPED LOOP KERNEL
       .line  29
          NOP            2
```

```
   [!A2]   STW     .D1T2   B8,*+A0[A4]       ;   ^ |121|
||         ADD     .D2     1,B7,B7           ; @Define a twin register
||         ADD     .S1     A7,A9,A4          ; @
||         MV      .L1     A9,A3             ; @

           CMPLT   .L2X    A4,B6,B0          ; @|118|
||         ADD     .L1     1,A9,A9           ; @|127|
||         LDW     .D2T2   *B5++,B8          ; @|121|
||         LDW     .D1T1   *+A0[A3],A3       ; @ ^ |121|

   [!A2]   LDW     .D1T1   *+A0[A1],A3       ; |125|
|| [!B0]   ADD     .S1     A6,A4,A4          ; @|118|
||         CMPLT   .L1     A9,A5,A1          ; @|127|

   [!A1]   ZERO    .D2     B7                ; @|128|

           MV      .D1     A3,A1             ; @Inserted to split a long life
|| [!A1]   MV      .S1X    B7,A9             ; @Define a twin register

           NOP             1

   [ B1]   B       .S2     L2                ; |129|
|| [ B1]   SUB     .D2     B1,1,B1           ; @|129|
||         INTSP   .L2     B8,B8             ; @|121|
||         MPYSP   .M1     A8,A3,A3          ; @ ^ |121|

           SPTRUNC .L1     A3,A3             ; |125|
           NOP             2
           ADDSP   .L2X    B8,A3,B8          ; @ ^ |121|
        .line  44

   [ A2]   SUB     .D1     A2,1,A2           ;
|| [!A2]   STW     .D2T1   A3,*B4++          ; |125|

;** -----------------------------------------------------------------------------*
L3:     ; PIPED LOOP EPILOG
           NOP             1
           MVC     .S2     B9,CSR            ; interrupts on
           STW     .D1T2   B8,*+A0[A4]       ; (E) @ ^ |121|
           LDW     .D1T1   *+A0[A1],A0       ; (E) @|125|
           NOP             4

           SPTRUNC .L1     A0,A0             ; (E) @|125|
||         B       .S2     B3                ; |133|

           NOP             3
           STW     .D2T1   A0,*B4++          ; (E) @|125|
        .line  48
           STW     .D2T1   A9,*+DP(_comb_state) ; |132|
           ; BRANCH OCCURS                   ; |133|
        .endfunc       133,000000000h,0
```

First, it is noticed that the loop was not disqualified from pipelining. Heuristically, the presence of parallel bars (||) in the kernel indicates greater utilization of the functional units. However, the presence of NOPs indicate that more optimization work can be done.

Recall that ii (iteration interval) is the number of cycles in the kernel of the loop. For medium to large loop counts, ii represents the average cycles for one iteration of the loop, i.e., the total cycle count for all iterations of the for() loop can be approximated by (`sampleCount` * ii).

From the Software Pipeline Feedback, **ii=14 cycles** for a performance **improvement of ~79%.** (In actuality, the improvement is much greater since the call to the run-time support library was not considered.)

By examining the original C source code, we can determine the theoretical minimum (best) performance for this loop. By looking at the original C source code, we see that the following operations are performed on each iteration of the loop:

- 1 floating-point multiply (M unit)
  -- delayPtr[read_ndx] * (comb_gain15/32768.)
- 1 floating-point add (L unit)
  -- result of above multiply + (float)*inPtr++
- 2 integer adds (L, S, or D unit)
  -- read_ndx + comb_delay
  -- comb_state += 1
  - 2 array loads (D unit)
  -- delayPtr[read_ndx] and inPtr
- 2 array stores (D unit)
  -- delayPtr[write_ndx] and outPtr

From this analysis, it can be determined that the D unit is the most heavily used as it has to perform 4 operations per iteration (2 loads and 2 stores). Since there are two D units, the theoretical minimum (best) performance for this loop due to resource constraints is approximately 2 cycles/iteration. More work can be done to further improve the performance of this loop to get closer to this theoretical best.

The Software Pipeline Feedback can guide future optimization efforts. Recall that ii can be constrained in two ways, either by **resources** or by **dependencies** in the code. To find the constraint by **resources**, find the line with the * next to it:

```
;*       Bound(.L .S .D .LS .LSD)      5*        3
```

This line indicates that the number of operations that need to be performed on an L or S or D unit is 5 per iteration. However, we only have three of these units per side of the CPU (1 L, 1 S and 1 D), so it will require at least two cycles to execute these five operations. Therefore, the minimum ii can be due to resources is 2.

To find the constraint imposed by **dependencies**, the following line needs to be examined:

```
;*       Loop Carried Dependency Bound(^) : 14
```

The constraint imposed by dependencies is that ii cannot be smaller than this number. Since ii = 14, it can be determined that ii is constrained by dependencies in the code and not constrained by lack of resources. In the actual assembly code, the lines marked with ^ are part of the dependency constraint (part of the loop carried dependency path).

If no further code optimizations are necessary, that is, the performance is good enough, the compiler can be set to the high level of aggressiveness for the final code. Remember to perform a sanity check on the code after the high level of aggressiveness is set.

I would like to thank George Mock for his contribution of the material in this appendix.

C

Cache Optimization in DSP and Embedded Systems

A cache is an area of high-speed memory linked directly to the embedded CPU. The embedded CPU can access information in the processor cache much more quickly than information stored in main memory. Frequently-used data is stored in the cache.

There are different types of caches but they all serve the same basic purpose. They store recently-used information in a place where it can be accessed very quickly. One common type of cache is a disk cache. This cache model stores information you have recently read from your hard disk in the computer's RAM, or memory. Accessing RAM is much faster than reading data off the hard disk and therefore this can help you access common files or folders on your hard drive much faster. Another type of cache is a processor cache which stores information right next to the processor. This helps make the processing of common instructions much more efficient, and therefore speeding up computation time.

There has been historical difficulty in transferring data from external memory to the CPU in an efficient manner. This is important for the functional units in a processor as they should be kept busy in order to achieve high performance. However, the gap between memory speed and CPU speed is increasing rapidly. RISC or CISC architectures use a memory hierarchy in order to offset this increasing gap and high performance is achieved by using data locality.

Principle of locality

The principle of locality says a program will access a relatively small portion of overall address space at any point in time. When a program reads data from address N, it is likely that data from address N+1 is also read in the near future (spatial locality) and that the program reuses the recently read data several times (temporal locality). In this context, locality enables hierarchy. Speed is approximately that of the uppermost level. Overall cost and size is that of the lowermost level. A memory hierarchy from the top to bottom contains registers, different levels of cache, main memory, and disk space, respectively (Figure C.1)

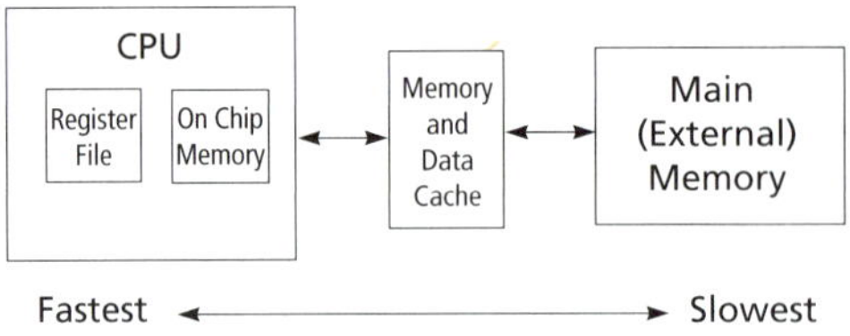

Figure C.1 Memory hierarchy for DSP devices

There are two types of memories available today. Dynamic Random Acces Memory (DRAM) is used for main memory and is cheap but slow. Static Random Access Memory (SRAM) is more expensive, consumes more energy, and is faster.

Vendors use a limited amount of SRAM memory as a high-speed cache buffered between the processors and main memory to store the data from the main memory currently in use. Cache is used to hide memory latency which is how quickly a memory can respond to a read or write request. This cache can't be very big (no room on chip), but must be fast. Modern chips can have a lot of cache, including multiple levels of cache:

- First level cache (L1) is located on the CPU chip
- Second level (L2) also located on the chip
- Third level cache (L3) is external to the CPU is larger

A general comparison of memory types and speed is shown in Figure C.2. Code that uses cache efficiently show higher performance. The most efficient way of using cache is through a block algorithm which is described later.

Memory Type	Speed
Registers	2 ns
L1 on chip	4 ns
L2 on chip	5 ns
L3 on chip	30 ns
Memory	220 ns

Figure C.2 Relative comparison of memory types and speed

Caching schemes

There are various caching schemes used to improve overall processor performance. Compiler and hardware techniques attempt to *look ahead* to ensure the cache is full of useful (not stale) information. Cache is refreshed at long intervals using several techniques. Random replacement will throw out any current block at random. First-in/first-out (FIFO) replace the information that has been in memory the longest. Least recently used (LRU) replace the block that was last referenced the furthest in the past. If a request is in the cache it's a cache hit. The higher hit rate the better performance. A request that is not in the cache is a miss.

As an example, if a memory reference were satisfied from L1 cache 70% of the time, L2 cache 20% of the time, L3 cache 5% of the time and main memory 5% of the time, the average memory performance would be:

(0.7 * 4) + (0.2 * 5) + (0.05 * 30) + (0.05 * 220) = 16.30 ns

Caches work well when a program accesses the memory sequentially;

```
do i = 1, 1000000
      sum = sum = sum + a(i)   ! accesses successive
      data element (unit stride)
enddo
```

The performance of the following code is not good!

```
do i = 1, 1000000, 10
      sum = sum = sum + a(i) ! accesses large data structure
enddo
```

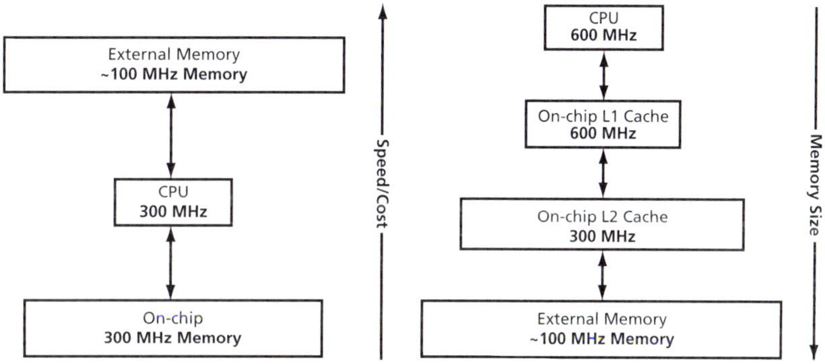

Figure C.3 Flat vs. hierarchical memory architecture (courtesy of Texas Instruments)

The diagram in Figure C.3 (the model on the left) shows a flat memory system architecture. Both CPU and internal memory run at 300 MHz. A memory access penalty will only occur when CPU accesses external memory. There are no memory stalls that will occur for accesses to internal memory. But what happens if we increase the CPU clock to 600 MHz? We would experience wait-states! We also would need a 600 MHz memory. Unfortunately, today's available memory technology is not able to keep up with increasing processor speeds. A same size internal memory running at 600 MHz would be far too expensive. Leaving it at 300 MHz is not possible as well, since this would effectively reduce the CPU clock to 300 MHz as well (Imagine in a kernel with memory access every EP, every EP will suffer 1 cycle stall, effectively doubling the cycle count and thus canceling out the double clock speed).

The solution is to use a memory hierarchy, with a fast, but small and expensive memory close to the CPU that can be accessed without stalls. Further away from the CPU the memories become larger but slower. The memory levels closest to the CPU typically act as a cache for the lower level memories.

So, how does a cache work in principle and why does it speed-up memory access time?

Let's look at the access pattern of a FIR filter, a 6-tap FIR filter in this case. The required computations are shown here. To compute an output we have to read in 6 data samples (

we also need 6 filter coefficients, but these can be neglected here since it's relatively little data compared to the sample data) from an input data buffer x[]. The numbers denote in which order the samples are accessed in memory. When the first access is made, the cache controller fetches the data for the address accessed and also the data for a certain number of the following addresses into cache. This range of addresses is called a *cache line*. [Fetching the line from the slower memory is likely cause a few CPU stall cycles.] The motivation for this behavior is that accesses are spatially local, that is if a memory location was accessed, a neighboring location will be accessed soon as well. And it is true for our FIR filter, the next 5 samples are required as well. This time all accesses will go into fast cache instead of slow lower level memory[, without causing any stall cycles]. Accesses that go to neighboring memory locations are called *spatially local.*

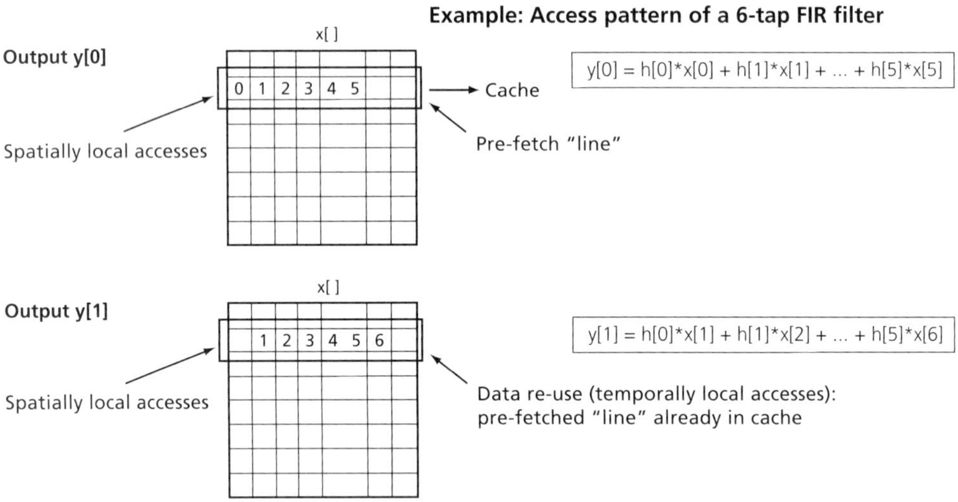

Figure C.4 Principle of locality (courtesy of Texas Instruments)

Let's see what happens if we calculate the next output. The access pattern is shown in Figure C.4. Five of the samples are being re-used and only one sample is new, but all of them are already held in cache. Also no CPU stalls occur. This illustrates the principle of temporal locality: the same data that was used in the previous step is being used again for processing;

- Spatial locality – If memory location was accessed, neighboring location will be accessed as well.
- Temporal locality – If memory location was accessed, it will be accessed soon again [look-up table]

Cache builds on the fact that data accesses are spatially and temporally local. Accesses to a slower lower level memory are minimal, and the majority of accesses can be serviced at CPU speed from the high level cache memory. As an example:

N = 1024 output data samples, 16-tap filter: 1024 * 16 / 4 = 4096 cycles

Stall cycles to bring 2048 byte into cache: about 100 cycles (2.5 cycles per 64-byte line), about 2.4%.

In other words, we have to pay with 100 cycles more, and in return we are getting double the execution speed, that's still in the end 1.95x speed-up!

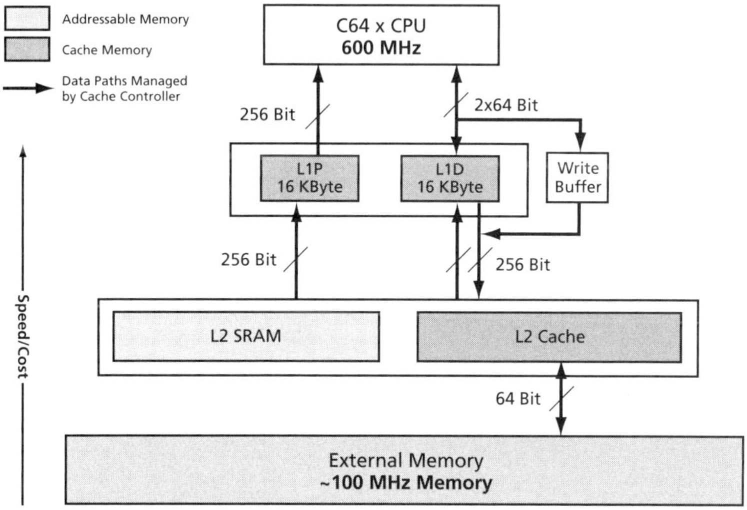

Figure C.5 C64x cache memory architecture (courtesy of Texas Instruments)

The C64x memory architecture (Figure C.5) consists of a 2-level internal cache-based memory architecture plus external memory. Level 1 cache is split into program (L1P) and data (L1D) cache, each 16 KBytes. Level 1 memory can be accessed by the CPU without stalls. Level 1 caches cannot be turned off. Level 2 memory is configurable and can be split into L2 SRAM (addressable on-chip memory) and L2 Cache that caches external memory addresses. On a TI C6416 DSP for instance, the size of L2 is 1 MByte, but only one access every two cycles can be serviced. Finally, external memory can be up to 2 GBytes, the speed depends on the memory technology used but is typically around 100 MHz. All caches (red) and data paths shown are automatically managed by the cache controller.

Mapping of addressable memory to L1P

First. let's have a look at how direct-mapped caches work. We will use the L1P of C64x as an example for a direct-mapped cache. Shown in Figure C.6 are the addressable memory (e.g., L2 SRAM), the L1P cache memory and the L1P cache control logic. L1P Cache is 16 KBytes large and consists of 512 32-byte lines. Each line always maps to the same fixed addresses in memory. For instance addresses 0x0 to 0x19 (32 bytes) will always be cached in line 0, addresses 3FE0h to 3FFFh will always be cached in line 511. Then, since we exhausted the capacity of the cache, addresses 4000h to 4019h map to line 0 again, and so forth. Note here that one line contains exactly one instruction fetch packet.

Access, invalid state of cache, tag/set/offset

Now let's see what happens if the CPU accesses address location 20h. Assume that cache has been completely invalidated, meaning that no line contains cached data.

The valid state of a line is indicated by the valid bit. A valid bit of 0 means that the corresponding cache line is invalid, i.e., doesn't contain cached data. So, when the CPU makes a request to read address 20h, the cache controller takes the address and splits it up into three portions: the offset, the set and the tag portion. The set portion tells the controller to which set the address maps to (in case of direct caches a set is equivalent to a line). For the address 20h the set portion is 1. The controller then checks the tag and the valid bit. Since we assumed that the valid bit is 0, the controller registers a *miss*, i.e., the requested address is not contained in cache.

Miss, line allocation

A miss means that the line containing the requested address will be allocated in cache. That is the controller fetches the line (20h-39h) from memory and stores the data in set 1. The tag portion of the address is stored in the tag RAM and the valid bit is set to 1. The fetched data is also forwarded to the CPU and the access is complete. Why we need to store a tag portion becomes clear when we access address 20h again.

Re-access same address

Now, let's assume some time later we access address 20h again. Again the cache controller takes the address and splits it up into the three portions. The set portion determines the set. The stored tag portion will be now compared against the tag portion of the address requested. This is comparison is necessary since multiple lines in memory map to the same set. If we had accessed address 4020h which also maps to the same set, but the tag portions would be different, thus the access would have been a miss. In this case the tag comparison is positive and also the valid bit is 1, thus the controller will register a *hit* and forward the data in the cache line to the CPU. The access is completed.

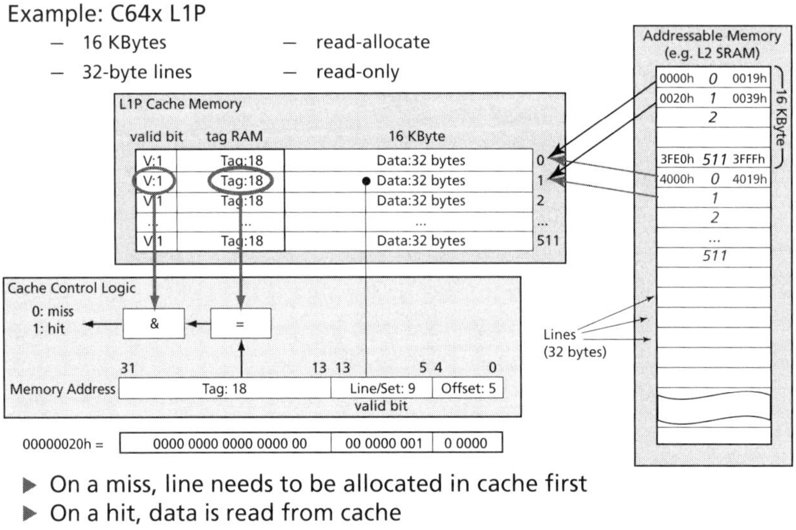

Figure C.6 Direct mapped cache architecture (courtesy of Texas Instruments)

At this point it is useful to remind ourselves what the purpose of a cache is. The purpose of a cache is to reduce the average memory access time. For each miss we have to pay a penalty to get (allocate) a line of data from memory into cache. So, to get the highest returns for what we "paid" for, we have to re-use (read and/or write accesses) this line as much as possible before it is replaced with another line. Re-using the line but accessing different locations within that line improves the spatial locality of accesses, re-using the same locations of a line improves the temporal locality of accesses. This is by the way the fundamental strategy of optimizing memory accesses for cache performance.

The key is here re-using the line *before it gets replaced*. Typically the term eviction is used in this context: The line is evicted (from cache). What happens if a line is evicted, but then is accessed again? The access misses and the line must first be brought into cache again. Therefore, it is important to avoid eviction as long as possible. So, to avoid evictions we must know what causes an eviction. Eviction is caused by conflicts, i.e., a memory location is accessed that maps to the same set as a memory location that was accessed earlier. The newly accessed location will cause the previous line held at that set to be evicted and to be allocated in its place. Another access to the previous line will now cause a miss.

This is referred to as a *conflict miss*, a miss that occurred because the line was evicted due to a conflict before it was re-used. It is further distinguished whether the conflict occurred because the capacity of the cache was exhausted or not. If the capacity was exhausted then the miss is referred to as a *capacity miss*. Identifying the cause of a miss, helps to choose the appropriate measure for avoiding the miss. If we have conflict misses that means the data accessed fits into cache, but lines get evicted due to conflicts.

In this case we may want to change the memory layout so that the data accessed is located at addresses in memory that do not conflict (i.e., map to the same set) in cache. Alternatively, from a hardware design point of view, we can create sets that can hold two or more lines. Thus, two lines from memory that map to the same set can *both* be allocated in cache without having one evict the other. How this works will see shortly on the next slide.

If we have capacity misses, we may want to reduce the amount of data we are operating on at a time. Alternatively, from a hardware design point of view, we could increase the capacity of the cache.

There is a third category of misses, the *compulsory misses* or also called first reference misses. They occur when the data is brought in cache for the first time. As opposed to the other two misses above, they cannot be avoided, thus they are compulsory.

An extension of direct-mapped caches are so-called set-associative caches (Figure C.7). For instance the C6x's L1D is a 2-way set associative cache, 16 KBytes capacity and has 64-byte lines. Its function shall now be explained. The difference to a direct-mapped cache is that here one set consists of two lines, one line in way 0 and one line in way 1, i.e., one line in memory can be allocated in either of the two lines. For this purpose the cache memory is split into 2 ways, each way consisting of 8 KBytes.

Hits and misses are determined the same as in a direct-mapped cache, except that now two tag comparisons are necessary, one for each way, to determine in which way the requested data is kept. If there's a hit in way 0, the data of the line in way 0 is accessed, if there's a hit in way 1 the data of the line in way 1 is accessed.

If both ways miss, the data needs to be allocated from memory. In which way the data gets allocated is determined by the LRU bit. An LRU bit exists for each set. The LRU bit can be thought of as a switch: if it is 0 the line is allocated in way 0, if it is 1 the line gets allocated in way 1. The state of the LRU bit changes whenever an access (read or write) is made to a cache line: If the line in way 0 is accessed the LRU bit switches to 1 (causing the line in way 1 to be replaced) and if line 1 is accessed the LRU bit switches to 0. This has the effect that always the Least-recently-used (or accessed) line in a set will be replaced, or in other words the LRU bit switches to the opposite way of the way that was accessed, as to "protect" the most-recently-used line from replacement. Note that the LRU bit is only consulted on a miss, but its status is updated every time a line in the set is accessed regardless whether it was a hit or a miss, a read or a write.

As the L1P, the L1D is a read-allocated cache, that is new data is allocated from memory on a read miss only. On a write miss, the data goes through a write buffer to memory bypassing L1D cache. On a write hit the data is written to the cache but not immediately to memory. This type of cache is referred to as write-back cache, since data that was modified by a CPU write access is written back to memory at a later time. So, when is the data written back?

First of all we need to know which line was modified and needs to be written back to lower level memory. For this purpose every cache line has a dirty bit (D) associated to it. It's called dirty because it tells us if the corresponding line was modified. Initially the dirty bit will be zero. As soon as the CPU writes to a line, the corresponding dirty bit will be set to 1. So again, when is a dirty line written back? It's written back on a read miss that will cause new data to be allocated in a dirty line. Let's assume the line in set 0, way 0 was written to by the CPU, and the LRU bit indicates that way 0 is to be replaced on the next miss. If the CPU now makes a read access to a memory location that maps to set 0, the current dirty data contained in the line is first written back to memory, then the new data is allocated in that line. A write-back can also be initiated by the program by sending a write back command to the cache controller. This is however, not usually required.

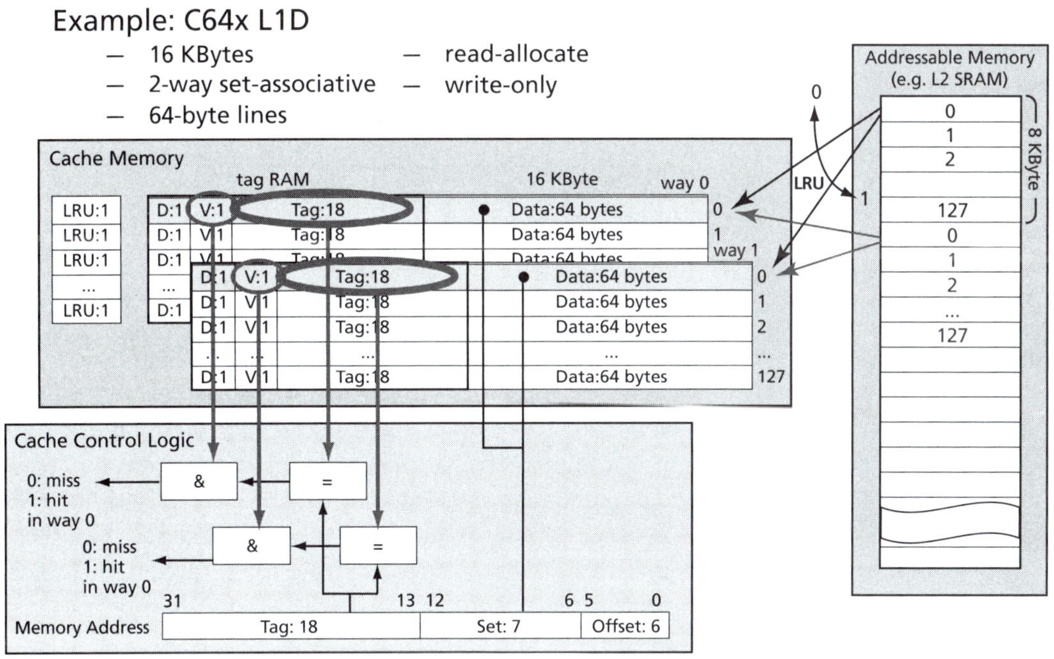

Figure C.7 Set-associative cache architecture (courtesy of Texas Instruments)

Now let's start looking at cache coherence. So, what do we mean by cache coherence? If multiple devices, such as the CPU or peripherals, share the same cacheable memory region, cache and memory can become incoherent. Let's assume the following system. Suppose the CPU accesses a memory location which gets subsequently allocated in cache (1). Later, a peripheral is writing data to this same location which is meant to be read and processed by the CPU (2). However, since this memory location is kept in cache, the memory access hits in cache and the CPU will read the old data instead of the new one (3). The same problem occurs if the CPU writes to a memory location that is cached, and the data is to be written out by a peripheral. The data only gets updated in cache but not in memory from where the peripheral will read the data. The cache and the memory is said to be "incoherent."

How is this problem addressed? Typically a cache controller is used that implements a cache coherence protocol that keeps cache and memory coherent. Let's see how this is addressed in the C6x memory system.

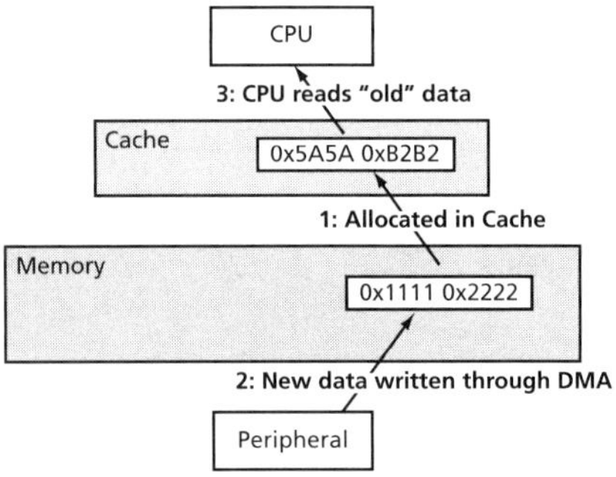

Figure C.8 Strategy for optimizing cache performance (courtesy of Texas Instruments)

A good strategy for optimizing cache performance is to proceed in a top-down fashion (Figure C.8), starting on application level, moving to procedural level, and if necessary considering optimizations on algorithmic level. The optimization methods for application level tend to be straightforward to implement and typically have a high impact on overall performance improvement. If necessary, fine tuning can then be performed using lower level optimization methods. Hence the structure of this chapter reflects the order in which one may want to address the optimizations.

To illustrate the coherency protocols, let's assume a peripheral is writing data to an input buffer located in L2 SRAM, (Figure C.9) then the CPU reads the data processes it and writes it back to an output buffer, from which the data is written to another peripheral. The data is transferred by the DMA controller. We'll first consider a DMA write, i.e., peripheral fills input buffer with data, the next slide will then show the DMA read, i.e., data in the output buffer is read out to a peripheral.

1. The peripheral requests a write access to line 1 (lines mapped form L1D to L2SRAM) in L2 SRAM. Normally the data would be committed to memory, but not here.
2. The L2 Cache controller checks it's local copy of L1D tag ram if the line that was just requested to be written to is cached in L1D by checking the valid bit and the tag. If the line is not cached in L1D no further action needs to be taken and the data is written to memory.
3. If the line is cached in L1D, the L2 controller sends a SNOOP-INVALIDATE command to L1D. This sets the valid bit of the corresponding line to zero, i.e., invalidates the line. If the line is dirty it is written back to L2 SRAM, then the new data from the peripheral is written.
4. The next time the CPU accesses this memory location, the access will miss in L1D and the line containing the new data written by the peripheral is allocated in L1D and read by the CPU. If the line had not been invalidated, the CPU would have read the "old" value that was cached in L1D.

Aside, the L2 controller sends an INVALIDATE command to L1P. This is necessary in case we want to load program code. No data needs to be written back in this case since data in L1P is never modified.

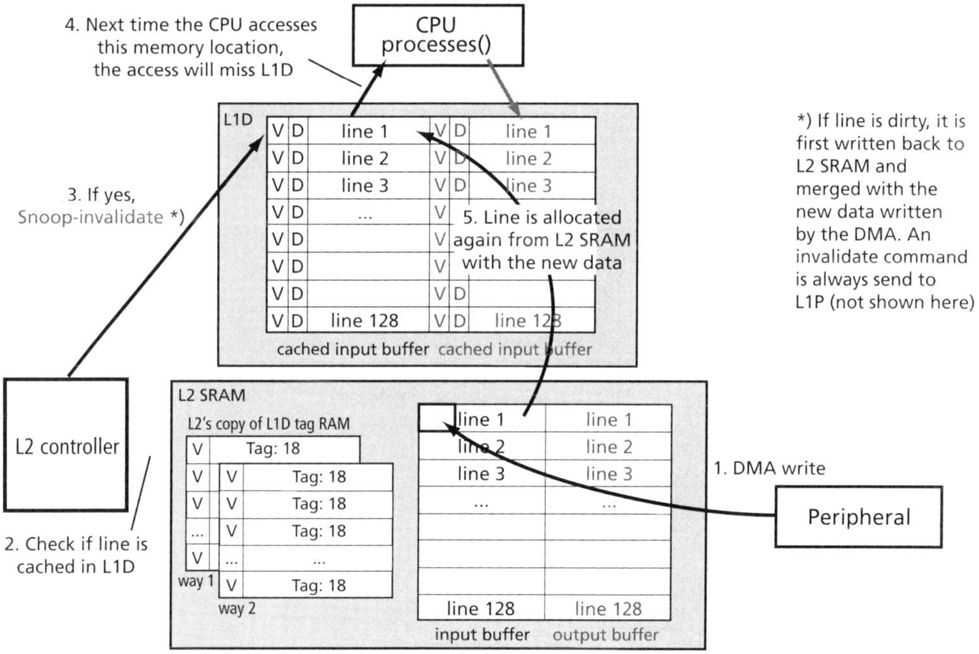

Figure C.9 DMA Write operations. This is the recommended use: Stream data to L2SRAM not external memory. (courtesy of Texas Instruments)

Having described how a DMA write and read to L2 SRAM works, we will see next how everything plays together in the example of a typical double-buffering scheme (Figure C.10). Let's assume we want read in data from one peripheral, process it, and write it out through another peripheral—a structure of a typical signal processing application. The idea is that while the CPU is processing data accessing one set of buffers (e.g., InBuff A and OutBuff A), the peripherals are writing/reading data using the other set of buffers such that the DMA data transfer occurs in parallel with CPU processing.

Let's start off assuming that InBuffA has been filled by the peripheral.

1. Transfer is started to fill InBuffB.
2. CPU is processing data in InBuffA. The lines of InBuffA are allocated in L1D. Data is processed by CPU and written through the write buffer (remember that L1D is read-allocated) to OutBuffA.
3. Buffers are then switched, and CPU is reading InBuffB and writing OutBuffB. InBuffB gets cached in L1D.
4. At the same time the peripheral fills InBuffA with new data. The L2 cache controller takes automatically care of invalidating the corresponding lines in L1D through Snoop-Invalidates so that the CPU will allocated the line again from L2 SRAM with the new data rather reading the cached line containing the old data.
5. Also, the other peripheral reads OutBuffA. However, since this buffer is not cached in L1D no Snoops are necessary here.

It's always a good idea to make the buffers fit into a multiple of cache lines, in order to get the highest return (in terms of cached data) for every cache miss.

Here's a code example how such a double buffering scheme could be realized. It uses CSL calls to DAT_copy and DAT_wait.

This is recommended over DMA double buffering in external memory.

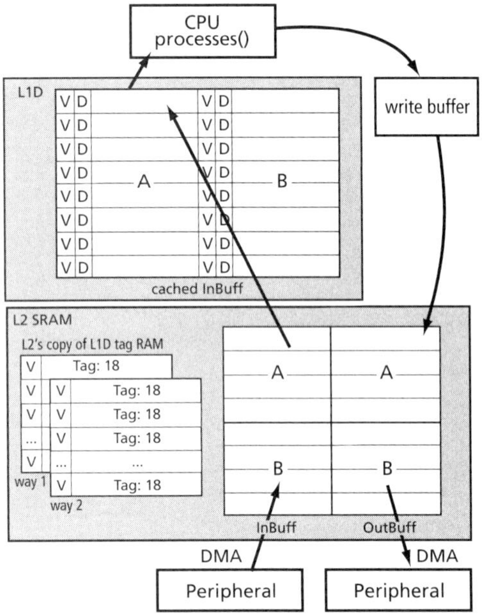


```
fOR [1=0] 1<(DATASIZE/BUFSIZE)-2; i+=2)
{
 /* ------------------------------------------------------------------ */
 /*  inBuffA -> OutBuffA Processing                                     */
 /* ------------------------------------------------------------------ */
 DMA_transfer(peripheral, InBuffB,    BUFSIZE);
 DMA_transfer(InBuffB,    peripheral, BUFSIZE);

 process(InBuffA, OutBuffA, BUFSIZE);

 /* ------------------------------------------------------------------ */
 /*  inBuffB -> OutBuffB Processing                                     */
 /* ------------------------------------------------------------------ */
 DMA_transfer(peripheral, InBuffA,    BUFSIZE);
 DMA_transfer(InBuffA,    peripheral, BUFSIZE);

 process(InBuffB, OutBuffB, BUFSIZE);
}
```

Figure C.10 DMA double buffering in coherent memory (courtesy of Texas Instruments)

Now let's look at the same double buffering scenario (Figure C.11), but now with the buffers located in external memory. Since the cache controller does not automatically maintain coherence, it is the responsibility of the programmer to maintain coherence manually. Again, the CPU reads in data from a peripheral processes it and writes it out to another peripheral via DMA. But now the data is additionally passed through L2 Cache.

Let's assume that transfers already have occurred and that both InBuff and OutBuff are cached in L2 Cache and also that InBuff is cached in L1D. Further let's assume that CPU has completed consuming inBuffB and filled OutBuffB and is now about to start processing the A buffers. Before we call the process function, we want to initiate the transfers that bring in new data into the InBuffB and commits the data in OutBuffB just written by the CPU to the peripheral.

We already know from the previous example what the L2 cache controller did to keep L2 SRAM coherent with L1D. We have to do exactly the same here to ensure that external memory is kept coherent with L2 Cache and L2 Cache with L1D. In the previous example, whenever data was written to an input buffer, the cache controller would invalidate the corresponding line in the cache. Similarly, here we have to invalidate all the lines in L2 Cache AND in L1D that map to the external memory input buffer before we initiate the transfer (or after the transfer is completed). This way the CPU will re-allocated these lines from external memory next time the input buffer is read, rather than accessing the previous data that would still be in cache if we hadn't invalidated. How can we invalidate the input buffer in cache?

Again the Chip Support Library (CSL) provides a set of routines that allow the programmer to initiate those cache operations. In this case we use CACHE_control(CACHE_L2, CACHE_INV, InBuffB, BUFSIZE); before the transfer starts. We need to specify the start address of the buffer in external memory and its number of bytes.

Similarly, before OutBuffB is transferred to the peripheral, the data first has to be written back from L2 Cache to external memory. This is done by issuing a CACHE_control(CACHE_L2, CACHE_WB, OutBuffB, BUFSIZE);. Again, this is necessary since the CPU writes data only to the cached version if OutBuffB.

Before we move on to the next slide which shows a summary of the available L2 cache operations: To prevent unexpected incoherence problems, it is important that we align all buffers at a L2 cache line size make their size a multiple of cache lines. It's also a good idea to place the buffers contiguously in memory to prevent evictions.

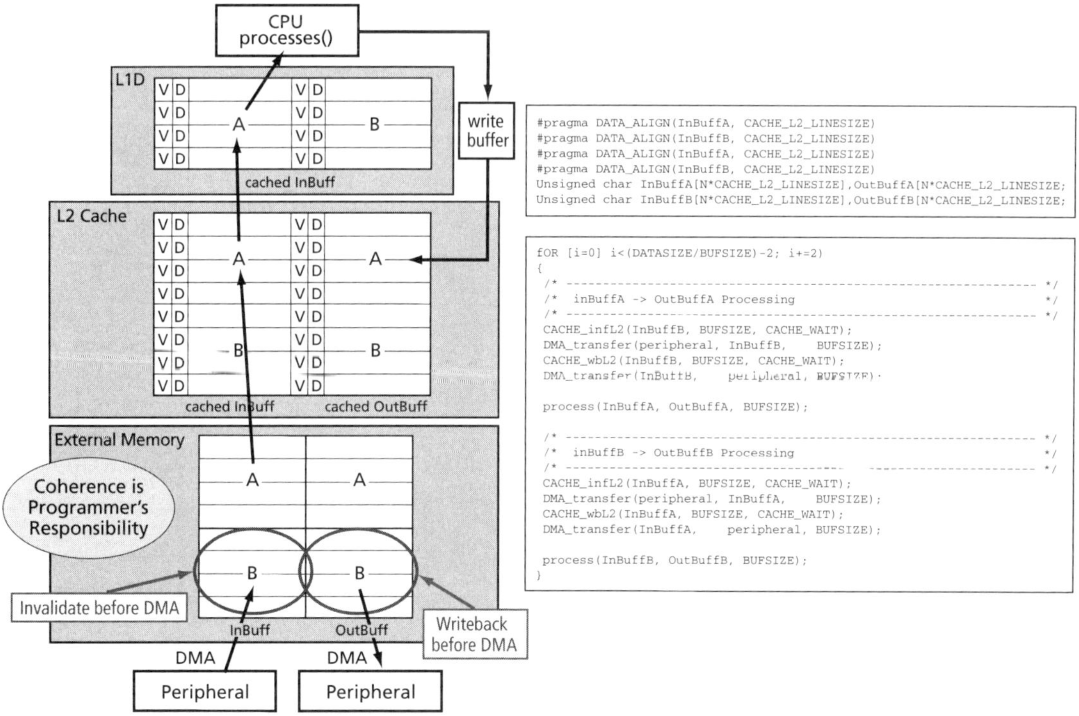

Figure C.11 DMA double buffering in incoherent memory (courtesy of Texas Instruments)

I used the double buffering examples to show how and when to use cache coherence operations. Now, when in general do I need to use them?

Only if CPU and DMA controller share a cacheable region of external memory. By share we mean the CPU reads data written by the DMA and vice versa. Only in this case I need to maintain coherence with external memory manually.

The safest rule is to issue a Global Writeback-Invalidate before any DMA transfer to or from external memory. However, the disadvantage here is that we'll probably operate on cache lines unnecessarily and get a relatively large cycle count overhead.

A more targeted approach is more efficient. First, we need to only operate on blocks of memory that we know are used as a shared buffer. Then we can also distinguish between the following three scenarios: The first two are familiar, we used them for our double buffering example. (1) If the DMA reads data written by the CPU, we need to use a L2 Writeback before the DMA starts, (2) If the DMA writes data that is to be read by the CPU, then we need to invalidate before the DMA starts. Then, a third case, the DMA may modify data that was written by the CPU that is consequently to be read back by the CPU. This is the case if the CPU initializes the memory first (e.g. sets it to zero) before a peripheral or else writes to the buffer. In this case we first need to commit the initialization data to external memory and then invalidate the buffer. This can be achieved with the Writeback-Invalidate command (On the C611/6711 an invalidate operation is not available. Instead a Writeback-Invalidate operation is used.)

When to use cache coherence control operations

The CPU and DMA share a cacheable region in external memory. The safest approach is to use *writeback-invalidate all cache* before any DMA transfer to/from external memory. The disadvantage is a large overhead for the operation. Overhead can be reduced by only operating on buffers used for DMA and distinguish between three possible scenarios shown in Figure C.12.

DMA reads data written by the CPU	Writeback before DMA
DMA writes data that is to be read by the CPU	Invalidate before DMA *)
DMA modifies data written by the CPU that is to be read back by the CPU	Writeback-Invalidate before DMA

Figure C.12 Three scenarios for using DMA to reduce cache misses

A good strategy for optimizing cache performance is to proceed in a top-down fashion, starting on application level, moving to procedural level, and if necessary considering optimizations on algorithmic level. The optimization methods for application level tend to be straightforward to implement and typically have a high impact on overall performance improvement. If necessary, fine tuning can then be performed using lower level optimization methods. Hence the structure of this chapter reflects the order in which one may want to address the optimizations.

Application level optimizations

There are various application level optimizations that can be performed in or to improve cache performance.

For signal processing code, the control and data flow of DSP processing are well understood and more careful optimization possible. Use the DMA for streaming data into on-chip memory to achieve the best performance. On-chip memory is closer to the CPU, and therefore latency is reduced. Cache coherence may be automatically maintained as well. Use L2 cache for rapid-prototyping applications but watch out for cache coherence related issues.

For general-purpose code, the techniques are a little different. General-purpose code has a lot of straight-line code and conditional branching. There is not much parallelism, and execution largely unpredictable In these situations, use L2 cache as much as possible.

Some general techniques to reduce the number of cache misses include maximizing the cache line re-use. Access *all* memory *locations* within a cached line The *same* memory *locations* within a cached line should be *re-used* as often as possible. Also avoid eviction of a line as long as it is being re-used. There are a few ways to do this;

- Prevent eviction: Don't exceed number of cache ways
- Delay eviction: Move conflicting accesses apart
- Controlled eviction: Utilize LRU replacement scheme

Optimization techniques for cache-based systems

Before we can begin to discuss techniques to improve cache performance, we need to understand the different scenarios that may exist in relation to the cache and our software application. There are three main scenarios we need to be concerned about:

Scenario 1 – All data/code of the working set fits into cache. There are no capacity misses in this scenario by definition, but conflict misses do occur. In this situation, the goal is to eliminate the conflict misses through contiguous allocation.

Scenario 2 – The data set is larger than cache. In this scenario, no capacity misses occur because data is not reused. The data is contiguously allocated, but conflict misses occur. In this situation, the goal is to eliminate the conflict misses by interleaving sets.

Scenario 3 – The data set is larger than cache. In this scenario, capacity misses occur because data is reused. Conflict misses will also occur. In this situation, the goal is to eliminate the capacity and conflict misses by splitting up the working set

I will show you an example of each of these.

Scenario 1

The main goal for scenario 1 is to allocate function contiguously in memory. Figure C.13a shows two functions allocated in memory in overlapping cache lines. When these functions are read into the cache, because of the conflict in memory mapping of the two functions, the cache will be trashed as each of these functions is called (C.13b and c). A solution to this problem is to allocate these functions contiguously in memory as shown in Figure C.13d.

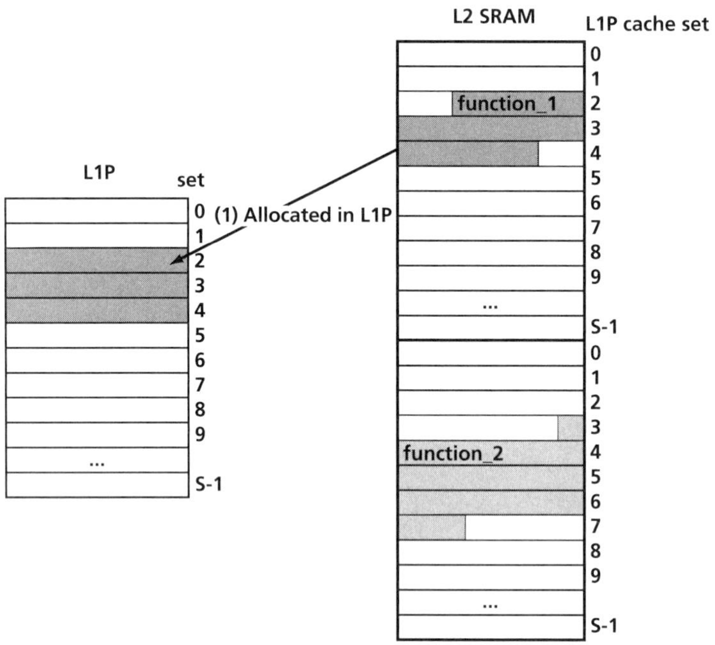

Figure C.13(a) Two function vying for cache space (courtesy of Texas Instruments)

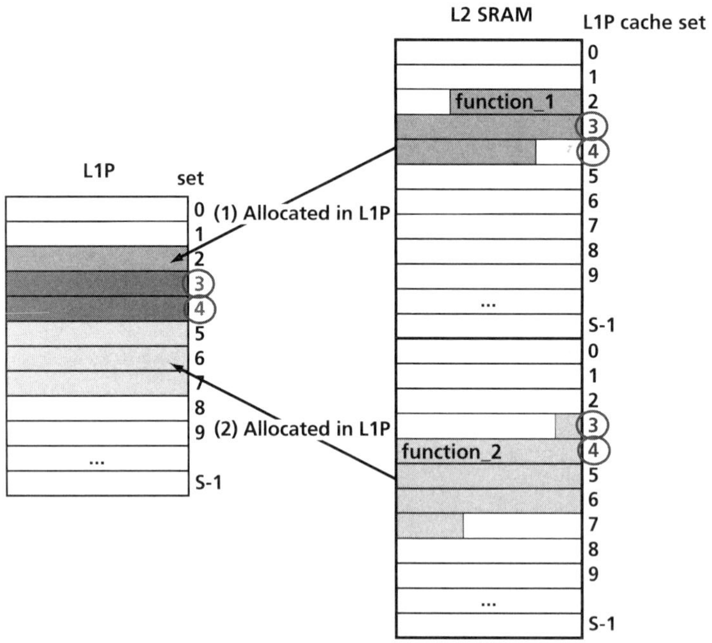

Figure C.13(b) Cache conflicts (lines 3 and 4) in two different functions (courtesy of Texas Instruments)

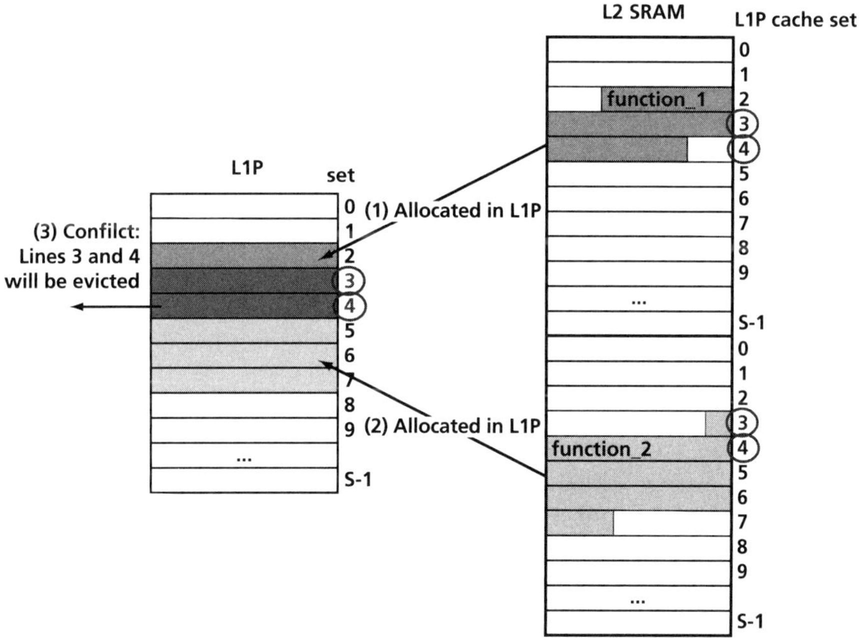

Figure C.13(c) Eviction due to cache conflicts (courtesy of Texas Instruments)

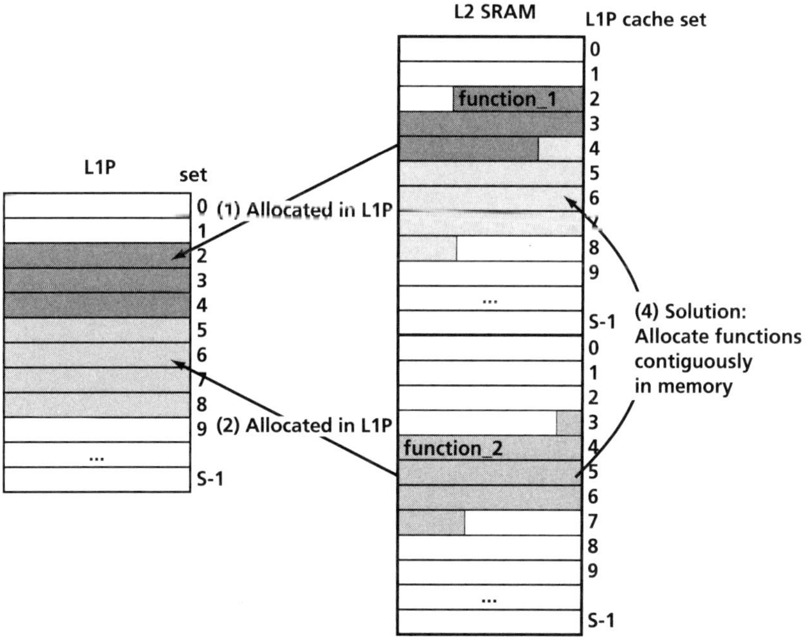

Figure C.13(d) Allocating arrays contiguously in memory prevents cache conflicts (courtesy of Texas Instruments)

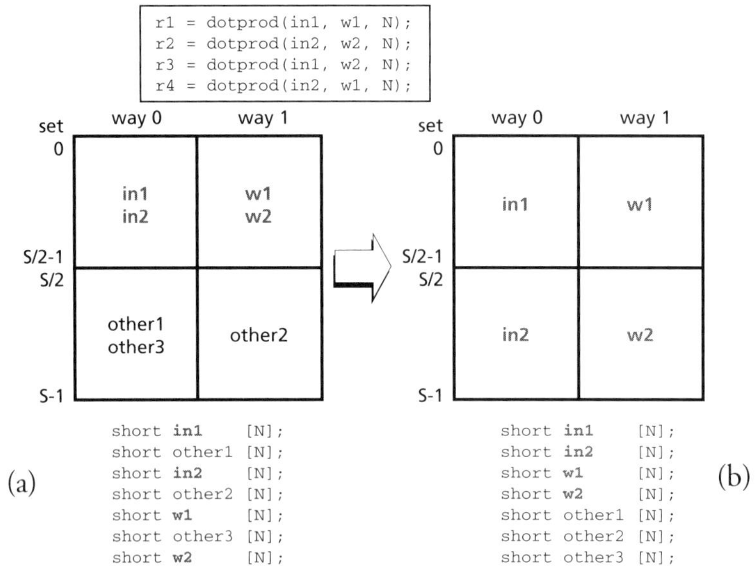

Figure C.14 Declaration of data that leads to cache conflicts (a) vs. one that prevents cache conflict (b) in a two-way associative cache (courtesy of Texas Instruments)

Scenario 2

Scenario 2 is the situation where the data set is larger than cache. No capacity misses occur in this situation because data is not reused. Data in memory is contiguously allocated but conflict misses occur. In this situation, the goal is to eliminate conflict misses by interleaving sets. Thrashing occurs if arrays are multiple of the size of one way. As an example consider Arrays w[ ], x[ ] and h[ ] map to the same sets in Figure C.15. This will cause misses and reduced performance. By simply adding a pad word, the offset for array h is now different from the other arrays and gets mapped into a different way in the cache. This improves overall performance.

Figure C.14a shows an example of a set of variables being declared in an order that causes an inefficient mapping onto the architecture of a two-way set associative cache. A simple rearrangement of these declarations will result in a more efficient mapping into the two-way set associative cache and eliminate potential thrashing (Figure C.14b).

Way0	Way1	LRU	Read Access to
		0	w[0]
w		1	x[0]
w	x	0	h[0]
h	x	1	w[1]
h	w	0	x[1]
x	w	1	h[1]

Every access misses!

```
short w   [N];
short x   [N];
char  pad [CACHE_L1D_LINESIZE];
short h   [N];
```

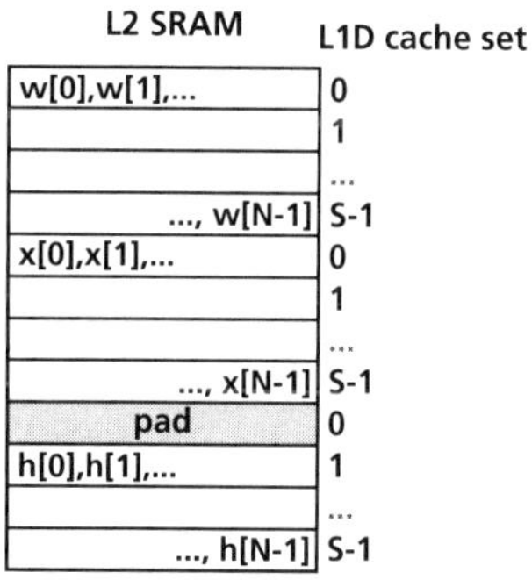

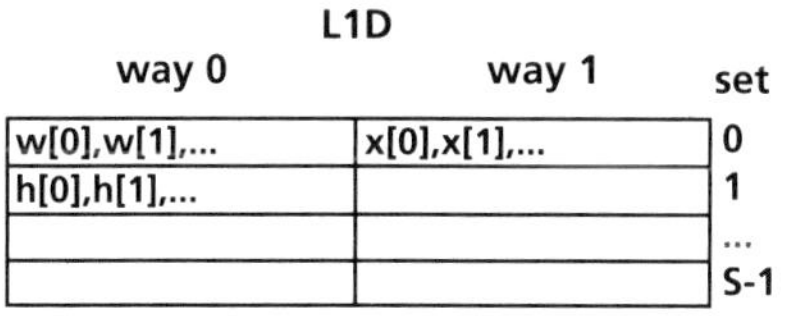

Figure C.15 Avoiding cache thrashing by padding the arrays! (courtesy of Texas Instruments)

Scenario 3

Scenario 3 occurs when the data set is larger than the cache. Capacity misses occur because data is reused and conflict misses can occur as well. In this situation, we must eliminate the capacity and conflict misses by splitting up the working set. Consider the example in Figure C14.b where the arrays exceed cache capacity. We can solve this with a technique called *blocking*. Cache blocking is a technique to optimally structure application data blocks in such a way that they fit into a cache memory block. This is a method of controlling data cache locality, which in turn improves performance. I'll show an algorithm to do this in a moment.

```
r1 = dotprod(in1, w, N);
r2 = dotprod(in2, w, N);
r3 = dotprod(in3, w, N);
r4 = dotprod(in4, w, N);
```

```
for (i=0; i<4; i++)
{
  o = i * N/4;
  dotprod(in1+o, w+o, N/4);
  dotprod(in2+o, w+o, N/4);
  dotprod(in3+o, w+o, N/4);
  dotprod(in4+o, w+o, N/4);
}
```

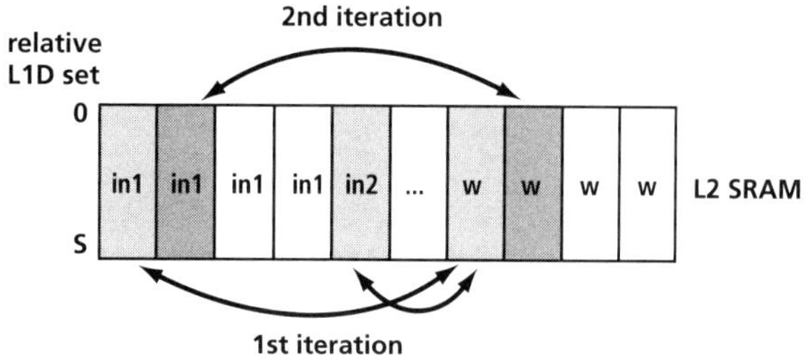

Figure C.16 Example of arrays exceeding the cache capacity (courtesy of Texas Instruments)

Processing chains

Before we consider the different cases, let's consider the concept of processing chains.

Often the output of one algorithms forms the input of the next algorithm. This is shown in Figure C.17. An input buffer is read by func1 and writes its results to an temporary buffer which is used as input to func2. Func2 then writes its output to an output buffer. If we can keep tmp_buf allocated in L1D, then func2 will not see any read misses. The data is written to L1D and read from L1D, avoiding any data fetches from L2 memory!

So, read misses will only for the first function in the processing chain. All following functions can operate…

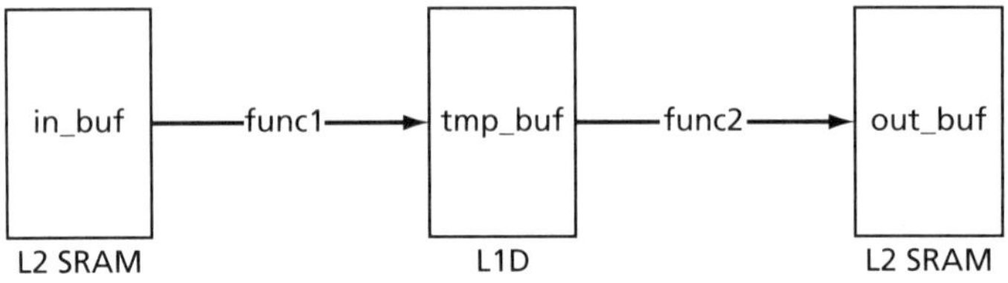

Figure C.17 Processing chains (courtesy of Texas Instruments)

As another example, consider Figure C.18. There is a Read miss only for the first function in the chain. All following functions can operate directly on cached buffers: In this situation there is no cache overhead! The last function writes through the write buffer.

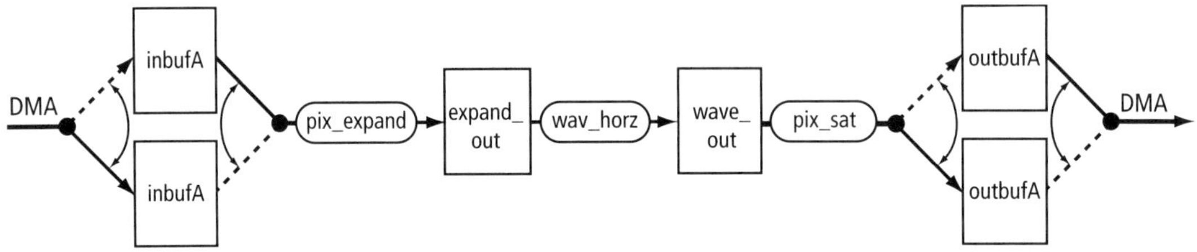

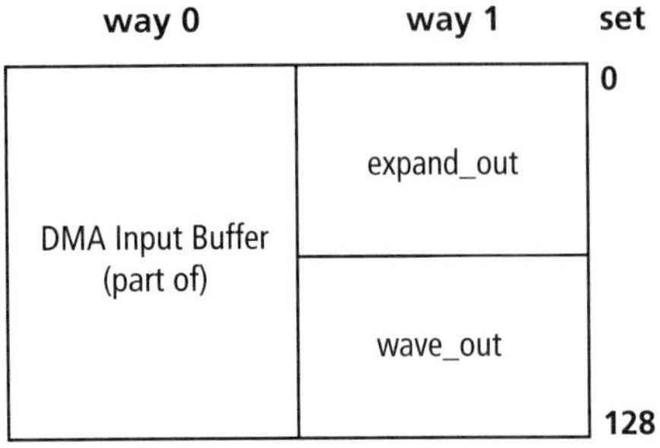

	DMA and Cache Opt.	DMA but No Cache Opt.	L2 Cache with Cache Opt.
L1D Stalls	10,042	219,615	1,008,224
L1P Stalls	130	633	1,489
Touch Execute	4,192	0	0
DMA Manag.	5,120	6,120	0
DMA Rel. Stalls	2,882	2,246	0
Loop Control	1,341	1,341	0
Total Overhead	23,707	228,955	1,009,713
Execute Cycles	650,875	650,875	650,875
Rel. Overhead	3.6%	35.2%	155%

Figure C.18 C64x Example: DMA processing chain (courtesy of Texas Instruments)

We have to keep the effects of processing chains in mind when we benchmark algorithms and applications. In this example the first function is a FIR filter which is followed by a dot product. The table below shows **cycle count** numbers for these algorithms. The execute cycles are shown, which are raw CPU execution cycles, below that L1D stalls that are predominately caused by L1D misses, and L1P stalls caused by L1P misses. The cycle counts are split up into 1st and 2nd iteration.

You see for the FIR filter the 1st and 2nd L1D stalls are about the same because we read new data from L2 SRAM, i.e., compulsory misses. On the other hand, once the temp buffer is allocated in L1D there are NO misses for the iterations after the 1st for dot prod (provided we have performed some optimization which I will describe in the next slides). We see in the 1st iteration we had 719 stalls (again compulsory misses) and in the 2nd we have only 6. Ideally this is 0, but we may get a few misses caused by stack accesses on function calls etc. The 719 stalls mean an overhead of 148%. This seems very large, but it's really only for the first iteration. If you run the processing loop only 4 times, the combined overhead is only 1.9%!

So, when you benchmark an algorithm on a cache-based memory architecture you have to keep this in mind and consider the algorithm in the context how you use it in the application.

What happens if FIR and DOTPROD are reversed (Figure C.19)? The overhead would be higher, but not by much, it would be 2.5%. They key is having a algorithm that does a high amount of processing compared to the number of memory accesses, or has high amount of data re-use.

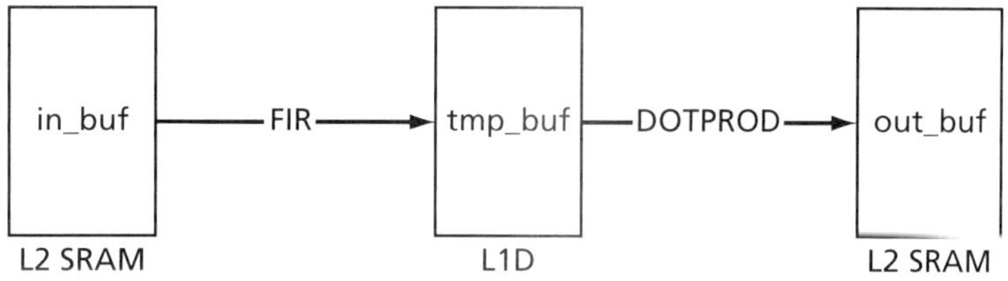

C64x	1st Iteration		2nd Iteration		Total (4 Iterations)
	fir	dotprod	fir	dotprod	
Execute Cycles	31,766	520	31,766	520	520
L1D Stalls	396	719	396	6	6
L1P Stalls	58	53	0	0	0
Total	32,220	1,292	32,160	526	526
Cache Overhead	1.5%	148%	1.2%	1.1%	1.1%

Figure C.19 processing chains (courtesy of Texas Instruments)

Interpretation of cache overhead benchmarks

Now, let's look at how to use and configure cache correctly. Whereas Level 1 cache cannot be switched off, L2 Cache can be configured in various sizes. The diagram on the left shows what configurations are possible. L2 Cache can be switched off, set to 32 KBytes, 64, 128 or 256 Kbytes. After reset L2 Cache is switched off. The rest is L2 SRAM which always starts at address 0x0. Let's say we want to use 256K of L2 as cache. First of all we need to make sure that no data or code is being linked into the L2SRAM portion to be used as cache. Figure C.20 is an example of how an appropriate linker command file could look like. L2 SRAM runs starts at 0 and is hex C0000 bytes large, which is 768K, then Cache starts at C0000 and is hex 40000, which is 256Kbytes in decimal. The last memory is external memory. We make sure nothing is linked into it by NOT declaring it under the MEMORY directive. In the program we need to call this sequence of CSL commands to enable caching of external memory locations into L2. The first command initializes the CSL, then we enable caching of the external memory space CE00 which corresponds to the first 16MB in external memory. Finally, we set L2 Cache to 256K. That's it.

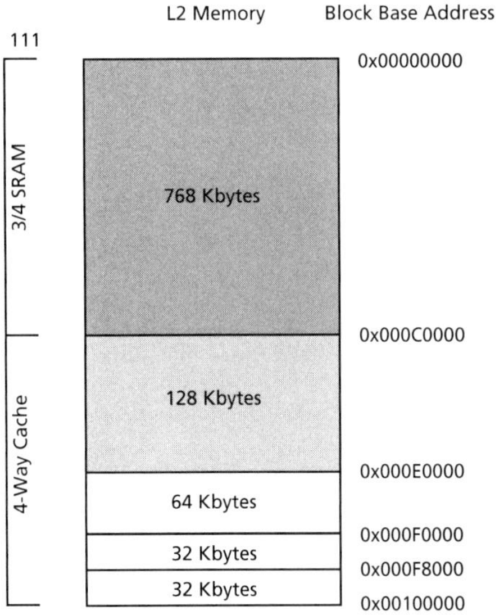


```
MEMORY
{
    /= Do NOT specify L2 Cache memory here! */
    L2SRAM:      o = 00000000h   1 = 000C0000b
    CEO:         o = 80000000h   1 = 01000000h
}
SECTIONS
{
    .cinit      >       L2SRAM
    .text       >       L2SRAM
    .stack      >       L2SRAM
    .bss        >       L2SRAM
    .const      >       L2SRAM
    .data       >       L2SRAM
    .far        >       L2SRAM
    .sysmem     >       L2SRAM
    .external   >       CEO
```

```
#include <csl.h>
#include <csl_cache.h>
...

CSL_init();
CACHE_enableCaching(CACHE_EMIFA_CE00);
CACHE_setL2Mode(CACHE_256KCACHE);
```

Figure C.20 Configuring L2 Cache (C64x) (courtesy of Texas Instruments)

Software transformation for cache optimization

There are various compiler optimizations that will improve cache performance. There are both instruction and data optimization opportunities using an optimizing compiler. An example of an instruction optimization is the reordering of procedures that reduce the cache thrashing that may occur. Reordering procedures can be done by first profiling the application and then using changes in the link control file to reorder these procedures.

The compiler may also perform several types of data optimizations including:

- Merging arrays
- Loop interchange
- Cache blocking
- Loop distribution

As an example of the compiler merging arrays, consider the array declarations below:

```
int array1[ array_size ];
int array2[ array_size ];
```

Restructuring the arrays as shown below will improve cache performance by improving spatial locality.

```
struct merged_arrays
{
        int array1;
        int array2;
} new_array[ array_ size ]
```

As an example of a loop interchange, consider the code snippet below:

```
for (i=0; i<100; i=i+1)
        for (j=0; j<200; j=j+1)
                for (k=0; k<10000; k=k+1)
                        z[ k ][ j ] = 10 * z[ k ][ j ];
```

By interchanging the second and third nested loops, we move to more of a sequential access instead of a stride of 100 words which will improve spatial locality.

```
for (i=0; i<100; i=i+1)
        for (k=0; k<10000; k=k+1)
                for (j=0; j<200; j=j+1)
                        z[ k ][ j ] = 10 * z[ k ][ j ];
```

As mentioned earlier, cache blocking is a technique to optimally structure application data blocks in such a way that they fit into a cache memory block. This is a method of controlling data cache locality which in turn improves performance. This can be done in the application or, in some cases, by using the compiler. The technique breaks large arrays of data into smaller blocks that can be repeat ably accessed during processing. Many video and image processing algorithms have inherent structures that allow them to be operated on this way. How much performance gains are available with this technique depend on the cache size, the data block size, and some other factors including how often the cached data can be accessed before being replaced. As an example, consider the snippet of code below:

```
for i = 1, 100
  for j = 1, 100
    for k = 1, 100
      A[i, j, k] = A[i, j, k]  +  B[i, k, j]
    end
  end
end
```

by changing this to a blocking structure like the one shown below, we can get maximum use of a data block while it is in cache before moving on to the next block. In this example, the cache block size is 20. The goal is to read in a block of data into the cache and perform the maximum amount of processing while the data is in the cache before moving on to the next block. In this case, there are 20 elements of array A and 20 elements of array B in the cache at the same time.

```
for x = 1, 100, 20
  for y = 1, 100, 20
    for k = x, x+19
      for j = y, y+19
        for i = 1, 100
          A[i, j, k] = A[i, j, k]  +  B[i, k, j]
        end
      end
    end
  end
end
```

Loop distribution is an optimization technique that attempts to split a large loop into smaller ones. These smaller loops not only increase the chance to be pipelined or vectorized by the compiler, but also increase the chance for the loop to fit into a cache in its entirety. This prevents the cache from trashing as the loop executes and improves performance. The code snippet below shows a for loop with a number of function calls inside the loop. By slitting up the loops,

```
*** code snippet before loop distribution
for (i=0; i< limit; i++) {
  a[i] = func1(i);
  b[i] = func2(i);
     :
     :
  z[i] = func10(i);
}

*** Code snippet after loop distribution
for (i=0; i< limit; i++) {
  a[i] = func1(i);
  b[i] = func2i);
     :
     :
  m[i] = func5(i);
}
for (i=0; i< limit; i++) {
  n[i] = func6(i);
  o[i] = func7(i);
     :
     :
  z[i] = func10(i);
}
```

Application level cache optimizations

There are also some application-level optimizations that the developer should be aware of that will improve overall application performance. The optimizations will vary depending on whether the application is signal processing software of control code software. For signal processing software, the control and data flow of DSP processing are well understood. We can therefore perform more careful optimizations. One specific technique to improve performance is to use the DMA for streaming data into on-chip memory to achieve best performance. On-chip memory is closer to the CPU, and therefore the latency is reduced. Cache coherence may be automatically maintained by using this on chip memory. L2 cache should also be used for rapid-prototyping applications (but watch out for potential cache coherence issues!).

General-purpose DSP software is characterized by straight-line code and a lot of conditional branching. There is not much parallelism in this software and the execution s largely unpredictable. This is when the developer should configure the DSP to use L2 cache, if available.

Techniques overview

In summary, the DSP developer should consider the following cache optimizations for embedded DSP applications;

1. Reduce the number of cache misses
   - Maximizing cache line re-use
     - Access *all* memory *locations* within a cached line
     - The *same* memory *locations* within a cached line should be *re-used* as often as possible.
   - Avoiding eviction of a line as long as it is being re-used
     - Prevent eviction: Don't exceed number of cache ways
     - Delay eviction: Move conflicting accesses apart
     - Controlled eviction: Utilize LRU replacement scheme
2. Reduce the number of stall cycles per miss
   - Miss pipelining

Figure C.21 is an example of a visual mapping of cache hits and misses over time. Both hits and misses form patterns when viewed visually that can alert the viewer to cache performance problems. Reordering data and program placement as well as using the other techniques we have been discussing will help eliminate some of these miss patterns.

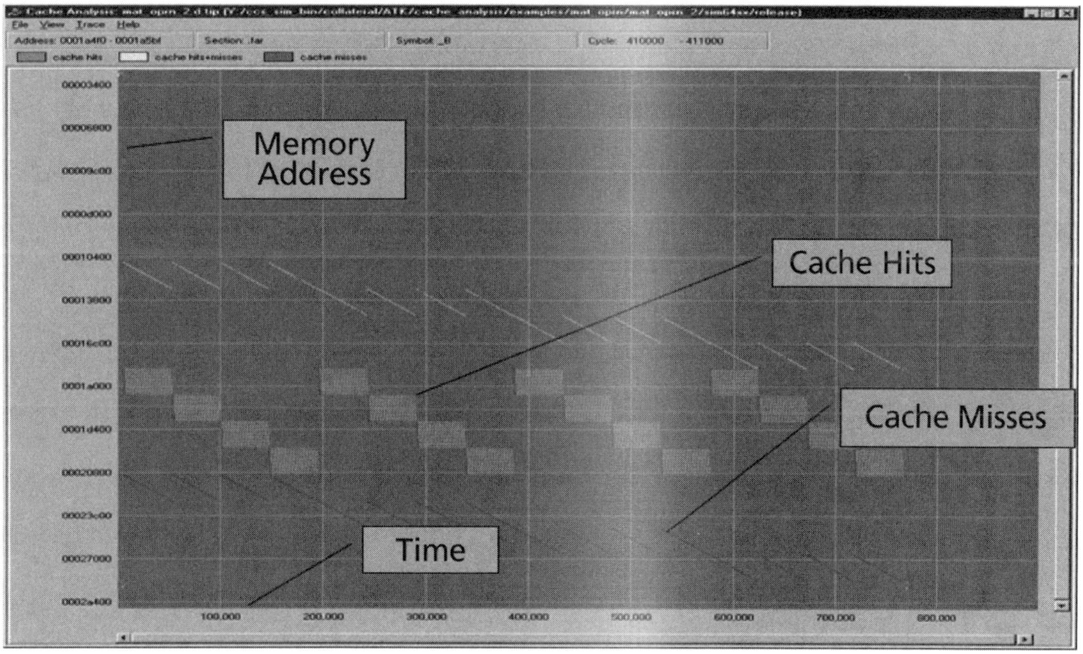

Figure C.21 Cache Analysis lets you determine how the placement of your code in memory is impacting code execution

I would like to thank Oliver Sohm for his significant contributions to this appendix.

References

1. Hennessy, J.L. and Patterson, D.A. *Computer Architecture: A Quantitative Approach.* Morgan Kaufmann Publishers, San Francisco CA, 1996.
2. Texas Instruments. C621x/C671x Two-Level Internal Memory Reference Guide, 2002.
3. Texas Instruments. C64x Two-Level Internal Memory Reference Guide, 2002.
4. Texas Instruments, C6000 Cache User's Guide, 2003.

D

Specifying Behavior of Embedded DSP Systems

What Makes a Good Requirement?

The criticality of correct, complete, testable requirements is a fundamental tenet in software engineering. Both functional and financial success is affected by the quality of requirements. So what is a requirement? It may range from a high-level abstract statement of a service or of a system constraint to a detailed mathematical functional specification. Requirements are needed for several reasons:

- Specify external system behavior.
- Specify implementation constraints.
- Serve as reference tool for maintenance.
- Record forethought about the life cycle of the system i.e., predict changes.
- Characterize responses to unexpected events.

The system designer must understand requirements and be able to organize them. A technical background and an understanding of the user are both required. Before design can start, each requirement must be understood in terms of significance and priority in the solution strategy. Because both the developer and the customer must understand the requirements, they are usually written in natural language. But natural language is a poor medium for communicating requirements. The use of natural language to specify complex requirement has at least two problems; those of ambiguity and inaccuracy. Many words and phrases have dual meanings and can be altered depending on the context in which they are used. A word that means one thing to one person can mean something entirely different to someone else. This is referred to as *semantic bypass.* For example the interpretation of the word "Bridge" can have completely different meanings depending on whether you are a dentist, a civil engineer, an electrical engineer, or someone who just retired early!

Other disciplines require a more exact language for communicating specifications. For example an architect blueprint or a circuit board schematic are formal specifications that define what is needed but not necessarily the implementation details of how to get there. Software requirements should be the same way. Software requirements specifications should not contain implementation details. These specifications should describe in sufficient detail what the software will do but not how. A good set of requirements has the following characteristics:

- *Correct*: Meets the need.
- *Unambiguous*: Only one possible interpretation.
- *Complete*: Covers all the requirements.
- *Consistent*: No conflicts between requirements.
- *Ranked for importance.*
- *Verifiable*: A test case can be written.
- *Traceable*: Referring to requirements easy.
- *Modifiable*: Easy to add new requirements.

Trying to define a large multidimensional capability of a complex embedded system within the limitations of a linear two-dimensional structure of a document becomes almost impossible. At the other end of the scale, the use of a programming language is too detailed. This is nothing more than "after the fact specification" which is just documenting what was implemented rather than what was required

Developing a good set of requirements to specify a system can be hard. In many cases the stakeholders don't know what they really want. The various stakeholders also tend to express requirements in their own terms and different stakeholders may have conflicting requirements that need to be resolved. Organizational and political factors may also influence the system requirements and the requirements change during the analysis process as new stakeholders emerge. The term "requirements engineering" has emerged that addresses the transformation by which vague and often unrelated customer requirements are transformed into detailed and precise requirements needed for system implementation.

Successful designs are usually the result of significant rethinking and reworking. Several iterations of design is the norm, rather than the exception. Design alternatives should be considered. In general, designs should get simpler rather than more difficult.

It can be difficult to specify the total behavior of a complex system because of the total number of possible uses of the system. But this is precisely what needs to be done in order to ensure completeness and consistency in our designs. As James Kowal describes:

"If the systems planners and customer do not specify what is expected in all types of interactions with the system, i.e., the behavior of the system, someone else will.

That someone else is most likely the programmer when he or she is coding the ELSE option of some IF statement. There is a very low probability that the programmer's guesses as to the expected behavior will be what the customer expects."

The hard part is being able to determine all the possible types of interaction with the system. Use cases are used to explore and elicit requirements and can help determine certain types of interaction to expect between the actors (external stimuli to the system) and the system. But use cases may not always combine to describe *complete* and *consistent* behavior. Once use cases are used to explore the problem and for front end domain analysis, other techniques can then be used to fully specify the solution strategy. I want to explain one approach that has worked well in our embedded system project, called sequence enumeration.

Sequence Enumeration

Sequence enumeration is a way of specifying stimuli and responses of an embedded system. This approach considers *all* permutations of input stimuli. Sequence enumerations consist of a list of prior stimuli and current stimuli as well as a response for that particular stimuli given the prior history. Equivalent histories are used to map certain responses. This technique maps directly to a state machine implementation. The strength of sequence enumerations is that the technique requires the developer to consider the obscure sequences that are usually overlooked.

As an example, I will consider the cell phone shown in Figure D.1. A very simplified set of natural language requirements for this system is shown in Table D.1.

Figure D.1 A cell phone

Tag Number	Requirement
1	When the "POWER" button is activated, the display light comes on and the initial screen is displayed.
2	If the cell phone is on and the "POWER" button is pressed, the cell phone turns off, and a "good bye" message is displayed on the screen.
3	After each digit button is pressed, the character is echoed on the display.
4	If there is a valid phone number entered into the display, when the "GO" button is pressed, a "calling" message is displayed on the display and the number is dialed.
5	If there are less than four digits on the display, the "GO" button is ignored.
6	If the phone is in a call, when the "STOP" button is pressed, a "terminating call" message is displayed on the screen and the call is terminated.
7	If the phone is not in a call but there are digits on the screen, when the "STOP" button is pressed, the screen is cleared.
8	If there are no digits on the screen, the "STOP" button is ignored.
9	This cell phone requires a 4-digit code for a phone number.

Table D.1 Simplified set of natural language requirements for the cell phone machine

Assumptions

- All four-digit combinations are valid phone numbers.
- The cell phone will only work when the power has been applied. This is a prerequisite for the system and not a requirement. Because power must be applied for the system to be operational, it is not considered a stimulus to the system. System stimuli must impact the functioning system in order to produce a response.

To start with, a use case can be developed that describes one type of interaction with the system. A use case is a story about the usage of a system told from an end users' perspective. It consists of:

- *Preconditions*: What must be available and done before the performance of the use case.
- *Description*: The story.
- *Exceptions*: Exceptions in the normal flow of the story.
- *Illustrations*: Figures to help understand the use case.
- *Postconditions*: The state of the system and actors after the use case has been performed.

The motivation for use cases are:

- Tool to analyze and improve functional requirements.
- Tool to model the functionality of the system as soon as possible.
- To allow the customer to understand the operation of the system.

Use cases must specify the most important functional requirements. A use case depicts a typical way of using the system but nothing more. In general, a use case should not try to define all possible ways if performing a task. Certain important exception conditions are described in "secondary" use cases or in exceptions within a use case. A simple use case for the coke machine can be as follows:

- *Precondition*: The cell phone has been turned on and is displaying the default screen.
- *Description*: A user decides to make a phone call using the cell phone. The battery is already in the cell phone, the on button is pressed, four digits are pressed to dial a number, the go button is pressed to connect the call, and the on switch is accidentally hit again. The system ignores the on button press since the phone is already on (the assumption is that only a "stop" or "off" button press can terminate the call.
- *Exceptions*: None.
- *Postconditions*: The cell phone is displaying the default screen.

Other use cases can be developed depending on the various business cases for this system. Once we have a general idea of what the system should do (which use cases are useful for determining), we can then perform a sequence enumeration to more completely map all possible combinations of stimuli to a response. To start with, it's a good idea to represent the system as a "black box," showing the stimuli and the responses that are possible. The black box for the coke machine is shown in Figure D.2. Table D.2 summarizes the stimuli and responses for the system.

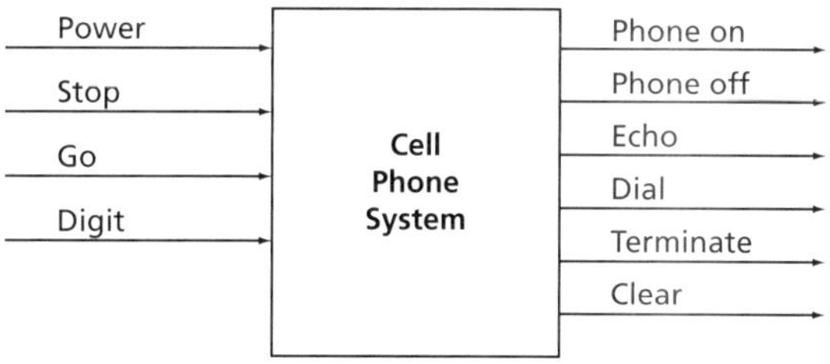

Figure D.2 Cell phone machine black box

Cell Phone Stimuli:

Stimulus	Description	Requirement Trace Number
Turn ON	Display light → on Initial screen is displayed	1
Turn OFF	"Good bye" message	2
Digit	A digit that is part of the phone number is pressed. The character is echoed on the display	3
Go	Initiate a phone call → "calling" message	4
Stop	Terminate a phone call or other entry / "terminating call"	5
Battery charge	Power has been applied to the phone	6

Cell Phone Responses:

Response	Description	Requirement Trace Number
Screen light ON	Screen illuminated	1
Screen light OFF	Screen not illuminated	2
Initial screen	Displays the initial screen	1
"Calling"	Screen Display "calling" message	4
"Terminating call"	Screen Display "terminating call"	5
"Good bye"	Display "good bye"	2
Digit echo	Display digit pressed	3
Cell phone has energy	Power has been applied to the phone	6

Table D.2 Cell phone machine stimuli and responses

A sequence enumeration can be done in a simple excel spreadsheet. Enumeration starts at evaluating a single stimuli to the system and determining the appropriate response for each stimuli. Table D.3 shows the enumeration for the system. As the table shows, each stimuli is separately evaluated. Starting at the beginning of the "length one" stimuli enumerations, the enumeration shows that a "Battery Charged" (B) stimuli must occur (the power is applied and the system is initialized) before operation of the cell phone, and a derived requirement (D6) documents this behavior. If any other stimuli occurs the response is considered an illegal sequence (you cannot use the cell phone if the power is not on, for example). The result from performing this first step is "B" is required before anything else can be done and we must remember that this event occurred. Therefore, the "B" stimuli is used in the next level of enumeration, level 2.

The level 2 enumeration uses all stimuli sequences that were not illegal or did not have a shorter equivalent sequence in the previous enumeration level as the first stimuli in the sequence. So, as the length 2 section in Table D.3 illustrates, the first stimuli is "B" and the second stimuli is each of possible stimuli to the system (the simple cross product). As an example, the level 2 sequence enumeration "B N" shows, if the system receives a power on signal, followed by the "ON" signal, the system will respond with turning the "screen light" on and displaying the "initial screen". The sequence " B D" describes that the sequence of Power on followed by the pressing of a single digit produces no response and is equivalent to the shorter sequence "B" (in other words, the sequence B D is equivalent to the sequence B—in other words, the D stimulus is ignored while the power is off.

Since the sequence "B N" is the only sequence at level 2 that does not have a shorter equivalent sequence and is not illegal, it must be evaluated at sequence length n+2, or in this case, sequence length 3.

In a similar fashion, the sequence B N is mapped to a cross product of all input stimuli again and yields the results shown in sequence length 3 in the table. In this case, we can see that the sequence " B N N" is equivalent to the shorter sequence "B N". The sequence "B N F" (which implies the battery is enabled, the on switch is turned on, and the "off" switch is then enabled, the system will display the "good bye" screen and turn the screen light off. The sequence "B N D" (battery on, system turned on, digit pressed) echoes a digit to the screen. The sequence "B N G" (battery on, system on, go button pressed) is equivalent to the shorter sequence "B N" (hitting the go button with having dialed any digits is ignored in this system).

This process continues until all behavior is eventually mapped to a shorter equivalent sequence (it must always end or the software system may never terminate!). Table 3 shows this process terminating in length 9.

The specific use case described earlier can be seen in the length 4 sequence "P D D S" which effectively describes a case of powering the phone on, entering two digits, and then pressing the Stop button which takes the system back to an initial state. The second part of the user case is relected in the length 7 enumeration "P D D D D G S", which completed the call and returns the system to the initial state "P" with the power on and the system in the initial state.

Stimuli

N – Turn ON

F – Turn OFF

D – Digit

G – Go

S – Stop

B – Battery charge

Sequence Enumeration

Sequence	Response	Equivalence	Requirement Trace Number
Length zero			
Empty	Null		Initially the cell phone is deactivated and has no power

Sequence	Response	Equivalence	Requirement Trace Number
Length one			
N	Illegal		D1,D6; Power should be applied to the phone, before Turn On
F	Illegal		D1; The cell phone is initially OFF
D	Illegal		D1; None key has effect while the cell phone is off
G	Illegal		D1; None key has effect while the cell phone is off
S	Illegal		D1; None key has effect while the cell phone is off
B (carry to next level)	Battery Charged		D6; First step should be: apply power to the phone

Sequence	Response	Equivalence	Requirement Trace Number
Length two			
BN (carry to next level)	screen light on / initial screen		1
BF	null	B	D1; The cell phone is initially OFF
BD	null	B	D1; None key has effect while the cell phone is off
BG	null	B	D1; None key has effect while the cell phone is off
BS	null	B	D1; None key has effect while the cell phone is off
BB	null	B	D6; Cell phone has already been charged

Sequence	Response	Equivalence	Requirement Trace Number
Length three			
BNN	null	BN	D1; Cell Phone has already been Turned on
BNF	"good bye" / screen light off		1,2
BND	Digit echo		3,7; needs 4-digit code
BNG	null	BN	7; 4-digit code must be entered before "Go"
BNS	null	BN	D5; In this case S has no effect
BNB	null	BN	D6; Cell Phone has already been charged

Sequence	Response	Equivalence	Requirement Trace Number
Length four			
BNFN	screen illuminated / Initial screen	BN	1
BNFF	null. Stimuli is ignored.	BF	D1; None key has effect while the cell phone is off
BNFD	null. Stimuli is ignored.	BD	D1; None key has effect while the cell phone is off
BNFG	null. Stimuli is ignored.	BG	D1; None key has effect while the cell phone is off
BNFS	null. Stimuli is ignored.	BS	D1; None key has effect while the cell phone is off
BNFB	null. Stimuli is ignored.	B	D1; None key has effect while the cell phone is off

Sequence	Response	Equivalence	Requirement Trace Number
Length four			
BNDN	null	BN	D7; Assumption: if "On", "Go" or "Stop" are pressed before complete 4-digit code, the cell phone will be reset.
BNDF	screen light off / "good bye"	BNF	
BNDD	Digit echo		7, 4-digit code is needed
BNDG	null	BN	D7; Assumption: if "On", "Go" or "Stop" are pressed before complete 4-digit code, the cell phone will be reset.
BNDS	null	BN	D7; Assumption: if "On", "Go" or "Stop" are pressed before complete 4-digit code, the cell phone will be reset.
BNDB	null	BND	Assumption: More power will not reset the cell phone. Ignoring stimuli.

Sequence	Response	Equivalence	Requirement Trace Number
Length five			
BNDDN	null	BN	D7; Assumption: if "On", "Go" or "Stop" are pressed before complete 4-digit code, the cell phone will be reset.
BNDDF	screen light off / "good bye"	BNF	2
BNDDD	Digit echo		7, 4-digit code is needed
BNDDG	null	BN	D7; Assumption: if "On", "Go" or "Stop" are pressed before complete 4-digit code, the cell phone will be reset.
BNDDS	null	BN	D7; Assumption: if "On", "Go" or "Stop" are pressed before complete 4-digit code, the cell phone will be reset.
BNDDB	null. Ignoring stimuli.	BNDD	D7; Assumption: More power will not reset the cell phone. Ignoring stimuli.

Sequence	Response	Equivalence	Requirement Trace Number
Length six			
BNDDDN	null	BN	D7; Assumption: if "On", "Go" or "Stop" are pressed before complete 4-digit code, the cell phone will be reset.
BNDDDF	screen light off / "good bye"	BNF	
BNDDDD			7, 4-digit code is needed
BNDDDG	null	BN	D7; Assumption: if "On", "Go" or "Stop" are pressed before complete 4-digit code, the cell phone will be reset.
BNDDDS	null	BN	D7; Assumption: if "On", "Go" or "Stop" are pressed before complete 4-digit code, the cell phone will be reset.
BNDDDB	null	BNDDD	Assumption: More power will not reset the cell phone. Ignoring stimuli.

Sequence	Response	Equivalence	Requirement Trace Number
Length seven			
BNDDDDN	null	BN	D7; Assumption: if "On", "Go" or "Stop" are pressed before complete 4-digit code, the cell phone will be reset
BNDDDDF	screen light off / "good bye"	BNF	
BNDDDDD	null	BN	D7; Assumption: if more than 4-digit code are pressed, the cell phone will be reset
BNDDDDG	"calling"		
BNDDDDS	null	BN	D7; Assumption: if "On", "Go" or "Stop" are pressed before complete 4-digit code, the cell phone will be reset
BNDDDDB	null	BNDDDD	Assumption: More power will not reset the cell phone. Ignoring stimuli.

Sequence	Response	Equivalence	Requirement Trace Number
Length eight			
BNDDDDGN	null (ignoring stimuli)	BNDDDDG	D5; Assumption: While making a call, only "Stop" or "Off" can finish the call.
BNDDDDGF	screen light off / "good bye"	BNF	2
BNDDDDGD	null (ignoring stimuli)	BNDDDDG	D5; Assumption: While making a call, only "Stop" or "Off" can finish the call.
BNDDDDGG	null (ignoring stimuli)	BNDDDDG	D5; Assumption: While making a call, only "Stop" or "Off" can finish the call.
BNDDDDGS	"terminating call"		
BNDDDDGB	null (ignoring stimuli)	BNDDDDG	D5; Assumption: While making a call, only "Stop" or "Off" can finish the call.

Sequence	Response	Equivalence	Requirement Trace Number
Length nine			
BNDDDDGSN	null (ignoring stimuli)	BN	The cell phone is already ON
BNDDDDGSF	screen light off / "good bye"	BNF	2
BNDDDDGSD	Digit echo	BND	3
BNDDDDGSG	null (ignoring stimuli)	BN	D7; Assumption: if "On", "Go" or "Stop" are pressed before complete 4-digit code, the cell phone will be reset
BNDDDDGSS	null (ignoring stimuli)	BN	D7; Assumption: if "On", "Go" or "Stop" are pressed before complete 4-digit code, the cell phone will be reset
BNDDDDGSB	null (ignoring stimuli)	BN	Assumption: More power will not reset the cell phone. Ignoring stimuli.

Table D.3 Cell phone sequence enumeration

You may have noticed that the enumeration tables have a "Requirement" column that is used to trace the requirement as we define the behavior. This is a simple easy way to trace all requirements to behavior. You will also notice the derived requirements (indicated as D1, D2, etc). A derived requirements is a lower level requirement that is determined to be necessary for a higher level requirement to be met. In general, a derived requirement is more specific and directed toward some sub-element of the project. Derived requirements often occur through the process of analysis.

There are a couple important conclusions to draw from this process. We can say that this process produced a set of behavior that is:

- *Complete*: The enumeration considered, in effect, every possible combination of stimuli to the system for an infinite length of possible stimuli. This ensures that we have considered every possible behavior condition. The requirement trace also ensures that we have covered all possible requirements.
- *Consistent*: Because we have considered every combination of stimuli only once, we have ensured consistency. In other words we have not mapped a stimuli combination to more than one possible output sequence.
- *Correct*: The mapping from stimuli to responses have been properly specified in the judgement of the domain experts.

Once the sequence enumeration is complete, it is a fairly simple step to develop the finite state machine that represents the behavior we have just specified. For one thing, I can determine how many states in the finite state machine there will be by simply counting the lines of enumeration that are not illegal and do not have equivalent sequences. These are referred as the *canonical* sequences. From Table D.3, these are those sequences that are not "illegal" and do not have an equivalence at that level of enumeration. These are listed in Figure D.5. State data can be invented that will encapsulate the behavior described in these canonical sequences. Figure D.6 shows that state data that I invented to encapsulate this relevant system behavior.

Canonical Sequence	State Variables (Power, Phone, Screen_Status)	Value before current stimulus	Value after current stimulus
B	Power	No	Yes
BN	Power Phone Screen_Status	Yes Off Light off	Yes On Light on / initial sc.
BNF	Power Phone Screen_Status	Yes On Light on	Yes On "good bye"
BND	Power Phone Screen_Status	Yes On Light on / initial sc.	Yes On d1
BNDD	Power Phone Screen_Status	Yes On d1	Yes On d1d2
BNDDD	Power Phone Screen_Status	Yes On d1d2	Yes On d1d2d3
BNDDDD	Power Phone Screen_Status	Yes On d1d2d3	Yes On d1d2d3d4
BNDDDDG	Power Phone Screen_Status	Yes On d1d2d3d4	Yes On "calling"
BNDDDDGS	Power Phone Screen_Status	Yes On "calling"	Yes On "terminating call"

State Mapping

Tag #	Current State	Response	State Update	Sequence Trace
1	Phone = ON Number = NONE Status = NOT_CONNECTED	Phone on	Phone = ON	P
2	Phone = ON Number = ONE Status = NOT_CONNECTED	Character echo	Number = ONE	P D
3	Phone = ON Number = TWO Status = NOT_CONNECTED	Character echo	Number = TWO	P D D
4	Phone = ON Number = THREE Status = NOT_CONNECTED	Character echo	Number = THREE	P D D D
5	Phone = ON Number = FOUR Status = NOT_CONNECTED	Character echo	Number = FOUR	P D D D D
6	Phone = ON Number = FOUR Status = CONNECTED	Dial	Status = CONNECTED	P D D D D G

Figure D.5 Canonical sequences for the cell phone machine

State variable	Range	Initial Value
Power	{Yes / No}	No
Phone	{On / Off}	Off
Screen Status	{Digit echo, "Calling", "Terminating", "Good bye"}	None

Figure D.6 State data for the cell phone machine and state data mapping

Based on the state data just invented and the behavior described by the enumeration sequences, the state machine for the cell phone is easily produced (Figure D.7). Notice the seven states corresponding to the seven canonical states produced during the enumeration process.

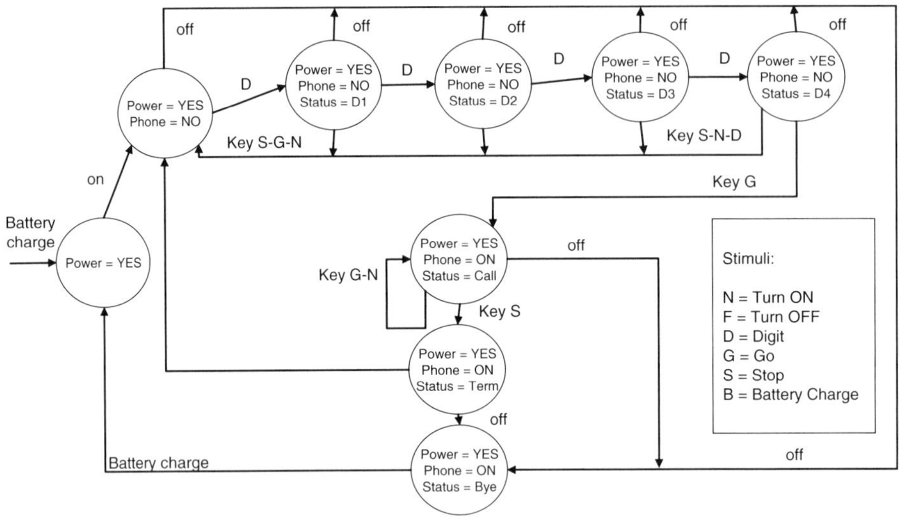

Figure D.7 State machine for the coke machine system

It's Really All Just Math!

The process described can be scaled to larger systems by using abstractions for the stimuli and responses. These abstractions can be decomposed and expanded at lower levels of behavior definition, as more and more details of the system are exposed. In effect, what we have done is define a rule for a function for this software system. This process is rooted in the mathematics of function theory, which maps a domain (valid inputs to the system) to a range (the correct results) within a co-domain (all *possible* results) (Figure D.8). A relation, on the other hand, is a more general mapping from domain to range and allows the same domain element to map to more than one range element (similar to what would happen in a software system with an un-initialized variable—different behavior for the same input sequence!).

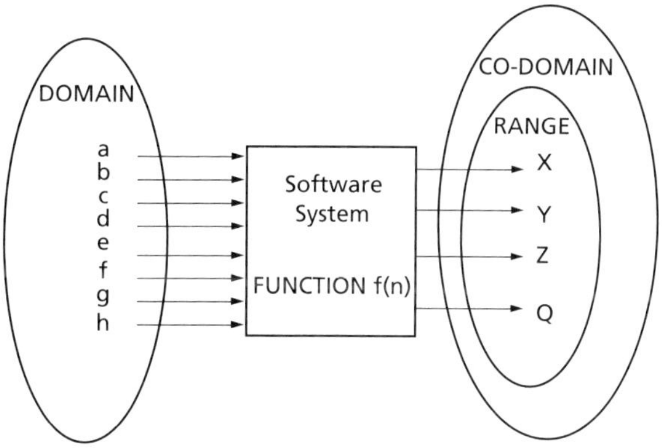

Figure D.8 A function maps a domain to a range within a co-domain

Referring back to the use case described earlier, you will notice that we can map the behavior from this use case to parts, but not all, of the enumeration we just completed. There are certain parts of the enumeration that describe the story being told by the use case. The sequence enumeration goes a step further, defining the complete behavior that was originally described in the story told in the use case. This is where the synergy between use cases and sequence enumeration becomes apparent. Use cases and scenarios are an effective tool to decide what we have to do for the various stakeholders of the system. The sequence enumeration drives home the completeness and consistency that will give your developers a clear picture of what to do in all circumstances of use as well as resolve any inconsistencies between the stakeholders.

E

Analysis Techniques for Real-Time DSP Systems

Introduction

DSP-based real-time systems have become more and more complex, causing designers to look at trade-offs between hardware and software when they define and analyze initial requirements. Today, engineers find themselves trading off system flexibility, speed, cost, schedules, and available tools. Plus, with newly available devices, functionality and requirements can be implemented in hardware, software, or a mixture of both.

Many system designers will execute a life cycle of hardware/software co-design, when both the hardware and the software are being developed simultaneously. In these environments, it is essential that hardware and software designers interact in the requirements definition and preliminary design process. Understanding the relationship between hardware and software functionality and the boundaries between the two helps to ensure requirements are designed and implemented completely and correctly.

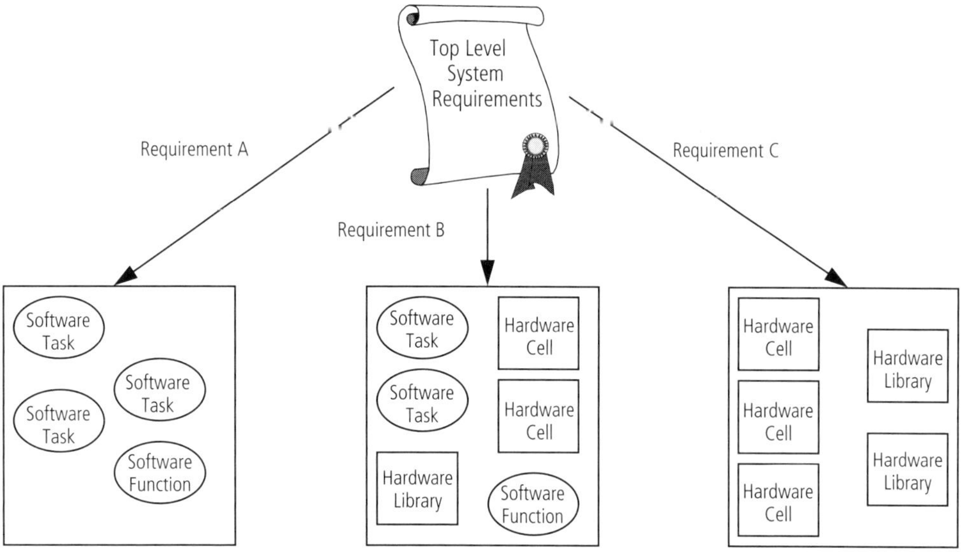

Figure E.1 Allocation of requirements to hardware and software

Early in the requirements definition and analysis phase, system developers, in close cooperation with the hardware and software design engineers, allocate requirements to hardware or software, and sometimes both (Figure E.1). This allocation is based on early system simulation, prototyping, and behavioral modeling results, as well as experience and other trade-offs mentioned earlier. Once this allocation has been made, detailed design and implementation begins. Various analysis techniques are applied to real-time micro-based systems development when both the hardware and the software are being designed concurrently:

- Hardware and software simulation.
- Hardware/Software emulation.
- Rate monotonic analysis.
- Prototyping and incremental development.

In some cases, existing products are available off-the-shelf to perform various levels of simulation and emulation. In other cases, some development effort is necessary to allow the proper analysis and risk management. In any case, system development environments are supported by several levels of modeling (Figure E.2) to support the designers throughout the development life cycle.

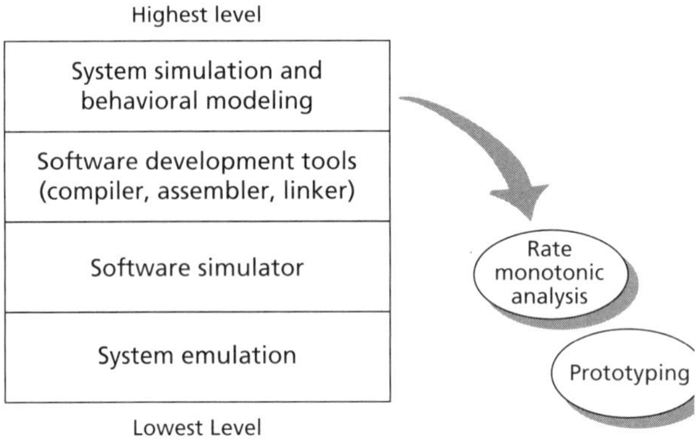

Figure E.2 Levels of system development environment

Hardware and Software Simulation and Modeling

Simulation of hardware/software systems is used to model the behavior of the system of interacting hardware and software parts. Until the advent of integrated simulation packages, modeling was mainly a static, nonexecutable analysis of a system. Much of the behavior of the system being modeled was subject to widely varying human interpretation. Simulation packages today allow executable models to be developed and analyzed. Simulation is used early in the design process to determine the functionality that must be performed in hardware and software. There are three types of simulation that can be performed:

- Functional (data and algorithms),
- Behavioral (process sequencing),
- Performance (resource utilization, throughput and timing).

Simulation also is used to make early evaluations of performance. Simulation can be performed at various levels of abstraction, depending on the goal of the particular simulation. Low level simulations are used to model bus bandwidths and data flows and are useful for evaluating performance. High level simulations can address interactions of functions and to perform hardware/software trade-off studies and validate designs. Using simulation, complex systems can be abstracted down to fundamental components and activities, allowing the designer to understand how these components and activities fit together.

Using integrated simulation tools available on the market today designers can develop executable system models to analyze the functional behavior as well as architectural performance.

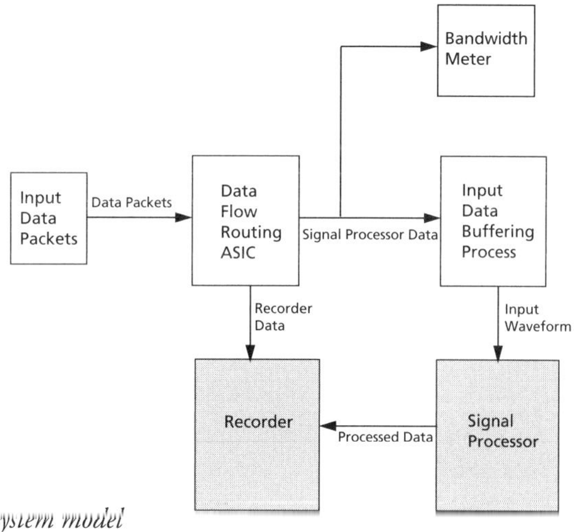

Figure E.3 A simple system model

Simulation models represent general system behavior typically using data flow diagrams (DFD). These DFDs are composed of nodes (system processes which represent units of work), arcs (system data and control flows) and data stores (Figure E.3). Simulators also provide library elements to allow the designer to monitor important system parameters such as bandwidth, temperature, or any other derived parameter. These types of models can be both satirically analyzed and dynamically executed.

Information obtained from initial modeling efforts allow hardware and software designers to begin more detailed analysis and design. For example, if a system simulation of the simple model in Figure E.3 showed that data would be arriving at the processor once every two milliseconds, then the task running on the processor responsible for processing that data would have two milliseconds to complete its work. The effective period has been established for that task. Likewise, data flowing from the processor to the recorder every 50 milliseconds establishes the period of the task associated with outputting data. These tasks can then be looked at more closely with other analysis techniques such as rate monotonic analysis (discussed later) to verify schedulability in worst case scenarios.

Although simulation and modeling are a powerful and effective way to abstract complex systems, they have several limitations. First of all, simulations are usually approximations or abstractions of the actual system behavior. Many simulations take significant amounts of time to run and normally require fast computers to run on. But as a first step in building up system models and partitioning hardware and software, simulations are effective.

Hardware/Software Emulation

Emulation is used to verify software and hardware co-design. Emulation allows various parts of system verification before the hardware is ready for actual implementation. Software emulation also permits software engineers to fine-tune the system before going to the target platform. Software emulations are done at various levels. One example is shown in Figure E.4. Here a software application interfaces with an operating system application programming interface (API).

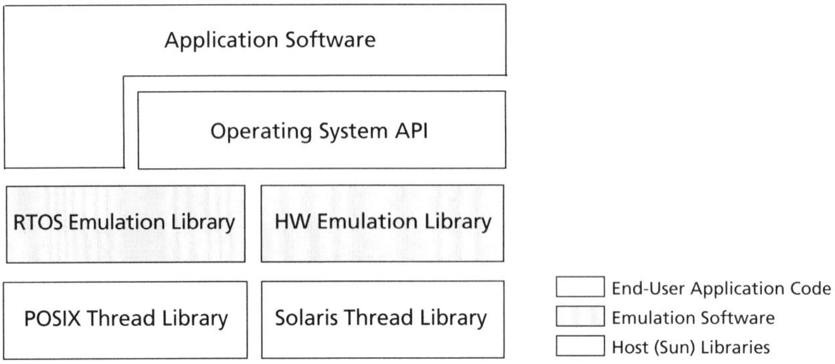

Figure E.4 System emulation block diagram.

The actual real-time operating system calls are emulated with POSIX and Solaris threads. Thus, software can be developed on larger, faster computers, such as a Sun system, using the target operating system calls, and development can begin before the hardware and the operating system are delivered. A hardware emulation library of functions also are available. These functions can include both DMA and I/O controllers and interrupt mechanisms. Emulation of the hardware and operating system interfaces permits host software development well before transitioning to a prototype or the real system.

Using system emulation, software design can proceed despite the lack of hardware and other software interfaces. We have developed a system emulation as a risk mitigation strategy. Hardware and commercial off-the-shelf (COTS) software schedule and delivery problems were causing concern. With the ability to develop the software in an environment that looked exactly like the target environment, we were able to continue development without the need to wait for other vendors.

Rate Monotonic Analysis

Real-time systems must do more than run fast. The system must meet its deadlines or failure can occur. In other words, the system may be correct functionally, but not temporally. The algorithms running on the processor may be implemented correctly, but the system could still fail. In real-time micro-based systems, many jobs are time-critical. The deadlines that face the tasks executing can be "soft" or "hard." Soft real-time systems can miss their deadline and still be marginally useful to the system.

In a hard real-time system, failure of a time-critical task to meet its deadline is considered a fatal fault. In some cases, a late response can be worse than no response! Meeting all of the required deadlines means that a task set is schedulable. Scheduling algorithms are usually used to determine schedulability. A task set is schedulable if the algorithm always generates a valid schedule and all of the task deadlines are met.

One of the biggest problems in software system design is deciding how to schedule tasks in the system. Determining which tasks should be the most important (the task priority) is usually subjective. How to respond to external events such as interrupts is a difficult deliberation. Designers often use simulations to see how the system will run. However, these simulations take a lot of time to develop. Simulations also are hard to change and modify if the designer wants to experiment with different scheduling policies and priority assignments. An easier way to model task schedulability is through rate monotonic analysis (RMA).

Theory of RMA

RMA is a model for predicting whether or not a system will meet its timing as well as throughput requirements when the system is operational. To build systems that are reliable and deterministic, the timing behavior of the tasks in a program must be predictable. If the system is predictable, it can be formally analyzed. RMA allows the designer to determine ahead of time whether the system will meet its timing requirements. The rate monotonic approach to task schedulability was developed in the 70's and is still popular and widely used. The main scheduling model assigns priorities based on shortest task period. Thus, in the set of tasks that are ready at any particular time, the one with the shortest period will be executed first (Figure E.5 and Table 1). Because the task period must be known in advance, this approach is considered to be a static-priority algorithm.

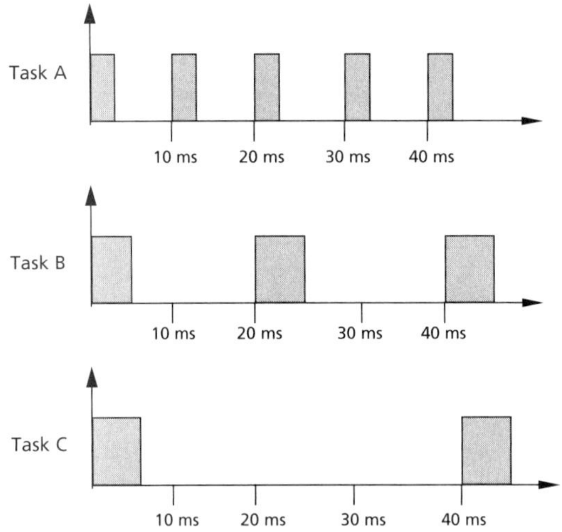

Figure E.5 Set of three tasks with different periods

Periodic Task	Execution Time	Period	Deadline
Task A	1 ms	10 ms	10 ms
Task B	5 ms	20 ms	20 ms
Task C	10 ms	40 ms	40 ms

Table E.1 Task characteristics of Figure E.1

So for the example above, we would get:

$$\frac{1}{10} + \frac{5}{20} + \frac{10}{40} <= U(3) = 3(2 ** 1/3 - 1)$$

$0.10 + 0.25 + 0.25 = 0.6 => 60\% <= 77.9\%$.

The algorithm for determining if a set of tasks will run is:

$$\frac{C1}{T1} + \ldots + \frac{Cn}{Tn} <= U(n) = n\,(2 ** 1/n - 1)$$

where: C(i) = worst case task execution time of task(i),
T(i) = period of Task(i),
U(n) = utilization bound for n tasks.

The set of three tasks is less than the utilization bound of 77.9%, which means that these three tasks, regardless of how and when they are scheduled, will always meet their deadlines. Because this algorithm is based on the worst case execution time of each task, it is considered a "pessimistic" algorithm. As a matter of fact, for an infinite number of tasks, the utilization bound is only around 70%, which means that we are effectively leaving 30% CPU unused. The algorithm provides a sufficient schedulability condition in the sense that it can always schedule the given system task set if the total task utilization does not exceed the utilization bound. The system is guaranteed to be schedulable if the test is passed.

The Rate Monotonic (RM) algorithm can be expanded to include blocking. Blocking leads to priority inversion which occurs when a lower priority task blocks a higher priority task because the lower priority task has a resource locked that is needed by the higher priority task. Priority inversion is a threat in real-time systems. Another problem in real-time systems is task switching. Whenever tasks are stopped in order to execute other tasks, a task switch is performed, which requires time. This extra time must be added to the task execution time.

The RM algorithm also is useful for tasks that occur at a periodic rate. However, real-time systems have many aperiodic events; e.g., hardware interrupts. These interrupts can occur at random and can be devastating if not anticipated and handled correctly. Using RMA, these aperiodic events are modeled using devices called periodic servers. A periodic server is assigned to each aperiodic task and is responsible for executing that task.

Tools to support RMA

There are numerous tools available to help the software developer prototype software and support application development. However, there are few tools available to help the system designer to verify if timing constraints are being met in hard real time systems. In most real time systems design, the application software is developed and then the timing constraints are verified using informal ad hoc methods. Usually in integration and test, many of the real time problems are found. In this phase, they are the most costly to fix. Simulations are used for real time analysis, but these tend to be large, costly, and time consuming to develop, when not integrated into the prototyping environment

Rate Monotonic Analysis tools use analytical approaches to verify real time constraints based on scheduling theory. These tools analyze the system abstractly, as a model. The system designer develops the model and can then use the framework to determine schedulability based on different scheduling algorithms.

With these tools, designers can analyze, validate, and evaluate real time systems and experiment with alternative scheduling and resource management strategies. Several different scheduling protocols are available. These tools support the rate monotonic (RM) and deadline monotonic (DM) scheduling algorithms, as well as the priority ceiling protocol (PCP), stack based protocol (SBP), earliest deadline first (EDF) and cyclic executive priority assignment algorithms.

The advantage of these tools are that they provide the system designer with an interactive design environment. They allowss the user to study the effects and relationships of various parameters in the system that directly affect the schedulability of the entire task set, and also determine if the system task set is schedulable, and also identifies and informs the designer when and where the failure to meet deadlines are most likely. The designer can modify easily a number of system parameters to improve the schedulability. The tools can also predicts the timing effects of changes to parameters at the hardware and operating system software level. Since these tools are based on analytical computation, results are generally fast, allowing the designer to quickly perform "what if" scenarios.

Rate monotonic analysis is a modeling approach to verifying schedulability that is different than simulation. Simulation needs to run for a certain length of time, which may or may not be known, to look for convergence or divergence of certain conditions. Modeling, such as RMA, allows you to run the analysis once to see what the worst case phasing would be without trying to guess what the time frame would be to see worst case scheduling. Modeling with RMA also lets one do the "what-if" scenarios much quicker than changing a simulation and then running it for a few hours or days to see the results. Some RMA tools on the market today include some rudimentary simulation capabilities.

Some RMA tools will also allow you to run a model with one of several scheduling algorithms. The scheduling algorithm can be changed easily and the analysis re-run quickly. Multiple node analysis and End-to-End analysis are also possible with some tools. This is especially useful in distributed applications or in modeling LAN/WANs (using end-to-end scheduling).

RMA tools can (and simulators usually cannot) identify blocking conditions and allow the user to change the blocking conditions easily without changing the physical architecture. Blocking and preemption are the most common reasons for missing deadlines and are one of the main focuses of most RMA tools.

RMA throughout the software development life cycle

Rate monotonic analysis can be used throughout the software life cycle with other methods or by itself. We have used RMA with other techniques during the various phases of software development to develop a constantly improving estimate of our task periods and system architecture. Figure E.7 shows a life cycle model for incorporating rate monotonic analysis in with the other techniques described earlier.

In the early stages of development (software requirements analysis),to determine the task execution times and deadlines, we use historical data if it exists and customer specifications. More in-depth analysis requires a discrete-event simulation of the system to be built. The results from a system level simulation may provide initial numbers to use for task periods and execution time.

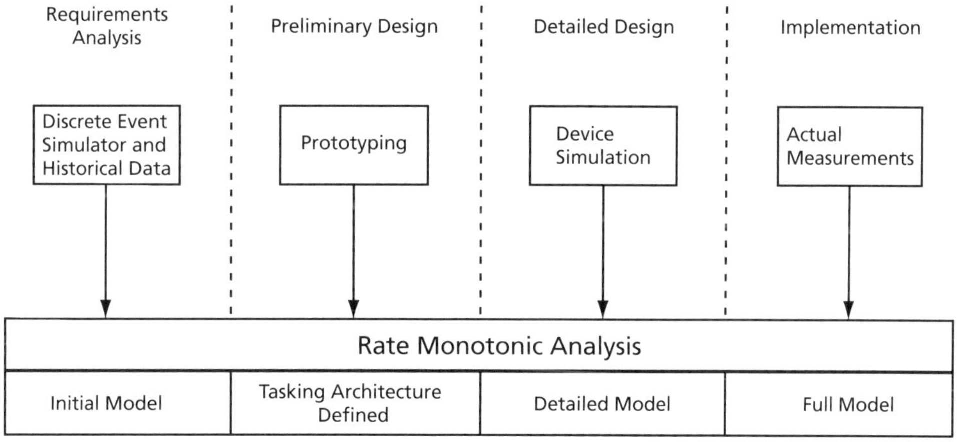

Figure E.6 Life cycle model of rate monotonic analysis

Preliminary design is the phase where the software tasking architecture (the number of tasks and how they will be interacting) must be defined. Each software task and its priority must be determined. Synchronization between tasks is also determined. Using a RMA tool during this phase is very useful for determining the tasking architecture. As Figure E.6 shows, prototyping can provide estimates of task execution times and resource requirements for use in updating the RMA model. Tasks can be created and modified quickly in the model. Different scheduling algorithms can be selected for the model, depending on the operating system you are using and the scheduling policy desired. Task priorities can be changed quickly. This flexibility is useful when performing sensitivity analysis to determine the best scheduling approach for a given set of tasks and resources. The exit criteria for the preliminary design phase is a software architecture that is schedulable using RMA; although you may not know just how the task phasings will be, being schedulable in RMA tells you even under the worst case, the system is schedulable. This is good news for the customer at this point in the development cycle.

The detailed design phase is typically where the system architecture is completely designed. Information to complete the design can come from several sources:

- Prototyping
- Instruction-accurate simulation
- Cycle accurate device simulation
- Hardware design/analysis

The development team can use these sources of information to more accurately determine:

- Resource requirements
- Task execution times
- I/O requirements
- Synchronization requirements

The details of task synchronization, inter-process communication, and resource management are decided. The schedulability model is updated to reflect these new conditions. Using a RMA tool, adding resources, tasks, and dependencies is not difficult (however, adding many blocking terms can be tedious). Adding the details of the resource requirements into the model allows more detailed analysis of the system. For example, blocking conditions, priority inversion conditions, and other bottlenecks can be discovered. Device simulation (for a new processor, for example), if available, can generate detailed estimates of tasks and resources. Problems can be corrected before the developers begin integration and test. The developers can use the results of the schedulability analysis to fix some of these problems early.

In the past, we never knew how the system was really going to run until we got to integration and test. Problems with system scheduling at that time are difficult to find and expensive to fix. Even developing a discrete event simulation of the system wasn't always good enough because we never knew how long to run the simulation. If a scheduling problem happens only during a certain task phasing situation, a simulation may never catch it. A RMA tool, based on mathematics of schedulability, will look for the worst case scenario, and give you an answer in seconds.

RMA can be useful even in the integration and test phase. The RMA model can be updated with actual timing measurements obtained in the lab. The RMA model can be used to assess the impact of proposed changes. One can analyze the effects of moving functionality from a task to an interrupt service routine, or re-allocating resources in the model before changing the software. Existing systems that are being upgraded or enhanced can benefit from using a model of the tasking architecture to determine if the proposed upgrades or enhancements will work.

Prototyping and Incremental Development

One approach to software product development that is gaining momentum is incremental development. With this approach, each increment contains end to end functionality. Each of the increments contains the full functionality of the previous increment, plus additional functionality. The final increment in the series is, effectively, the final deliverable product. Each increment should be able to demonstrate operational user functions. Each of the software increments should be executed in the operational environment. In general, the most stable requirements are implemented in the earlier increments, resulting in a minimal system. These requirements will probably not change and can be implemented without concern for change later (Figure E.8) .

The more unstable requirements can be implemented in later increments. By waiting until later, many of these requirements will have had time to stabilize. Incremental development allows early assessment of software reliability for managers, feedback of information to the developers concerning risk areas, and early looks at functionality for the customers and users. Increment planning is first performed as the requirements are flowed down from the system level to the individual software products.

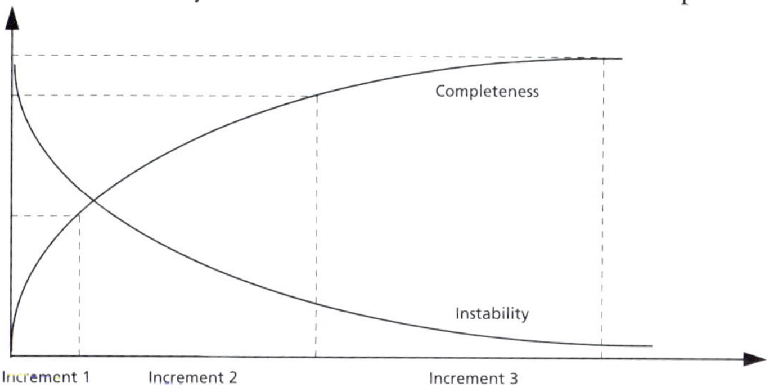

Figure E.7 Increment planning for evolving requirements

Determining increment content

Software increments should be planned based on several factors:

- *Available resources* – Larger and more complex increments require more time and resources and need to be planned accordingly. In some cases, parallel development efforts may be necessary to allow increments to be completed on time and integrated into the system.
- *Availability of hardware* – The increments should execute in the system environment if at all possible. This is so that end to end user functionality can be demonstrated easily.
- *Requirements functionality* – Functionally, increments should be designed such they seamlessly integrate into the final system. Each successive increment should build on the previous increment without having to make changes to the earlier increments.
- *Type and number of requirements* – The top level approach to developing and integrating increments allows the developers to build up the framework for the system first and then enhance and evolve that framework in later increments. This is an approach that is used in other engineering disciplines.
- *Operational usage profiles* – Increments should be planned based on expected usage of the system. Those functions will high expected usage are candidates for incorporation into earlier increments. Since these functions will be used most often by the user, they should be tested more completely.
- *Reliability and risk considerations* – As more and more customers expect software that meets predetermined reliability measures, software components that contribute most to the reliability of the system can be developed first. More testing is then performed on these components, increasing the overall reliability.

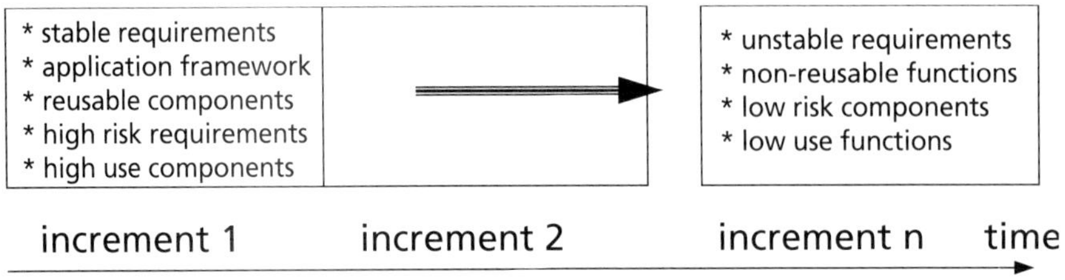

Figure E.8 Allocating software requirements to increments

Figure E.8 summarizes the parameters to consider when allocating requirements to the different increments in a software system. In this diagram the darker shades indicate a stronger parameter. For example, the "high" risk requirements are candidates for earlier increments and the "low" reuse requirements are candidates of requirements for later increments.

Summary

Several techniques were described that allows system designers to model systems, perform hardware/software trade-offs, and start developing software before the hardware and other vendor software is ready. System modeling and simulation tools aid in developing behavior models of complex systems. These models can be analyzed statically or executed to allow analysis of architectural performance.

Emulation can allow various parts of system verification before the hardware is ready for actual implementation. Software emulation allows the software engineers to fine tune the system before going to the target platform.

Rate monotonic analysis is a method for determining schedulability of real-time software systems. The building of reliable real-time systems requires predictable the timing behavior of the tasks. RMA can predict whether or not a system will meet its timing and throughput requirements. RMA can guarantee a system will be schedulable if a set of requirements is met. Other methods of determining schedulability, such as simulations, cannot reliably predict if the set of tasks will, under all conditions, be schedulable. These tools allow rapid prototyping and extensive "what if" analysis to be performed.

RMA tools will also point the designer to the problem areas. Blocking problems and other resource bottlenecks are indicated graphically to the designer. Previously, many of these problems were not discovered until system integration and test, where hours were spent using logic and bus analyzers to find subtle, nonrepeatable timing problems that caused many schedule delays. With end-to-end analysis capabilities, allow designers to look at an isolated task-set on a specific processor or entire embedded systems and networks.

Incremental development allows early assessment of software reliability for managers, feedback of information to the developers concerning risk areas, and early looks at functionality for the customers and users. Increment planning is first performed as the requirements are flowed down from the system level to the individual software products. With this approach, each increment contains end to end functionality. Each of the increments contains the full functionality of the previous increment, plus additional functionality. The final increment in the series is, effectively, the final deliverable product.

References

Nu-Thena Systems, Foresight Training Material, September 1996.

Liu, C. and J. Layland. "Scheduling Algorithms for Multiprogramming in a Hard Real Time Environment," *Journal of the ACM*, January 1973.

A Practitioners Handbook for Real Time Analysis, by Mark Klein, Thomas Ralya, Bill Pollak, and Ray Obenza, Kluwer Academic Publishers, 1993.

Adams, Jay K. and Donald E. Thomas, "Design Automation for Mixed Hardware-Software Systems," *Electronic Design*, March 1997.

Oshana, Rob, "Rate-monotonic analysis keeps real-time systems on schedule," *EDN*, September 1, 1997.

Trammel, C.J., P.A.Hausler and C.E. Galbraith, "Incremental Implementation of Cleanroom practices," *Proceedings of the 25th Hawaii International Conference on System Sciences*, 1992.

Programming digital signal processors with high-level languages, By Robert Jan Ridder, *DSP Engineering*, Summer 2000

1. Smith, Connie U. "Performance Engineering for Software Architectures," 21st Annual Computer Software and Applications Conference, 1997, p 166-167.
2. Baker, Michelle and Warren Smith, "Performance Prototyping: A Simulation Methodology for Software Performance Engineering," Proceeding of the Computer Systems and Software Engineering," 1992, p 624-629.
3. Obenza, Ray, "Rate monotonic analysis for real-time systems," March 1993

Liu, C. and J. Layland. "Scheduling Algorithms for Multiprogramming in a Hard Real Time Environment," *Journal of the ACM*, January 1973.

A Practitioner's Handbook for Real-Time Analysis, by Mark Klein, Thomas Ralya, Bill Pollak, and Ray Obenza, Kluwer Academic Publishers, 1993.

1. Hakkarainen, Harri, "Evaluating the TMS320C62xx for Comm Applications," *Communication Systems Design*, October 1997.
2. Blalock, Garrick, "General-purpose uPs for DSP applications: consider the trade-offs," *EDN*, October 23, 1997.
3. Levy, Markus, "Virtual Processors and the Reality of Software Simulation," *EDN*, January 15, 1998.
4. Mayer, John H., "Sophisticated tools bring real-time DSP applications to market," *Military and Aerospace Electronics*, January 1998.
5. Stearns, Samuel D. and Ruth David, "Signal Processing Algorithms in MATLAB," Prentice Hall 1996.

Practical Software Requirements, A Manual of Content and Style, Ben Kovitz, Manning pubishers

[1] Erwin, Harry R., "Characteristics of Performance Engineers," *Computer Measurement Group Transactions*, Spring, 1988, (pp 77-80).

[2] Connie U. Smith, *Performance Engineering of Software Systems*, Addison-Wesley, 1990.

[3] Major Joseph B., "Software Performance Engineering: Rules of Thumb and Tradeoffs," *Computer Measurement Group Transactions*, December 1992, pp 87-98).

[4] International Business Machines, Inc., *TPF System Performance and Measurement Program Reference Manual (GH20-7505-02)*, September 1991.

[1] Jeff Bier, Phil Lapsley, Edward A. Lee, Franz Weller, *DSP Design Tools and Methodologies*, Berkeley Design Technology, Inc.: Fremont, California. 1995.

DSP Engineering / Summer 2000 Copyright 2000 / All rights reserved.

Programming Digital Signal Processors with High-level Languages by Robert Jan Ridder

EEDESIGN January 17, 2001 Next-Generation Model-Based Design Tools for Automotive System Design, Model-Based Approach Simplifies Embedded Application Design, By Paul Bernard, Simulink Application Manager, The MathWorks Inc.

F

DSP Algorithmic Development—Rules and Guidelines

Digital signal processors are often programmed like "traditional" embedded microprocessors. They are programmed in a mix of C and assembly language, they directly access hardware peripherals, and, for performance reasons, almost always have little or no standard operating system support. Thus, like traditional microprocessors, there is very little use of commercial off-the-shelf (COTS) software components for DSPs. However, unlike general-purpose embedded microprocessors, DSPs are designed to run sophisticated signal processing algorithms and heuristics. For example, they may be used to detect important data in the presence of noise, to or for speech recognition in a noisy automobile traveling at 65 miles per hour. Such algorithms are often the result of many years of research and development. However, because of the lack of consistent standards, it is not possible to use an algorithm in more than one system without significant reengineering. This can cause significant time to market issues for DSP developers.

This appendix defines a set of requirements for DSP algorithms that, if followed, allow system integrators to quickly assemble production-quality systems from one or more such algorithms. Thus, this standard is intended to be used for DSP algorithm developers to more rapidly develop and deploy systems with these algorithms, to enable in house reuse programs for DSP algorithms, to enable a rich COTS marketplace for DSP algorithm technology and to significantly reduce the time-to-market for new DSP-based products.

Requirements of a DSP Algorithm Development Standard

This section lists the required elements of a DSP Algorithm Standard. These requirements are used throughout the remainder of the document to motivate design choices. They also help clarify the intent of many of the stated rules and guidelines.

- Algorithms from multiple vendors can be integrated into a single system.
- Algorithms are framework-agnostic. That is, the same algorithm can be efficiently used in virtually any application or framework.

- Algorithms can be deployed in purely static as well as dynamic run-time environments.
- Algorithms can be distributed in binary form.
- Integration of algorithms does not require recompilation of the client application, although reconfiguration and re-linking may be required.

A huge number of DSP algorithms are needed in today's marketplace, including modems, vocoders, speech recognizers, echo cancellation, and text-to-speech. It is not possible for a product developer, who wants to leverage this rich set of algorithms, to obtain all the necessary algorithms from a single source. On the other hand, integrating algorithms from multiple vendors is often impossible due to incompatibilities between the various implementations. This is why an algorithm standard is needed.

DSP algorithms and standards can be broken down into general and specific components as shown in Figure F.1.

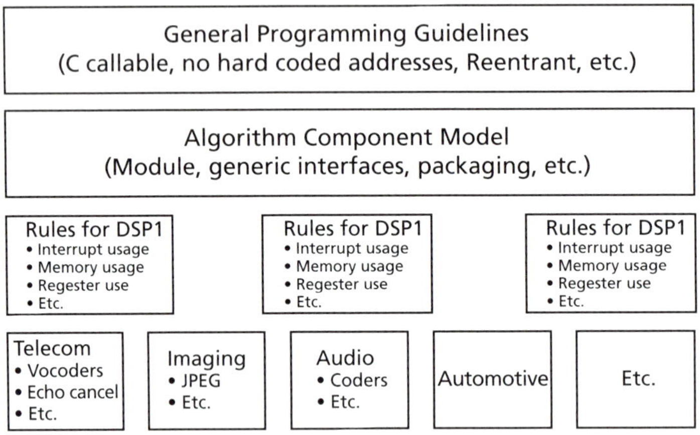

Figure F.1 A model of a DSP algorithm standard (courtesy of Texas Instruments)

Dozens of distinct DSP frameworks exist. Each framework optimizes performance for an intended class of systems. For example, client systems are designed as single-channel systems with limited memory, limited power, and lower-cost DSPs. As a result, they are quite sensitive to performance degradation. Server systems, on the other hand, use a single DSP to handle multiple channels, thus reducing the cost per channel. As a result, they must support a dynamic environment. Yet, both client-side and server-side systems may require exactly the same

vocoders. It is important that algorithms be deliverable in binary form. This not only protects the algorithm intellectual property; it also improves the reusability of the algorithm. If source code were required, all clients would require recompilation. In addition to being destabilizing for the clients, version control for the algorithms would be close to impossible.

DSP system architecture

Many modern DSP system architectures can be partitioned along the lines depicted in Figure F.2 below.

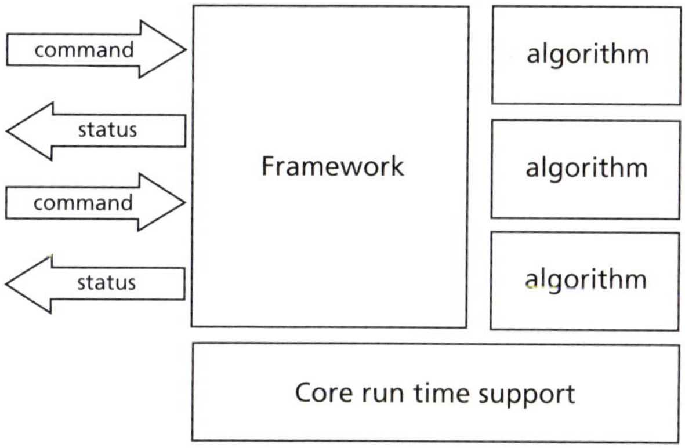

Figure F.2 DSP Software Architecture (courtesy of Texas Instruments)

Algorithms are "pure" data transducers; i.e., they simply take input data buffers and produce some number of output data buffers. The core run-time support includes functions that copy memory, and functions to enable and disable interrupts. The framework is the "glue" that integrates the algorithms with the real-time data sources and links using the core run time support, to create a complete DSP sub-system. Frameworks for the DSP often interact with the real-time peripherals (including other processors in the system) and often define the I/O interfaces for the algorithm components.

Unfortunately, for performance reasons, many DSP systems do not enforce a clear line between algorithm code and the system-level code (i.e., the framework). Thus, it is not possible to easily reuse an algorithm in more than one system. A DSP algorithm standard should clearly define this line in such a way that performance is not sacrificed and algorithm reusability is significantly enhanced.

DSP frameworks

Frameworks often define a device independent I/O sub-system and specify how essential algorithms interact with this sub-system. For example, does the algorithm call functions to request data or does the framework call the algorithm with data buffers to process? Frameworks also define the degree of modularity within the application; i.e., which components can be replaced, added, removed, and when can components be replaced (compile time, link time, or real-time). Even within the telephony application space, there are a number of different frameworks available and each is optimized for a particular application segment (e.g., large volume client-side products and low volume high-density server-side products).

DSP algorithms

Careful inspection of the various frameworks in use reveals that, at some level, they all have algorithm components. While there are differences in each of the frameworks, the algorithm components share many common attributes.

- Algorithms are C callable
- Algorithms are reentrant
- Algorithms are independent of any particular I/O peripheral
- Algorithms are characterized by their memory and MIPS requirements

In many frameworks, algorithms are required to simply process data passed to the algorithm. The others assume that the algorithm will actively acquire data by calling framework-specific, hardware-independent, I/O functions. In all cases, algorithms should be designed to be independent of the I/O peripherals in the system. In an effort to minimize framework dependencies, the standard should require that algorithms process data that is passed to them via parameters. It seems likely that conversion of an "active" algorithm to one that simply accepts data in the form of parameters is straightforward and little or no loss of performance will be incurred.

It is important to realize that each particular implementation of, say a video codec, represents a complex set of engineering trade-offs between code size, data size, MIPS, and quality. Moreover, depending on the system designed, the system integrator may prefer an algorithm with lower quality and smaller footprint to one with higher quality and larger footprint (e.g., a digital camera vs. a cell phone). Thus, multiple implementations of exactly the same algorithm sometimes make sense; there is no single best implementation of many algorithms. Unfortunately, the system integrator is often faced with choosing all algorithms from a single vendor to ensure compatibility between the algorithms and to minimize the overhead of managing disparate APIs.

By enabling system integrators to plug or replace one algorithm for another, we reduce the time to market because the system integrator can chose algorithms from multiple projects, if the standard is used in house, or vendors if the standard is applied more broadly. The DSP program can effectively create a huge catalog of interoperable parts from which any system can be built.

Core run-time support

In order to enable algorithms to satisfy the minimum requirements of reentrancy, I/O peripheral independence, and debuggability, algorithms must rely on a core set of services that are always present. Since many algorithms are still produced using assembly language, many of the services provided by the core must be accessible *and* appropriate for assembly language. The core run-time support should include a subset of interrupt functions of the DSP operating system to support atomic modification of control/status registers (to set the overflow mode, for example). It also includes a subset of the standard C language run-time support libraries; e.g., memcpy, strcpy, etc.

General Programming Guidelines for DSP Algorithms

We can apply a set of general guidelines to DSP algorithm development. In this section, we develop programming guidelines that apply to all algorithms on all DSP architectures, regardless of application area.

Use of C language

All algorithms will follow the run-time conventions imposed by the C programming language. This ensures that the system integrator is free to use C to "bind" various algorithms together, control the flow of data between algorithms, and interact with other processors in the system easily.

Rule 1

All algorithms must follow the run-time conventions imposed by an implementation of the C programming language. This does *not* mean that algorithms must be written in the C language. Algorithms may be implemented entirely in assembly language. They must, however, be callable from the C language and respect the C language run-time conventions. Most significant algorithms are not implemented as a single function; like any sophisticated software, they are composed of many interrelated internal functions. Again, it is important to note that these internal functions do *not* need to follow the C language conventions; only the top-most interfaces must obey the C language conventions. On the other hand, these internal functions must be careful not to cause the top-most function to violate the C run-time conventions; e.g., no called function may use a word on the stack with interrupts enabled without first updating the stack pointer.

Threads and reentrancy

Because of the variety of frameworks available for DSP systems, there are many differing types of threads, and therefore, reentrancy requirements. The goal of this section is to precisely define the types of threads supported by a standard and the reentrancy requirements of algorithms.

Threads

A thread is an encapsulation of the flow of control in a program. Most people are accustomed to writing single-threaded programs; i.e., programs that only execute one path through their code "at a time." Multithreaded programs may have several threads running through different code paths "simultaneously." In a typical multithreaded program, zero or more threads may actually be running at any one time. This depends on the number of CPUs in the system in which the process is running, and on how the thread system is implemented. A system with *n* CPUs can, intuitively run no more than *n* threads in parallel, but it may give the appearance of running many more than *n* "simultaneously," by sharing the CPUs among threads. The most common case is that of n equal to one; that is, a single CPU running all the threads of an application.

Why are threads interesting? An OS or framework can schedule them, relieving the developer of an individual thread from having to know about all the other threads in the system. In a multi-CPU system, communicating threads can be moved among the CPUs to maximize system performance without having to modify the application code. In the more common case of a single CPU, the ability to create multithreaded applications allows the CPU to be used more effectively; while one thread is waiting for data, another can be processing data.

Virtually all DSP systems are multithreaded; even the simplest systems consist of a main program and one or more hardware interrupt service routines. Additionally, many DSP systems are designed to manage multiple "channels" or "ports," i.e., they perform the same processing for two or more independent data streams.

Preemptive vs. nonpreemptive multitasking

Nonpreemptive multitasking relies on each thread to voluntarily relinquish control to the operating system before letting another thread execute. This is usually done by requiring threads to periodically call an operating system function, say yield(), to allow another thread to take control of the CPU or by simply requiring all threads to complete within a specified short period. In a nonpreemptive multithreading environment, the amount of time a thread is allowed to run is determined by the thread, whereas in a preemptive environment, the time is determined by the operating system *and* the entire set of tasks that are ready to run.

Note that the difference between those two flavors of multithreading can be a very big one; for example, under a nonpreemptive system, you can safely assume that no other thread executes while a particular algorithm processes data using on-chip data memory. Under preemptive execution, this is not true; a thread may be preempted while it is in the middle of processing. Thus, if your application relies on the assumption that things do not change in the middle of processing some data, it might break under a preemptive execution scheme.

Since preemptive systems are designed to *preserve* the state of a preempted thread and restore it when its execution continues, threads can safely assume that most registers and all of the thread's data memory remain unchanged. What would cause an application to fail? Any assumptions related to the maximum amount of time that can elapse between any two instructions, the state of any global system resource such as a data cache, or the state of a global variable accessed by multiple threads, can cause an application to fail in a preemptive environment.

Nonpreemptive environments incur less overhead and often result in higher performance systems; for example, data caches are much more effective in nonpreemptive systems since each thread can control when preemption (and therefore, cache flushing) will occur.

On the other hand, nonpreemptive environments require that either each thread complete within a specified maximum amount of time, or explicitly relinquish control of the CPU to the framework (or operating system) at some minimum periodic rate. By itself, this is not a problem since most DSP threads are periodic with real-time deadlines. However, this minimum rate is a function of the *other* threads in the system and, consequently, nonpreemptive threads are not completely independent of one another; they must be sensitive to the scheduling requirements of the other threads in the system. Thus, systems that are by their nature multirate and multichannel often require preemption; otherwise, all of the algorithms used would have to be rewritten whenever a new algorithm is added to the system.

If we want all algorithms to be framework-independent, we must either define a framework-neutral way for algorithms to relinquish control, or assume that algorithms used in a nonpreemptive environment always complete in less than the required maximum scheduling latency time. Since we require documentation of worst-case execution times, it is possible for system integrators to quickly determine if an algorithm will cause a nonpreemptive system to violate its scheduling latency requirements.

Since algorithms can be used in both preemptive and nonpreemptive environments, it is important that all algorithms be designed to support both. This means that algorithms should minimize the maximum time that they can delay other algorithms in a nonpreemptive system.

Reentrancy

Reentrancy is the attribute of a program or routine that allows the same copy of the program or routine to be used concurrently by two or more threads. Reentrancy is an extremely valuable property for functions. In multichannel systems, for example, any function that can be invoked as part of one channel's processing must be reentrant; otherwise, that function would not be usable for other channels. In single channel multirate systems, any function that must be used at two different rates must be reentrant; for example, a general digital filter function used for both echo cancellation and pre-emphasis for a vocoder. Unfortunately, it is not always easy to determine if a function is reentrant.

The definition of reentrant code often implies that the code does not retain "state" information. That is, if you invoke the code with the same data at different times, by the same or other thread, it will yield the same results. This is not always true, however. How can a function maintain state and still be reentrant? Consider the rand() function. Perhaps a better example is a function with state that protects that state by disabling scheduling around its critical sections. These examples illustrate some of the subtleties of reentrant programming.

The property of being reentrant is a function of the threading model; after all, before you can determine whether multiple threads can use a particular function, you must know what types of threads are possible in a system. For example, if threads are not preemptive, a function may freely use global variables if it uses them for scratch storage only; i.e., it does not assume these variables have any values upon entry to the function. In a preemptive environment, however, use of these global variables must be protected by a critical section or they must be part of the context of every thread that uses them.

Although there are exceptions, reentrancy usually requires that algorithms:

- only modify data on the stack or in an instance "object"
- treat global and static variables as read-only data
- never employ self modifying code

These rules can sometimes be relaxed by disabling *all* interrupts (and therefore, disabling all thread scheduling) around the critical sections that violate the rules above. Since algorithms are not permitted to directly manipulate the interrupt state of the processor, the allowed DSP operating system functions (or equivalent implementations) must be called to create these critical sections.

Rule 2

All algorithms must be reentrant within a preemptive environment (including time-sliced preemption).

Example

In the remainder of this section we consider several implementations of a simple algorithm, digital filtering of an input speech data stream, and show how the algorithm can be made reentrant and maintain acceptable levels of performance. It is important to note that, although these examples are written in C, the principles and techniques apply equally well to assembly language implementations.

Speech signals are often passed through a pre-emphasis filter to flatten their spectrum prior to additional processing. Pre-emphasis of a signal can be accomplished by applying the following difference equation to the input data:

yn = xn – xn-1 + 13/32 c xn-12

The following implementation is not reentrant because it references and updates the global variables z0 and z1. Even in a nonpreemptive environment, this function is not reentrant; it is not possible to use this function to operate on more than one data stream since it retains state for a particular data stream in two fixed variables (z0 and z1).

```
int z0 = 0, z1 = 0; /* previous input values */
void PRE_filter(int input[], int length)
{
int i, tmp;
for (i = 0; i < length; i++) {
tmp = input[i] - z0 + (13 * z1 + 16) / 32;
```

```
z1 = z0;
z0 = input[i];
input[i] = tmp;
}
}
```

We can make this function reentrant by requiring the caller to supply previous values as arguments to the function. This way, PRE_filter1 no longer references any global data and can be used, therefore, on any number of distinct input data streams.

```
void PRE_filter1(int input[], int length, int *z)
{
int i, tmp;
for (i = 0; i < length; i++) {
tmp = input[i] - z[0] + (13 * z[1] + 16) / 32;
z[1] = z[0];
z[0] = input[i];
input[i] = tmp;
}
}
```

This technique of replacing references to global data with references to parameters illustrates a general technique that can be used to make virtually any code reentrant. One simply defines a "state object" as one that contains all of the state necessary for the algorithm; a pointer to this state is passed to the algorithm (along with the input and output data).

```
typedef struct PRE_Obj { /* state obj for pre-emphasis alg */
int z0;
int z1;
} PRE_Obj;
void PRE_filter2(PRE_Obj *pre, int input[], int length)
{
int i, tmp;
for (i = 0; i < length; i++) {
tmp = input[i] - pre->z0 + (13 * pre->z1 + 16) / 32;
pre->z1 = pre->z0;
pre->z0 = input[i];
input[i] = tmp;
}
}
```

Although the C code looks more complicated than our original implementation, its performance is comparable, it is fully reentrant, and its performance can be configured on a "per data object" basis. Since each state object can be placed in any data memory, it is possible to place some objects in on-chip memory and others in external memory. The pointer to the state object is, in effect, the function's private "data page pointer." All of the function's data can be efficiently accessed by a constant offset from this pointer.

Notice that while performance is comparable to our original implementation, it is slightly larger and slower because of the state object redirection. Directly referencing global data is often more efficient than referencing data via an address register. On the other hand, the decrease in efficiency can usually be factored out of the time-critical loop and into the loop-setup code. Thus, the incremental performance cost is minimal and the benefit is that this same code can be used in virtually any system—independent of whether the system must support a single channel or multiple channels, or whether it is preemptive or nonpreemptive.

"We should forget about small efficiencies, say about 97% of the time: premature optimization is the root of all evil." —Donald Knuth "Structured Programming with go to Statements," Computing Surveys, Vol. 6, No. 4, December, 1974, page 268.

Data memory

The large performance difference between on-chip data memory and off-chip memory (even 0 wait-state SRAM) is so large that every algorithm designer designs their code to operate as much as possible within the on-chip memory. Since the performance gap is expected to continue to increase in the coming years, this trend will continue for the foreseeable future. The TMS320C6000 series DSP, for example, incurs a 25 wait state penalty for external SDRAM data memory access. While the amount of on-chip data memory may be adequate for each algorithm in isolation, the increased number of MIPS available on modern DSPs encourages systems to perform multiple algorithms concurrently with a single chip. Thus, some mechanism must be provided to efficiently share this precious resource among algorithm components from one or more third parties.

Memory spaces

In an ideal DSP, there would be an unlimited amount of on-chip memory and algorithms would simply always use this memory. In practice, however, the amount of on-chip memory is very limited and there are even two common types of on-chip memory with very different performance characteristics: dual-access memory which allows simultaneous read and write operations in a single instruction cycle, and single access memory that only allows a single access per instruction cycle.

Because of these practical considerations, most DSP algorithms are designed to operate with a combination of on-chip and external memory. This works well when there is sufficient on-chip memory for all the algorithms that need to operate concurrently; the system developer simply dedicates portions of on-chip memory to each algorithm. It is important, however, that no algorithm assume specific region of on-chip memory or contain any "hard coded" addresses; otherwise the system developer will not be able to optimally allocate the on-chip memory among all algorithms.

Rule 3

Algorithm data references must be fully relocatable (subject to alignment requirements). That is, there must be no "hard-coded" data memory locations. Note that algorithms *can* directly access data contained in a static data structure located by the linker. This rule only requires that all such references be done symbolically; i.e., via a relocatable label rather than a fixed numerical address.

In systems where the set of algorithms is not known in advance or when there is insufficient on-chip memory for the worst-case working set of algorithms, more sophisticated management of this precious resource is required. In particular, we need to describe how the on-chip memory can be shared at run-time among an arbitrary number of algorithms.

Scratch vs. persistent

In this section, we develop a general model for sharing regions of memory among algorithms. This model is used to share the on-chip memory of a DSP, for example. This model is essentially a generalization of the technique commonly used by compilers to share CPU registers among functions.

Compilers often partition the CPU registers into two groups: "scratch" and "preserve." Scratch registers can be freely used by a function without having to preserve their value upon return from the function. Preserve registers, on the other hand, must be saved prior to being modified and restored prior to returning from the function. By partitioning the register set in this way, significant optimizations are possible; functions do not need to save and restore scratch registers, and callers do not need to save preserve registers prior to calling a function and restore them after the return. Consider the program execution trace of an application that calls two distinct functions, say a() and b().

```
Void main()
{
... /* use scratch registers r1 and r2 */
/* call function a() */
a() {
... /* use scratch registers r0, r1, and r2 */
}
/* call function b() */
b() {
... /* use scratch registers r0 and r1*/
}
}
```

Notice that both a() and b() freely use some of the same scratch registers and no saving and restoring of these registers is necessary. This is possible because both functions, a() and b(), agree on the set of scratch registers and that values in these registers are indeterminate at the beginning of each function.

By analogy, we partition all memory into two groups: scratch and persistent.

- Scratch memory is memory that is freely used by an algorithm without regard to its prior contents, i.e., no assumptions about the content can be made by the algorithm and the algorithm is free to leave it in any state.
- Persistent memory is used to store state information while an algorithm instance is not executing.

Persistent memory is any area of memory that an algorithm can write to assume that the contents are unchanged between successive invocations of the algorithm within an application. All physical memory has this behavior, but applications that share memory among multiple algorithms may opt to overwrite some regions of memory (e.g., on-chip DARAM).

A special variant of persistent memory is the write-once persistent memory. An algorithm's initialization function ensures that its write-once buffers are initialized during instance creation and that all subsequent accesses by the algorithm's processing to write-once buffers are strictly read-only. Additionally, the algorithm can link its own statically allocated write-once buffers and provide the buffer addresses to the client. The client is free to use provided buffers or allocate its own. Frameworks can optimize memory allocation by arranging multiple instances of the same algorithm, created with identical creation parameters, to share write-once buffers.

Note that a simpler alternative to declaring write-once buffers for sharing statically initialized read-only data is to use global statically linked constant tables and publish their alignment and memory space requirements in the required standard algorithm documentation. If data has to be computed or relocated at run-time, the write-once buffers approach can be employed.

The importance of making a distinction between scratch memory and persistent memory is illustrated in Figure F.3

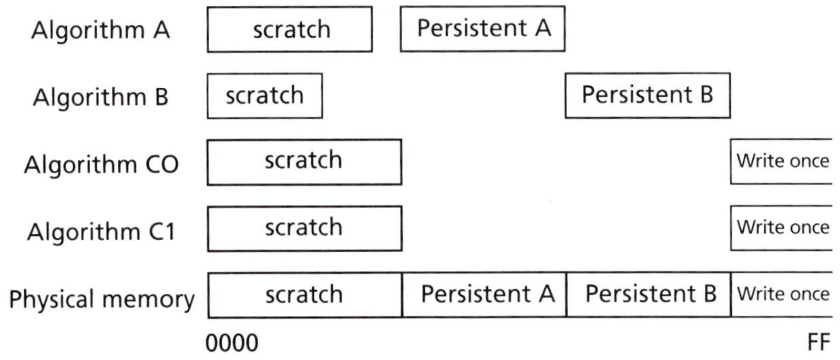

Figure F.3 Scratch vs. persistent memory allocation (courtesy of Texas Instruments)

All algorithm scratch memory can be "overlaid" on the same physical memory.

Without the distinction between scratch and persistent memory, it would be necessary to strictly partition memory among algorithms, making the total memory requirement the sum of all algorithms' memory requirements. On the other hand, by making the distinction, the total memory requirement for a collection of algorithms is the sum of each algorithm's distinct persistent memory, plus any shared write-once persistent memory, plus the *maximum* scratch memory requirement of any of these algorithms.

Guideline 1

Algorithms should minimize their persistent data memory requirements in favor of scratch memory. In addition to the types of memory described above, there are often several memory spaces provided by a DSP to algorithms.

- *Dual-access memory (DARAM)* is on-chip memory that allows two simultaneous accesses in a single instruction cycle.
- *Single-access memory (SARAM)* is on-chip memory that allows only a single access per instruction cycle.
- *External memory* is memory that is external to the DSP and may require more than zero wait states per access.

These memory spaces are often treated very differently by algorithm implementations; in order to optimize performance, frequently accessed data is placed in on-chip memory, for example. *The scratch vs. persistent attribute of a block of memory is independent of the memory space.* Thus, there are six distinct memory classes; scratch and persistent for each of the three memory spaces described above.

Algorithm vs. application

Other than a memory block's size, alignment, and memory space, three independent questions must be answered before a client can properly manage a block of an algorithm's data memory.

- Is the block of memory treated as scratch or persistent by the algorithm?
- Is the block of memory shared by more than one algorithm?
- Do the algorithms that share the block preempt one another?

The first question is determined by the implementation of the algorithm; the algorithm must be written with assumptions about the contents of certain memory buffers. There is significant benefit to distinguish between scratch memory and persistent memory, but it is up to the algorithm implementation to trade the benefits of increasing scratch, and decreasing persistent memory against the potential performance overhead incurred by re-computing intermediate results.

The second two questions regarding sharing and preemption, can only be answered by the client of a DSP algorithm. The client decides whether preemption is required for the system and the client allocates all memory. Thus, only the client knows whether memory is shared among algorithms. Some frameworks, for example, never share any allocated memory among algorithms whereas others always share scratch memory.

There is a special type of persistent memory managed by clients of algorithms that is worth distinguishing: *shadow* memory is unshared persistent memory that is used to shadow or save the contents of shared registers and memory in a system. Shadow memory is not used by algorithms; it is used by their clients to save the memory regions shared by various algorithms.

Figure F.4 illustrates the relationship between the various types of memory.

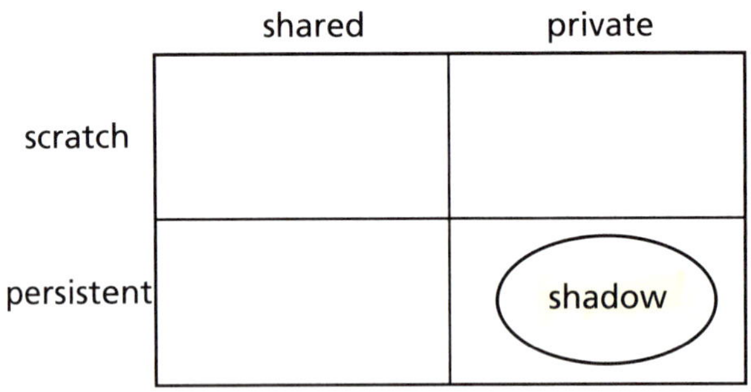

Figure F.4. Data memory types (courtesy of Texas Instruments)

Program memory

Like the data memory requirements described in the previous section, it is important that all DSP algorithms are fully relocatable; i.e., there should never be any assumption about the specific placement of an algorithm at a particular address. Alignment on a specified page size should be permitted, however.

Rule 4

All algorithm code must be fully relocatable. That is, there can be no hard coded program memory locations. As with the data memory requirements, this rule only requires that code be relocated via a linker. For example, it is *not* necessary to always use PC-relative branches. This requirement allows the system developer to optimally allocate program space to the various algorithms in the system. Algorithm modules sometimes require initialization code that must be executed prior to any other algorithm method being used by a client. Often this code is only run once during the lifetime of an application. This code is effectively "dead" once it has been run at startup. The space allocated for this code can be reused in many systems by placing the "run-once" code in data memory and using the data memory during algorithm operation. A similar situation occurs in "finalization" code. Debug versions of algorithms, for example, sometimes implement functions that, when called when a system exits, can provide valuable debug information; e.g., the existence of objects or objects that have not been properly deleted. Since many systems are designed to never exit (i.e., exit by power-off), finalization code should be placed in a separate object module. This allows the system integrator to avoid including code that can never be executed.

Guideline 2

Each initialization and finalization function should be defined in a separate object module; these modules must not contain any other code. In some cases, it is awkward to place each function in a separate file. Doing so may require making some identifiers globally visible or require significant changes to an existing code base. Some DSP C compilers support a pragma directive that allows you to place specified functions in distinct COFF output sections. This pragma directive may be used in lieu of placing functions in separate files. The table below summarizes recommended section names and their purpose.

Section	Name	Purpose
Text:init	Run once	Initialization code
Text:exit	Run once	Finalization code
Text:create	Run time	Object creation
Text:delete	Run time	Object deletion

ROM-ability

There are several addressing modes used by algorithms to access data memory. Sometimes the data is referenced by a pointer to a buffer passed to the algorithm, and sometimes an algorithm simply references global variables directly. When an algorithm references global data directly, the instruction that operates on the data often contains the address of the data (rather than an offset from a data page register, for example). Thus, this code cannot be placed in ROM without also requiring that the referenced data be placed in a fixed location in a system. If a module has configuration parameters that result in variable length data structures *and* these structures are directly referenced, such code is not considered ROM-able; the offsets in the code are fixed and the relative positions of the data references may change. Alternatively, algorithm code can be structured to always use offsets from a data page for all fixed length references and place a pointer in this page to any variable length structures. In this case, it is possible to configure and locate the data anywhere in the system, provided the data page is appropriately set.

Rule 5

Algorithms must characterize their ROM-ability; i.e., state whether or not they are ROM-able. Obviously, self-modifying code is not ROM-able. We do not require that no algorithm employ self-modifying code; we only require documentation of the ROM-ability of an algorithm. It is also worth pointing out that if self-modifying code is used, it must be done "atomically," i.e., with all interrupts disabled; otherwise this code would fail to be reentrant.

Use of peripherals

To ensure the interoperability of DSP algorithms, it is important that algorithms never directly access any peripheral device.

Rule 6

Algorithms must never directly access any peripheral device. This includes but is not limited to on-chip DMAs, timers, I/O devices, and cache control registers. In order for an algorithm to be framework-independent, it is important that no algorithm directly calls any device interface to read or write data. All data produced or consumed by an algorithm must be explicitly passed to the algorithm by the client. For example, no algorithm should call a device-independent I/O library function to get data; this is the responsibility of the client or framework.

DSP Algorithm Component Model

This section should describe in detail, the rules and guidelines that apply to all algorithms on all DSP architectures regardless of application area.

- Interfaces and Modules
- Algorithms
- Packaging

These rules and guidelines enable many of the benefits normally associated with object-oriented and component-based programming but with little or no overhead. More importantly, these guidelines are necessary to enable two different algorithms to be integrated into a single application without modifying the source code of the algorithms. The rules include naming conventions to prevent duplicate external name conflicts, a uniform method for initializing algorithms, and specification of a uniform data memory management mechanism.

There is a lot that can be written concerning this level of the standard, and it makes sense at this point to reference the details of a particular Algorithm Component Model. Please reference the following URL for the details of one DSP Algorithm Component Model

http://focus.ti.com/lit/ug/spru352e/spru352e.pdf

A summary of the rules from this standard is listed at the end of this appendix.

DSP Algorithm Performance Characterization

In this section, we need to define the performance information that should be provided by algorithm components to enable system integrators to assemble combinations of algorithms into reliable products.

The only resources consumed by most DSP algorithms are MIPS and memory. All I/O, peripheral control, device management, and scheduling should be managed by the application—not the algorithm. Thus, we need to characterize code and data memory requirements and worst-case execution time. There is one important addition, however. It is possible for an algorithm to inadvertently disrupt the scheduling of threads in a system by disabling interrupts for extended periods. Since it is not possible for a scheduler to get control of the CPU while interrupts are disabled, it is important that algorithms minimize the duration of these periods and document the worst-case duration. It is important to realize that, due to the pipeline of modern DSPs, there are many situations where interrupts are implicitly disabled; e.g., in some zero-overhead loops. Thus, even if an algorithm does not explicitly disable interrupts, it may cause interrupts to be disabled for extended periods. The main areas to define and document are:

- Data Memory
- Program Memory
- Interrupt Latency
- Execution Time

Data memory

All data memory for an algorithm falls into one of three categories:

- *Heap memory* – data memory that is potentially (re)allocated at run-time;
- *Stack memory* – the C run-time stack; and
- *Static data* – data that is fixed at program build time.

Heap memory is bulk memory that is used by a function to perform its computations. From the function's point of view, the location and contents of this memory may persist across functions calls, may be (re)allocated at run-time, and different buffers may be in physically distinct memories. Stack memory, on the other hand, is scratch memory whose location may vary between consecutive function calls, is allocated and freed at run-time, and is managed using a LIFO (Last In First Out) allocation policy. Finally, static data is any data that is allocated at design-time (i.e., program-build time) and whose location is fixed during run-time. In this section, we should define performance metrics that describe an algorithm's data memory requirements.

Program memory

Algorithm code can often be partitioned into two distinct types: frequently accessed code and infrequently accessed code. Obviously, inner loops of algorithms are frequently accessed. However, like most application code, it is often the case that a few functions account for most of the MIPS required by an application.

Interrupt latency

In most DSP systems, algorithms are started by the arrival of data and the arrival of data is signaled by an interrupt. It is very important, therefore, that interrupts occur in as timely a fashion as possible. In particular, algorithms should minimize the time that interrupts are disabled. Ideally, algorithms would never disable interrupts. In some DSP architectures, however, zero overhead loops implicitly disable interrupts and, consequently, optimal algorithm efficiency often requires some interrupt latency.

Execution time

It is also important to define the execution time information that should be provided by algorithm components to enable system integrators to assemble combinations of algorithms into reliable products.

MIPS is not enough

It is important to realize that a simple MIPS calculation is far from sufficient when combining multiple algorithms. It is possible, for example, for two algorithms to be "unschedulable" even though only 84% of the available MIPS are required. In the worst case, it is possible for a set of algorithms to be unschedulable although only 70% of the available MIPS are required!

Suppose, for example, that a system consists of two tasks A and B with periods of 2 ms and 3 ms respectively (Figure F.5). Suppose that task A requires 1 ms of the CPU to complete its processing and task B also requires 1 ms of the CPU. The total percentage of the CPU required by these two tasks is approximately 83.3%; 50% for task A plus 33.3% for task B.

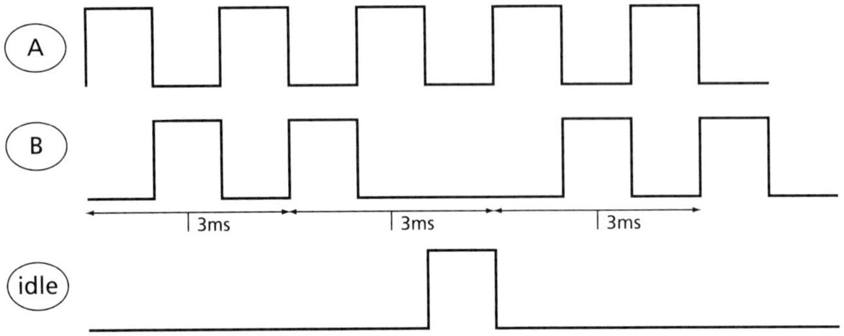

Figure F.5 Execution timeline for two periodic tasks (courtesy of Texas Instruments)

In this case, both task A and B meet their deadlines and we have more than 18% (1 ms every 6 ms) of the CPU idle. Suppose we now increase the amount of processing that task B must perform very slightly, say to 1.0000001 ms every 3 ms. Notice that task B will miss its first deadline because task A consumes 2 ms of the available 3 ms of task B's period. This leaves only 1 ms for B but B needs just a bit more than 1 ms to complete its work. If we make task B higher priority than task A, task A will miss its deadline line because task B will consume more than 1 ms of task A's 2 ms period. In this example, we have a system that has over 18% of the CPU MIPS unused but we cannot complete both task A and B within their real-time deadlines.

Moreover, the situation gets worse if you add more tasks to the system.

Liu and Layland proved that in the worst case you may have a system that is idle slightly more than 30% of the time that still can't meet its real-time deadlines! The good news is that this worst-case situation does not occur very often in practice. The bad news is that we can't rely on this not happening in the general situation. It is relatively easy to determine if a particular task set will meet its real-time deadlines *if* the period of each task is known *and* its CPU requirements during this period are also known. It is important to realize, however, that this determination is based on a mathematical model of the software and, as with any model, it may not correspond 100% with reality.

Moreover, the model is dependent on each component accurately characterizing its performance; if a component underestimates its CPU requirements by even 1 clock cycle, it is possible for the system to fail. Finally, designing with worst-case CPU requirements often prevents one from creating viable combinations of components. If the average case CPU requirement for a component differs significantly from its worst case, considerable CPU bandwidth may be wasted.

DSP-Specific Guidelines

DSP algorithms are often written in assembly language and, as a result, they will take full advantage of the instruction set. Unfortunately for the system integrator, this often means that multiple algorithms cannot be integrated into a single system because of incompatible assumptions about the use of specific features of the DSP (e.g., use of overflow mode, use of dedicated registers, etc.). This section should cover those guidelines that are specific to a particular DSP instruction set. These guidelines are designed to maximize the flexibility of the algorithm implementers, while at the same time ensure that multiple algorithms can be integrated into a single system. The standard should include things like;

CPU register types

The standard should include several categories of register types.

- *Scratch register* – These registers can be freely used by an algorithm, cannot be assumed to contain any particular value upon entry to an algorithm function, and can be left in any state after exiting a function.
- *Preserve registers* – These registers may be used by an algorithm, cannot be assumed to contain any particular value upon entry to an algorithm function, but must be restored upon exit from an algorithm to the value it had at entry.
- *Initialized register* – These registers may be used by an algorithm, contain a specified initial value upon entry to an algorithm function (as stated next to the register), and must be restored upon exit from the algorithm.
- *Read-only register* – These registers may be read but must not be modified by an algorithm.

In addition to the categories defined above, all registers should be further classified as being either local or global. Local registers are thread specific; i.e., every thread maintains its own copy of this register and it is active whenever this thread is running. Global registers, on the other hand, are shared by all threads in the system; if one thread changes a global register then all threads will see the change.

Use of floating-point

Referencing the float data type in an algorithm on a fixed-point DSP causes a large floating-point support library to be included in any application that uses the algorithm.

Endian byte ordering

DSP families support both big and little endian data formats. This support takes the form of "boot time" configuration. The DSP may be configured at boot time to access memory either as big endian or little endian and this setting remains fixed for the lifetime of the application.

The choice of which data format to use is often decided based on the presence of other processors in the system; the data format of the other processors (which may not be configurable) determines the setting of the DSP data format. Thus, it is not possible to simply choose a single data format for DSP algorithms.

Data models

DSP C compilers support a variety of data models; one small model and multiple large model modes. Fortunately, it is relatively easy to mix the various data memory models in a single application. Programs will achieve optimal performance using small model compilation. This model limits, however, the total size of the directly accessed data in an application to 32K bytes (in the worst case). Since algorithms are intended for use in very large applications, all data references should be far references.

Program model

DSP algorithms must never assume placement in on-chip program memory; i.e., they must properly operate with program memory operated in cache mode. In addition, no algorithm may ever directly manipulate the cache control registers. It is important to realize that compliant algorithms *may* be placed in on-chip program memory by the system developer. The rule above simply states that algorithms must not *require* placement in on-chip memory.

Register conventions

There must also be rules and guidelines that apply to the use of DSP on-chip registers. There are several different register types. Only those registers that are specifically defined with a programming model are allowed to be accessed. The list below is an example of a register set that must be defined with a programming model (register use rules):

- Address mode register Init (local)
- General-purpose Scratch (local)
- General-purpose Preserve (local)
- Frame Pointer Preserve (local)
- Data Page pointer Preserve (local)
- Stack Pointer Preserve (local)
- Control and Status Register Preserve
- Interrupt clear register Not accessible (global)
- Interrupt enable register Read-only (global)
- Interrupt flag register Read-only (global)
- Interrupt return pointer Scratch (global)
- Interrupt set register Not accessible (global)
- Interrupt service table pointer Read-only (global)
- Nonmaskable Interrupt return pointer Read-only (global)
- Program counter Read-only
- Floating-point Control register

Two examples are the status register and interrupt latency;

Status register

DSPs, like other embedded processors, contain status registers. This status register is further divided into several distinct fields. Although each field is often thought of as a separate register, it is not possible to access these fields individually. For example, in order to set one field it may be necessary to set *all* fields in the same status register. Therefore, it is necessary to treat the status registers with special care; if any field of a status register is of type Preserve or Read-only, the entire register must be treated as a Preserve register, for example.

Interrupt latency

Although there are no additional rules for DSP algorithms that deal with interrupt latency, it may be necessary to understand some important architectural impacts that DSPs have with respect to algorithmic execution. For example, on some DSPs, all instructions in the delay slots of branches are noninterruptible; i.e., once fetched, interrupts are blocked until the branch completes. Since these delay slots may contain other branch instructions, care must be taken to avoid long chains of noninterruptible instructions. In particular, tightly coded loops often result in unacceptably long noninterruptible sequences. Note that the C compiler has options to limit the duration of loops. Even if this option is used, you must be careful to limit the length of loops whose length is not a simple constant.

Use of the DMA Resource for Algorithm Development

The direct memory access (DMA) controller performs asynchronously scheduled data transfers in the background while the CPU continues to execute instructions. A good DSP algorithm standard should include rules and guidelines for creating compliant DSP algorithms that utilize the DMA resources.

The fact is that some algorithms require some means of moving data in the background of CPU operations. This is particularly important for algorithms that process and move large blocks of data; for example, imaging and video algorithms. The DMA is designed for this exact purpose and algorithms need to gain access to this resource for performance reasons.

A DSP algorithm standard should outline a model to facilitate the use of the DMA resources for these DSP algorithms.

A DSP algorithm standard should look upon algorithms as pure "data transducers." They are, among other things, not allowed to perform any operations that can affect scheduling or memory management. All these operations must be controlled by the framework to ensure easy integration of algorithms, possibly from different vendors. In general, the framework must be in command of managing the system resources, including the DMA resource.

Algorithms cannot access the DMA registers directly, nor can they be written to work with a particular physical DMA channel only. The framework must have the freedom to assign any available channel, and possibly share DMA channels, when granting an algorithm a DMA resource.

Requirements for the use of the DMA resource

Below is a list of requirements for DMA usage in DSP algorithms. These requirements will help to clarify the intent of the stated rules and guidelines in this chapter.

1. All physical DMA resources must be owned and managed by the framework.
2. Algorithms must access the DMA resource through a handle representing a logical DMA channel abstraction. These handles are granted to the algorithm by the framework using a standard interface.
3. A mechanism must be provided so that algorithms can ensure completion of data transfer(s).
4. The DMA scheme must work within a preemptive environment.
5. It must be possible for an algorithm to request multiframe data transfers (two-dimensional data transfers).
6. The framework must be able to obtain the worst-case DMA resource requirements at algorithm initialization time.
7. The DMA scheme must be flexible enough to fit within static and dynamic systems, and systems with a mix of static and dynamic features.

8. All DMA operations must complete prior to return to caller. The algorithm must synchronize all DMA operations before return to the caller from a framework-callable operation.
9. It must be possible for several algorithms to share a physical DMA channel.

For details of an existing algorithm standard that details the use of the DMA in algorithm develop, see the previously mentioned URL.

Rules and Guidelines for DSP Algorithm Development

This section will summarize the general rules, DSP specific rules, performance characterization rules, and DMA use rules for an existing DSP algorithm standard as well as some guidelines. This can be tailored accordingly.

General Rules

Rule 1 All algorithms must follow the run-time conventions imposed by the implementation of the C programming language.

Rule 2 All algorithms must be reentrant within a preemptive environment (including time-sliced preemption).

Rule 3 All algorithm data references must be fully relocatable (subject to alignment requirements). That is, there must be no "hard coded" data memory locations.

Rule 4 All algorithm code must be fully relocatable. That is, there can be no hard coded program memory locations

Rule 5 Algorithms must characterize their ROM-ability; i.e., state whether they are ROM-able or not.

Rule 6 Algorithms must never directly access any peripheral device. This includes but is not limited to on-chip DMAs, timers, I/O devices, and cache control registers.

Note, however, algorithms can utilize the DMA resource by implementing the appropriate interface.

Rule 7 All header files must support multiple inclusions within a single source file.

Rule 8 All external definitions must be either API identifiers or API and vendor prefixed.

Rule 9 All undefined references must refer either to the operations specified in the appropriate C runtime support library functions or the appropriate library functions.

Rule 10 All modules must follow compliant naming conventions for those external declarations disclosed to clients of those algorithms.

Rule 11 All modules must supply an initialization and finalization method.

Rule 12 All algorithms must implement the appropriately defined interfaces.

Rule 13 Each of the methods implemented by an algorithm must be independently relocatable.

Rule 14 All abstract algorithm interfaces must derive from a standard interface definition.

Rule 15 Each DSP algorithm must be packaged in an archive which has a name that follows a uniform naming convention.

Rule 16 Each DSP algorithm header must follow a uniform naming convention.

Rule 17 Different versions of a DSP algorithm from the same vendor must follow a uniform naming convention.

Rule 18 If a module's header includes definitions specific to a "debug" variant, it must use a common symbol to select the appropriate definitions; the symbol is defined for debug compilations and only for debug compilations.

Examples of DSP Specific Rules

Rule 25 All DSP algorithms must be supplied in little-endian format.

Rule 26 All DSP algorithms must access all static and global data as far data.

Rule 27 DSP algorithms must never assume placement in on-chip program memory; i.e., they must properly operate with program memory operated in cache mode.

Rule 28 On processors that support large program model compilation, all function accesses to independently relocatable object modules must be far references. For example, intersection function references within algorithm and external function references to other DSP modules must be far on the specific DSP type; i.e., the calling function must push both the XPC and the current PC.

Rule 29 On processors that support large program model compilation, all independently relocatable object module functions must be declared as far functions; for example, on the specific DSP, callers must push both the XPC and the current PC and the algorithm functions must perform a far return.

Rule 30 On processors that support an extended program address space (paged memory), the code size of any independently relocatable object module should never exceed the code space available on a page when overlays are enabled.

Rule 31 All DSP algorithms must document the content of the stack configuration register that they follow.

Rule 32 All DSP algorithms must access all static and global data as far data; also the algorithms should be instantiable in a large memory model.

Rule 33 DSP algorithms must never assume placement in on-chip program memory; i.e., they must properly operate with program memory operated in instruction cache mode.

Rule 34 All DSP algorithms that access data by B-bus must document: the instance number of the IALG_MemRec structure that is accessed by the B-bus (heapdata), and the data-section name that is accessed by the B-bus (static-data).

Rule 35 All DSP algorithms must access all static and global data as far data; also, the algorithm should be instantiable in a large memory model.

Performance Characterization Rules

Rule 19 All algorithms must characterize their worst-case heap data memory requirements (including alignment).

Rule 20 All algorithms must characterize their worst-case stack space memory requirements (including alignment).

Rule 21 Algorithms must characterize their static data memory requirements.

Rule 22 All algorithms must characterize their program memory requirements.

Rule 23 All algorithms must characterize their worst-case interrupt latency for every operation.

Rule 24 All algorithms must characterize the typical period and worst-case execution time for each operation.

DMA Rules

DMA Rule 1 All data transfer must be completed before return to caller.

DMA Rule 2 All algorithms using the DMA resource must implement a standard interface.

DMA Rule 3 Each of the DMA methods implemented by an algorithm must be independently relocateable.

DMA Rule 4 All algorithms must state the maximum number of concurrent DMA transfers for each logical channel.

DMA Rule 5 All algorithms must characterize the average and maximum size of the data transfers per logical channel for each operation. Also, all algorithms must characterize the average and maximum frequency of data transfers per logical channel for each operation.

DMA Rule 6 For certain DSP (like the C6x) algorithms must not issue any CPU read/writes to buffers in external memory that are involved in DMA transfers. This also applies to the input buffers passed to the algorithm through its algorithm interface.

DMA Rule 7 If a DSP algorithm has implemented the DMA interface, all input and output buffers residing in external memory and passed to this algorithm through its function calls, should be allocated on a cache line boundary and be a multiple of the cache line length in size. The application must also clean the cache entries for these buffers before passing them to the algorithm.

DMA Rule 8 When appropriate, all buffers residing in external memory involved in a DMA transfer should be allocated on a cache line boundary and be a multiple of the cache line length in size.

DMA Rule 9 When appropriate, algorithms should not use stack allocated buffers as the source or destination of any DMA transfer.

DMA Rule 10 When appropriate, algorithms must request all data buffers in external memory with 32-bit alignment and sizes in multiples of 4 (bytes).

DMA Rule 11 When appropriate, algorithms must use the same data types, access modes and DMA transfer settings when reading from or writing to data stored in external memory, or in application-passed data buffers.

General Guidelines

Guideline 1 Algorithms should minimize their persistent data memory requirements in favor of scratch memory.

Guideline 2 Each initialization and finalization function should be defined in a separate object module; these modules must not contain any other code.

Guideline 3 All modules that support object creation should support design-time object creation.

Guideline 4 All modules that support object creation should support run-time object creation.

Guideline 5 Algorithms should keep stack size requirements to a minimum.

Guideline 6 Algorithms should minimize their static memory requirements.

Guideline 7 Algorithms should never have any scratch static memory.

Guideline 8 Algorithm code should be partitioned into distinct sections and each section should be characterized by the average number of instructions executed per input sample.

Guideline 9 Interrupt latency should never exceed 10 ~s. (or other appropriate measure)

Guideline 10 Algorithms should avoid the use of global registers.

Guideline 11 Algorithms should avoid the use of the float data type.

Guideline 12 When appropriate, all algorithms should be supplied in both little- and big-endian formats.

Guideline 13 On processors that support large program model compilations, a version of the algorithm should be supplied that accesses all core run-time support functions as near functions and all algorithms as far functions (mixed model).

Guideline 14 When appropriate, all algorithms should not assume any specific stack configuration and should work under all the three stack modes.

DMA Guidelines

Guideline 1 The data transfer should complete before the CPU operations executing in Parallel.

Guideline 2 All algorithms should minimize channel (re)configuration overhead by requesting a dedicated logical DMA channel for each distinct type of DMA transfer it issues.

Guideline 3 To ensure correctness, All DSP algorithms that use DMA need to be supplied with the internal memory they request from the client application.

Guideline 4 To facilitate high performance, DSP algorithms should request DMA transfers with source and destinations aligned on 32-bit byte addresses (when appropriate).

Guideline 5 DSP algorithms should minimize channel configuration overhead by requesting a separate logical channel for each different transfer type.

References

This appendix lists sources for additional information.

Dialogic, *Media Stream Processing Unit; Developer's Guide*, 05-1221-001-01 September 1998.

ISO/IEC JTC1/SC22 N 2388 dated January 1997, *Request for SC22 Working Groups to Review DIS 2382-07.*

Intermetrics, Mwave Developer's Toolkit, DSP Toolkit User's Guide, 1993.

Liu, C.L.; Layland, J.W. "Scheduling Algorithms for Multi-Programming in a

Hard Real-Time Environment," *JACM* 20, 1, (January 1973): 40-61.

Massey, Tim and Iyer, Ramesh. *DSP Solutions for Telephony and Data/Facimile Modems*, SPRA073, 1997.

Texas Instruments, *TMS320C54x Optimizing C Compiler User's Guide*, SPRU103C, 1998.

Texas Instruments, *TMS320C6x Optimizing C Compiler User's Guide*, SPRU187C, 1998.

Texas Instruments, *TMS320C62xx CPU and Instruction Set*, SPRU189B, 1997.

Texas Instruments, *TMS320C55x Optimizing C/C++ Compiler User's Guide*, SPRU281, 2001.

Texas Instruments, *TMS320C2x/C2xx/C5x Optimizing C Compiler User's Guide*, SPRU024, 1999.

Texas Instruments, *TMS320C28x Optimizing C/C++ Compiler User's Guide*, SPRU514, 2001.

URLs

http://www.faqs.org/faqs/threads-faq/part1/

Glossary

Abstract Interface An interface defined by a C header whose functions are specified by a structure of function pointers. By convention these interface headers begin with the letter 'i' and the interface name begins with 'I'. Such an interface is "abstract" because, in general, many modules in a system implement the same abstract interface; i.e., the interface defines abstract operations supported by many modules.

Algorithm Technically, an algorithm is a sequence of operations, each chosen from a finite set of well-defined operations (e.g. computer instructions), that halts in a finite time, and computes a mathematical function. In the context of this specification, however, we allow algorithms to employ heuristics and do not require that they always produce a correct answer.

API Acronym for Application Programming Interface i.e., a specific set of constants, types, variables, and functions used to programmatically interact with a piece of software

Asynchronous System Calls Most system calls block (or "suspend") the calling thread until they complete, and continue its execution immediately following the call. Some systems also provide asynchronous (or *nonblocking*) forms of these calls; the kernel notifies the caller through some kind of out-of-band method when such a system call has completed. Asynchronous system calls are generally much harder for the programmer to deal with than blocking calls. This complexity is often outweighed by the performance benefits for real-time compute intensive applications.

Client The term client is often used to denote any piece of software that uses a function, module, or interface; for example, if the function a() calls the function b(), a() is a client of b(). Similarly, if an application App uses module MOD, App is a client of MOD.

COFF Common Output File Format. The file format of the files produced by the TI compiler, assembler, and linker.

Concrete Interface An interface defined by a C header whose functions are implemented by a single module within a system. This is in contrast to an abstract interface where multiple modules in a system may implement the same abstract interface. The header for every module defines a concrete interface.

Context Switch A context switch is the action of switching a CPU between one thread and another (or transferring control between them). This may involve crossing one or more protection boundaries.

Critical Section A critical section of code is one in which data that may be accessed by other threads are inconsistent. At a higher level, a critical section can be viewed as a section of code in which a guarantee you make to other threads about the state of some data may not be true. If other threads can access these data during a critical section, your program may not behave correctly. This may cause it to crash, lock up, produce incorrect results, or do just about any other unpleasant thing you care to imagine.

Other threads are generally denied access to inconsistent data during a critical section (usually through use of locks). If some of your critical sections are too long, however, it may result in your code performing poorly.

Endian Refers to which bytes are most significant in multibyte data types. In big-endian architectures, the leftmost bytes (those with a lower address) are most significant. In little-endian architectures, the rightmost bytes are most significant. HP, IBM, Motorola 68000, and SPARC systems store multibyte values in big-endian order, while Intel 80x86, DEC VAX, and DEC Alpha systems store them in little-endian order. Internet standard byte ordering is also big-endian. The TMS320C6000 is bi-endian because it supports both systems.

Frame Algorithms often process multiple samples of data at a time. This set of samples is sometimes referred to as a frame. In addition to improving performance, some algorithms require specific minimum frame sizes to properly operate.

Framework A framework is that part of an application that has been designed to remain invariant while selected software components are added, removed, or modified. Very general frameworks are sometimes described as application specific operating systems.

Instance The specific data allocated in an application that defines a particular object.

Interface A set of related functions, types, constants, and variables. An interface is often specified via a C header file.

Interrupt Latency The maximum time between when an interrupt occurs and its corresponding Interrupt Service Routine (ISR) starts executing.

Glossary of Terms

D-3 glossary

Interrupt Service Routine (ISR) An ISR is a function called in response to an interrupt detected by a CPU.

Method The term method is a synonym for a function that can be applied to an object defined by an interface.

Module A module is an implementation of one (or more) interfaces. In addition, all modules follow certain design elements that are common to *all* standard-compliant software components. Roughly speaking, a module is a C language implementation of a C++ class. Since a module is an implementation of an interface, it may consist of many distinct object files.

Multithreading Multithreading is the management of multiple concurrent uses of the same program. Most operating systems and modern computer languages also support multithreading.

Preemptive A property of a scheduler that allows one task to asynchronously interrupt the execution of the currently executing task and switch to another task; the interrupted task is *not* required to call any scheduler functions to enable the switch.

Protection Boundary A protection boundary protects one software subsystem on a computer from another, in such a way that only data that is explicitly shared across such a boundary is accessible to the entities on both sides. In general, all code within a protection boundary will have access to all data within that boundary. The canonical example of a protection boundary on most modern systems is that between processes and the kernel. The kernel is protected from processes, so that they can only examine or change its internal state in certain strictly defined ways. Protection boundaries also exist between individual processes on most modern systems. This prevents one buggy or malicious process from wreaking havoc on others. Why are protection boundaries interesting? Because transferring control across them is often expensive; it takes a lot of time and work. Most DSPs have no support for protection boundaries.

Reentrant Pertaining to a program or a part of a program in its executable version, that may be entered repeatedly, or may be entered before previous executions have been completed, and each execution of such a program is independent of all other executions.

Run to Completion A thread execution model in which all threads run to completion without ever synchronously suspending execution. Note that this attribute is completely independent of whether threads are preemptively scheduled. Run to completion threads may be preempt on another (e.g., ISRs) and nonpreemptive systems may allow threads to synchronously suspend

Glossary of Terms

D-4 execution Note that only one run-time stack is required for all run to completion threads in a system.

Scheduling The process of deciding what thread should execute next on a particular CPU. It is usually also taken as involving the context switch to that thread.

Scheduling Latency The maximum time that a "ready" thread can be delayed by a lower priority thread.

Scratch Memory Memory that can be overwritten without loss; i.e., prior contents need not be saved and restored after each use.

Scratch Register A register that can be overwritten without loss; i.e., prior contents need not be saved and restored after each use.

Thread The program state managed by the operating system that defines a logically independent sequence of program instructions. This state may be as little as the Program Counter (PC) value but often includes a large portion of the CPU's register set.

About the Author

Robert Oshana is an Engineering Manager for the Software Development Organization of Texas Instruments's DSP Systems group. He has over 23 years of real-time embedded software development experience in several industries including military and commercial. He is also an adjunct professor at Southern Methodist University where he teaches graduate software engineering as well as embedded and real-time systems courses. He speaks regularly at the Embedded Systems conference and has numerous publications in the areas of software engineering and real-time systems. Robert is a licensed professional engineer and a Senior Member of IEEE.

Index

ELSEVIER SCIENCE CD-ROM LICENSE AGREEMENT

PLEASE READ THE FOLLOWING AGREEMENT CAREFULLY BEFORE USING THIS CD-ROM PRODUCT. THIS CD-ROM PRODUCT IS LICENSED UNDER THE TERMS CONTAINED IN THIS CD-ROM LICENSE AGREEMENT ("Agreement"). BY USING THIS CD-ROM PRODUCT, YOU, AN INDIVIDUAL OR ENTITY INCLUDING EMPLOYEES, AGENTS AND REPRESENTATIVES ("You" or "Your"), ACKNOWLEDGE THAT YOU HAVE READ THIS AGREEMENT, THAT YOU UNDERSTAND IT, AND THAT YOU AGREE TO BE BOUND BY THE TERMS AND CONDITIONS OF THIS AGREEMENT. ELSEVIER SCIENCE INC. ("Elsevier Science") EXPRESSLY DOES NOT AGREE TO LICENSE THIS CD-ROM PRODUCT TO YOU UNLESS YOU ASSENT TO THIS AGREEMENT. IF YOU DO NOT AGREE WITH ANY OF THE FOLLOWING TERMS, YOU MAY, WITHIN THIRTY (30) DAYS AFTER YOUR RECEIPT OF THIS CD-ROM PRODUCT RETURN THE UNUSED CD-ROM PRODUCT AND ALL ACCOMPANYING DOCUMENTATION TO ELSEVIER SCIENCE FOR A FULL REFUND.

DEFINITIONS

As used in this Agreement, these terms shall have the following meanings:

"Proprietary Material" means the valuable and proprietary information content of this CD-ROM Product including all indexes and graphic materials and software used to access, index, search and retrieve the information content from this CD-ROM Product developed or licensed by Elsevier Science and/or its affiliates, suppliers and licensors.

"CD-ROM Product" means the copy of the Proprietary Material and any other material delivered on CD-ROM and any other human-readable or machine-readable materials enclosed with this Agreement, including without limitation documentation relating to the same.

OWNERSHIP

This CD-ROM Product has been supplied by and is proprietary to Elsevier Science and/or its affiliates, suppliers and licensors. The copyright in the CD-ROM Product belongs to Elsevier Science and/or its affiliates, suppliers and licensors and is protected by the national and state copyright, trademark, trade secret and other intellectual property laws of the United States and international treaty provisions, including without limitation the Universal Copyright Convention and the Berne Copyright Convention. You have no ownership rights in this CD-ROM Product. Except as expressly set forth herein, no part of this CD-ROM Product, including without limitation the Proprietary Material, may be modified, copied or distributed in hardcopy or machine-readable form without prior written consent from Elsevier Science. All rights not expressly granted to You herein are expressly reserved. Any other use of this CD-ROM Product by any person or entity is strictly prohibited and a violation of this Agreement.

SCOPE OF RIGHTS LICENSED (PERMITTED USES)

Elsevier Science is granting to You a limited, non-exclusive, non-transferable license to use this CD-ROM Product in accordance with the terms of this Agreement. You may use or provide access to this CD-ROM Product on a single computer or terminal physically located at Your premises and in a secure network or move this CD-ROM Product to and use it on another single computer or terminal at the same location for personal use only, but under no circumstances may You use or provide access to any part or parts of this CD-ROM Product on more than one computer or terminal simultaneously.

You shall not (a) copy, download, or otherwise reproduce the CD-ROM Product in any medium, including, without limitation, online transmissions, local area networks, wide area networks, intranets, extranets and the Internet, or in any way, in whole or in part, except that You may print or download limited portions of the Proprietary Material that are the results of discrete searches; (b) alter, modify, or adapt the CD-ROM Product, including but not limited to decompiling, disassembling, reverse engineering, or creating derivative works, without the prior written approval of Elsevier Science; (c) sell, license or otherwise distribute to third parties the CD-ROM Product or any part or parts thereof; or (d) alter, remove, obscure or obstruct the display of any copyright, trademark or other proprietary notice on or in the CD-ROM Product or on any printout or download of portions of the Proprietary Materials.

RESTRICTIONS ON TRANSFER

This License is personal to You, and neither Your rights hereunder nor the tangible embodiments of this CD-ROM Product, including without limitation the Proprietary Material, may be sold, assigned, transferred or sub-licensed to any other person, including without limitation by operation of law, without the prior written consent of Elsevier Science. Any purported sale, assignment, transfer or sublicense without the prior written consent of Elsevier Science will be void and will automatically terminate the License granted hereunder.

TERM

This Agreement will remain in effect until terminated pursuant to the terms of this Agreement. You may terminate this Agreement at any time by removing from Your system and destroying the CD-ROM Product. Unauthorized copying of the CD-ROM Product, including without limitation, the Proprietary Material and documentation, or otherwise failing to comply with the terms and conditions of this Agreement shall result in automatic termination of this license and will make available to Elsevier Science legal remedies. Upon termination of this Agreement, the license granted herein will terminate and You must immediately destroy the CD-ROM Product and accompanying documentation. All provisions relating to proprietary rights shall survive termination of this Agreement.

LIMITED WARRANTY AND LIMITATION OF LIABILITY

NEITHER ELSEVIER SCIENCE NOR ITS LICENSORS REPRESENT OR WARRANT THAT THE INFORMATION CONTAINED IN THE PROPRIETARY MATERIALS IS COMPLETE OR FREE FROM ERROR, AND NEITHER ASSUMES, AND BOTH EXPRESSLY DISCLAIM, ANY LIABILITY TO ANY PERSON FOR ANY LOSS OR DAMAGE CAUSED BY ERRORS OR OMISSIONS IN THE PROPRIETARY MATERIAL, WHETHER SUCH ERRORS OR OMISSIONS RESULT FROM NEGLIGENCE, ACCIDENT, OR ANY OTHER CAUSE. IN ADDITION, NEITHER ELSEVIER SCIENCE NOR ITS LICENSORS MAKE ANY REPRESENTATIONS OR WARRANTIES, EITHER EXPRESS OR IMPLIED, REGARDING THE PERFORMANCE OF YOUR NETWORK OR COMPUTER SYSTEM WHEN USED IN CONJUNCTION WITH THE CD-ROM PRODUCT.

If this CD-ROM Product is defective, Elsevier Science will replace it at no charge if the defective CD-ROM Product is returned to Elsevier Science within sixty (60) days (or the greatest period allowable by applicable law) from the date of shipment.

Elsevier Science warrants that the software embodied in this CD-ROM Product will perform in substantial compliance with the documentation supplied in this CD-ROM Product. If You report significant defect in performance in writing to Elsevier Science, and Elsevier Science is not able to correct same within sixty (60) days after its receipt of Your notification, You may return this CD-ROM Product, including all copies and documentation, to Elsevier Science and Elsevier Science will refund Your money.

YOU UNDERSTAND THAT, EXCEPT FOR THE 60-DAY LIMITED WARRANTY RECITED ABOVE, ELSEVIER SCIENCE, ITS AFFILIATES, LICENSORS, SUPPLIERS AND AGENTS, MAKE NO WARRANTIES, EXPRESSED OR IMPLIED, WITH RESPECT TO THE CD-ROM PRODUCT, INCLUDING, WITHOUT LIMITATION THE PROPRIETARY MATERIAL, AN SPECIFICALLY DISCLAIM ANY WARRANTY OF MERCHANTABILITY OR FITNESS FOR A PARTICULAR PURPOSE.

If the information provided on this CD-ROM contains medical or health sciences information, it is intended for professional use within the medical field. Information about medical treatment or drug dosages is intended strictly for professional use, and because of rapid advances in the medical sciences, independent verification f diagnosis and drug dosages should be made.

IN NO EVENT WILL ELSEVIER SCIENCE, ITS AFFILIATES, LICENSORS, SUPPLIERS OR AGENTS, BE LIABLE TO YOU FOR ANY DAMAGES, INCLUDING, WITHOUT LIMITATION, ANY LOST PROFITS, LOST SAVINGS OR OTHER INCIDENTAL OR CONSEQUENTIAL DAMAGES, ARISING OUT OF YOUR USE OR INABILITY TO USE THE CD-ROM PRODUCT REGARDLESS OF WHETHER SUCH DAMAGES ARE FORESEEABLE OR WHETHER SUCH DAMAGES ARE DEEMED TO RESULT FROM THE FAILURE OR INADEQUACY OF ANY EXCLUSIVE OR OTHER REMEDY.

U.S. GOVERNMENT RESTRICTED RIGHTS

The CD-ROM Product and documentation are provided with restricted rights. Use, duplication or disclosure by the U.S. Government is subject to restrictions as set forth in subparagraphs (a) through (d) of the Commercial Computer Restricted Rights clause at FAR 52.22719 or in subparagraph (c)(1)(ii) of the Rights in Technical Data and Computer Software clause at DFARS 252.2277013, or at 252.2117015, as applicable. Contractor/Manufacturer is Elsevier Science Inc., 655 Avenue of the Americas, New York, NY 10010-5107 USA.

GOVERNING LAW

This Agreement shall be governed by the laws of the State of New York, USA. In any dispute arising out of this Agreement, you and Elsevier Science each consent to the exclusive personal jurisdiction and venue in the state and federal courts within New York County, New York, USA.